ESSAYS ON MATHEMATICAL AND PHILOSOPHICAL LOGIC

ESSAYS ON MATHEMATICAL AND PHILOSOPHICAL LOGIC

PROCEEDINGS OF THE FOURTH SCANDINAVIAN LOGIC SYMPOSIUM AND OF THE FIRST SOVIET-FINNISH LOGIC CONFERENCE, JYVÄSKYLÄ, FINLAND, JUNE 29–JULY 6, 1976

Edited by

JAAKKO HINTIKKA, ILKKA NIINILUOTO,
and ESA SAARINEN

D. REIDEL PUBLISHING COMPANY

DORDRECHT : HOLLAND / BOSTON : U.S.A.

LONDON : ENGLAND

Library of Congress Cataloging in Publication Data

Scandinavian Logic Symposium, 4th, Jyväskylä, Finland, 1976.
Essays on mathematical and philosophical logic.

(Synthese library ; v. 122)
Includes bibliographies and index.
1. Logic—Congresses. I. Hintikka, Kaarlo Jaakko Juhani, 1929– II. Niiniluoto, Ilkka. III. Saarinen, Esa, 1953– IV. Soviet-Finnish Logic Conference, 1st, Jyväskylä, Finland, 1976. V. Title.
BC5.S28 1976 160 78-14736
ISBN 90-277-0879-7

Published by D. Reidel Publishing Company,
P.O. Box 17, Dordrecht, Holland

Sold and distributed in the U.S.A., Canada, and Mexico
by D. Reidel Publishing Company, Inc.
Lincoln Building, 160 Old Derby Street, Hingham,
Mass. 02043, U.S.A.

Printed in The Netherlands

PREFACE

The Fourth Scandinavian Logic Symposium and the First Soviet–Finnish Logic Conference were held in Jyväskylä, Finland, June 29–July 6, 1976. The Conferences were organized by a committee which consisted of the editors of the present volume. The Conferences were supported financially by the Ministry of Education of Finland, by the Academy of Finland, and by the Division of Logic, Methodology, and Philosophy of Science of the International Union of History of Science. The Philosophical Society of Finland and the Jyväskylä Summer Festival gave valuable help in various practicalities.

35 papers by authors representing 10 countries were presented at the two meetings. Of those papers 24 appear here.

THE EDITORS

TABLE OF CONTENTS

PART I

PROOF THEORY

G. KREISEL

SOME FACTS FROM THE THEORY OF PROOFS AND SOME FICTIONS FROM GENERAL PROOF THEORY

INTRODUCTION

A working definition of the distinction intended in the title is this. Proof theory is principally interested in what is traditionally called the *essence* or, equivalently, 'defining property' of proofs, namely their being valid arguments. This property of validity, which – like most notions and questions of traditional philosophy – occurs to us at a very early stage of experience, is regarded as basic in proof theory. And general proof theory develops such refinements as the distinction between different kinds of validity, for example, logical or constructive validity (and others familiar from the foundational literature). In contrast, the theory of proofs questions the utility of these distinctions compared to taking for granted the validity at least of currently used principles. Instead, this theory studies such structural features as the length of proofs, and especially relations between proofs and other things, so to speak, the 'role' of proofs. With this different emphasis, the classifications needed in the theory of proofs are often demonstrably in conflict with the basic distinctions of proof theory, and not, as has sometimes been thought, refinements of the latter.

The switch from proof theory to the theory of proofs fits in with the scepticism of the *silent majority* of mathematicians towards traditional foundational questions, but not with so-called antiphilosophical doctrines such as logical positivism (which are also negative about traditional questions). Specifically, at least as I see this switch, it is needed not because traditional notions are particularly imprecise and difficult to formalize, let alone 'meaningless' or the result of 'bewitchment' by our language. The issue is elsewhere. There is doubt whether a precise analysis of traditional notions or indeed of anything remotely like them is at all rewarding for the intended purpose, namely for an understanding of the phenomena of mathematical reasoning.

N.B. The silent majority has no doubt about the occasional *heuristic* value, to suitably gifted people, of some of the traditional questions; 'heuristic' because in general that value does not derive from the originally

J. Hintikka, I. Niiniluoto, and E. Saarinen (eds.), Essays on Mathematical and Philosophical Logic, 3–23.

intended purpose, and so will be judged by criteria appropriate to the *new* purposes; cf. the appendix to this article. In particular, any *faithful* analysis of the traditional notions or, more generally, any logic-chopping is not an end in itself, but may have heuristic value (for the people involved).

Summary. Section 1 lists a number of arguments in the literature which have interpretations in terms of both (i) proof theory and (ii) of the theory of proofs; more specifically, in terms of (i) the metamathematical principles used and (ii) structural properties of the proofs mentioned (in the 'object language') and of the operations applied to these proofs. By confronting two such interpretations one can then judge the *adequacy* of the basic notions of (i), resp. (ii) – at least, as far as formulating the particular arguments is concerned: Section 2 complements Section 1 by listing a number of general assumptions of (i) which are unquestionably natural on more or less limited logical experience, for example, if one thinks primarily of purely logical reasoning (when the latter is trivially 'fundamental'), but *obviously contradicted* by our existing experience (and so the assumptions are quite implausible w.r.t. realistic possibilities of mathematical reasoning).

Discussion. The kind of detail given in Sections 1–2 is probably necessary for effective progress in the theory of proofs or, for that matter, in other 'heirs' of traditional proof theory such as those mentioned in the appendix. But it is also possible to look at the switch from proof theory to a theory of proofs as parallel to other, incidentally *familiar chapters in the history of foundations*, for example, *axiomatic foundations of geometry*. Unquestionably, the latter was originally intended to contribute to the question:

(*) What is space?

(tacitly, given a few preliminary distinctions, for example, between physical and visual space). Looking back we can say *now* that substantial progress w.r.t. (*) was possible; specifically in Einstein's theory of relativity. But assuming only that something like this theory is correct, nothing like the massive work of axiomatic geometry, say on euclidean geometry and its subsystems, provides a substantial contribution to (*) itself. For one thing, the most famous independence result, of the parallel postulate, is a mere corollary of any theory for whcih the *postulate is* *false*, that is, *decided*. But, more radically, the most rewarding progress does not look 'inwards', at the (kinematic) essence of space or even of space-time, but relates the latter to 'other things', to inertial and electromagnetic phenomena (to mention a neglected aspect of so-called unified theories). Furthermore, a really convincing presentation of the theory relies on unfamiliar phenomena occurring when objects move near the speed of light. All this is far removed from the (foundational) questions which first occur to us when looking 'inwards' – or to be more specific, which occurred to Eudoxus. Without suggesting that proofs are spati(o-tempor)al objects, I am impressed by the parallel between different stages in the *study* of (*), and in the study of:

(**) What is proof?

(which is the starting point of proof theory). Naturally, a principal question here is this: Has our experience w.r.t. (**) reached anything like the stage which was patently necessary for convincing progress w.r.t. (*)? This question is considered at the end of Kreisel [1977B]with due respect for the possibility that matters 'might' be different w.r.t. (*) and (**).

On the other hand, however limited the value of *axiomatic geometry* for (*) itself may have been, its *heuristic* value for the whole of algebra can hardly be exaggerated (cf. (c) of the Appendix for a recent parallel). As a corollary it is hardly surprising that, for example, the notion of *quadratic form* has a greater permanent value within algebra than the (corresponding) *euclidean metric* within a theory of (*).[1]

Remarks on the literature. The reasons given above why many traditional philosophical questions are (meaningful but) unrewarding were previously set out in a chatty style in Kreisel [1976B], especially around p. 184, and in Section 2 of Kreisel [A], by comparison with answers to: What is matter? and: What is a (literal) family resemblance? A more compact formulation is in (i)–(iii) of Note 7 to Kreisel [1977A]. Instead of comparing proof theory to axiomatic geometry, I have previously compared it to Sturm's algebraic theory of real polynomials intended (by him) to remove the use of philosophically doubtful infinitesimals; cf. the 'Historical' remark on p. 102 of Kreisel [1976A]. This comparison seems to me pedagogically bad for my present purpose because most readers are not familiar with Sturm's original (published) intentions while they are familiar with axiomatic geometry; of course, the comparison is good if one looks for a little history of mathematics which is not altogether hackneyed.

1. SOME RESULTS FROM THE THEORY OF PROOFS

The selection in (a) and (b) below revolves around *length* of proofs, (here) measured by the number of nodes or, equivalently, of inferences. The selection in (c) derives from so-called E-theorems, also involved in (b). The general points which particular results are intended to illustrate will be stated at the beginning of each subsection.

(a) We begin with two results which have their sources in traditional proof theory. Specifically, the proof theoretic literature gave elaborate, so-called syntactic or more elementary proofs of theorems which are model theoretically almost immediate. Thus the elaborate proofs are unnecessary and pedantic for the results *stated* (unless model theoretic arguments are rejected for ideological reasons). At the same time the proofs have a certain 'raw' interest. We set ourselves the problem of analyzing this interest more adequately by stating new results also established by those proofs; in other words, we reinterpret the proofs. These restatements require concepts not prominent in traditional proof theory; in fact, concepts which are trivial for proof theoretic aims. In this sense those aims are *inappropriate for our experience*, specifically, for our (mathematical) experience *extended* by the new proofs.

N.B. We have here an instance of a general scheme for *testing aims* by reference to (extended) experience, as one says, empirically – and not only by internal coherence or plausibility (relative to limited experience), the two pillars of traditional philosophical tests of aims.

(i) Model theoretically it is obvious that $\vdash(A \vee A) \Rightarrow \vdash A$ for a (propositionally) complete set of rules. Nevertheless, several proof theoretic texts give 'syntactic' proofs, often even after having established completeness. The new proofs involve precise and reasonably sensible concepts such as invertibility of rules and the like. But the import of those concepts is lost unless the final result is restated. Let $\vdash^n$ mean: derivable by means of a proof figure containing $\leqslant n$ nodes. We now ask (for the rules considered): Do we have

(*) $\qquad \vdash^n(A \vee A) \Rightarrow \vdash^n A$?

The answer (which is positive for the usual cut-free rules) is certainly not obvious from completeness itself.

N.B. The reader should consider refinements which involve not only the *lengths of derivations*, say d and d', of $A \vee A$, resp. A, but also the length of the *process* converting d into d'. (*Hint.* Consider a proof of (*) by induction on n, involving a search through d, compared with a proof which, roughly speaking, simply replaces '$A \vee A$' by 'A' in d.)

(ii) In Kreisel and Tait [1961] there is a short model theoretic and a longer combinatorial proof of the decidability of a certain (complete) equation calculus.[2] The difference between the results obtained from these two proofs which is actually *stated* (in the corollaries on p. 30 and p. 37, resp.) refers to recursive, resp. primitive recursive decision procedures. This is inadequate because well-known general metamathematical results such as Herbrand's theorem (alluded to at the end of Section 3 on p. 31) allow one to extract a primitive recursive decision procedure from the short proof too. A convincing distinction was given only recently in Statman [1977] by *refining* the analysis of bounds *within* the class of primitive recursive functions (exponential *versus* quadratic bounds). Statman's analysis should be compared to the half-hearted attempt in Kreisel and Tait [1961], p. 36, 1.–12 to 1.–9, of stating the interest of the longer proof, namely as providing (a complete family of) solutions for the equations considered.

N.B. The inadequate analysis loc. cit. (p. 30 and p. 37) is by no means an accidental aberration, but the direct consequence of proof theoretic aims. Since here the *whole* of PRA (primitive recursive arithmetic) is accepted, subdivisions within PRA are simply trivial. In contrast, for the theory of proofs or, incidentally, for all mathematical practice, *only* subdivisions of PRA are significant as far as bounds are concerned; in fact, the difference between exponential and non-recursive bounds (for procedures) is practically trivial. As a corollary: since PRA is demonstrably 'natural' (for the many authors who have written on PRA), it provides an example of the conflict mentioned at the end of the first paragraph of this paper.

(b) Though the measures of length used in (a) above, both for derivations and for operations on derivations, are not only precise but also superficially plausible, closer inspection shows that their range of applicability is not easy to judge.

N.B. In this respect those measures are similar to the familiar measure of the *complexity of decision problems* by the number of steps needed to compute the decision by means of a *universal* Turing machine. This measure ignores what is surely the single most important factor of existing computer technology, the introduction of *chips* which replace a whole sub-routine; in other words, *speed-up by hardware* whose behavior in unit time can be simulated by a universal Turing machine, but only in a large number of steps. Besides, as I have repeated *ad nauseam*, even a realistic measure of length (of derivations or terms, that is, programs) is often less significant than less hackneyed data such as *genus* in Statman [1974] or loops in flow diagrams.

To *establish* some (range of) significance of – our measure of – length we use the time-honoured method of deriving implications of this measure formulated in terms of familiar, so to speak 'tested' concepts. In particular, the length of a derivation of an existential formula, $\exists xA$, is shown to provide a bound for the least n such that $A[x/s^n0]$ holds. Below we consider *classical* systems and restrict ourselves to Σ_1^0-formulae $\exists xA$.[3]

(i) For *cut-free* analysis, the length itself is a bound. More precisely, for the systems of cut-free analysis formulated in Kreisel and Takeuti [1974] we have this bound, by pp. 25–28 loc. cit.

(ii) For analysis *with* cut, Chapter 3 of Statman [1974] contains a (much larger) bound obtained by means of a suitable cut-elimination or, equivalently, normalization procedure. 'Suitable' means this: the procedure should of course preserve end formulae, and a bound on the length of the normal form of a derivation d should be determined by the length of d. (The measure of length satisfies the condition that, for a given length λ, there are only finitely many logical forms of derivations $\leq \lambda$, but infinitely many distinct derivations since, for example, an arbitrary substitution for a variable is a single inference.) Granted this, we have a reduction to (i).

Comparison with proof theoretic interpretations. (i) was established to show that the Σ_1^0-reflection principle for cut-*free* analysis is provable within this system, and that all *literal Henkin sentences* are refutable (loc. cit., p. 21, 2.7); concerning the interest of these results, see p. 9, l. −14 to l. −6. As to (ii), Statman's use of (his particular procedure of) normalization provides clear *adequacy conditions* for choosing such procedures, namely that they be 'suitable' in the sense described above. Proof theoretic reasons for the choice of normalization procedures usually appeal to some kind of *operational semantics* (or to the metamathematical methods needed to prove the convergence of the procedure, for example, by means of transfinite induction). This is the subject of Section 2(b) and Section 2(c) below.

(c) Inspection of the proofs of (i) and (ii) in (b) above shows that more is established than asserted; not merely in the trivial sense in which this applies to practically every proof but in that we can *say* what is shown: we obtain bounds not only for the least n for which $A[x/s^n0]$ holds, but for the value of the particular term t_d provided by the given derivation d of $\exists xA$. Evidently, the significance of this sharpening of (b) depends on the significance of the relation

t is the realization of A provided by the derivation d.

For traditional proof theoretic aims this relation is marginal since the *basic* purpose of establishing reflection principles (involved here, as mentioned in (b) above) is simply to 'justify' the principles of proof (used in d); to show that $\exists xA$ holds at all. Which t realizes A is a minor matter as long as there is some t, and that this fact has been 'properly' established. Naturally, for other aims, say for mechanical operations on derivations, the relation above is absolutely central, in fact, the map

$$d \mapsto t_d$$

is central because this is the operation which will be performed. The (metamathematical) methods of proof establishing that $A[x/t_d]$ actually holds are immaterial here provided there is no genuine doubt that they are correct.

Digression. As a matter of empirical fact, in these circumstances the proof theorist steeped in traditional (proof theoretic) aims is tempted to go farther: not only to question the significance of the relation considered, which is usually thought to be a 'delicate' or 'subjective' matter or what have you, but to search for *ambiguities* or some other lack of precision. To avoid leading anybody into temptation, I introduced certain *formal analogues* of the relation above which are relative to operations familiar in the proof theoretic literature, such as functional interpretations and normalization procedures.

For interpretations I in the literature we consider the relation:

t is the realization of A provided by I applied to the derivation of $\exists xA$.

For the usual I, the interpretation of $\exists xA$ is $\exists x\exists s\forall tA_0$ where A_0 is quantifier-free and $\exists s\forall tA_0$ is the interpretation of A. Applied to d, I provides a pair of terms (x_d, s_d) such that $\forall tA_0[x/x_d, s/s_d]$ holds, and x_d is t_d.

If $\exists xA$ contains the free variables **a**, so does x_d. (The same applies to the 'usual' normalization procedures for logical systems such as predicate logic, but not *directly* for, say, arithmetic.)

(i) Mints [1975] shows that several popular interpretations provide, in the sense sketched above, the *same* x_d (up to appropriate convertibility), and so does normalization where applicable. Diller, in lectures, pinpointed the relevant more general property of the operations (of interpretation and normalization) involved, namely that they *commute*.

'Popular' does not include Hilbert's ϵ-substitution method (for arithmetic), which I happened to use for establishing the first published functional interpretation, called 'no-counterexample-interpretation' (n.c.i.) in Kreisel [1951]. As is usual in such circumstances, I found Hilbert's method very *natural* (it having once been the *only* method which I understood): I had second thoughts in connection with our present topic.

(ii) In general, Hilbert's ϵ-substitution method applied to a derivation d of (a Σ_1^0-formula) $\exists xA$ does *not* provide the intended term t_d nor, equivalently, the term provided by the interpretations considered in (i).

This was pointed out to me by Mints in reply to my question. A simple example is this. Let d be a derivation of $\exists x(x = x)$ where the only critical ϵ-formula in d is, say,

$$5 = 5 \rightarrow \epsilon_x(x = x) = \epsilon_x(x = x).$$

Hilbert's method (for arithmetic, not predicate logic) searches for the *least* $n : n = n$, that is, replaces $\epsilon_x(x = x)$ by 0. But clearly, the realization of $x = x$ provided by *this* d is 5 and not 0.

N.B. This defect (for the present purpose) concerns my particular proof of the n.c.i. (by reference to Ackermann's consistency proof for number theory), which can also be proved 'directly', for example, in the style of Gödel [1958]. The difference is that now only definition principles for functionals of *lowest* type are used such as the rule of bar recursion in Tait [1965]. *Presumably*, the map: $d \mapsto t_d$ (for derivations d of Σ_1^0-formulae) which is implicit in this more 'direct' proof agrees with (i) above.

Remarks on the proof theoretic literature. The list of alternative proofs, of known results, which are said to be more 'direct', is long, for example, one may avoid the passage to *prenex forms* for the n.c.i. or for Herbrand's theorem (which can be formulated as an n.c.i. provided one has this notion); again, one may avoid the familiar detour *via*, say, the double negation interpretation of a classical system S in a *corresponding intuitionistic system* S^- (on which, e.g., Gödel [1958] relies). Now these more direct proofs demonstrably *do not constitute a significant improvement for* (anything like) *traditional proof theoretic aims* since the passages involved are unsurpassably elementary. The situation is totally and obviously different for the theory of proofs:

Those passages mess up the structural properties of proofs,

including some of those properties which are crucial for intelligibility: certainly double negations make one's head spin. The facts mentioned suggest the following:

Test for adequacy. Go over the literature and use more direct proofs for (sharpened) results in terms of the theory of proofs. Then compare this sharpening with existing applications of more direct proofs in the literature, for example, when authors state sharper ordinal bounds (say, in the case of *k*-consistency).

2. SOME ASSUMPTIONS OF GENERAL PROOF THEORY

(This is in the sense indicated in Prawitz [1971] and especially Prawitz [1973][4] which I assume readers to know.) (a) and (b) below concern the assumptions, found in Prawitz's writings, that logical reasoning and his style of normalization (involved in most variants of so-called operational semantics of the logical particles) are privileged. The assumption considered in (c), concerning the special role of what is called 'the ordinal of a formal system' (for example, ϵ_0 in the case of first order arithmetic), dominates the rest of the proof theoretic literature. Incidentally, this assumption can also be given a 'semantic' dress by use of the idea that principles of proof are 'justified from below', an idea formulated in terms of so-called autonomous progressions.

In (a)–(c) the assumptions above are confronted with a number of detailed results. This seems to me useful pedagogically: the aim of refuting the assumptions is a peg on which to hang a lot of otherwise isolated results. But as already discussed at the end of the Introduction, a much simpler, more general reason for suspecting the assumptions comes from comparison with work on 'grand' questions (What is space? and – here even better – What is matter?). Each assumption takes *one* aspect of proofs as basic or fundamental. This is parallel to the fundamental theories in physics; but with an important difference. The basic elements of these fundamental theories are far removed from ordinary experience, and so there is a reasonable chance that relatively simple laws about a few such elements have significant implications for – what, on our ordinary view, is – an endless variety of phenomena. In contrast, the aspects of proofs involved in each of the assumptions (a)–(c) are quite close to our untutored view of (mathematical) experience, much closer than are, say, atoms (in the sense of modern physics) to what we see with the naked eye. Put differently, the 'unity' or 'system' which those assumptions are, evidently, intended to bring into our thinking about proofs is at best superficial; at worst, they blind us to other aspects of proofs which can be studied much more profitably at the present stage of knowledge.

(a) The first assumption of general proof theory can be explained very simply in the case of purely logical reasoning, formulated, say, in the language of first order predicate calculus. We know, from Gödel's completeness theorem, that every logically valid theorem, F, has a purely logical proof (by means of the usual rules). In particular, the *ranges of the variables* in F, that is, the domains of individuals, and the realizations of the non-logical symbols act as parameters: in the proofs we use nothing of what we may know about the 'nature' of the ranges. This is in contrast to non-logical proofs of F, where one establishes

$$\mathrm{Val}(F)\text{: validity of } F \text{ defined set-theoretically}$$

using more or less sophisticated properties of *sets* (entering into the definition of: Val). The issue is not whether there is a precise and otherwise appealing definition of 'logical proof', but in which way, if any, *logical proofs* (of F) *are privileged among all proofs of* F.[5]

(i) As has often been pointed out, modern mathematics constantly uses non-logical proofs of – assertions evidently expressed by – logical theorems, for example, on ordered fields, where a polynomial $p(x)$ of degree n has at most n changes of sign. This can be proved (i) by embedding ordered fields in real closed fields where $p(x)$ has a zero at any change of sign or (ii) by using Lagrange's identity

$$p(x) = \sum_0^n p(x_i) \prod_{j \neq i} \left(\frac{x - x_j}{x_i - x_j} \right) \text{ for } x_0 < x_1 \ldots < x_n < x,$$

and showing that $\bigwedge p(x_i)p(x_{i+1}) < 0 \rightarrow p(x)p(x_n) > 0$ by a logical proof. (For specialists: Which logical proof is obtained by normalizing (i) above?)

(ii) As is less well known (but can be read off some discussions by Wittgenstein), 'reductions' of arithmetic to logic show convincingly that we *need* arithmetic concepts to recognize logical validity; cf. Section 1(a′) of the Appendix to Kreisel [A].

Certainly, if – like operational semantics – one wants to analyze proofs in terms of what we do or even of what we should do, it is necessary to consider what we *can* do: (i) and (ii) are clearly relevant here.

(b) We now come to the second assumption of proof theory (or at least the part which is based on operational semantics), namely that *normal* derivations and *normalization procedures* are in fact used to determine the meaning of the logical particles. By (a) it would be idle to take this literally since the particles are understood when we recognize as *evident* the set theoretic or number theoretic properties which are formulated by logically compound definitions, and used in the shorter (and reliable!) proofs in, say, a(i)–a(ii) above. But there remains the possibility that the assumption has heuristic value at least within the general program of normalization, and that *here* logical reasoning is basic (specifically, for choosing *between* equivalent axiomatizations of (first order) *mathematical theories*, say *T*):

Such *T* are presented by a set $\mathfrak{A}(T)$ of axioms. *If* the analysis of logically valid arguments, say in Prawitz [1973], is at all basic for mathematical reasoning (and not 'trivial' compared to the choice of axioms), it *should* help us decide on (i) appropriate rules $\mathfrak{R}_{\mathfrak{A}}$ which, together with normal rules, generate *T* and on (ii) a corresponding normalization procedure (for derivations built up by adding the rule of 'cut'). Here 'decide on' means: find such rules $\mathfrak{R}_{\mathfrak{A}}$, etc., or else show that $\mathfrak{A}(T)$ does not lend itself to this style of axiomatization. Since the answer depends on $\mathfrak{A}(T)$ and not only on *T*, we have here a means of choosing between different axiomatizations of *T*.

The evidence for the heuristic value considered here is weak. This is shown by the literature on the systems (i) and (ii) below which have been considered in general proof theory though the systems are actually dubious for operational semantics. (The systems are of course elegant and familiar from set theoretic and traditional intuitionistic semantics.)

(i) Even for classical impredicative predicate calculus of second order, Prawitz's analysis has not led to a correct normalization procedure; cf. Statman [1974]. In fact, Statman's own procedure, σ, was found by a quite different requirement, namely that σ be *suitable* for preserving equality of logical form needed in Section 1b(ii) above.

(ii) For the intuitionistic impredicative theory of species with (some usual formulation of) the axiom of choice, no suitable procedure is known at all – though the procedure of Prawitz's [1971] *is* suitable for the theory of species itself, by the corollary to Proposition 4 on p. 106 of Statman [1974] as mentioned already in Note 3.

On present evidence – and in accordance with what was said earlier about the pretentious character of any would-be 'systematic' analysis of proofs available at the present time – we do much better by looking for *relatively few procedures* which have *relatively many applications*; for examples in the case of derivations, see III.2 of Kreisel and Takeuti [1974], esp. p. 41 under the slogan of: normalization according to (arithmetic) content, and in the case of (normalization of) terms, see the review of Feferman [1971]. Actually the bulk of normalization or, equivalently, of cut-elimination procedures in the literature were introduced for the sake of

consistency proofs by means of transfinite induction,

the requirement being that the convergence of the procedure be established by elementary (= quantifier-free) recursion. This matter is considered separately in (c) below.

(c) *Ordinals*, say $o_{\mathfrak{F}}$, have been associated to *formal systems* $\mathfrak{F}$ in several ways; first by Gentzen in the case of first order arithmetic, $\mathfrak{P}$, where $o_{\mathfrak{P}} = \epsilon_0$.

(i) $o_{\mathfrak{F}}^{\mathrm{CON}}$ is the least ordinal of some well-ordering with the *natural definition* $<_{\mathfrak{F}}$, such that CON $\mathfrak{F}$ can be proved by *elementary transfinite induction* on $<_{\mathfrak{F}}$. (The expressions in italics have been made formally precise in the case of $\mathfrak{P}$, and $o_{\mathfrak{P}}^{\mathrm{CON}} = \epsilon_0$; cf. for example, App. I of Kreisel [1977C].)

(ii) $o_{\mathfrak{F}}^{\mathrm{PW}}$ is the least upper bound of the Σ_1^0-orderings which can be proved in $\mathfrak{F}$ to be well-founded, so-called 'provable well orderings'. (In the case of $\mathfrak{P}$ one first passes to the two-sorted theory of arithmetic properties on pp. 339–340 of Kreisel [1968] where well-foundedness can be expressed directly.) As is well known, $o_{\mathfrak{F}}^{\mathrm{PW}}$ is not sensitive to the particular choice of Σ_1^0-orderings: adapting the proof that all well-founded Σ_1^1-orderings have recursive ordinals too, $o_{\mathfrak{F}}^{\mathrm{PW}}$ also bounds the ordinals of all Σ_1^1-orderings which can be proved in $\mathfrak{F}$ to be well founded, cf. Kreisel [1968] bottom of p. 340.

(iii) $o_{\mathfrak{F}}^{\infty}$ is the least upper bound of the generally infinite derivations d_∞ which, in Brouwer's terminology, *analyze fully* the (proofs expressed by) the derivations d of $\mathfrak{F}$; here the ordinal of the familiar tree ordering of d_∞ is meant. (We have $o^\infty_{\mathfrak{P}} = \epsilon_0$ if a 'fully analyzed' proof is a cut-*free* derivation with the ω-rule in place of induction, and the 'analysis' of a given derivation of $\mathfrak{P}$ is taken to be provided by one of the cut-elimination procedures in the literature, for example Lopez-Escobar [1975].)

(iv) $o^{\text{RED}}_{\mathfrak{F}}$ is the least upper bound for the *reduction diagrams* used in 'consistency' proofs in the style of Gentzen's original publications (where $o^{\text{RED}}_{\mathfrak{P}} = \epsilon_0$).

N.B. Usually, the reduction diagram is not defined explicitly in the literature, but must be read off the 'ordinal assignments' used as tools for the consistency proof in question.

Unquestionably, the principal attraction of those $o_{\mathfrak{F}}$, especially of $o^{\text{CON}}_{\mathfrak{F}}$ and $o^{\text{PW}}_{\mathfrak{F}}$, is that they provide a *universal measure* for all reasonably 'rich' formal systems, at least modulo the notions of 'natural definition of well orderings' and 'direct expression of well foundedness' in $o^{\text{CON}}_{\mathfrak{F}}$, resp. $o^{\text{PW}}_{\mathfrak{F}}$ above.[6] (The ideas of a 'full analysis' and '(natural) reduction' used in $o^\infty_{\mathfrak{F}}$ and $o^{\text{RED}}_{\mathfrak{F}}$, seem to be less seductive.) But on present evidence, there are very few, quite isolated problems to which the 'universal' measure is *relevant*, and this applies already in those cases where the notions of 'natural definitions' and 'direct expressions' *are* made precise. Furthermore, the bigger the $o_{\mathfrak{F}}$ (where the former notion may be dubious), the less likely are applications: after all, it is a virtue of impredicative definitions to avoid a laborious analysis in stages even if the latter is possible and heuristically useful; cf. Jensen [1972] on the fine structure of L.

Remark. It seems instructive to compare the universal measures $o_{\mathfrak{F}}$ with another chapter in what Tarski somewhere called *theory of theories*, namely, his program of characterizing (natural) syntactic classifications of sets $\mathfrak{A}$ of axioms by use of (familiar) algebraic operations on the class of models of $\mathfrak{A}$; the $\mathfrak{A}$ are intended to correspond to our formal systems, and the algebraic or 'mathematical' operations to our ordinals (or, in terms of category theory, ordinal structures). It is a fact that people tend to be preoccupied with possible ambiguities in Tarski's program, and forget that the program *was* carried out in the intended way for many $\mathfrak{A}$; the defect is elsewhere: only very few, simple results have turned out to be relevant in logic and mathematics (and those few, incidentally very pretty applications concern singularly isolated phenomena: Why can a field be ordered just in case it satisfies a suitable set of purely universal conditions?

specifically, those expressing that the field is formally real); for a parallel in the case of $o_{\mathfrak{F}}$, see p. 121 of Kreisel [1976A]. All this is in accordance with the general discussion on pp. 11–12 which leads one to expect that measures which are both *universal* and *on the surface* will generally be unrewarding.

Finally, it should be observed that work on $o_{\mathfrak{F}}^{\infty}$ can also be interpreted in terms of the *theory of proofs* (since it involves structural properties of the d and d_∞). In fact, instead of taking the *cardinal* of the set of nodes as a measure of length as was done in Section 1, we now have an *ordinal*. Granted this, the *efficiency* of the cut elimination procedures, say ν and ν', involved in the definition of $o_{\mathfrak{F}}^{\infty}$, is crudely compared by:

$$|d_\infty| < |d'_\infty|, \text{ averaging over all } d \text{ in } \mathfrak{F},$$

where $|d_\infty|$ is the ordinal of the tree ordering of d_∞. The comparison is crude in the sense that the *trivial* procedure ν^o is *almost optimal* by Mints [1976]: for all ν, $|d_\infty^o| < \omega|d_\infty|$. More precisely, ν^o which is independent of all details of d except its end formula A_d, associates to d, primitive recursively, a *canonical cut free ω-derivation tree of A_d* (which is well founded if and only if A_d is ω-valid). In fact, $|d_\infty^o| \leq \omega^2$ for arithmetic A_d, and the tree d_∞^o can be mapped primitive recursively into $\omega \cdot d_\infty$ preserving order. In other words, the procedures ν of the literature are *not* distinguished by efficiency for the ordinal measure above.

N.B. Some of the ν *are* distinguished (from ν^o) by being suitable in the sense of Section 1b(ii) while ν^o evidently is not. Thus those ν are more clearly relevant to the aims of Section 1(b) than to the proof theoretic ordinal analysis where they were first discovered.

Reminder concerning the nature of the satisfaction to be derived from a detailed critique of the assumptions of general proof theory (such as those in (a)–(c) above). It is a kind of intimate pedagogy. Contrary to a view popular among philosophers, it seems (to me) to have little value for instruction, and even less philosophical *interest* (attributed to its 'clarifying our ideas') though it is undoubtedly a philosophical *activity*. Sure, it 'clarifies'. But unless it helps – some readers, if not the author – to make positive progress, the ritual of 'clarification' is unconvincing. For one thing, there is a more radical way of achieving clarity, by simply putting aside ideas which are very much in need of clarification. More seriously, once progress has been made, one usually can go back to halfbaked general ideas, and 'clarify' them almost automatically.

APPENDIX. OTHER HEIRS OF CURRENT PROOF THEORY

The juxtaposition of *general proof theory* and the *theory of proofs* in this article is not intended to suggest that these two exhaust all possibilities of developing such 'current' proof theory as I tried to 'survey' in Kreisel [1968], or [SPT] for short, and Kreisel [1971] (nor, incidentally, do I suggest that there are no fictions in the theory of proofs; cf. the beginning of Section 1(b) above). To avoid all misunderstanding, let me give a very small selection of realistic possibilities.

(a) While it was clear that there was some overlap between proof theory (especially of intuitionistic systems) and category theory, both of which make explicit the operations which are 'hidden' in logical symbolism ([SPT], p. 352), nothing like the results on *closed* categories established in Kelley and McLane [1971] and extended in Mints [1977] was anticipated. As Mints remarks in effect, the derivations of a certain fragment of so-called 'relevance' logic, which he had studied in Mints [1972], constitute a *free* closed category (where the normal derivations are the constructible maps, and equality of normal forms determines identity of morphisms). It may fairly be said that relevance logic is not *logically* central, and the fragment considered even less; while closed categories arise naturally in the elementary theory of tensor products, a well-established part of mathematics. Since it is usually easier to make progress with tested notions, closed categories seem promising for the aim, stated in [SPT], p. 384, l. 1–2 of 'formulating the mathematical methods used in proof theory abstractly'; more promising than, for example, the fragments of propositional logic proposed *loc. cit.* for this purpose.

(b) While it has always been clear that the 'logic of forcing' is not classical, it appears that some notions suggested (for example, to A. Robinson) by 'forcing' are also useful for classical model theory. In particular, some notions, concerning a theory T, are *not* invariant for classically equivalent sets of axioms for T, and so some choice *among* its axiomatizations $\mathfrak{A}(T)$ is needed. The possibility of using experience with normalization for this choice was raised in Section 1(a). Another direction is suggested by the extension of the 'double negation translation' to set theory Powell [1975] where also a choice has to be made between classically equivalent axiomatizations. The (hypothetical) use of such work for forcing in model theory is likely to be less dubious than its more usual interpretation for 'constructivity'.

(c) Subsystems of (the more familiar – and necessarily incomplete – systems of) analysis, type theory and set theory were introduced into current proof theory in order to verify quickly that the bulk of the theorems proved in mathematical practice can already be proved in such weak systems, by using higher order language, but weak (existential) axioms. This *refinement of* PM (in the terminology of Kreisel [1976A], p. 108) corrected for example the impression that the more abstract principles of analytic number theory are 'needed (logically)' since the principles actually used in the refinement are obviously conservative over (subsystems of) first order arithmetic. The issues involved, and the devices used for reducing the 'strength' of abstract principles are discussed in pp. 327–328 of [SPT]; for a more careful formalization, see for example Takeuti [1978]. Though this is hardly needed to correct the false impression mentioned above (which refers to provability, not to proofs) it is, for example, necessary in order to settle the status of earlier publications, by Goodstein [1961] and Lorenzen [1965] where arguments of ordinary practice are transcribed into (their) unfamiliar languages of 'recursive', resp. 'ramified' analysis. Specifically, one might apply some normalization procedure to Takeuti's formalizations, and see if the results agree more or less with the transcriptions mentioned (cf. also the Exercise in Section 1a(i) above): this use of normalization for stating facts in the psychology of mathematics, of the authors cited, seems quite plausible. Another, more familiar example of weak systems is the theory KP, with its variants, in the literature on admissible sets which are the well-founded models of KP. But though the original interest in, and some critical contributions to, those weak systems required (people with) experience in current proof theory, the further development did not.

It seems to me that a much more essential use of such experience is needed for the filigree work reported in Friedman [1976], which establishes close relations between (i) systems familiar from the proof theoretic literature such as those on p. 327 of [SPT] and (ii) systems particularly adapted to special chapters of elementary mathematics; the first example of this kind of thing, described on p. 118 of Kreisel, Mints and Simpson [1975], involved the distinction between finitely and semi-finitely branching trees in König's lemma. The relation between (i) and (ii) may be compared to the pretty work in axiomatic geometry where one relates (i) such algebraic properties of the space of coordinates as commutativity to (ii) superficially quite different geometric propositions such as Pappus' theorem, all this being relative to axioms for (i) fields, resp. (ii) projective geometry. To be precise, Friedman's systems have a more *algebraic* look than those of (i). These, unlike the usual algebraic systems, are of *unbounded logical complexity* because they contain the full induction schema; his systems contain only the induction axiom (of the theory of *arithmetic properties* on pp. 339–340)[7]. They still differ from algebraic systems which are usually only 'about' a *few* operations ($+$, $\times$ and those built up by composition), in contrast to the full schema of definition by primitive recursion; perhaps this can be avoided.

The facts above suggest (to me) that the collection of systems in Friedman [1976] may be a *source for* – one or two – *useful new algebraic systems* in mathematics, especially on the border of category theory. Specifically, those systems are two-sorted, with variables for individuals and functions, and with axioms involving both sorts. These axioms may serve as a model for theories of *categories together with* (categories of) *certain functors between them* – in place of the algebraically artificial practice of giving a privileged place to the category of *all* functors (between given categories). *Warning*. The modest use suggested here is not affected by the doubts about the so-called empirical version of Hilbert's program which are set out on pp. 108–109 of Kreisel [1976A]. Work like Friedman [1976] is indeed relevant to some points in the empirical variant, but not at all to Hilbert's original intention: he was not merely interested in – limitations of – any current practice, but wanted to establish the finitist nature of *all* mathematical reasoning; cf. the abstract and p. 98 of Kreisel [1976A]. This 'grand' aim made Hilbert's program glamorous.

Perhaps more interestingly: as the reader will have noticed, the type of careful formalization in Friedman [1976] promises to be *useful for the theory of proofs provided only the emphasis is changed*, for example *from* (i) conservation, that is, the fact that a theorem F in the language of arithmetic is provable here only if it is also provable in some familiar arithmetic system $\mathfrak{P}^-$, *to* (ii) the possibility that the structural features of some proofs of F in the new systems are significantly different from those of purely arithmetical proof of F, for example, if *lower* bounds for speed-up are large.

(d) A quite different development, not at all anticipated in [SPT], comes from the business about 'elementary' consistency proofs by transfinite induction (as in Section 2c(i) of this article), where induction is applied only to 'elementary' predicates, say $\mathscr{E}$. In other words, the metamathematical results involved hold for all (appropriate) structures in which non-empty subsets $\in\mathscr{E}$ have a first element, not necessarily all sets.

An obviously interesting class of such structures are 'proof' figures which are built up according to the usual rules and which are $\mathscr{E}$-founded without being well-founded; interesting because, roughly speaking, exactly those formulae are 'provable' on such figures which are valid in all models given by data $\in\mathscr{E}$; cf. Kreisel, Mints and Simpson [1975]. *Correction.* Some details of this paper are unsatisfactory because it considers languages containing so-called Henkin constants, but establishes the generalized form of completeness just described, only for the formulae without such constants.

Department of Philosophy,
Stanford University

REFERENCES

S. Feferman, 'Ordinals and functionals in proof theory', in vol. 1, pp. 229–233 of *Actes du Congrès Int. Math. 1970*, Gauthier-Villars, Paris, 1971. Reviewed in *JSL* **40** (1975), 625–626.

H. Friedman, 'Programs and results in logic I', *Notices A.M.S.* **22** (1975), 476.

H. Friedman, 'Systems of second order arithmetic with restricted induction' (abstracts), *JSL* **41** (1976), 557–559.

K. Gödel, 'Über eine bisher noch nicht benützte Erweiterung des finiten Standpunktes', *Dialectica* **12** (1958), 280–287.

R. L. Goodstein, *Recursive Analysis*, North-Holland, Amsterdam, 1961.

R. B. Jensen, 'The fine structure of L', *Ann. Math. Logic* **4** (1972), 229–308.

G. M. Kelley and S. MacLane, 'Coherence in closed categories', *J. Pure Appl. Algebra* **1** (1971), 97–140.

G. Kreisel, 'On the interpretation of non-finitist proofs', *JSL* **16** (1951), 241–267.

G. Kreisel, 'A survey of proof theory', *JSL* **33** (1968), 321–388.[8]

G. Kreisel, 'A survey of proof theory II', in *Proc. Second Scand. Logic Symp.*, ed. J. E. Fenstad, North-Holland, Amsterdam, 1971.

G. Kreisel, 'What have we learnt from Hilbert's second problem?', *Proc. Symp. Pure Math.* **28** (1976), 93–130. (A)

G. Kreisel, 'Der unheilvolle Einbruch der Logik in the Mathematik', *Acta Phil. Fennica* **28** (1976), 166–187.[9] (B)

G. Kreisel, 'L. E. J. Brouwer, *Collected Works*' (review), *Bull. A.M.S.* **83** (1977), 86–93. (A)

G. Kreisel, 'On the kind of data needed for a theory of proofs', in R. O. Gandy and J. E. M. Hyland (eds.), *Logic Colloquium '76*, North-Holland, Amsterdam, 1977. (B)

G. Kreisel, 'Wie die Beweistheorie zu ihren Ordinalzahlen kam und kommt', *Jber. d. Dt. Math.-Verein.* **78** (1977), 177–223. (C)

G. Kreisel, 'The motto of "Investigations" and the philosophy of proofs and rules' (to appear: A).

G. Kreisel and J.-L. Krivine, *Elements of Model Theory*, 2nd. ed. North-Holland, Amsterdam, 1971.

G. Kreisel, G. E. Mints and S. G. Simpson, 'The use of abstract language in elementary metamathematics: some pedagogic examples', *Springer Lecture Notes* **453** (1975), 38–131.

G. Kreisel and W. W. Tait, 'Finite definability of number theoretic functions and parametric completeness of equation calculi', *Z. Math. Logik Grundlagen Math.* **7** (1961), 28–38.

G. Kreisel and G. Takeuti, 'Formally self-referential propositions for cut-free analysis and related systems', *Diss. Math.* **118** (1974).

E. G. K. Lopez-Escobar, 'On an extremely restricted ω-rule', *Fund. Math.* **90** (1976), 159–172.

P. Lorenzen, *Differential und Integral; Eine konstruktive Einführung in die klassische Analysis*, Akad. Verlagsges., Frankfurt, 1965.

G. E. Mints, 'A theorem on cut elimination for relevant logic', *Zap. Naučn. Sem. LOMI* **32** (1972), 90–97. This appears in translation in *J. Soviet Math.* **6** (1976), 422–428.

G. E. Mints, 'Finite investigation of infinite derivations', *ibid.* **49** (1975), 67–122.

G. E. Mints, 'Universality of canonical trees', *Dokl. Akad. Nauk SSSR* **227** (1976), 808–812.

G. E. Mints, 'Closed categories and the theory of proofs', *Zap. Naučn. Sem. LOMI* **68** (1977), 83–114.

W. C. Powell, 'Extending Gödel's negative interpretation to ZF', *JSL* **40** (1975), 221–229

D. Prawitz, 'Ideas and results in proof theory', in *Proc. Second Scand. Logic Symp.*, ed. J. E. Fenstad, North-Holland, Amsterdam, 1971.

D. Prawitz, 'Towards a foundation of general proof theory', in *Logic, Methodology and Philosophy of Science IV*, North-Holland, Amsterdam, 1973.

R. Statman, *Structural complexity of proofs*, Ph.D.Thesis, Stanford University, Stanford, 1974.

R. Statman, 'Herbrand's theorem and Gentzen's notion of direct proof' in J. Barwise (ed.), *Handbook of Mathematical Logic*, North-Holland, Amsterdam, 1977.

W. W. Tait, 'The substitution method', *JSL* **30** (1965), 175–192.

G. Takeuti, 'Two applications of logic to mathematics' (to appear, A).

NOTES

[1] (*Added in proof*) The speculations in the text, concerning (Hilbert's) original intentions for the axiomatic analysis of geometry, can now be replaced by references to the literature. Specifically, Mr. Judson Webb of Boston University has drawn my attention to a transcript, at the Widener Library (Harvard), of Hilbert's lectures which were later condensed into: *Grundlagen der Geometrie*. In the lectures (but not in the book) Hilbert explicitly proposed his axiomatic analysis as a (more fruitful) substitute for traditional philosophical questions about (*). – Statistically, he may have been right; but not for Einstein who at about that time was pondering (*) along traditional (positivist) lines – to better effect than all of axiomatic geometry put together.

[2] The principal aim of the paper was a *model theoretic* characterization of the class of recursive functions; cf. footnote on p. 29. Note that the model theoretic proof in Section 3 (p. 31) is an instance of a more general result on *free* models, given in Ex. 8 on p. 48 of Kreisel and Krivine [1971].

[3] For the record, this result holds for *all* existential formulae $\exists xA$ (with the numerical quantifier x) in the so-called theory of species, that is, analysis restricted to Heyting's logical rules. For normal derivations the proof is somewhat simpler than the argument of Kreisel and Takeuti [1974], by a remark of Statman who also showed that the normalization procedure for the theory of species in Prawitz [1971] is suitable; cf. Cor. to Prop. 4 on p. 106 of Statman [1974]. It seems open whether the result holds when (one of the usual formalizations of) the axiom of choice is added to the theory of species.
[4] but in contrast to the sense of 'general proof theory' used in my earlier writings. Since this was apparently misunderstood (in the last paragraph on p. 236 of Prawitz [1971]), I now use 'theory of proofs' for my original meaning of 'general proof theory'.
[5] The *heuristic* value (40 years ago) of the vague general idea of logical proof is not in doubt: it leads immediately to the *compactness theorem* for predicate logic which has been widely exploited during the last 40 years. Here we are concerned with the proofs themselves, for example, their reliability.
[6] for example, if well foundedness is expressed in terms of transfinite induction, $o^{PW}_{\mathfrak{F}}$ depends – in general; not for $\mathfrak{F} = \mathfrak{P}!$ – on the logical complexity of the predicates to which induction is applied; cf. p. 342(3) of Kreisel [1968]. In the case of Heyting's arithmetic, $\mathfrak{H}$, $o^{PW}_{\mathfrak{H}}$ is remarkably stable for different expressions of 'well foundedness' by Friedman [1975].
[7] This is in contrast to the asymmetry between comprehension and induction referred to on p. 327 of [SPT]. The reason is this. The 'informal analytic arguments' meant in l. 15–16, loc. cit., are not those occurring in a systematic exposition of elementary mathematics, but those imaginative ones which introduce notions far removed from the objects mentioned in the theorem to be proved, an aspect which suggested (to me) that *notions* or, equivalently, predicates should be treated differently from *objects* (sets and functions). But, given the preoccupation with proof theoretic 'strength' in Kreisel [1968], the principal objection to restricting the induction schema (to the axiom) at the cost of adding instances of the comprehension schema is a different asymmetry: more comprehension automatically implies new instances of induction, but not conversely. To take an extreme example, add to 'elementary' analysis (with both induction and comprehension restricted to quantifier-free formulae, but containing function parameters) (i) full comprehension, and (ii) full induction. Then (i) → (ii), and so we have full (impredicative) analysis, but (ii) is conservative over *first* order arithmetic.
[8] *Addenda* to this paper appeared in Kreisel [1971], 166–167. Further work has appeared since then:

p. 332. The open question in footnote 8 is settled in Kreisel and Takeuti [1974], p. 49, l. 22.

p. 334, l. 12, read: $\forall n\, \exists m\, \mathrm{Prov}(m, \ulcorner \exists y\, \mathrm{Prov}(y, \ulcorner \mathrm{Con}\, S_n \urcorner) \urcorner)$.

p. 335, l. 11–12 has been improved, by H. Friedman, to a sharper (model theoretic) result in terms of ω-consequence. For any bounded extension EA^+ of elementary analysis there is an ω-model of EA^+ for which some instance of WF($<\cdot$) fails.

p. 356, l. -15 to -13 have been refuted by H. Friedman, 'Adding propositional connectives to countable infinitary logic', *Math. Proc. Cambridge Phil. Soc.* **77** (1975), 1–6.

p. 357, l. 4–6. The question is settled positively by J. Gregory, 'On a finiteness condition for infinitary languages', *Springer Lecture Notes* **492** (1975), 143–206.

p. 373, l. 5 and l. 8. The open problems have been settled negatively by J. Steel, 'Forcing with tagged trees', *Notices A.M.S.* **21** (1974), 627–628.

p. 383, Footnote 43. Friedman's proof (mentioned in the *addition*) of my conjecture in the same footnote on p. 382, l. −7 to l. −5 has now been published, 'Uniformly defined descending sequences of degrees', *JSL* **41** (1976), 363–367. The sketch of my proof of my other conjecture (last 7 lines of footnote 43) is elaborated in the contribution of C. Smorynski 'The incompleteness theorems' on pp. 821–865 of J. Barwise (ed.), *Handbook of Mathematical Logic*, North-Holland, Amsterdam, 1977. (*Added December* 1976): A consequence of the proof, more interesting than the conjecture itself, was pointed out by H. Friedman: For any formula A of predicate logic (from which enough concatenation theory can be developed to express the consistency of A adequately), $A \cup \{\neg \text{Con } A\}$ has a uniform A-model in the sense of p. 364 b(i) in Kreisel [1968]. This consequence is a special case of an unpublished, very general result of Friedman's on a class of relative consistency proofs, of A' relative to A, which fits in very well with the theme of the present paper. The class involves explicit bounds for (hypothetical) derivations of $\neg A$, resp. $\neg A'$, with metamathematical methods restricted to A, as in my abstract *JSL* **41** (1976), 285–286; the bounds involve the number of characters and a variant of the 'cut-degree' of the derivations, so that for a fixed value of this measure there are relatively few derivations – and not the number of nodes, in accordance with the introduction to Section 1(b) of this paper.

[9] *Correction*. In l. −2 of footnote 9 on p. 121 replace 'the principle of the least upper bound' by 'completeness of the ordering of $\mathbb{R}$ (in the sense of (*) on p. 101)'. (*Added December* 1976): The type of analysis of relative consistency proofs stressed in l.–13 on p. 100 of Kreisel [1976 A] is very well illustrated by the general result of Friedman mentioned at the end of Note 8 of this paper.

DAG PRAWITZ

PROOFS AND THE MEANING AND COMPLETENESS OF THE LOGICAL CONSTANTS

In this lecture I shall discuss the notions of meaning and proof and how they depend on each other. In conclusion, I shall deal with the question of the completeness of the usual sentential operations.

I. THE NATURE OF THE SUBJECT

I.1. *The Weakness of Some Common Views*

There are two different views on the nature of proofs which seem to have great popular support and which I want to oppose in this lecture. They seem to be the ones most often articulated, and are connected each with their own philosophy of mathematics.

One view is the *formalistic* one according to which one may speak about proofs only relative a logical calculus with its given set of inference rules. On this view – and this is its obvious weakness – one cannot even ask the question whether the proofs of a given formal system adequately represent real proofs. A formal system has to be looked upon, I think, as a codification of an already existing deductive practice, and to be able to discuss whether a codification is adequate, it must be meaningful to speak about proofs not only relative to an already given system.

According to a second view, which we may call the *two-layers-view*, one has first to clarify the notions of meaning and truth independently of the notion of proof, and once this is done, the notion of proof is easily tackled. A most influential theory of meaning is the *Platonistic* one, which identifies the meaning of a sentence with its truth-conditions: semantics becomes here essentially constituted by a theory of truth, which is taken to be prior to questions of how we get to know the truths (e.g. questions of proofs). This view is often connected with the idea that when the notions of truth and logical truth are settled in this way, a proof (from some hypotheses), or more generally, a valid argument (from hypotheses) can be understood as a sequence of sentences where each sentence (is either an hypothesis or) follows logically from the preceding sentences.

J. Hintikka, I. Niiniluoto, and E. Saarinen (eds.), Essays on Mathematical and Philosophical Logic, 25–40.

The latter idea is obviously untenable (although it seems to be encouraged by many textbooks of logic). It is true that the question whether something is a proof must depend on the meaning of the sentences involved. And obviously, a valid argument must preserve truth. But the preservance of truth is clearly not a sufficient condition for validity; nobody would consider e.g. Peano's axioms followed by Fermat's last theorem as a proof, even if in fact Fermat's last theorem follows from these axioms. As every examiner stresses, it is not enough that the steps of a proof happen to follow from the preceding ones, it must also be *seen* that they follow; and it is this last requirement that must be attacked in any real analysis of the notion of proof.

I shall soon return to some difficulties that the platonistic meaning theory itself runs into, and I shall suggest that the notions of meaning and proof are intertwined with each other. But it should be clear already here that there is no easy route from such a platonistic theory of meaning to a theory of proofs.

I.2. *Theory and Practice*

Although relevant for the mathematicians, questions about meaning and proofs do not necessarily lead to interesting mathematical theories. Semantics and logic, as I understand them, are not abstract, mathematical theories but disciplines with a concrete subject matter, namely the empirically given linguistic and deductive practices that we are already engaged in. This is of course not to say that these disciplines are empirical in the sense of predicting linguistic and deductive behaviour, nor that this behaviour exists as a fixed object of study uninfluenced by our theories about it. On the contrary, theoretical reflections on the now existing practice often lead us to modify this practice. Philosophical investigations aim at making the practice more systematic and intelligible, and a modification of the practice may be the best or only way of explaining it. But these practices are nevertheless the object of study in philosophical disciplines such as semantics and logic.

This means that in one sense we already know what is to count as a proof and what it is to grasp the meaning of an expression. That is, in practice, we are able to tell whether something is a proof and whether somebody has grasped the meaning of a given expression. What semantics and logic have to do is to explain this practice by giving a systematic account of it, and by doing this, our implicit knowledge may be improved and become explicit to some extent (although it seems that the explanation will usually still have to presuppose some implicit knowledge of the same kind). The understanding of such a phenomenon as proofs will normally proceed by connecting it with other phenomena. As I want to argue, the connection between proofs and meaning is especially important here.[1]

1.3. *A Test of Meaning Theories*

The task of a meaning theory is thus to explain our linguistic practice. This means that we have a certain test of the adequacy of any proposed meaning theory, viz. that it really explains the behaviour that it should explain. Since we already know in practice what is to count as grasping the meaning of an expression, an adequate meaning theory has to identify the linguistic or other kind of behaviour that is counted as a sign of grasping meaning and has to establish a deductive connection between assumptions about knowledge of meaning and the behaviour in question. More precisely, we have an *adequacy condition* on a theory T about the meaning of a sentence A where B_A is a kind of behaviour counted as a sign of grasping the meaning of A: if T is to be adequate, it must be possible to derive in T the implication

if P knows the meaning of A, then P shows behaviour B_A.

Furthermore, T should contain nothing more than is needed to derive these implications.

I shall return in a moment to the kind of behaviour B_A that is appropriate in these tests, but it is clear that in general, it will be a question of dispositions to act in a certain way in connection with the *use* of the sentence A.

This adequacy condition is a methodological version of the slogan 'meaning is use' especially as it has been understood by Dummett (see e.g. Dummett [1] and Prawitz [7] for a discussion of this slogan). According to the adequacy condition, a theory of meaning is, roughly speaking, to contain neither more nor less than is needed to account for certain uses of these expression, viz. the use that depends on a grasp of their meaning.

As formulated here, this condition is nothing else than a special case of a general methodological requirement on theoretical terms such as meaning. A sentence to the effect that somebody knows the meaning of an expression is of course not directly the report of an observation, and general requirement of empirical import of theoretical terms demands that such a theoretical sentence is connected with appropriate observation sentences as required above; and again, this requirement is nothing else than an expression of the conviction that theoretical terms such as meaning are introduced just to account for certain linguistic behaviour.

II. DATA RELEVANT FOR A THEORY OF MEANING

II.1. *Conditions for the Making of Correct Assertions*

A person who knows the meaning of a sentence *A* knows in particular in what kind of situations it is correct or appropriate to assert *A* and in what situations this assertion is incorrect or inappropriate. This knowledge of the condition for correctly asserting *A* has to be understood as a disposition which shows itself in certain behaviour in certain situations such as actually asserting *A* in certain situations satisfying the condition and criticizing the assertion of *A* in situations when the condition is not satisfied. These kinds of behaviour thus constitute test-cases for an adequate theory of meaning as suggested above. In short: knowledge of the meaning of *A* is typically demonstrated by the ability to apply *A* correctly; in an adequate meaning theory, one should consequently be able to derive that a person knows the condition for correctly asserting a sentence *A* from the assumption that he knows the meaning of *A*.

It is important to note the difference between the truth-condition of a sentence and the condition for correctly asserting it. Truth is a property of sentences and is essentially non-temporal (provided the sentence is not elliptical or indexical), but the question whether an assertion is correct depends essentially on the situation in which it was made. That a sentence is true is certainly not sufficient for the correctness of asserting it in any situation. For instance, Fermat's last theorem may turn out to be true, but we may still doubt that Fermat had enough information to be right in asserting the theorem.

The weakness of the platonistic principle of meaning that to know the meaning of a sentence is to know its truth-conditions should now be clear: knowledge of the truth-condition of a sentence does not directly imply knowledge of the condition for correctly asserting it; and indeed, it is difficult to see how such an implication is to be obtained at all. It is even obscure what is to be meant by knowing the truth-condition of a sentence when the condition is not decidable or known explicitely (and as pointed out by Dummett [1], knowledge of meaning has in the end to be implicit). The inductive definition of truth seems to be of no help here. What is it to know that the truth of *all* instances of a universal sentence is the condition of the truth of the universal sentence? (To say that one knows some universal fact, say $\forall x A(x)$, is certainly not the same as saying that for each t, one knows that $A(t)$; but rather that one *knows* that for each t, one knows that $A(t)$; i.e., the question of what it is to know a universal fact cannot be reduced in this way.)

In other words, the platonistic theory of meaning does not seem to satisfy the adequacy condition of establishing a deductive connection between the knowledge of the meaning of a sentence and a kind of behaviour in which such knowledge is typically displayed, namely the correct application of the sentence in different situations. This criticism of a platonistic meaning theory is essentially a reformulation of Dummett's criticism in [1] (using the adequacy condition stated above instead of Dummett's stronger requirement that implicit knowledge is to be fully manifest in behaviour).

II.2. *Commitments Made by Asserting*

There are two general aspects of the use of an expression (both stressed by Dummett [1]): on the one hand, the conditions for its correct application or uttering, and on the other hand, the reactions or consequences of uttering it. In the case of sentences, this second aspect concerns the commitments made by a speaker when asserting the sentence, or in other words, what conclusions the hearer is entitled to draw from the assertion, what inferences he has the right to make. That a person understands a sentence, knows its meaning, shows itself not only in his knowledge of the condition for asserting it but of course also in his correct reactions to the assertion of the sentence, i.e. in his ability to draw the correct conclusions, to make the correct inferences.

What is to be meant by a correct inference here? The point of an inference is to allow us to assert the conclusion when we can correctly assert the premisses. This is not to say that a situation in which the premisses can be correctly asserted has to be a situation in which also the conclusion is correctly assertable. On the contrary, an inference is essentially a method which brings about a situation in which the conclusion can be asserted when a situation in which the premisses are assertable is given. An inference is thus correct in a given situation if and only if a method is known with the property that when the method is applied to a fulfilment of the conditions for asserting the premisses, it yields a fulfilment of the condition for asserting the conclusion. Hence, if we know the conditions for correctly asserting the sentences involved in an inference, we have a criterion for the correctness of the inference.

I am here making a distinction between, on the one hand, the act of inferring and its correctness, and, on the other hand, the relation of logical consequence – a distinction which is parallel with that between the correctness of an assertion and the truth of a sentence. It is the first notion, the correctness of an inference (the act of infering), that is reduced to the notion of the correctness of an assertion.

II.3. *Identification of Meaning and Conditions for Asserting. The Interdependence of Proofs and Meaning*

One may argue that the use of language that shows our grasp of its meaning consists just in our acceptance and rejection of assertions and inferences in various situations. Since the correctness of an inference is reduced to that of an assertion, one may then simply identify knowledge of meaning with knowledge of the conditions of correct assertions. I shall not directly discuss this question any further in this paper but shall now somewhat elaborate such an identification and its consequences.

A first question is the more precise nature of the condition for correctly asserting a sentence. In the case of mathematics, there is an obvious answer to this question: the construction of a proof of the sentence. My discussion so far is supposed to hold quite generally, but in the sequel, I shall restrict myself to the context of mathematics where the relevance of proofs is quite clear; I shall leave open the interesting question to what extent the rest of the discussion is applicable also outside mathematics.

The answer that the construction of a proof is the condition for asserting a sentence is not controversial, classical and intuitionistic mathematicians e.g. will completely agree here, but the answer is also very general until we have said what it is to have constructed a proof.

The view of meaning advocated here, however, gives rise to the following circle: on the one hand, the meaning of a sentence will depend on what is counted as a proof of it, but on the other hand, as already pointed out, whether something is a proof must depend on the meaning of the sentences involved. The concepts of proof and meaning must thus be clarified simultaneously.

III. ARGUMENTS AND PROOFS

III.1. *Arguments*

The syntactical form of a proof I call an *argument* (Prawitz [5] and [6]).

An argument is a tree of formulas where a formula that is immediately inferred from some formulas stands immediately below these formulas. An argument is *open* if some of its top-formulas stand as *free* assumptions, i.e. assumptions which are not *bound* in the course of the argument, or if some of its formulas are open and their parameters (free variables) are not *bound* in the course of the argument. Typical examples of inferences that bind assumptions or parameters are implication introduction and universal introduction, written

$$(1) \qquad \frac{\begin{array}{c}[A]\\ \Sigma\\ B\end{array}}{A \to B} \qquad \text{and} \qquad \frac{\begin{array}{c}\Sigma(a)\\ A(a)\end{array}}{\forall x A(x)}$$

where some set $[A]$ of assumptions of the form A free in the argument $\begin{array}{c}\Sigma\\ B\end{array}$ are bound in the resulting argument for $A \to B$, and the parameters of the form a free in the argument $\begin{array}{c}\Sigma(a)\\ A(a)\end{array}$ are bound in the resulting argument for $\forall x A(x)$, respectively.

Quite generally an argument is a tree of formulas with a specification of (i) which top formulas are assumptions, (ii) at what places the assumptions are bound (if bound at all), and (iii) at what places the parameters are bound (if bound at all). When a parameter is bound it must not occur in an assumption that is still free.

III.2. *Correctness of Arguments*

An argument may be correct or incorrect. It is correct if all its inferences are correct. As we have seen, an inference is correct, if a method is known which applied to the fulfilment of the conditions for asserting the premisses yields a fulfilment of the condition for asserting the conclusion, or, in other words (since the condition for asserting is a proof), if a method M is known which applied to proofs of the premisses yields a proof of the conclusion. Not an argument alone but an argument together with such methods M associated with the various inferences of the argument is what constitutes a proof.

The notion of proof thus depends on the notion of correctness of an inference, which again depends on the notion of proof. But this circle is now broken by observing that certain inferences are correct directly in virtue of the meaning of the sentence inferred. This occurs exactly in the case when the condition for asserting the conclusion of the inference requires just proofs of the premisses of the inference. In this case, the method which makes the inference valid is nothing else than the application of the inference.

III.3. *Specification of the Conditions for Asserting*

As a typical example of a situation of this kind, we may consider the condition for asserting $A \wedge B$. Quite uncontroversially, we require here that we either have already constructed a proof of A and a proof of B or know methods for finding such proofs. In the first case, we get a proof of $A \wedge B$ of the form

$$(2) \qquad \frac{\begin{matrix}\Sigma_1 & \Sigma_2 \\ A & B,\end{matrix}}{A \wedge B}$$

and the last inference is here obviously correct in virtue of the condition for asserting $A \wedge B$.

(2) is said to be in *canonical form* if the arguments for A and B are in canonical forms. In the second case, when we only know methods for finding proofs of A and B, we may of course also form a proof of $A \wedge B$ of the form (2) by first applying the methods in question. (It is true that the methods for finding proofs of A and B may depend on another proof of $A \wedge B$ that is not in canonical form, and the methods in question then amount to a transformation of the non-canonical or indirect proof of $A \wedge B$ to a proof in the form (2).) We can therefore conclude (given that the same holds for A and B) that the condition for asserting $A \wedge B$ is that we either have a canonical proof of $A \wedge B$ (i.e. know the correctness of an argument for $A \wedge B$ in canonical form) or know a method for finding one.

Similarly, to assert $A \rightarrow B$ or $\forall x A(x)$, we require a method which applied to a proof of A or to a term t yields a proof of B or a proof of $A(t)$, respectively, and an argument for $A \rightarrow B$ or $\forall x A(x)$ is therefore said to be in canonical form if it is in the form shown in (1) above, which obviously offers a method of the required kind when correct. Again, we can conclude that the condition for asserting $A \rightarrow B$ or $\forall x A(x)$ is that we either have a canonical proof of $A \rightarrow B$ or $\forall x A(x)$ or know a method for finding one.

The conditions for asserting sentences of these forms and the meaning of the logical constants for conjunction, implication and universal quantification are thus given by specifying the forms of the canonical arguments for these sentences. The intuitionistic meaning of disjunction and existence are in the same way determined by the canonical forms of arguments for $A_1 \vee A_2$ and $\exists x A(x)$, which are indicated by the figures

$$\frac{\begin{matrix}\Sigma \\ A_i\end{matrix}}{A_1 \vee A_2} \quad \text{and} \quad \frac{\begin{matrix}\Sigma \\ A(t)\end{matrix}}{\exists x A(x)}$$

($i = 1$ or 2), while the classical meaning of disjunction and existence is determined by specifying other canonical forms, which are at the same time the forms of canonical arguments for $\neg(\neg A \wedge \neg B)$ and $\neg\forall x\neg A(x)$ i.e. the forms

$$\frac{\begin{matrix}[\neg A][\neg B] \\ \Sigma \\ \bot\end{matrix}}{A \vee B} \qquad \frac{\begin{matrix}[\neg A(t)] \\ \Sigma \\ \bot\end{matrix}}{\exists x A(x)}.$$

where $\bot$ stands for falsehood.

III.4. *Introductions and Eliminations*

Inferences which give rise to canonical arguments in the way described above are called *introductions*. In other words, an introduction is an inference such that the result of applying it to given proofs of the premisses is, as it stands, a sufficient condition for asserting the conclusion. An inference which does not in itself produce a sufficient condition for asserting the conclusion in this way but which is correct only in virtue of a method for transforming proofs of the premisses to a proof of the conclusion (different from the trivial method of simply applying the inference) is called an *elimination*. This classification agrees with Gentzen's terminology for his system of natural deduction. It can furthermore be seen that the methods in virtue of which the eliminations are correct are just the reduction procedures in normalizations for natural deductions which I described in [3]. The proof-theoretical work reported there and in [4] thus fits into the broader semantical framework described here; or rather, this semantics is a generalization of ideas discussed in [4], [5] and [6].

IV. QUESTIONS OF COMPLETENESS

There are two natural questions about completeness that now arise. Are all logical operations whose meanings are given by conditions for asserting sentences of the corresponding form (i.e. sentences whose principal sign is the operation in question) definable by means of the usual logical operations? And are all logically correct inferences derivable from the usual inference rules?

If the semantical analysis suggested here is at all correct, every meaningful logical constant must of course get its meaning from the conditions for asserting sentences of the corresponding form, and all inferences that are seen to be correct must be correct in virtue of conditions for making assertions in the sense described above. I have argued above for the view that this must be so. But the answer to the quite different questions whether the usual logical constants and inference rules are complete may of course be negative.

In fact, even if we restrict ourselves to sentential operations, the answer to the question of completeness is obviously negative. The situation is quite parallel to the one of ordinary classical semantics, where there are many sentential operations that are not definable in sentential logic but whose meaning may be claimed to be determined by truth-conditions (to the extent that this claim makes sense at all); the modal operations being an example. The question thus arises whether the sentential operations that are definable in sentential logic are determined by some special kind of conditions for making assertions and whether they are complete with respect to that kind of conditions (comparable to the result in classical semantics that all sentential operations determined by truth-conditions of the truth-functional kind are definable in classical sentential logic). Below, I shall give an answer to this question.

I think that the answer to the other question whether all logically correct inferences are derivable from the usual inference rules is positive. But besides some comments in Section 2 below, I shall not deal with this question here. It requires a definition of logical correctness, which I am not quite sure how to give.

IV.1. *The General Form of Schematic Introductions*

By a *schematic introduction rule for an n-ary sentential operation* φ I shall understand an introduction rule given by r ($r \geqslant 0$) rule schemata of the form

$$\frac{\begin{matrix} & & [H^i_{j,1}][H^i_{j,2}]\ldots[H^i_{j,h^i_j}] & \\ \vdots & \vdots & \vdots & \vdots \\ G^i_1 & G^i_2 \;\ldots & G^i_j \;\ldots & G^i_{p_i} \end{matrix}}{\varphi(F_1, F_2, \ldots, F_n)}$$

for $i = 1, 2, \ldots, s$, where F_i, G^i_j, and $H^i_{j,k}$ ($j = 1, 2, \ldots, p_i$; $k = 1, 2, \ldots, h^i_j$) are formula schemata built up of syntactical variables for sentences and sentential operators. An inference according to such a rule schema infers a formula $\varphi(A_1, A_2, \ldots, A_n)$ from p_i premisses $B^i_1, B^i_2, \ldots, B^i_{pi}$ and may bind assumptions of the forms $C^i_{j,1}, C_{i,2}, \ldots, C^i_{j,h^i}$ that occur above the j:th premiss B^i_j, where the A:s, B:s, and C:s are formulas obtained from the corresponding F:s, G:s, and H:s by a simultaneous substitution of formulas for the syntactical variables.

Whether something is an inference according to a schematic introduction rule is thus determined only by the form of the premisses of the inference and the form of the assumptions bound by the inference, and the relation between these forms and the form of the conclusion is determined uniformly in some of the subformulas.

Example 1 (*of schematic introduction rules*). The ordinary introduction rules for $\wedge$, $\vee$, and $\rightarrow$ are obviously schematic:

for $\wedge$, we have $s = 1$, $p_1 = 2$, $h^1_1 = h^1_2 = 0$, and $G^i_j = F_j$ ($j = 1, 2$);

for $\vee$, we have $s = 2$, $p_1 = p_2 = 1$, $h^1_1 = h^2_1 = 0$, and $G^i_1 = F_i$ ($i = 1, 2$);

for $\rightarrow$, we have $s = p_1 = h^1_1 = 1$, $G^1_1 = F_2$, and $H^1_{1,1} = F_1$.

The *empty* introduction rule for a 0-place operational constant, i.e. the one with $s = 0$ and $n = 0$, is the schematic introduction rule for $\perp$; in other words, there is no introduction inference with $\perp$ as conclusion.

Example 2 (*of non-schematic introduction rules*). The introduction rule for the modal operation for necessity, N, or for strict implication, $\prec$, allowing inferences of the form

$$\frac{\begin{matrix}\Sigma \\ A\end{matrix}}{NA} \qquad \text{and} \qquad \frac{\begin{matrix}[A] \\ \Sigma \\ B\end{matrix}}{A \prec B}$$

provided all the free assumptions in the resulting argument are modal (see e.g. Prawitz [3], pp. 74–75) is not schematic since the restrictions put on the arguments for the premisses concern not only the form of the premisses and the assumptions bound by the inference but also the form of the assumptions that remain free. Similarly, the introduction rule for relevant implication (Prawitz [3], p. 82) is not schematic since it is required here that some assumption is bound in every application of the rule.

A schematic inference rule is said to be *independent* if all the formula schemata (all the F:s, G:s and H:s) consist of just syntactical variables for sentences[2]; otherwise, i.e. when some logical constants occur in the formula schemata, the rule is said to be *dependent* (*on* the set of constants occurring in the schemata). The introduction rules for $\wedge$, $\vee$, and $\rightarrow$ are obviously independent.

Example 3 (*of a dependent rule*). The schematic inference rule given by the one schema

$$\frac{\begin{matrix}[\neg A_1] & [\neg A_2] \\ \cdot & \cdot \\ \cdot & \cdot \\ \multicolumn{2}{c}{\cdot} \\ \multicolumn{2}{c}{\bot}\end{matrix}}{A_1 \vee A_2}$$

(appropriate for classical disjunction, cf. Section III.3) is dependent.

A schematic introduction rule is said to be *explicit* if the schemata $F_1, F_2, \ldots,$ and F_n consist of syntactical variables for sentences and the G:s and H:s contain no other syntactical variables than these, and if the principal sign φ of the conclusion does not occur in the G:s and H:s; otherwise it is said to be *implicit*. All the introduction rules of examples 1 and 3 are obviously explicit.

Example 4 (*of an implicit rule*). The introduction rules for constructible falsity stated in Prawitz [3], p. 97, which is given by four rule schemata, two for the negations of conjunction, one for the negation of implication, and one for the negation of negation, would be schematic if we relaxed the requirement that the conclusions in the different rule schemata are to be the same but would still be implicit since the defined operation, say $-$, occurs also in the schema for the premiss; e.g. one rule schema is

$$\frac{-A}{-(A \wedge B)}$$

IV.2. *The Elimination Rule Corresponding to a Schematic Introduction Rule*
Gentzen [2] suggested that "it should be possible to display the elimination

rules as unique functions of the corresponding introduction rules on the basis of certain requirements". One has indeed a strong feeling that the elimination rules are obtained from the corresponding introduction rules in a uniform way. For the introduction rules of the schematic type, we can now easily describe how the elimination rules are obtained uniformly from the corresponding introduction rule.

But a solution to the problem suggested by Gentzen should also establish that these elimination rules are the strongest possible elimination rules corresponding to the given introduction rules, i.e. the strongest possible elimination rules that are correct in virtue of the meaning of the logical constant stated by the condition for asserting sentences of the form in question. In other words, a solution of the completeness problem for the introduction and elimination rules as suggested in the introduction to this Section IV is required. It seems obvious that there can be no stronger correct elimination rules for e.g. $\wedge$, $\vee$, and $\rightarrow$ than the usual ones (but it does not seem obvious how this is to be proved). Similarly, the elimination rule now to be stated is conjectured to be the strongest elimination rule corresponding to the introduction rule for φ stated in the section above, i.e. the strongest rule whose correctness can be seen merely from the meaning of φ.

The *elimination rule* for an n-ary sentential operation φ whose introduction rule is the one stated in Section IV.1 above is given by one rule schema of the form

$$\frac{\varphi(F_1, F_2, \ldots, F_n) \quad \begin{matrix}\Gamma_1 \\ A\end{matrix} \ \ldots \ \begin{matrix}\Gamma_i \\ A\end{matrix} \ \ldots \ \begin{matrix}\Gamma_{pi} \\ A\end{matrix}}{A}$$

with r minor premisses of the form A and where assumptions occurring above the i:th ($i \leqslant r$) minor premiss may be bound when they are of one of the forms shown in Γ_i

$$\Gamma_i = \bigwedge_{\substack{i \\ k \le h_1}} H^i_{1,k} \rightarrow G^i_1, \bigwedge_{\substack{i \\ k \le h_2}} H^i_{2,k} \rightarrow G^i_2, \ldots, \bigwedge_{\substack{i \\ k \le h_{p_i}}} H^i_{p_i,k} \rightarrow G^i_{p_i}$$

($\bigwedge$ being a sign for iterated conjunction). If h^i_j is 0, then the iterated conjunction in question is just to be left out and the implication is replaced by its consequent. If $p_i = 0$, then there are no assumptions occurring above the i:th minor premiss bound by the inference, in which case the whole rule is without any deductive power. If $r = 0$, there is only one premiss, namely $\varphi(F_1, F_2, \ldots, F_n)$.

The usual elimination rules for ∨ and ⊥ have exactly the form described here, while those for ∧ and → have a slightly different form, which is easily seen to be deductively equivalent to the one given here.

It is not difficult to see that all inferences according to this elimination rule are correct, and that there are reductions of introduction inferences for ϕ followed by such elimination inferences, which give rise to the usual normalizability results.

IV.3. *Completeness of the Set* {⊥, ∧, ∨, →}

The set {⊥, ∧, ∨, →} is complete within intuitionistic sentential logic with respect to all sentential operations that have explicit schematic introduction rules.[3] More precisely, each n-ary sentential operation with such an introduction rule depending on a set Δ of logical constants is definable within intuitionistic sentential logic in the sense that there exists a formula schema $\Phi(F_1, F_2, \ldots, F_n)$ built up of only the syntactical variables $F_1, F_2, \ldots, F_n$ for sentences and logical constants from the set $\{\bot, \wedge, \vee, \rightarrow\} \cup \Delta$ such that for any sentences $A_1, A_2, \ldots, A_n$

$$\varphi(A_1, A_2, \ldots, A_n) \equiv \Phi(A_1, A_2, \ldots, A_n)$$

is provable in ordinary intuitionistic sentential logic extended with the introduction rule for φ and the corresponding elimination rule described in Sections IV.1 and IV.2.[4]

For the proof, consider first some special cases. When the introduction rule is the empty one, i.e. when $r = 0$, then φ is a logical constant for falsehood and

$$\varphi(A_1, A_2, \ldots, A_n) \equiv \bot$$

as required.

When $r > 0$ but one of the rule schemata has no premisses, i.e. some $p_i = 0$, then φ is a logical constant for truth and e.g.,

$$\varphi(A_1, A_2, \ldots, A_n) \equiv (\bot \rightarrow \bot)$$

is provable as required.

(In other words, there are only two 0-place sentential operations with independent schematic introduction rules, namely truth and falsehood.)

In the main case, when $r > 0$ and $p_i > 0$ for each $i \leq r$, then the equivalence

$$(3) \qquad \varphi(A_1, A_2, \ldots, A_n) \equiv \bigvee_{i \leq r} (\bigwedge_{j \leq p_i} (\bigwedge_{k \leq h^i_j} C^i_{j,k} \rightarrow B^i_j))$$

is provable as required ($\bigwedge$ and $\bigvee$ being signs for iterated conjunctions and disjunctions), where the B:s and C:s are obtained from the corresponding G:s and H:s by substituting $A_1, A_2, \ldots,$ and A_n for $F_1, F_2, \ldots,$ and F_n, respectively.

It follows that if a set of sentential operations with dependent, explicit, and schematic introduction rules can be well-ordered in such a way that for each constant φ, the corresponding introduction rule depends only on the set of constants preceding φ in the well-ordering, then all the constants of the set can be defined using only the constants $\perp$, $\wedge$, $\vee$, and $\rightarrow$.

When the introduction rule is schematic and implicit, an equivalence of the same form (3) is provable except that the A:s, B:s, and C:s are then obtained by a simultaneous substitution in the corresponding F:s, G:s and H:s. But the equivalence does not then amount to an explicit definition since φ is then not necessarily defined in all contexts (only the ones of the form $\varphi(F_1, F_2, \ldots, F_n)$) and φ may occur also in the right member of the equivalence.

Department of Philosophy,
University of Stockholm

REFERENCES

Dummett, Michael, 'The philosophical basis of intuitionistic logic', in *Logic Colloqium '73* (H. E. Rose and J. C. Shepherdson, eds.), Amsterdam, 1975.

Gentzen, Gerhard, 'Untersuchungen über das logische Schliessen', *Mathematische Zeitschrift* **39** (1935), pp. 176–210.

Prawitz, Dag, *Natural Deduction. A Proof-Theoretical Study*, Stockholm, 1965.

Prawitz, Dag, 'Ideas and results in proof theory', in *Proceedings of the Second Scandinavian Logic Symposium* (ed. J. E. Fenstad), Amsterdam, 1975, pp. 235–250.

Prawitz, Dag, 'Towards a foundation of general proof theory', in *Logic, Methodology, and the Philosophy of Science IV* (eds. P. Suppes *et al.*), Amsterdam, 1973, pp. 63–77.

Prawitz, Dag, 'On the idea of a general proof theory', *Synthese* **27** (1974), pp. 63–77.

Prawitz, Dag, 'Meaning and proofs: on the conflict between classical and intuitionistic logic', *Theoria* **43**, 2–40.

Zucker, Jeffrey, *The adequacy problem for inferential logic*, Department of Mathematics, University of Utrecht, preprint nr. 37, 1976.

NOTES

[1] That understanding of e.g. proofs must proceed not by concentrating on the question 'what is a proof' in isolation but by finding connections between proofs and other things is well described by Kreisel's contribution to this volume. However, somewhat surprisingly, he seems nowadays to think that the things with which the proofs shall be revealingly connected are some measures of the complexity of these very same proofs (such as their length and genus).

[2] It can then be assumed without any loss of generality that no other variables than $F_1, F_2, \ldots, F_n$ occur as G:s and H:s: if G^i_j was a variable different from $F_1, F_2, \ldots,$ and F_n, then any instance of this rule would remain a correct instance if the argument foı G^i_j was replaced by a correct argument for a provable formula (e.g. $\bot \to \bot$), and hence the deductive power of the inference rule would not change by deleting G^i_j (and its accompanying H:s); similarly, if $C^i_{j,k}$ was a variable different from $F_1, F_2, \ldots,$ and F_n, then any instance of this rule would remain a correct instance if the argument for G^i_j was replaced by the correct argument

$$\frac{\bot}{G^i_j}$$

and the assumption $\bot$ was bound by the inference as being of the form $H^i_{j,k}$, and hence the deductive power of the inference rule would again not be changed by deleting G^i_j (and its accompanying H:s).

[3] J. Zucker [8] has independently obtained a similar result but within a somewhat different semantical framework.

[4] It follows that no proper subset of $\{\bot, \vee, \vee, \to\}$ has this property within intuitionistic first order predicate logic (Prawitz [3], p. 59) but that $\{\to, \forall\}$ is sufficient within intuitionistic second order predicate logic (ibid., p. 67). Within classical sentential logic $\{\bot, \to\}$ is of course complete with respect to all sentential operations with explicit schematic introduction rules.

V. A. SMIRNOV

THEORY OF QUANTIFICATION AND $\mathscr{E}$-CALCULI

There are two types of systems of natural deduction: the first is based on an indirect rule for the elimination of the existential quantifier; and the second is based on a direct rule for the elimination of the existential quantifier.

We shall call the first calculus NC (Natural Classical); the second, N$\mathscr{E}$C (Natural Classical with $\mathscr{E}$-symbol). NC is equivalent, as we all know, to the axiomatic calculus of predicates HC. But N$\mathscr{E}$C is not equivalent to HC. It is equivalent to H$\mathscr{E}$C (classical axiomatic calculus with $\mathscr{E}$-symbol). N$\mathscr{E}$C is not an inessential extension of NC, but only a conservative extension of NC. Then we shall construct an intuitionistic natural system of the second type – system N$\mathscr{E}$I.

I

Let us construct the system H$\mathscr{E}$C.

The language has two kinds of variables: free – $v, w, u, \ldots$ and bound – $x, y, z. \ldots$ There are predicate constants and logical constants: &, $\vee$, $\supset$, $\neg$, $\forall$, $\exists$, $\mathscr{E}$. We define with the help of simultaneous induction 'quasiterm with the list of free variables V and the list of bound variables W' and 'quasiformula with the list of free variables V and the list of bound variables W':

1. If t is a free individual variable, then t is a quasiterm with the list of free variables t and the empty list of bound variables.
2. If t is a bound individual variable, then t is a quasiterm with the empty list of free variables and the list of bound variables t.
3. If $t_1, \ldots, t_k$ are quasiterms with the lists of free variables $V_1, \ldots, V_k$ and with the lists of bound variables $W_1, \ldots, W_k$ respectively and if P is a k-place predicate symbol (constant), then $P(t_1, \ldots, t_k)$ is a quasiformula with the list of free $V_1 \ldots V_k$ and the list of $W_1 \ldots W_k$ bound variables respectively.
4. If A is a quasiformula with the list of free V and of bound W variables, then $(\neg A)$ is a quasiformula with the same lists of variables.

J. Hintikka, I. Niiniluoto, and E. Saarinen (eds.), Essays on Mathematical and Philosophical Logic, 41–47.

5. If A and B are quasiformulae with the lists V_1 and V_2 of free and lists W_1 and W_2 of bound variables, then $(A \& B)$, $(A \wedge B)$, $(A \supset B)$ are quasiformulae with the list of free V_1V_2 and of bound W_1W_2 variables.

6. If A is a quasiformula with the list of free V and of bound W variables, and x is a bound variable, then $\forall xA$ and $\exists xA$ are quasiformulae with the list of free V and of bound W_x variables, where W_x is the result of crossing out all occurrences x in W.

7. If A is a quasiformula with the list of free V and of bound W_x variables, then $\mathscr{E}xA$ is a quasiterm with the list of free V and of bound W_x variables.

A term is a quasiterm with an empty list of bound variables. A formula is a quasiformula with an empty list of bound variables. A sentence is a formula with an empty list of free variables.

Axiom schemata:

propositional axiom schemata

$\forall xAx \supset At$

$At \supset \exists xAx$

$\exists xAx \supset A(\mathscr{E}xAx)$

$A(\mathscr{E}x \neg Ax) \supset \forall xAx$

where t is a term, that is, a free variable or $\mathscr{E}$-term.

Inference rules

modus ponens $\dfrac{A \quad A \supset B}{B}$

the rule of generalization $\dfrac{Aw}{\forall xAx}$

Deduction is a sequence of formulae each of which is either an assumption, or an axiom, or is immediately deduced according to the rules of inference from the previous formulae.

It is not difficult to prove that H$\mathscr{E}$C is equivalent to Hilbert's $\mathscr{E}$-calculus (with the definitions of quantifiers). H$\mathscr{E}$C is an extension of HC that is: HC $\subseteq$ H$\mathscr{E}$C.

But is the calculus H$\mathscr{E}$C really an inessential extension of HC? No, it isn't. Since I have not found any proof of this theorem in the literature I shall give now a simple proof of it.

As is well known, theory T_1 is known as the conservative extension of T

iff T_1 is an extension of T and for any $A \in LT_1$ (LT_1 is the language of theory T_1) if $T_1 \vdash A$, then $T \vdash A$.

It is not difficult to see that if $T \subseteq T_1$, $T_1 \subseteq T_2$ and T_2 is a conservative extension of T then T_1 is a conservative extension of T.

T_1 is called an inessential extension of T iff

(I) T_1 is a conservative extension of T and

(II) There is the function φ which transforms formulae LT_1 into formulae L_T so that

(1) if $A \in L_T$, then $\varphi(A)$ is A

(2) $T_1 \vdash A \equiv \varphi(A)$

We shall need a Lemma for our proof.

LEMMA. If the theories T_1 and T_2 are formulated in the same language ($L_{T1} = L_{T2}$), T_2 is a conservative extension of T, $T \subseteq T_1$ and $T_1 \subset T_2$ ($T_1 \neq T_2$), then T_1 is not an inessential extension of T.

The proof is simple.

Proof:

1.	$\exists\varphi\forall C(T_1 \vdash C \equiv \varphi(C))$	assumption
2.	$\forall C(T_1 \vdash C \equiv \varphi(C))$	$\exists_E$; 1
3.	$T_1 \subset T_2$ and $T_1 \neq T_2$	condition
4.	$\forall C(T_2 \vdash C \equiv \varphi(C))$	2,3
5.	$\exists C(T_2 \vdash C$ and $T_1 C \nvdash)$	3
6.	$T_2 \vdash C$	5
7.	$T_1 \nvdash C$	5
8.	$T_2 \vdash \varphi(C)$	4,6
9.	$T \vdash \varphi(C)$	T_2 is a conservative extension of T; 8
10.	$T \subseteq T_1$	condition
11.	$T_1 \vdash \varphi(C)$	9,10
12.	$T_1 \vdash C$	2,11
13.	Contradiction.	

Consequently, T_1 is not an inessential extension of T.

Now we shall use this lemma as a proof that H$\mathscr{E}$C is not an inessential extension of HC.

Let H$\mathscr{E}$C$^+$ be an extension of H$\mathscr{E}$C at the expense of an addition to the axiom schema

$$\forall x(Ax \equiv Bx) \supset (F_{\mathscr{E}xAx}^{\;w}\, C \supset F_{\mathscr{E}xBx}^{\;w}\, C)$$

or $\forall x(Ax \equiv Bx) \supset \mathscr{E}xAx = \mathscr{E}xBx$, if we deal with a calculus with an identity.

As is well known, H$\mathscr{E}$C$^+$ is a conservative extension of HC and H$\mathscr{E}$C$^+$ is the proper extension of H$\mathscr{E}$C.

On the basis of the given lemma we get the following theorem.

THEOREM. H$\mathscr{E}$C is not an inessential extension of HC.

Thus H$\mathscr{E}$C is a generalization of the standard theory of quantification as well as the theory with quantifiers of Henkin, $Q\alpha$ of Mostowski and others.

It seems to me that with the help of the $\mathscr{E}$-symbol, it is possible to reconstruct all the examples in the analysis of which it is necessary for Professor Hintikka to have Henkin-quantifiers.

N$\mathscr{E}$C is a natural calculus built by the method of subordinate proofs. Quantifier rules are:

$$\forall_J \frac{Aw}{\forall xAx} \qquad \forall_E \frac{\forall xAx}{At}$$

$$\exists_J \frac{At}{\exists xAx} \qquad \exists_E \frac{\exists xAx}{A(\mathscr{E}xAx)}$$

N$\mathscr{E}$C is equivalent to H$\mathscr{E}$C.

Sometimes the $\mathscr{E}$-symbol is not used explicitly during the construction of the system of natural deduction with a direct rule of elimination of the existential quantifier. But $\mathscr{E}$-terms are present implicitly. Thus, instead of description $\mathscr{E}xA(x, y_1, \ldots, y_n)$ it is possible to use a constant with indexes $a_{y_1, \ldots, y_n}$. In this case different letters correspond to different descriptions. This method is developed in the book by Slupecki and Borkowski. This is another method of fixation of a description.

Thus the system H$\mathscr{E}$C is equivalent to the system of natural deduction with the direct rule of elimination of the existential quantifier. Hence the system of natural deduction with the direct rule of elimination of the existential quantifier is not equivalent to the standard calculus of predicates, as is often stated. It is only a conservative extension of the predicate calculus.

II

The method of natural deduction with the direct rule of elimination of the existential quantifier is very elegant. It is useful to apply it to the construction of intuitionistic and other systems of logic.

The first attempt of this kind was made by Borkowski in 1958 (Borkowski and Slupecki, 1958). Intuitionistic logic is constructed by him with the aid of changing the propositional part of the classical natural system. But unfortunately, this method is not adequate. Indeed, in this system the provable formulae

$$\exists x(\exists yAy \supset Ax),\ (\exists xAx \supset \exists xBx) \supset \exists x(Ax \supset Bx)$$

are not accepted by intuitionists.

For instance

1.	|	$\exists yAy$
2.	|	$A(\mathscr{E}yAy)$
3.		$\exists yAy \supset A(\mathscr{E}yAy)$
4.		$\exists x(\exists yAy \supset Ax)$

In 1971 in my article 'Elimination des termes dans la logique intuitioniste' (*Revue international de philosophie*) and in the book *Formal Deduction and Logical Calculus* I proposed to use the natural intuitionistic calculus with the $\mathscr{E}$-term.

Quantifier rules are:

$$\forall_E\ \frac{\forall xAx}{Aa} \qquad \forall_J\ \frac{Aw}{\forall xAx}$$

$$\exists_E\ \frac{\exists xAx}{A(\mathscr{E}xAx)} \qquad \exists_J\ \frac{Aa}{\exists xAx}$$

$$\mathscr{E}_E\ \frac{B(\mathscr{E}xAx)}{\exists xAx} \qquad \mathscr{E}_J\ \frac{\forall xBx\ \exists yAy}{B(\mathscr{E}yAy)}$$

Here, where the term a is different from $\mathscr{E}$-term.

Some limitations and restrictions are imposed on the use of subsidiary rules: $\mathscr{E}$-symbols is not to occur in the eliminable assumptions and conclusion of subsidiary deduction.

Propositional rules are ordinary.

Let us call this calculus $N\mathscr{E}I_1$. $N\mathscr{E}I_1$ differs from the classical calculus $N\mathscr{E}C_1$ only by the propositional rules.

THEOREM. $N\mathscr{E}I_1(N\mathscr{E}C_1)$ is a conservative extension of NI (NC).

The proof is given in my works paper and my book (1971, 1972). Similar constructions (but evidently independent) were received by D. Leivant (1973). G. E. Mints (1974) has noted that my limitation on deduction (namely, the demand that $\mathscr{E}$-symbol should not occur in the eliminated assumption and conclusion of subsidiary deduction) is too hard. Other conservative extensions of intuitionistic quantifier theory were proposed by A. G. Dragalin (1974) and G. E. Mints (1974).

A. G. Dragalin considers intuitionistic calculus with identity supplemented by axiom schemata:

$$\exists t \;\&\; \forall x A x \supset A t$$
$$\exists t \;\&\; A t \supset \exists x A x$$
$$\exists x A x \supset \mathscr{E} x A x = \mathscr{E} y A y$$
$$\exists x A x \supset A(\mathscr{E} x A x)$$

where $\exists t$ is the abbreviation for $\exists x Bx$, if t is $\mathscr{E}xBx$, and $\exists t$ is $\exists x(x = x)$, if t is x. Let us call this calculus $H\mathscr{E}I_2$.

Without identity it is equivalent to $N\mathscr{E}I_2$. The rules of inference of $N\mathscr{E}I_2$ are

$$\forall_E \frac{\forall x A x}{A a} \qquad \forall_J \frac{A w}{\forall x A x}$$

$$\exists_E \frac{\exists x A x}{A(\mathscr{E} x A x)} \qquad \exists_J \frac{A a}{\exists x A x}$$

$$\mathscr{E}_E \frac{\exists x A x \;\; B(\mathscr{E} x A x)}{\exists x B x} \qquad \mathscr{E}_J \frac{\forall x B x \;\; \exists y A y}{B(\mathscr{E} y A y)}$$

Dragalin, with the help of model theory, proved that $H\mathscr{E}I_2$ is a conservative extension of HI. However, as Dragalin has shown, $H\mathscr{E}I_2$ with identity and supplemented by axiom schema

$$\exists x A x \;\&\; \exists x B x \;\&\; \forall x(Ax \equiv Bx) \supset \mathscr{E} x A x = \mathscr{E} x B x$$

is not a conservative extension of HI. Intuitionistic arithmetic with the formal analogue of Church's thesis in this system is inconsistent.

Above, I asked whether the system obtained from $N\mathscr{E}I_2$ after replacing $\mathscr{E}_J$ by $\dfrac{\forall x Bx}{B(\mathscr{E}xAx)}$ is a conservative extension of HI. Now I can give a negative answer.

Institute of Philosophy,
Academy of Science, Moscow

BIBLIOGRAPHY

L. Borkowski and J. Slupecki, 'A logical system based on rules', *Studia Logica* **7** (1958).

А. Г. Драгалин, 'Интуиционистская логика и $\mathscr{E}$-символ Гильберта,' *История и методология естественных наук*, вып. XYI, Математика, механика. МГУ, 1974.

A. C. Leisenring, *Mathematical logic and Hilbert's $\mathscr{E}$-symbol*, London, 1969.

Г. Е. Минц, ,Гейтинговское исчисление предикатов с эпсилонсимволом,, Записки научных семинаров ЛОМИ, т.40. *Исследования по конструктивной математике и математической логике*. YI. Ленинград, 1974.

В. Н. Оревков, ,Элиминация функциональных знаков без условия единственности конструктивные исчисления предикатов с $\mathscr{E}$-символом', *Теория логического вывода* (тезисы докладов Всесоюзного симпозиума, Москва, март 25-27, 1974), ч.I, М., 1974.

Е. Слупецкий, Л. Борковский, *Элементы математической логики и теория множеств*, М., 1965 (перевод с польского, 1963).

V. A. Smirnov, 'Elimination des termes $\mathscr{E}$ dans la logique intuitioniste', *Revue International de Philosophie* **98**, 1971.

В. А. Смирнов, *Формальный вывод и логические исчисления*, М., 1972.

D. Leivant, 'Existential instantiation in a system of natural deduction for intuitionistic arithmetic', Stichting Mathematisch Centrum, Amsterdam, 1973.

LARS SVENONIUS

TWO KINDS OF EXTENSIONS OF PRIMITIVE RECURSIVE ARITHMETIC

INTRODUCTION

The main object of study in this paper is a class of calculi of second-order recursive arithmetic, called 'calculi of Kind I'. These are very similar to the second-order calculi considered in Tait (1965). The main focus is on questions about 'translations' of various sorts between these calculi, and certain other calculi, called 'calculi of kind II'. These are more like first-order calculi, only with some built-in rules for definition by recursion.

The results on the relations between the two kinds of calculi might serve to illuminate the relation between second-order quantifier-free calculi, and first-order quantifier-free calculi.

Many of the main ideas in the paper can already be found in the paper by Tait.

PART I. ELEMENTARY CALCULI (FIRST-ORDER SYSTEMS OF RECURSIVE ARITHMETIC)

What we call elementary calculi are certain extensions of primitive recursive arithmetic, similar to those considered in Tait (1965).

We introduce some formal apparatus for discussing these calculi.

1. Elementary Symbols

The elementary symbols are

(i) A denumerable set of variables.

(ii) For each $k \geqslant 0$, k-ary function constants, predicate constants, and operation constants.

(iii) The propositional connectives $\neg$, $\vee$, $\wedge$, $\rightarrow$, $\leftrightarrow$.

The 0-ary function constants will be called *numerical constants*.

2. Definition of "Numerical Term", "Function Term", "Formula"

(i) Variables are numerical terms.

J. Hintikka, I. Niiniluoto, and E. Saarinen (eds.), Essays on Mathematical and Philosophical Logic, 49–94.

(ii) When A is a k-ary function constant (predicate constant, operation constant), and $t_0, \ldots, t_{k-1}$ are numerical terms, then $At_0, \ldots, t_{k-1}$ is a numerical term (formula, function term).

(Observe: k is allowed to be 0 in this rule.)

(iii) When χ is a function term, and t a numerical term, then $\{\chi\}(t)$ is a numerical term.

(iv) If A and B are formulas, then so are $\neg A$, $(A \vee B)$, $(A \wedge B)$, $(A \rightarrow B)$, and $(A \leftrightarrow B)$.

3. The Set $\mathscr{L}_0$ (of Elementary Function, Predicate, and Operation (Constants)

The set $\mathscr{L}_0$ is freely generated from a set of 'initial constants' (enumerated below) by the partially defined operations (i) – (vi) to be described. With each constant other than the initial constants is associated one or two 'defining axioms'.

A. Initial Constants

The elementary initial constants are the following: 0 (a numerical constant), Eq (a binary predicate constant), S and I (1-ary function constants).

Note: Eq is the identity symbol, and we write as usual $s = t$ for Eq(s, t).

B. The Generating Operations

(i) When f is an n-ary function constant in $\mathscr{L}_0$, and Π a mapping $n \rightarrow k$, then

$M_{n,k,\Pi}(f)$ is a k-ary function constant in $\mathscr{L}_0$.

When g is $M_{n,k,\Pi}(f)$, its defining axiom is

$$g(x_0, \ldots, x_{k-1}) = f(x_{\Pi(0)}, \ldots, x_{\Pi(n-1)}).$$

(ii) When g is an m-ary function constant, and h an $m + 1$-ary function constant, then $Su(h, g)$ is an m-ary function constant.

When f is $Su(h, g)$, its defining axiom is

$$f(\bar{x}) = h(\bar{x}, g(\bar{x})).$$

[We start here to use the convenient convention of writing $\bar{x}$ for the string of variables $x_0, \ldots, x_{m-1}$. The length of the string will always be clear from the context.]

(iii) When g is an m-ary function constant in $\mathscr{L}_0$, and h an $m + 2$-ary function constant in $\mathscr{L}_0$, then $Re(h, g)$ is an $m + 1$-ary function constant in $\mathscr{L}_0$.

When f is $Re(h, g)$, its defining axioms are:

$$\begin{cases} f(\bar{y}, 0) = g(\bar{y}); \\ f(\bar{y}, Sz) = h(\bar{y}, z, f(\bar{y}, z)). \end{cases}$$

(iv) When f is an m-ary function constant in $\mathscr{L}_0$, then Pred(f) is an m-ary predicate constant in $\mathscr{L}_0$.

When A is Pred(f), its defining axiom is

$$A(\bar{y}) \leftrightarrow f(\bar{y}) = 0.$$

(v) When f is an $m + 1$-ary function constant in $\mathscr{L}_0 (m \geqslant 0)$, then $\lambda(f)$ is an m-ary operation constant in $\mathscr{L}_0$.

When p is $\lambda(f)$, its defining axiom is

$$\{p(\bar{y})\}(z) = f(\bar{y}, z).$$

(vi) When R is a binary relation constant in $\mathscr{L}_0$, and p is an m-ary operation constant in $\mathscr{L}_0$, then $Fo(R, p)$ is an m-ary function constant in $\mathscr{L}_0$.

When h is $Fo(R, p)$, its defining axioms are:

$$z = h(\bar{y}) \rightarrow \langle \{p(\bar{y})\}(Sz), \{p(\bar{y})\}(z) \rangle \notin R;$$
$$z < h(\bar{y}) \rightarrow \langle \{p(\bar{y})\}(Sz), \{p(\bar{y})\}(z) \rangle \in R.$$

Here we start the habit of sometimes writing $\in$ instead of juxtaposition, i.e. $\langle s, t \rangle \in R$ instead of $R(s, t)$.

The predicate constant $<$ is one defined in the usual way: (see Section 7 below).

The operations mentioned in (i)–(v) are called the '*obligatory operations*' on the set $\mathscr{L}_0$. The operation Fo is an '*optional operation*'.

We define the following notion of '*immediate predecessor*' (i.p.) for constants in $\mathscr{L}_0$: (i) f is the i.p. of the constant $M_{n,k,\Pi}(f)$; (ii) g and h are the immediate predecessors of $Su(h, g)$ and also of $Re(h, g)$; (iii) f is the immediate predecessor of Pred(f), and of $\lambda(f)$; (iv) R and p are the i.p.s of $Fo(R, p)$.

4. *General Axioms, and Rules of Inference*

A. The general axioms

1. All tautologies;
2. $x = x$, $Sx \neq 0$, $Sx = Sy \rightarrow x = y$, $Ix = x$;
3. $x = y \rightarrow (A(x) \rightarrow A(y))$;

where $A(y)$ is obtained from $A(x)$ by replacement of y for some occurrence of x, in the usual sense.

B. Rules of Inference

$$\text{(i)}\ \frac{A(x);}{A(t)} \qquad \text{(ii)}\ \frac{A,\ A \to B;}{B} \qquad \text{(iii)}\ \frac{A(0),\ A(z) \to A(Sz)}{A(x)}.$$

(In (i), x is a variable, t a numerical term, $A(t)$ the result of substituting t for x in $A(x)$; in (iii), $A(0)$, $A(Sz)$, and $A(x)$ are the results of substituting, resp., 0, Sz and x, for z in $A(z)$).

5. *Elementary Calculi*

We define an *elementary calculus* to be a subset of $\mathscr{L}_0$, which is closed under the obligatory operations (i)–(v), and closed under the relation of immediate predecessor.

When C is an elementary calculus, we will refer to the following associated sets as follows:

(a) The 'set of constants of C' is C itself.

(b) The 'terms and formulas of C' are those elementary terms and formulas which contain no constants except those in C.

(c) A 'proof in C' is a proof from the general axioms and defining axioms using the rules of inference above, where all the formulas involved are formulas of C.

(d) A 'theorem of C' is a formula, for which there is a proof in C. We write $\vdash_C A$ for the statement 'A is a theorem of C'.

There is a minimal elementary calculus, namely the set of constants generated from the initial constants by the obligatory operations (i)–(v). This calculus is a formulation of ordinary primitive recursive arithmetic, and we will refer to it as PRA.

More generally, for any subset $\mathscr{M}$ of the set $\mathscr{L}_0$, there is a minimal elementary calculus C for which $\mathscr{M} \subseteq C$. We will denote this calculus $C[\mathscr{M}]$.

6. *Equivalence between Constants in a Calculus* C; *Translations* $C_1 \to C_2$.

Equivalence between constants in a calculus C is defined in the obvious way:

If f and g are two k-ary function constant, p and q two k-ary operation constants, and A and B two k-ary predicate constants, then

(i) f and g are equivalent in C, if $\vdash_C f(\bar{y}) = g(\bar{y})$;
(ii) p and q are equivalent in C, if $\vdash_C \{p(\bar{y})\}(z) = \{q(\bar{y})\}(z)$;
(iii) A and B are equivalent in C, if $\vdash_C A(\bar{y}) \leftrightarrow B(\bar{y})$.

When C_1 and C_2 are two elementary calculi, and M a mapping $C_1 \rightarrow C_2$, which preserves syntactic category (i.e., m-ary function-, predicate-, and operation constants of C_1 are associated with, resp., m-ary function-, predicate-, and operation constants in C_2), and if M^* is the induced mapping of formulas, then we call M a *translation* $C_1 \rightarrow C_2$ (and C_1 *translatable in* C_2), if

(i) $M(0) = 0$, $M(S) = S$; and
(ii) For each formula A of C_1, if $\vdash_{C_1} A$, then $\vdash_{C_2} M^*(A)$.

Two calculi C_1 and C_2 are called *equivalent* if each is translatable in the other.

When C_1 and C_2 are two calculi, M and M' two translations $C_1 \rightarrow C_2$, we call M and M' *equivalent translations*, if for every constant K of C_1, the constants $M(K)$ and $M'(K)$ are equivalent in C_2.

The following theorem states that translations are unique, up to equivalence.

THEOREM 1. If C_1 and C_2 are two elementary calculi, M and M' two translations $C_1 \rightarrow C_2$, then M and M' are equivalent.

We can prove the standard result on explicit definitions, i.e.:

THEOREM 2. Let C be any elementary calculus, t any numerical term in C in the variables $y_0, \ldots, y_{k-1}$. Then there is a function constant f in C, such that

$$\vdash_C f(\bar{y}) = t.$$

A constant in $\mathscr{L}_0$ obtained by the operation Fo, i.e., of form $Fo(R, p)$, is said to be 'defined by recursion on R'. When C is an elementary calculus, R a binary predicate constant in C, we say that 'C is closed under recursion on R', if for every operation constant p in C, C also contains $Fo(R, p)$.

7. *Some Special Constants Definable in PRA*

We introduce, in this section, notations for constants in PRA with some useful properties. The proof of the existence of such constants is routine.

1. We introduce constants $+$, $\cdot$, $\dot{-}$, $<$, min, max, exp, with the usual defining axioms (see Kleene, 1952, for instance). In some contexts we will write ω for $<$ (as in $\langle x, y\rangle \in \omega$, instead of $x < y$).

2. We choose our pairing function $\langle\rangle$, so that

$$\vdash_{\text{PRA}} \langle x, y\rangle = 2^x \cdot (2y + 1) \dot{-} 1$$

and we write L and R for the inverses, i.e., the constants for which

$$\vdash_{\text{PRA}} \langle L(x), R(x)\rangle = x.$$

We also want to find functions which give a convenient representation of finite sequences.

For this purpose, we first introduce a 3-ary auxiliary function constant mb, constructed so that the following holds in PRA:

$$\begin{cases} mb(0, s, 1) = s \dot{-} 1, \text{ if } s < 0; \\ mb(0, s, 2) = L(s); \\ mb(1, s, 2) = R(s); \\ m > 2 \wedge i < m \dot{-} 1 \rightarrow mb(i, s, m) = mb(i, L(s), m \dot{-} 1); \\ m > 2 \wedge i = m \dot{-} 1 \rightarrow mb(i, s, m) = R(s). \end{cases}$$

(We leave out the proof for the existence of this constant, although it is not completely trivial.)

We define the three functions lh, *, and Mb in terms of mb:

(Intuitively, the sense of these functions will be: $\text{lh}(s)$ is the length of the sequence s; $\text{Mb}(i, s)$ will be, for i less than the length of s, the i:th member of s; $s * x$ will be the sequence obtained from s by adding on an extra element x at the end.)

$$\begin{cases} \text{lh}(0) = 0; \\ \text{lh}(x) = R(x) + 1 \text{ if } x \neq 0; \end{cases}$$

$$\text{Mb}(i, s) = mb(i, L(s), \text{lh}(s));$$

$$s * x = \begin{cases} \langle x + 1, 0\rangle \text{ if } s = 0; \\ \langle\langle L(s) - 1, x\rangle, 1\rangle \text{ if } (s \neq 0 \wedge R(s) = 0); \\ \langle\langle L(s), x\rangle, R(s) + 1\rangle \text{ if } R(s) \neq 0. \end{cases}$$

We also define relation constants $\subseteq$ and $\subset$;
($s \subseteq t$ can be read: 's is an initial segment of t'; and $s \subset t$ as 's is a proper initial segment of t').

The definitions:

$$s \subseteq t \leftrightarrow \mathrm{lh}(s) > \mathrm{lh}(t) \wedge \forall j_{j < lh(s)}(\mathrm{Mb}(j, s) = \mathrm{Mb}(j, t));$$
$$s \subset t \leftrightarrow s \subseteq t \wedge \mathrm{lh}(s) \neq \mathrm{lh}(t).$$

THEOREM 1. The following formulas are provable (with the constants used defined as above).

(i) $\mathrm{lh}(0) = 0$;
(ii) $\mathrm{lh}(s * x) = \mathrm{lh}(s) + 1$;
(iii) $s \subset s * x$;
(iv) $\mathrm{Mb}(\mathrm{lh}(s), s * x) = x$;
(v) $s \subseteq t \wedge t \subseteq s \rightarrow s = t$;

We will also use the following notations:

When $\bar{s}$ and $\bar{t}$ represent strings of variables $s_0, \ldots, s_{k-1}$, and $t_0, \ldots, t_{k-1}$ of the same length, and u, v are two other variables, then

$\bar{s} \subseteq \bar{t}$ will abbreviate $s_0 \subseteq t_0 \wedge \ldots \wedge s_{k-1} \subseteq t_{k-1}$;
$\bar{s} \subset \bar{t}$ will abbreviate $s_0 \subset t_0 \wedge \ldots \wedge s_{k-1} \subset t_{k-1}$;
$(u; \bar{s}) \subseteq (v; \bar{t})$ will abbreviate $u \leqslant v \wedge \bar{s} \subseteq \bar{t}$;
$(u; \bar{s}) \subset (v; \bar{t})$ will abbreviate $u < v \wedge \bar{s} \subset \bar{t}$.

When $\bar{s}$ is a variable string $s_0, \ldots, s_{k-1}$, we write $\mathrm{lh}(\bar{s})$ for $\min(\mathrm{lh}(s_0), \ldots, \mathrm{lh}(s_{k-1}))$.

We will write Λ for 0, when we want to think of 0 as representing the empty sequence. We will use the common notation $(s)_z$ for $\mathrm{Mb}(z, s)$ (with the intended meaning: the z:th member of the sequence s).

When R is a binary predicate constant, and s is a variable, we write R-desc(s) for $\forall z_{z+1 < lh(s)}(\langle (s)_{(z+1)}, (s)_z \rangle \in R)$.

PART II. CALCULI OF KIND I (SECOND ORDER CALCULI)

1. *Symbols of Kind I*

The symbols of Kind I are:

(i) Denumerable set of numerical variables, and of function variables.

(ii) For each m, $n \geqslant 0$, function constants, predicate constants, and operation constants of type (m, n).

(iii) The propositional connectives $\neg$, $\vee -$, $\wedge$, $\rightarrow$, $\leftrightarrow$.

The function constants of type (0, 0) are called *numerical constants.*

2. Definition of 'Numerical Term', 'Function Term', and 'Formula'

(i) Numerical variables are numerical terms.

(ii) Function variables are function terms.

(iii) When A is a function (predicate, operation) constant of type (m, n), and $\chi_0, \ldots, \chi_{m-1}$ are function terms, and $t_0, \ldots, t_{n-1}$ are numerical terms, then $A\chi_0, \ldots, \chi_{m-1}, t_0, \ldots, t_{n-1}$ is a numerical term (formula, function term).

(iv) When χ is a function term and t a numerical term, then $\{\chi\}(t)$ is a numerical term.

(v) If A and B are formulas, then so are $\neg A$, $(A \vee B)$, $(A \wedge B)$, $(A \rightarrow B)$, and $(A \leftrightarrow B)$.

3. The Set $\mathscr{L}_I$ of Function, Predicate, and Operation Constants of Kind I

The set $\mathscr{L}_I$ is freely generated from a set of 'initial constants' (enumerated below), by the partially defined operations described in (i)–(ix) below. With each constant other than the initial constants is associated a 'defining axiom', or, in some cases, a pair of defining axioms.

A. Initial Constants

The initial constants (of Kind I) are the following: 0 (a numerical constant), Eq (a predicate constant of type (0, 0)), S and I (function constants of type (0, 1)), and Val (a function constant of type (1, 1).)

Note: Eq is the identity symbol, and we write as usual $s = t$ for Eq(s, t).

B. The Generating Operations for $\mathscr{L}_I$ (and the Defining Axioms)

1. The operation $\mathrm{Op}_{1,n,k,\Pi}$.

When G is a function constant in $\mathscr{L}_I$ of type (m, n), and Π a mapping $n \rightarrow k$, then

$\mathrm{Op}_{1,n,k,\Pi}(G)$ is a function constant in $\mathscr{L}_I$ of type (m, k). When F is $\mathrm{Op}_{1,n,k,\Pi}(G)$, then the defining axiom for F is

$$F(\bar{\phi}, y_0, \ldots, y_{k-1}) = G(\phi, y_{\Pi(0)}, \ldots, y_{\Pi(n-1)}).$$

2. The operation $\mathrm{Op}_{2,n,k,\Pi}$.

When G is a function constant in $\mathscr{L}_I$ of type (n, m), and Π a mapping $n \rightarrow k$, then

$\mathrm{Op}_{2,n,k,\Pi}(G)$ is a function constant in $\mathscr{L}_I$ of type (k, m).

When F is $\mathrm{Op}_{2,n,k,\Pi}(G)$, then the defining axiom for F is

$$F(\phi_0, \ldots, \phi_{k-1}, \bar{y}) = G(\phi_{\Pi(0)}, \ldots, \phi_{\Pi(n-1)}, \bar{y}).$$

3. The operation Op_3 (or $Su^{(0)}$)

When G is a function constant in $\mathscr{L}_I$ of type (m, n), and H a function constant of type $(m, n + 1)$, then

$Op_3(H, G)$ is a function constant in $\mathscr{L}_I$ of type (m, n).

When F is $Op_3(H, G)$, the defining axiom for F is

$$F(\bar{\phi}, \bar{y}) = H(\bar{\phi}, \bar{y}, G(\bar{\phi}, \bar{y})).$$

(We will sometimes write $Su^{(0)}(H, G)$ for $Op_3(H, G)$.)

4. The operation $Op_{4,m,k}$(or $Su^{(1)}_{m,k}$)

When Ψ is an operation constant of type (m, k), and H a function constant of type $(1 + m, k)$, then

$Op_{4,m,k}(H, \Psi)$ is a function constant in $\mathscr{L}_I$ of type (m, k).

When F is the constant $Op_{4,m,k}(H, \Psi)$, the defining axiom for F is

$$F(\bar{\phi}, \bar{y}) = H(\Psi(\bar{\phi}, \bar{y}), \bar{\phi}, \bar{y}).$$

(We will sometimes write $Su^{(1)}_{m,k}(H, \Psi)$ for $Op_{4,m,k}(H, \Psi)$.)

5. The operation Op_5(or Re)

When G is a function constant of type (m, n), and H a function constant of type $(m, n + 2)$, then

$Op_5(G, H)$ is a function constant of type $(m, n + 1)$.

When F is $Re(G, H)$, the defining axioms for F are

$$\begin{cases} F(\bar{\phi}, \bar{y}, 0) = G(\bar{\phi}, \bar{y}); \\ F(\bar{\phi}, \bar{y}, Sz) = H(\bar{\phi}, \bar{y}, z, F(\bar{\phi}, \bar{y}, z)). \end{cases}$$

(We will sometimes write $Re(G, H)$ for $Op_5(G, H)$.)

6. The operation Op_6(or Pred)

When F is a function constant of type (m, k), then

$Op_6(F)$ is a predicate constant of type (m, k).

When P is $Op_6(F)$, then the defining axiom for P is:

$$P(\bar{\phi}, \bar{y}) \leftrightarrow F(\bar{\phi}, \bar{y}) = 0.$$

7. The operation $Op_{7,m,k}$(or $\lambda_{m,k}$)

When G is a function constant of type $(m, k + 1)$, with $k \geqslant 0$, then $Op_{7,m,k}(G)$ is an operation constant of type (m, k).

When Ψ is $Op_{7,m,k}(G)$, then the defining axiom for Ψ is:

$$\{\Psi(\overline{\phi}, \overline{y})\}(z) = G(\overline{\phi}, \overline{y}, z).$$

(We will sometimes write $\lambda_{m,k}(G)$ or simply $\lambda(G)$ for $\mathrm{Op}_{7,m,k}(G)$.)

8. The operation Op_8(or $Fo^{(1)}$)
When R is a predicate constant of type (0, 2), then
$\mathrm{Op}_8(R)$ is a function constant of type (1, 0).
When F is the constant $\mathrm{Op}_8(R)$, the defining axioms for F are:

$$z = F(\phi) \rightarrow \langle\{\phi\}(z+1), \{\phi\}(z)\rangle \notin R;$$
$$z < F(\phi) \rightarrow \langle\{\phi\}(z+1), \{\phi\}(z)\rangle \in R.$$

9. The operation Op_9(or Fo)
When R is a predicate constant of type (0, 2), and p an operation constant of type (0, k), then
$\mathrm{Op}_9(R, p)$ is a function constant of type (0, k).
When h is $\mathrm{Op}_9(R, p)$, the defining axioms for h are:

$$z = h(\overline{y}) \rightarrow \langle\{p(\overline{y})\}(Sz), \{p(\overline{y})\}(z)\rangle \notin R;$$
$$z < h(\overline{y}) \rightarrow \langle\{p(\overline{y})\}(Sz), \{p(\overline{y})\}(z)\rangle \in R.$$

The operation Op_8 and Op_9 are called 'optional operations on $\mathscr{L}_I$'. The others (those mentioned in 1–7) are 'obligatory operations on $\mathscr{L}_I$'.

The notion of 'immediate predecessor' of a constant in $\mathscr{L}_I$ is defined in analogy with the case of constants of kind 0. The relation 'predecessor' is the ancestral of 'immediate predecessor'.

A function (predicate, operation) constant of type (m, k) is called *elementary*, if $m = 0$. A constant is called *strictly elementary*, if it is elementary, and all its predecessors are elementary.

4. *General Axioms, and Rules of Inference*

A. *Axioms*

1. All tautologies.
2. $x = x,\ Sx \neq 0,\ Sx = Sy \rightarrow x = y,\ Ix = x.$
3. $x = y \rightarrow (A(x) \leftrightarrow A(y)).$

where $A(x)$ is any formula, and $A(y)$ is obtained from $A(x)$ by replacement of y for some occurrence of x, in the usual sense.

4. $\mathrm{Val}(\phi, z) = \{\phi\}(z).$

B. Rules of Inference

(ia) $\dfrac{A(x);}{A(t)}$ (ib) $\dfrac{A(\phi);}{A(\chi)}$

(Here x and ϕ are, resp., a numerical variable and a function variable; and t and χ are, resp., a numerical term and a function term. The lower formula is obtained from the upper by the indicated substitution.)

(ii) $\dfrac{A,\ A \to B;}{B}$

(iii) $\dfrac{A(0),\ A(z) \to A(Sz).}{A(x)}$

5. *Calculi of Kind I*

We define a *calculus of kind I* to be a subset of $\mathscr{L}_I$ which contains the initial constants,
is closed under the obligatory operations (1)–(7), and closed under the relation of predecessor.

When $\mathscr{C}$ is a calculus of kind I, we will refer to the following associated sets as follows:

(a) The 'set of constants of $\mathscr{C}$' is $\mathscr{C}$ itself.

(b) The 'terms and formulas of $\mathscr{C}$' are those terms and formulas of kind I which contain no constants outside of $\mathscr{C}$.

(c) A 'proof in $\mathscr{C}$' is a proof from the general axioms and defining axioms, using the rules of inference above, where all the formulas involved are formulas of $\mathscr{C}$.

(d) A 'theorem of $\mathscr{C}$' is a formula, for which there is a proof in $\mathscr{C}$. We write $\vdash_{\mathscr{C}} A$ for the statement 'A is a theorem of $\mathscr{C}$'.

As in the elementary case, there is a minimal calculus of kind I, namely the set of constants generated from the initial constants by the obligatory operations on $\mathscr{L}_I$. This calculus is a formulation of second order primitive recursive arithmetic, and we will refer to it as $\mathrm{PRA}^{(1)}$.

More generally: For any subset $\mathscr{M}$ of the set $\mathscr{L}_I$, there is a minimal calculus (of kind I) which includes the set $\mathscr{M}$. We will denote this calculus $\mathscr{C}[\mathscr{M}]$. In the cases where $\mathscr{M}$ is a finite set $\{A_1, \ldots, A_k\}$, we may write $\mathscr{C}[A_1, \ldots, A_k]$ for $\mathscr{C}[\{A_1, \ldots, A_k\}]$.

Some useful constants definable in $PRA^{(1)}$. We can define a predicate constant $\supset$ of type (1, 1), and a function constant Segm of type (1, 1), for which it is provable:

(i) $\supset(\phi, s) \leftrightarrow \forall j_{j < lh(s)}((s)_j = \{\phi\}(j))$;

(ii) $\begin{cases} \text{Segm}(\phi, 0) = 0; \\ \text{Segm}(\phi, Sz) = \text{Segm}(\phi, z) * \{\phi\}(z); \end{cases}$

We will usually write '$s \subset \phi$' for $\supset(\phi, s)$.
($s \subset \phi$ can be read 's is an initial segment of ϕ'; Segm(ϕ, z) is the initial segment of ϕ of length z).

Theorem on explicit definition. We can prove for calculi of kind I the expected theorem on explicity definitions, i.e.,

THEOREM I. Let $\mathscr{C}$ be any calculus of Kind I, t any numerical term in $\mathscr{C}$ in the variables $\phi_0, \ldots, \phi_{k-1}, y_0, \ldots, y_{m-1}$. Then there is in $\mathscr{C}$ a function constant F of type (k, m), such that $\vdash_{\mathscr{C}} F(\bar{\phi}, \bar{y}) = t$.

6. *Equivalence between Constants in $\mathscr{C}$. Translations $\mathscr{C}' \to \mathscr{C}$, and Translations $C \to \mathscr{C}$, with $\mathscr{C}'$ Another Calculus of Kind I, C an Elementary Calculus*

Equivalence between constants in a calculus $\mathscr{C}$ is defined in the obvious way.

If $\mathscr{C}$ is a calculus of Kind I, F_1 and F_2 two function constants, A_1, A_2 two predicate constants, Ψ_1, Ψ_2 two operation constants, all of the same type (m, k), then

(i) F_1 and F_2 are equivalent in $\mathscr{C}$, if $\vdash_{\mathscr{C}} F_1(\bar{\phi}, \bar{y}) = F_2(\bar{\phi}, \bar{y})$;

(ii) A_1 and A_2 are equivalent in $\mathscr{C}$, if $\vdash_{\mathscr{C}} A_1(\bar{\phi}, \bar{y}) \leftrightarrow A_2(\bar{\phi}, \bar{y})$;

(iii) Ψ_1 and Ψ_2 are equivalent in $\mathscr{C}$, if $\vdash_{\mathscr{C}} \{\Psi_1(\bar{\phi}, \bar{y})\}(z) = \{\Psi_2(\bar{\phi}, \bar{y}\}(z)$.

When $\mathscr{C}$ and $\mathscr{C}'$ are two calculi of Kind I, and M is a mapping $\mathscr{C}' \to \mathscr{C}$, which preserves syntactic category and type (i.e., when F is a function constant in $\mathscr{C}'$ of type (m, k), then $M(F)$ is a function constant in $\mathscr{C}$ of type (m, k), etc.), and M^* is the induced mapping of formulas, then we call M a *translation* $\mathscr{C}' \to \mathscr{C}$ if

(i) M maps 0 on 0, and S on S, and

(ii) For each formula A of $\mathscr{C}'$, if $\vdash_{\mathscr{C}'} A$, then $\vdash_{\mathscr{C}} M^*(A)$.

When $\mathscr{C}$ is a calculus of Kind I, and C an elementary calculus, then a mapping $M: C \to \mathscr{C}$ is called a *translation* $C \to \mathscr{C}$, if

(i) M maps 0 on 0, S on S

(ii) M maps k-ary function (predicate, operation) constants on function (predicate, operation) constants of type $(0, k)$.

(iii) When M^* is the induced mapping of formulas, M^* preserves theoremhood, i.e., for each theorem A of C, the formula $M^*(A)$ is a theorem of $\mathscr{C}$.

If M_1 and M_2 are two translation $\mathscr{C}' \to \mathscr{C}$ (or two translations $C \to \mathscr{C}$, with C elementary), then we say that M_1 and M_2 are *equivalent translations*, if for every constant A in $\mathscr{C}'$ (every constant A in C), the two constants $M_1(A)$ and $M_2(A)$ are equivalent in $\mathscr{C}$.

Translations of the kind defined are unique up to equivalence, i.e.,

(a) If $\mathscr{C}$ and $\mathscr{C}'$ are two calculi of Kind I, and M_1 and M_2 are two translations $\mathscr{C}' \to \mathscr{C}$, then M_1 and M_2 are equivalent.

(b) If C is an elementary calculus, and $\mathscr{C}$ a calculus of Kind I, and M_1 and M_2 two translations $C \to \mathscr{C}$, then M_1 and M_2 are equivalent.

There is an obvious mapping $\mathscr{L}_0 \to \mathscr{L}_I$, which to every constant in the class $\mathscr{L}_0$ associates 'the corresponding' strictly elementary constant in $\mathscr{L}_I$. We will call this the 'canonical mapping' $\mathscr{L}_0 \to \mathscr{L}_I$, and denote it by $\mathscr{E}_{0,1}$. (We will often speak of this correspondence as an identity.)

7. *The Notion of an NDS-relation; Some Operations Producing NDS-relations*

Let $\mathscr{C}$ be a calculus of Kind I, R a binary elementary predicate constant. Then we call R an *NDS-relation in* $\mathscr{C}$ (NDS for 'no descending sequences'), if for some constant T of type (1, 0), it is provable in $\mathscr{C}$:

(i) $$z = T(\phi) \to \langle \{\phi\}(z + 1), \{\phi\}(z) \rangle \notin R.$$

When T and R are related as in (i), we call T an *NDS-functional for* R *in* $\mathscr{C}$.

We now define four operations on predicate constants, which will be seen to preserve the NDS-character.

DEFINITIONS (a) When R is any binary elementary predicate constant, we let Succ(R) be a binary elementary predicate constant R', for which it is provable in $\mathscr{C}[R]$:

$$\langle\langle x, y \rangle, \langle u, v \rangle\rangle \in R' \leftrightarrow ((y = v = 0 \wedge \langle x, u \rangle \in R) \vee (y = 0 \wedge \langle u, v \rangle = \langle 0, 1 \rangle));$$

(b) When R and S are binary elementary predicate constants, we let $R \oplus S$ be a binary elementary predicate constant Q, for which it is provable in $\mathscr{C}[R, S]$:

$$\langle\langle x, y\rangle, \langle u, v\rangle\rangle \in Q \leftrightarrow ((y = v = 0 \wedge \langle x, u\rangle \in R) \vee (y = v = 1 \wedge \langle x, u\rangle \in S) \vee (y = 0 \wedge v = 1));$$

(c) When R and S are binary elementary predicate constants, we let $R \otimes S$ be a binary predicate constant Q, for which it is provable in $\mathscr{C}[R, S]$

$$\langle\langle x, y\rangle, \langle u, v\rangle\rangle \in Q \leftrightarrow \langle y, v\rangle \in S \vee (y = v \wedge \langle x, u\rangle \in R);$$

(d) When R is a binary elementary predicate constant, we let $\Sigma(R)$ be a binary elementary predicate constant Q, for which it is provable in $\mathscr{C}[R]$:

$$\langle t, s\rangle \in Q \leftrightarrow s \subset t \wedge R\text{-desc}(s).$$

An observation on the above definitions. Since everything involved in these definitions is elementary, it will be easily seen that entirely analogous operations can be defined for binary predicate constants in elementary calculi, and also in the calculi of Kind II to be introduced later.

THEOREM 1. Let $\mathscr{C}$ be a calculus of Kind I. Let R and S be binary elementary predicate constants in $\mathscr{C}$. Then

(i) The constants Succ(R), $R \oplus S$, $R \otimes S$, and $\Sigma(R)$ belong to $\mathscr{C}$, and

(ii) If R and S are NDS-relations in $\mathscr{C}$, then so are Succ(R), $R \oplus S$, $R \otimes S$, and $\Sigma(R)$.

Proof: The proof involves the construction of NDS-functionals for the new constants, in terms of NDS-functionals for the given constants. The construction is not difficult.

8. *Calculi in Standard Form; Standardizations; the Elementary Part of a Calculus of Kind I*

A calculus $\mathscr{C}$ of Kind I is said to be in *standard form*, when it has the following closure property:

When R is an NDS-relation in $\mathscr{C}$, and p any elementary operation constant in $\mathscr{C}$, then $\mathscr{C}$ also contains the constant $Fo(R, p)$.

When $\mathscr{C}$ is not in standard form, there is an obvious way to form an extension of $\mathscr{C}$ which is in standard form, namely: by successively adding all the missing constants of form $Fo(R, p)$ (and, of course, closing under the obligatory operations). We call this extension $\mathscr{C}^+$, or, '*the standardization of* $\mathscr{C}$'. It is easy to verify, that the calculus $\mathscr{C}^+$ has property (b) below, as well as (a):

THEOREM 1. When $\mathscr{C}$ is any calculus of Kind I, then: (a) the standardization $\mathscr{C}^+$ is in standard form; (b) the calculi $\mathscr{C}$ and $\mathscr{C}^+$ are equivalent.

The set of strictly elementary constants in $\mathscr{C}^+$ ('regarded' as constants of $\mathscr{L}_0$ under the canonical identification mapping) forms an elementary calculus, which we will call *the elementary part of* $\mathscr{C}$, or $\mathscr{C}^0$.

Evidently, there is a translation from $\mathscr{C}^0$ to $\mathscr{C}$.

We will later be able to prove also (III, 12, 1), that $\mathscr{C}^0$ is, in the appropriate sense, the *maximal* elementary calculus translatable into $\mathscr{C}$.

9. *The Notions of 'Representing Triples', and 'Representing Quadruples', for Constants in a Calculus* $\mathscr{C}$

We assume that $\mathscr{C}$ is a calculus of Kind I.

When F is a function constant of type (m, k) in $\mathscr{C}$, then a triple $\langle Q, g, r\rangle$ of constants in $\mathscr{C}$ is called a *representing triple for F in* $\mathscr{C}$, if

(i) Q is an NDS-relation in $\mathscr{C}$.

(ii) g and r are elementary $m + k$-ary function constants.

(iii) (letting $\bar{s}$ and $\bar{t}$ be variable strings of length m, $\bar{y}$ a string of length k)

$$\vdash_{\mathscr{C}} r(\bar{t}, \bar{y}) = 0 \wedge \bar{s} \subset \bar{t} \rightarrow \langle g(\bar{t}_3, \bar{y}), g(\bar{s}, \bar{y})\rangle \in Q;$$

(iv) $\vdash_{\mathscr{C}} r(\bar{s}, \bar{y}) \neq 0 \wedge \bar{s} \subseteq \bar{t} \rightarrow r(\bar{t}, \bar{y}) = r(\bar{t}, \bar{y});$

(v) (writing $\mathrm{Segm}(\bar{\phi}, n)$ for $\mathrm{Segm}(\phi_0, n), \ldots, \mathrm{Segm}(\phi_{m-1}, n)$:)

$$\vdash_{\mathscr{C}} r(\mathrm{Segm}(\bar{\phi}, n), \bar{y}) \neq 0 \rightarrow F(\bar{\phi}, \bar{y}) = r(\mathrm{Segm}(\bar{\phi}, n), \bar{y}) \dot{-} 1.$$

When P is a predicate constant of type (m, k) in $\mathscr{C}$, of form $\mathrm{Pred}(F)$, with F a function constant, then $\langle Q, g, r\rangle$ is called a *representing triple for P in* $\mathscr{C}$, if it is a representing triple for F.

When Ψ is an operation constant of type (m, k) in $\mathscr{C}$ of form $\lambda(F)$, with F a function constant of type $(m, k + 1)$, then $\langle Q, g, r\rangle$ is called a *representing triple for* Ψ *in* $\mathscr{C}$, if it is a representing triple for F.

Remark: When f is an *elementary* k-ary function constant in $\mathscr{C}$, then a representing triple $\langle Q, g, r\rangle$ for f in $\mathscr{C}$ can be defined by the following conditions:

(i) Q is the empty relation;

(ii) $g(\bar{y}) = 0;$

(iii) $r(\bar{y}) = f(\bar{y}) + 1$.

A representing triple for f satisfying these conditions in $\mathscr{C}$ will be called a *straight representation for f in* $\mathscr{C}$. We can similarly define 'straight representations' for predicate constants and operations constants.

The following property of representing triples is immediate:

THEOREM I. If $\mathscr{C}$ is a calculus of Kind I, and if A and B are two function constants (two predicate constants, two operation constants) of the same type, then if a triple $\langle Q, g, r\rangle$ is a representing triple in $\mathscr{C}$ for both A and B, then A and B are equivalent in $\mathscr{C}$.

Potential representing triples. A triple $\langle Q, g, r\rangle$ of constants in a calculus $\mathscr{C}$, where Q is a binary elementary predicate constant, and for some m and k, g and r are $m + k$-ary elementary function constants which satisfy (in $\mathscr{C}$) the conditions (i)–(iv) above (with $\bar{s}$ and $\bar{t}$ strings of length m, and $\bar{y}$ a string of length k) will be called a *potential representing triple for type* (m, k) *in* $\mathscr{C}$.

Evidently, any representing triple for a function constant F of type (m, k) in $\mathscr{C}$ is also a potential representing triple for type (m, k). We also have the following converse:

THEOREM 2. When $\mathscr{C}$ is a calculus of Kind I, and $\langle Q, g, r\rangle$ is a potential representing triple in $\mathscr{C}$ for type (m, k), then there is a function constant F in $\mathscr{C}$ such that $\langle Q, g, r\rangle$ is a representing triple for F in $\mathscr{C}$.

Proof: Because of the assumed properties of Q, g and r, we can define a function T of type (m, k), such that

$$\vdash_{\mathscr{C}} r(\text{Segm}(\bar{\phi}, T(\bar{\phi}, \bar{y})), \bar{y}) \neq 0.$$

F is then defined by:

$$F(\bar{\phi}, \bar{y}) = r(\text{Segm}(\bar{\phi}, T(\varphi, \bar{y})), \bar{y}) \dot{-} 1.$$

It is easily verified that this function F has the desired properties. (And because of the previous theorem, it is also essentially the only function constant with the desired properties.)

Remark: It follows evidently, than when $\langle Q, g, r\rangle$ is a potential triple in $\mathscr{C}$ for type (m, k), we can construct a predicate constant P (of type (m, k)), with the given triple as a representing triple; and, when $k \neq 0$, we can construct an operation constant Ψ (of type $(m, k = 1)$), for which the given triple is a representing triple.

It turns out to be convenient, for some purposes, to extend the representing triples to 'representing quadruples'. The definition is as follows:

DEFINITION. When $\mathscr{C}$ is a calculus of Kind I, A a constant in $\mathscr{C}$, and $\langle Q, g, r\rangle$ is a representing triple for A in $\mathscr{C}$, (say, of type (m, k), so that A is either a function or a predicate constant of type (m, k), or, in case $k \neq 0$, an operation constant of type $(m, k-1)$), then the quadruple $\langle Q, g, r, B\rangle$ is called a *representing quadruple for* A *in* $\mathscr{C}$, if one of the following is true:

(1) A is a function constant (of type (m, k)), B a function constant of the same type as r, and B and r are related as follows:

$$[\mathrm{Fu}]\ \vdash_{\mathscr{C}} B(\bar{s}, \bar{y}) = r(\bar{t}, \bar{y}) \dot{-} 1$$

(2) A is a predicate constant (of type (m, k)), B a predicate constant of type $(0, m+k)$, and r and B are related as follows:

$$[\mathrm{Pred}]\ \vdash_{\mathscr{C}} B(\bar{s}, \bar{y}) \leftrightarrow r(\bar{s}, \bar{y}) = 1$$

(3) m is 0, $k \neq 0$, A is an operation constant (of type $(0, k-1)$), B an elementary k-ary function constant, and r and B are related as follows, (with $\bar{y}$ a variable string of length $k-1$):

$$[\mathrm{Op}_0]\begin{cases} \vdash_{\mathscr{C}} B(0, \bar{y}) = \Lambda \\ \vdash_{\mathscr{C}} B(z+1, \bar{y}) = B(z, \bar{y}) * (r(\bar{y}, z) \dot{-} 1) \end{cases}$$

(4) $m \neq 0, k \neq 0$, A is an operation constant of type $(m, k-1)$, B is an elementary $m+k-1$-ary function constant, and r and B are related as follows, (with $\bar{y}$ a variable string of length $k-1$):

$$[\mathrm{Op}_m]\begin{cases} \vdash_{\mathscr{C}} j < \mathrm{lhb}(\bar{s}, \bar{y})) \rightarrow (r(\bar{s}, \bar{y}, j) \neq 0 \wedge (B(\bar{s}, \bar{y}))_j = \\ r(\bar{s}, \bar{y}, j) \dot{-} 1) \\ \vdash_{\mathscr{C}} \mathrm{lh}(B(\bar{s}, \bar{y}) = \mathrm{lh}(\bar{s}) \vee \\ (\mathrm{lh}(B(\bar{s}, \bar{y})) < \mathrm{lh}(\bar{s}) \wedge r(\bar{s}, \bar{y}, \mathrm{lh}(B(\bar{s}, \bar{y})) = 0). \end{cases}$$

The notion of potential representing quadruple fitting a given type of constant is defined as follows:

DEFINITION. If $\langle Q, g, r\rangle$ is a potential representing triple in $\mathscr{C}$ of type (m, k), then

(a) $\langle Q, g, r, B\rangle$ is a *potential representing quadruple in* $\mathscr{C}$ *for function constants of type* (m, k), if r and B satisfy [Fu] above.

(b) $\langle Q, g, r, B\rangle$ is a *potential representing quadruple for predicate constants of type* (m, k) *in* $\mathscr{C}$, if r, B satisfy [Pred] above.

(c) If $m = 0$ and $k \neq 0$, then $\langle Q, g, r, B\rangle$ is a *potential representing quadruple in* $\mathscr{C}$ *for operation constants of type* $(0, k - 1)$, if r and B satisfy conditions [Op_0] above.

(d) If $m \neq 0$ and $k \neq 0$, then $\langle Q, g, r, B\rangle$ is a *potential representing quadruple for operation constants of type* $(m, k - 1)$ in $\mathscr{C}$ if r and B satisfy the conditions [Op_m] listed in (4) above.

10. The Definition of the 'Canonical Representing Mappings' α, σ, ρ, V*; and the Definition of the Relations* $\Delta^C{}_\theta$*, for Various Subscripts* $_\theta$

We want to define uniformly four functions α, σ, ρ and V on $\mathscr{L}_I$, in such a way that for any constant A in $\mathscr{L}_I$, the quadruple $\alpha(A)$, $\sigma(A)$, $\rho(A)$, $V(A)$ will be a representing quadruple for A in any calculus in standard form, which contains the constant A.

As a useful preliminary step we will define a class of relations between 4-tuples, one family of relations corresponding to each of the generating operations on $\mathscr{L}_I$.

We will say of a quadruple $\langle Q, g, r, B\rangle$ of constants, that it is 'of the right type for function (predicate, operation) constants of type (m, k)', when the quadruple satisfies the type requirements for representing quadruples for constants of that type (i.e.: Q must be an elementary binary predicate constant: to be right for function constants of type (m, k), g, r and B must be $m + k$-ary elementary function constants, etc.).

Throughout the following definition, we will assume that q_1, q_2, q_3 are quadruples of constants, and that

$$q_1 = \langle Q_1, g_1, r_1, B_1\rangle, q_2 = \langle Q_2, g_2, r_2, B_2\rangle \text{ and}$$
$$q_3 = \langle Q_3, g_3, r_3, B_3\rangle.$$

C is assumed to be an elementary calculus or a calculus of Kind I.

Definition of Various Relations Δ^C_ν *Corresponding to the Generating Operations for* $\mathscr{L}_I$

1. The relation $\Delta^C_{1,n,k,\pi}$.

When π is a mapping $n \to k$, then we say of two quadruples q_1 and q_2, that $\langle q_1, q_2\rangle \in \Delta^C_{1,n,k,\pi}$ if:

(a) q_1 and q_2 are of the right type for function constants of type (m, k) and (m, n), respectively;

(b) Q_1 and Q_2 are the same;
(c) $\vdash_C g_1(\bar{s}, y_0, \ldots, y_{k-1}) = g_2(\bar{s}, y_{\pi(0)}, \ldots, y_{\pi(n-1)})$;
(d) $\vdash_C r_1(\bar{s}, y_0, \ldots, y_{k-1}) = r_2(\bar{s}, y_{\pi(0)}, \ldots, y_{\pi(n-1)})$;
(e) $\vdash_C B_1(\bar{s}, \bar{y}) = r_1(\bar{s}, \bar{y}) \dot{-} 1$.

2. The relation $\Delta^C_{2,n,k,\pi}$.

When π is a mapping $n \to k$, then we say of two quadruples q_1 and q_2, that $\langle q_1, q_2\rangle \in \Delta^C_{2,n,k}$, if:

(a) q_1 and q_2 are of the right type for function constants of type (k, m) and (n, m), resp., for some m;
(b) Q_1 and Q_2 are the same;
(c) $\vdash_C g_1(s_0, \ldots, s_{k-1}, \bar{y}) = g_2(s_{\pi(0)}, \ldots, s_{\pi(n-1)}, \bar{y})$;
(d) $\vdash_C r_1(s_0, \ldots, s_{k-1}, \bar{y}) = r_2(s_{\pi(0)}, \ldots, s_{\pi(n-1)}, \bar{y})$;
(e) $\vdash_C B_1(\bar{s}, \bar{y}) = r_1(\bar{s}, \bar{y}) - 1$.

3. The relation Δ^C_3.

We say that $\langle q_1, q_2, q_3\rangle \in \Delta^C_3$ if, for some m and n:

(a) The types of q_1 and q_3 correspond to function constants of type (m, n), and the type of q_2 to function constants of type $(m, n + 1)$;
(b) Q_1 and $Q_2 \oplus Q_3$ are equivalent in C;
(c) $\vdash_C g_1(\bar{s}, \bar{y}) = \begin{cases} \langle g_3(\bar{s}, \bar{y}), 1\rangle \text{ if } r_3(\bar{s}, \bar{y}) = 0; \\ \langle g_2(\bar{s}, \bar{y}, B_3(\bar{s}, \bar{y})), 0\rangle \text{ if } r_3(\bar{s}, \bar{y}) \neq 0; \end{cases}$
(d) $\vdash_C r_1(\bar{s}, \bar{y}) = \begin{cases} 0 \text{ if } r_3(\bar{s}, \bar{y}) = 0; \\ r_2(\bar{s}, \bar{y}, B_3(\bar{s}, \bar{y})) \text{ if } r_3(\bar{s}, \bar{y}) \neq 0; \end{cases}$
(e) $\vdash_C B_1(\bar{s}, \bar{y}) = r_1(\bar{s}, \bar{y}) \dot{-} 1$.
($\bar{s}$ and $\bar{y}$ above are assumed to be strings of length m and n resp.)

4. The relation $\Delta^C_{4,m,k}$.

The definition of this relation has different forms for the two cases $m = 0$, and $m \neq 0$.

Case 1. $m = 0$.

We say that $\langle q_1, q_2, q_3\rangle \in \Delta^C_{4,0,k}$ if:
(a) q_1, q_2, q_3 are of types corresponding to, resp., (1) a function constant of type $(0, k)$, (2) a function constant of type $(1, k)$, and (3) an operation constant of type $(0, k)$;
(b) Q_1 is the empty relation (in C);
(c) $\vdash_C g_1(\bar{y}) = 0$;

(d) $\vdash_C r_2(B_3(z, \bar{y}), \bar{y}) \neq 0 \rightarrow B_1(\bar{y}) = r_2(B_3(z, \bar{y}), \bar{y}) \dot{-} 1;$
(e) $\vdash_C r_1(\bar{y}) = B_1(\bar{y}) + 1.$

Case 2. $m \neq 0$.
In this case we say that $\langle q_1, q_2, q_3 \rangle \in \Delta^C_{4,m,k}$ if:

(a) The types of q_1, q_2 and q_3 correspond, resp., to (1) a function constant of type (m, k), (2) a function constant of type $(1 + m, k)$, and (3) an operation constant of type (m, k);
(b) Q_1 and $Q_3 \otimes Q_2$ are equivalent in C;
(c) $\vdash_C g_1(\bar{s}, \bar{y}) = \langle g_3(\bar{s}, \bar{y}, \mathrm{lh}(B_3(\bar{s}, \bar{y}))), g_2(B_3(\bar{s}, \bar{y}), \bar{s}, \bar{y})\rangle;$
(d) $\vdash_C r_1(\bar{s}, \bar{y}) = r_2(B_3(\bar{s}, \bar{y}), \bar{s}, \bar{y});$
(e) $\vdash_C B_1(\bar{s}, \bar{y}) = r_1(\bar{s}, \bar{y}) \dot{-} 1.$

5. The relation Δ^C_5.
We say that $\langle q_1, q_2, q_3 \rangle \in \Delta^C$ if, for some m and k:

(a) The types of q_1, q_2, q_3 correspond, resp., to, a function constant of type $(m, k + 1)$, a function constant of type (m, k), and a function constant of type $(m, k + 2)$;
(b) Q_1 and $Q_3 \otimes \omega$ are equivalent in C;
(c) $\vdash_C r_1(\bar{s}, \bar{y}, 0) = r_2(\bar{s}, \bar{y});$
(d) $\vdash_C r_1(\bar{s}, \bar{y}, z + 1) = \begin{cases} 0 \text{ if } r_1(\bar{s}, \bar{y}, z) = 0; \\ r_3(\bar{s}, \bar{y}, z, r_1(\bar{s}, \bar{y}, z) \dot{-} 1) \text{ otherwise;} \end{cases}$
(e) $\vdash_C r_1(\bar{s}, \bar{y}, w) \neq 0 \wedge r_1(\bar{s}, \bar{y}, w + 1) = 0 \wedge w < z \rightarrow$
$g_1(\bar{s}, \bar{y}, z) = \langle g_3(\bar{s}, \bar{y}, w, r_1(\bar{s}, \bar{y}, w) \dot{-} 1), w \dot{-} z)\rangle;$
(f) $\vdash_C r_1(\bar{s}, \bar{y}, z) \neq 0 \rightarrow g_1(\bar{s}, \bar{y}, z) = 0;$
(g) $\vdash_C B_1(\bar{s}, \bar{y}, z) = r_1(\bar{s}, \bar{y}, z) \dot{-} 1.$

6. The relation Δ^C_6.
We say that $q_1, q_2 \in \Delta^C_6$, if, for some m and k:

(a) The types of q_1 and q_2 correspond, resp., to a predicate constant of type (m, k), and to a function constant of type (m, k);
(b) Q_1, g_1, r_1 are the same as Q_2, g_2, r_2;
(c) $\vdash_C B_1(\bar{s}, \bar{y}) \leftrightarrow r_1(\bar{s}, \bar{y}) = 1.$

7. The relation $\Delta^C_{7,m,k}$
We divide the definition in two cases:
Case 1: $m = 0$.
We say that $(q_1, q_2) \in \Delta^C_{7,0,k}$ if:

(a) The types of q_1 and q_2 correspond, resp., to operation constants of type $(0, k)$, and to function constants of type $(0, k+1)$;

(b) Q_1, g_1, r_1 are the same as Q_2, g_2, r_2;

(c) $\vdash_C \mathrm{lh}(B_1(z, \bar{y})) = z$;

(d) $\vdash_C x < z \rightarrow (B_1(z, \bar{y}))_x = B_2(\bar{y}, x)$.

Case 2. $m \neq 0$.
In that case we say that $\langle q_1, q_2 \rangle \in \Delta_{7,m,k}$ if:

(a) The types of q_1 and q_2 correspond, resp., to operation constants of type (m, k), and to function constants of type $(m, k+1)$;

(b) Q_1, g_1, r_1 are the same as Q_2, g_2, r_2;

(c) $\vdash_C z = \mathrm{lh}(B_1(\bar{s}, \bar{y})) \rightarrow (B_1(\bar{s}, \bar{y}))_z + 1 = r_2(\bar{s}, \bar{y}, z)$;

(d) $\vdash_C \mathrm{lh}(B_1(\bar{s}, \bar{y})) \leqslant \mathrm{lh}(\bar{s})$;

(e) $\vdash_C \mathrm{lh}(B_1(\bar{s}, \bar{y})) < \mathrm{lh}(\bar{s}) \rightarrow r_2(\bar{s}, \bar{y}, \mathrm{lh}(B_1(\bar{s}, \bar{y}))) = 0$.

8. The relation Δ_8^C.
We say that $\langle q_1, q_2 \rangle \in \Delta^C$ if:

(a) The types of q_1 and q_2 correspond, resp., to function constants of type $(1, 0)$, and to predicate constants of type $(0, 2)$;

(b) Q_1 and $\Sigma(B_2)$ are equivalent in C;

(c) $\vdash_C g_1(s) = s$;

(d) $\vdash_C B_2\text{-desc}(s) \leftrightarrow r_1(s) = 0$;

(e) $\vdash_C \neg B_2\text{-desc}(s) \rightarrow (z < r_1(s) \rightarrow (\langle (s)_{z+1}, (s)_z \rangle \notin B_2 \leftrightarrow z + 1 = r_1(s)))$;

(f) $\vdash_C B_1(s) = r_1(s) \dot{-} 1$.

9. The relation Δ_9^C.
We say that $\langle q_1, q_2, q_3 \rangle \in \Delta_9^C$, if, for some k,

(a) The types of q_1, q_2, q_3, correspond, resp., to function constants of type $(0, k)$, predicate constants of type $(0, 2)$, and operation constants of type $(0, k)$;

(b) Q_1 is empty, $\vdash_C g_1(s) = 0$;

(c) B_1 and $Fo(B_2, \lambda(r_3))$ are equivalent in C;

(d) $\vdash_C r_1(\bar{y}) = B_1(\bar{y}) + 1$.

The definitions above were chosen so that the following two lemmas would be true.

THEOREM 1. Let $\mathscr{C}$ be a calculus of Kind I. Let Op_ν be one of the operations listed in II, 3 as the 'generating operations' for $\mathscr{L}_I$, and let $\Delta_\nu^{\mathscr{C}}$ be the relation, among those defined above, with the same subscript. Then

(a) (In case Op_ν is a unary operation).

If $A_1 \in \mathscr{C}$, and $A_1 = \mathrm{Op}_\nu(A_2)$, and q_2 is a representing quadruple for A_2 in $\mathscr{C}$, then if q_1 is a quadruple, such that $\langle q_1, q_2 \rangle \in \Delta_\nu^{\mathscr{C}}$, then q_1 is a representing quadruple for A_1 in $\mathscr{C}$.

(b) (In case Op_ν is a binary operation).

If $A_1 \in \mathscr{C}$, and $A_1 = \mathrm{Op}_\nu(A_2, A_3)$, and q_2, q_3 are representing quadruples in $\mathscr{C}$, for A_2 and for A_3, then if q_1 is a quadruple, such that $\langle q_1, q_2, q_3 \rangle \in \Delta_\nu^{\mathscr{C}}$, then q_1 is a representing quadruple for A_1 in $\mathscr{C}$.

THEOREM 2. (On the existence of a quadruple q_1, such that $\langle q_1, q_2 \rangle \in \Delta_\nu^C$ (or: such that $\langle q_1, q_2, q_3 \rangle \in \Delta_\nu^C$).)

For each of the subscripts ν, used in the above definitions, the following holds:

If q_2 (and q_3) are quadruples of elementary constants of the right types for places 2 (and 3) in a relation Δ_ν^C, and if C is any elementary calculus which contains the constants of q_2 (and q_3), and which closed under recursion on Q_2 (and Q_3), then there is a quadruple q_1 in C, such that $\langle q_1, q_2) \in \Delta_\nu^C$ (such that $\langle q_1, q_2, q_3 \rangle \in \Delta_\nu^C$.

The following proposition states a simple property of the relations Δ_ν^C defined above, which is easily verified by inspection of the form of the definitions.

THEOREM 3. For each of the subscripts ν used above, it holds: If C and C' are two elementary calculi, J a translation $C \to C'$, $\langle q_1, q_2, q_3 \rangle$ quadruples of constants in C, and q_1', q_2', q_3' their translations by J, then if $\langle q_1, q_2, q_3 \rangle \in \Delta_\nu^C$, then $\langle q_1, q_2, q_3 \rangle \in \Delta_\nu^{C'}$. (This statement was for the case that Δ_ν^C is a 3-ary relation; the modification for the other case is obvious.)

From Theorem 2 we can derive a useful consequence on the existence of a quadruple of mappings $\mathscr{L}_I \to \mathscr{L}_0$.

THEOREM 4. There is a quadruple α, σ, ρ, V of mappings $\mathscr{L}_I \to \mathscr{L}_0$ with the properties below:

(First a piece of terminology: When A is an element of $\mathscr{L}_I$, we call $\langle \alpha(A), \sigma(A), \rho(A), V(A) \rangle$ the 'quadruple associated with A by $\langle \alpha, \sigma, \rho, V \rangle$'.)

(i) The quadruple associated to a constant A always belongs to the calculus $(\mathscr{C}[A])^0$.

(ii) For each of the initial constants A of $\mathscr{L}_I$, the associated quadruple is a representing quadruple for A in $\mathrm{PRA}^{(1)}$.

(iii) For each generating operation Op_ν, it holds:

When $A_1 = \mathrm{Op}_\nu(A_2, A_3)$ (when $A_1 = \mathrm{Op}_\nu(A_2)$), $\mathscr{C}$ any calculus of Kind I containing A_1, and C its elementary part, and if q_1, q_2 (and q_3) are the quadruples associated to A_1, A_2 (and A_3), then $\langle q_1, q_2, q_3\rangle \in \Delta_\nu^C$ (then $\langle q_1, q_2\rangle \in \Delta_\nu^C$).

DEFINITION. From now on we will use the symbols α, σ, ρ, V for a particular quadruple of mappings $\mathscr{L}_I \to \mathscr{L}_0$, which satisfies the conditions listed in Theorem 4. We will call these mappings 'the canonical representation mappings'. We will often write α_A, σ_A, ρ_A and V_A, instead of $\alpha(A)$, $\sigma(A)$, $\rho(A)$ and $V(A)$.

From the conditions on the quadruple $\langle \alpha, \sigma, \rho, V\rangle$ listed in Theorem 4, it follows easily by Theorem 1, that:

THEOREM 5. If $\mathscr{C}$ is a calculus of Kind I in standard form, and A a constant in $\mathscr{C}$, then

(a) the quadruple $\langle \alpha_A, \sigma_A, \rho_A, V_A\rangle$ is a representing quadruple for A in $\mathscr{C}$, which is, moreover, strictly elementary.

(b) The elementary portion of the conditions, making $\langle \alpha_A, \sigma_A, \rho_A, V_A\rangle$ a potential representing quadruple in $\mathscr{C}$ for the type of A, is provable in the elementary part $\mathscr{C}^0$ (by the 'elementary portion' we mean those formulas we used in characterizing $\langle \alpha_A, \sigma_A, \rho_A, V_A\rangle$ as a potential quadruple for the type in question, except the condition that α_A be an NDS-relation, a condition, which obviously cannot be stated in $\mathscr{C}^0$).

COROLLARY (of 5a). When $\mathscr{C}$ is a calculus of Kind I in standard form, then every elementary constant in $\mathscr{C}$ has a strictly elementary equivalent.

Proof: When A is an elementary constant in $\mathscr{C}$, then V_A can be seen to be a strictly elementary equivalent.

We will also need the following related result about provability in $\mathscr{C}^0$:

THEOREM 6. Let $\mathscr{C}$ be a calculus of Kind I, $\mathscr{C}^0$ its elementary part. Then if A is a strictly elementary constant in $\mathscr{C}$ (i.e., $A \in \mathscr{C} \cap \mathscr{C}^0$), the equivalence of A and V_A is provable in $\mathscr{C}^0$.

11. Relations between Interpretability, Strong Interpretability, and Translatability for Calculi of Kind I

When $\mathscr{C}_1$ and $\mathscr{C}_2$ are two calculi of Kind I, then by an *interpretation* $\mathscr{C}_1 \to \mathscr{C}_2$ we will mean a translation of the elementary parts J: $\mathscr{C}_1^0 \to \mathscr{C}_1^0$, which satisfies the condition:

(i) For every constant A in $\mathscr{C}_1$, the constant $J(\alpha_A)$ is an NDS-relation in $\mathscr{C}_2$.

We will call J a *strong interpretation* $\mathscr{C}_1 \to \mathscr{C}_2$, if, in addition to (i), it satisfies the stronger condition:

(ii) Whenever R, in $\mathscr{C}_1^0$, is an NDS-relation in $\mathscr{C}_1$, then $J(R)$ is an NDS-relation in $\mathscr{C}_2$.

(We will show right below, that for $\mathscr{C}_1$, $\mathscr{C}_2$ of Kind I, the notions of interpretation and of strong interpretation are equivalent. But until we show this, we have to treat them as separate notions.)

THEOREM 1. Let $\mathscr{C}_1$ and $\mathscr{C}_2$ be two calculi of Kind I, and let J be an interpretation $\mathscr{C}_1 \to \mathscr{C}_2$. Then

(a) There is a translation $\mathscr{C}_1 \to \mathscr{C}_2$; and
(b) J is a strong interpretation.

Proof: We introduce the following notation: When A is any constant in $\mathscr{C}_1$, then $\langle \alpha_A, \sigma_A, \rho_A, V_A \rangle$ is the canonical representing quadruple for A. These constants will belong to $\mathscr{C}_1^0$. We write α_A', σ_A', ρ_A', V_A' for the translations $J(\alpha_A)$, $J(\sigma_A)$, $J(\rho_A)$, $J(V_A)$.

Because of condition (i) on the translation J, the translation by J of a potential representing quadruple in $\mathscr{C}_1$ is still a potential representing triple in $\mathscr{C}_2$. By Theorem 9.2, there is then a way to construct a constant in $\mathscr{C}_2$, of the same type as A, for which $\langle \alpha_A', \sigma_A', \rho_A' \rangle$ is a representing triple in $\mathscr{C}_2$.

We write M for the mapping, which assigns to each constant A in this constant in $\mathscr{C}_2$, which has the 'corresponding' representation.

We want to show that the mapping M so defined is a translation. It will be sufficient to verify, that for each constant A in $\mathscr{C}_1$ the defining axioms for A get translated by M into theorems of $\mathscr{C}_2$. Let us consider for illustration only one case, namely the case where A (in $\mathscr{C}_1$) is a function constant of type (m, k), $m \neq 0$, and of form $\mathrm{Su}^{(1)}_{m,k}(H, \Psi)$ (i.e., $A = \mathrm{Op}_{4,m,k}(H, \Psi)$).

We assume $F = \mathrm{Op}_{4,m,k}(H, \Psi)$, with $m \neq 0$. We let $F' = M(F)$,

$H' = M(H)$, $\Psi' = M(\Psi)$. We let $C_1 = \mathscr{C}_1^0$, $C_2 = \mathscr{C}_2^0$; we write F^* for $\mathrm{Op}_{4,m,k}(H',\Psi')$. To prove that the defining axioms for F translate into theorems, it will be sufficient to show that F' and F^* are equivalent in $\mathscr{C}_2$.

Let q_1, q_2, q_3 be the canonical representing quadruples for F, H, and Ψ resp., and let q_1', q_2', q_3' be their translations by J. We know:

q_1', q_2', q_3' are representing quadruples for F', H' and Ψ' in $\mathscr{C}_2$.

We also know:

$$\langle q_1, q_2, q_3\rangle \in \Delta^{C_1}{}_{4,m,k}.$$

Then, by 10,3, we also have:

$$\langle q_1', q_2', q_3'\rangle \in \Delta^{C_2}{}_{4,m,k}.$$

But then, by 10.1, q_1' is a representing quadruple for F^*.

Then, by 9.1, F' and F^* are equivalent in $\mathscr{C}_2$.

We have then shown that M is a translation $\mathscr{C}_1 \to \mathscr{C}_2$. Clearly then, the mapping M will map NDS-relations in $\mathscr{C}_1$ on NDS-relations in $\mathscr{C}_2$. Using the fact that translations are essentially unique, we conclude that J is a strong interpretation.

PART III. CALCULI OF KIND II

1. *Symbols of Kind II*

A. Elementary Symbols

The elementary symbols (paralleling the categories in elementary calculi) are a denumerable set of numerical variables; for each $n \geqslant 0$, n-ary function-, predicate-, and operation constants; also, the connectives $\neg, \vee, \wedge, \rightarrow, \leftrightarrow$.

B. Special Constants

2. *Definition of 'Numerical Term', 'Formula', Function Term'*

The inductive rules for forming terms and formulas are the rules (i)–(viii) of I.2, with the addition of one rule:

(ix) When P is a special constant, and t_1, t_2 are numerical terms, then $P(t_1, t_2)$ is a formula. (We will often write $\langle t_1, t_2\rangle \in P$ instead of $P(t_1, t_2)$.)

3. *The Set $\mathscr{L}_{II}$ of Function-, Predicate-, and Operation Constants of Kind II, and of 'Special Constants'*

The set $\mathscr{L}_{II}$ is freely generated from a set of 'initial constants' described below, by the following set of partially defined operations.

A. The Initial Constants

The initial constants are: 0 (0-ary function constant), Eq (2-ary predicate constant), S and I (1-ary function constants), and ω (special constant).

B. The Generating Operations (and Associated 'Defining Axioms')

We take over the generating operations of I.3 subsections (i)–(vi), and add the following five operations:

(vii) If R is a special constant, then **Succ**(R) is a special constant. When R' is **Succ**(R), then the defining axiom for R' is:

$$\langle\langle x, y\rangle, \langle u, v\rangle\rangle \in R' \leftrightarrow ((y = v = 0 \wedge \langle x, u\rangle \in R) \vee (y = 0 \wedge \langle u, v\rangle = \langle 0, 1\rangle))$$

(viii) If R and S are special constant, then $R \oplus S$ is a special constant.

When Q is $R \oplus S$, the defining axiom for Q is:

$$\langle\langle x, y\rangle, \langle u, v\rangle\rangle \in Q \leftrightarrow ((u = v = 0 \wedge \langle x, u\rangle \in R) \vee (y = v = 1 \wedge \langle x, u\rangle \in S) \vee (y = 0 \wedge v = 1))$$

(ix) If R and S are special constants, then $R \oplus S$ is a special constant. When Q is $R \oplus S$, the defining axiom for Q is:

$$\langle\langle x, y\rangle, \langle u, v\rangle\rangle \in Q \leftrightarrow \langle y, v\rangle \in S \vee (y = v \wedge \langle x, u\rangle \in R)$$

(x) If R is a special constant, and p an m-ary operation constant, then $Fo^{(2)}(R, p)$ is an m-ary function constant.

When f is $Fo^{(2)}(R, p)$, the defining axioms for f are:

$$z = f(\bar{y}) \rightarrow \langle\{p(\bar{y})\}(z + 1), \{p(\bar{y})\}(z)\rangle \notin R$$
$$z < f(\bar{y}) \rightarrow \langle\{p(\bar{y})\}(z + 1), \{p(\bar{y})\}(z)\rangle \in R$$

We also have the following (optional) operation:

(xi) When Q is an ordinary binary predicate constant, then Sp(Q) is a special constant.

When R is Sp(Q), its defining axiom is:

$$R(x, y) \leftrightarrow Q(x, y)$$

Of the generating operations specified above, the operations *Fo* (of subsection (vi), and Sp (of subsection (xi)) are optional, the others obligatory.

The relation of 'predecessor' among constants in $\mathscr{L}_{II}$ is defined in obvious analogy with previous cases.

4. *The General Axioms; Rules of Inference*

A. Axioms

Group 1: Formally the same as the axioms and axiom schemata (i)–(v) as listed in I.4.
Group 2: $\langle x, y\rangle \in \omega \leftrightarrow x < y$.

B. Rules of Inference

Formally the same as those in I.4 (Substitution, Modus Ponens, Rule of induction).

5. *Calculi of Kind II*

A subset M of $\mathscr{L}_{II}$ is 'closed downwards', if whenever M contains a constant A, M also contains the predecessors of A. A subset of $\mathscr{L}_{II}$ is called a *calculus of Kind II*, if it is closed downwards, contains the initial constants, and is closed under the obligatory operations on $\mathscr{L}_{II}$.

When D is a calculus of Kind II, the *terms, formulas, proofs in D* are those terms, formulas, and proofs, all of whose constants are members of the set D. We write $\vdash_D A$ for the statement 'A has a proof in D'.

As in the case of calculi of Kind I and of elementary calculi, there is a minimal calculus D_0 of Kind II, namely the set of constants generated from the initial constants by the obligatory operations.

6. *The Notions of 'Embedding' and of 'NDS-Relation'*

When D is a calculus of Kind II, R a binary predicate constant or special constant in D, the abbreviation 'R-desc(s)' represents a formula in D as explained in I.7.

When R is a binary predicate constant, γ a unary function constant, and S either a binary predicate constant, or a special constant, we write 'Emb(γ, R, S) for the formula

$$R\text{-desc}(s) \ \&\ t \subset s \rightarrow \langle \gamma(s), \gamma(t)\rangle \in S$$

When D is a calculus of Kind II, we call R *embeddable into* S *in* D, if for some function constant γ, $\text{Emb}(\gamma, R, S)$ is provable in D.

And we call R an *NDS-relation in* D, if for some special constant S, R is embeddable into S in D.

THEOREM 1. When D is any calculus of Kind II, then if R, S and Q are binary predicate constants or special constants, then if Q is embeddable into R in D, and R is embeddable into S in D, then Q is embeddable into S in D.

The operations on binary predicate constants dealt with below were introduced in II.7. As was remarked there, their definitions can be taken over, without formal change, for calculi of Kind II.

THEOREM 2. Let D be a calculus of Kind II, A and B two binary predicate constants in D.

Then, if A and B are NDS-relations in D, then so are $\text{Succ}(A)$, $A \oplus B$, $A \otimes B$, and $\Sigma(A)$.

The proofs of Theorem 1 and Theorem 2 are straightforward, and are omitted.

THEOREM 3. Let D be a calculus of Kind II. Let A be an NDS-relation in D, and p an m-ary operation constant in D. Then there is a function constant f, satisfying

$$z = f(\bar{y}) \rightarrow \langle \{p(\bar{y})\}(z+1), \{p(\bar{y})\}(z) \rangle \notin A;$$
$$z < f(\bar{y}) \rightarrow \langle \{p(\bar{y})\}(z+1), \{p(\bar{y})\}(z) \rangle \in A.$$

(This means that the behaviour of NDS-relations is essentially the same as that of special constants.)

7. *Equivalence between Constants in a Calculus D; Translations D′ → D, and Translations C → D, with D′ Another Calculus of Kind II, and C an Elementary Calculus; Standard Form; the Elementary Part of a Calculus*

Two constants A and B in $\mathcal{L}_{II}$ are said to be *of the same type in the strict sense*, if either both are special constants, or, for some m, both are either m-ary function constants, or m-ary predicate constants, or m-ary operation constants.

Two constants A, B in $\mathcal{L}_{II}$ are said to be *of the same type in the wide sense* if they are either of the same type in the strict sense, or else one is a special constant and the other a binary predicate constant. (Binary predicate constants and special constants are both 'binary relation constants in the wide sense'.)

Equivalence between two constants relative to a calculus D, when the constants are of the same type in the wide sense, can be defined in obvious analogy to the definition in Section II.6.

When D is a calculus of Kind II, C an elementary calculus, then a mapping $M: C \to D$ is called a *translation* $C \to D$ if

(1) M preserves type in the sense that each binary predicate constant in C is mapped on a binary relation constant in the wide sense, and otherwise, each constant of C is mapped on a constant of the same type, and

(2) The induced mapping of formulas, M^*, preserves theoremhood.

(3) $M(0) = 0$, $M(S) = S$.

When D and D' are two calculi of Kind II, then a mapping $M{:}D' \to D$ is called a *translation* $D' \to D$ if

(1) M preserves type in the wide sense.

(2) M^* preserves theoremhood.

(3) $M(0) = 0$, $M(S) = S$.

(4) Whenever R is an NDS-relation in D', $M(R)$ is an NDS-relation or a special constant in D.

When D is a calculus of Kind II, and K is either elementary or of Kind II, then a translation $M: K \to D$ is called *monomorphic* if for A, B in K, $M(A)$ and $M(B)$ are equivalent in D only if A and B are equivalent in K. We say that D is a *conservative* extension of K, if there is a monomorphic translation $K \to D$.

When D is a calculus of Kind II, and K is either an elementary calculus, or another calculus of Kind II, then if M_1 and M_2 are two translations $K \to D$, we call M_1 and M_2 *equivalent translations* if, for each constant A in K, $M_1(A)$ and $M_2(K)$ are equivalent in D.

It is easy to prove:

THEOREM 1. (Uniqueness of translations)
When D is a calculus of Kind II, and K is either an elementary calculus or another calculus of Kind II, then if M_1 and M_2 are two translations $K \to D$, then M_1 and M_2 are equivalent translations.

Two calculi D_1 and D_2 of Kind II are called *equivalent* if each is translatable in the other.

All constants in $\mathscr{L}_{II}$ other than the special constants are called *elementary*. A constant in $\mathscr{L}_{II}$ is called *strictly elementary*, when it is elementary, and all its predecessors are elementary.

A calculus D of Kind II is said to be *in standard form*, if, for any strictly elementary p and R in D, such that p is an operation constant and R an NDS-relation in D, the calculus D contains the constant $Fo(R, p)$.

Given a calculus D which is not in standard form, there is an obvious method for extending it to a calculus in standard form, namely, the method of successively adding the missing constants of form $Fo(R, p)$, and closing w.r.t. the obligatory operations on $\mathscr{L}_{II}$.

We write D^+ for the calculus which we get by applying this procedure to D, and we call it the 'standardization' of D.

We can prove:

THEOREM 2. When D is any calculus of Kind II, then

(a) D and D^+ are equivalent.
(b) D^+ is in standard form.
(c) If D is in standard form, then D^+ is the same as D.

THEOREM 3. When D is a calculus of Kind II in standard form, each constant in D has a strictly elementary equivalent in D.

We let $\mathscr{E}_{0,2}$ be the obvious mapping of constants of $\mathscr{L}_0$ onto the set of strictly elementary constants of $\mathscr{L}_{II}$. Observe, that we have chosen the generating operations for $\mathscr{L}_0$ and $\mathscr{L}_{II}$ in such a way that this mapping is one-one.

When D is a calculus of Kind II, then by the *elementary part* of D we mean the elementary calculus consisting of the inverse image of D^+ by the mapping $\mathscr{E}_{0,2}$: we write D^0 for the elementary part of D. It is evident that D^0 will be translatable into D. It is also easy to verify:

THEOREM 4. When D is any calculus of Kind II, and D^0 its elementary part, then D is a conservative extension of D^0.

When D^0 is the elementary part of a calculus D (of Kind II), then a constant R in D^0 is said to be an *NDS-relation in* D, when its translation in D is an NDS-relation in D.

8. *The Notions of 'Universal Formula', 'Stable Term', 'Convergent Term' in a Calculus D of Kind II*

For the purposes of this and the following section, we will pick out and name certain sets of variables in $\mathscr{L}_{II}$, and make some definitions which will be dependant on the choice of sets.

The designated classes will be denoted W, Σ, Y. We assume the following about them:

$$Y = \{y_0, y_1, \ldots\}$$
$$\Sigma = \{s_0, s_1, \ldots\}$$
$$W = \{w\}$$

We also assume: Y, Σ, W are disjoint, and the variables $\Sigma' = \{s'_0, s'_1, \ldots\}$, and w' are disjoint from those and from each other.

We will call the variables Y 'numerical parameters', those in Σ 'sequence variables', and w the 'numerical indicator'.

Before introducing the interesting definitions, we have to explain some new notation:

We will use $\vdash$ as symbol for 'deducibility' in the following way:

(1) When F is a formula, and $\mathscr{M}$ a finite set of formulas, (and D a calculus), we write $\mathscr{M} \vdash_D F$ for $\vdash_D \wedge \mathscr{M} \rightarrow F$, where $\wedge \mathscr{M}$ is the conjunction of the members of $\mathscr{M}$.

(2) When F is a formula, and $\mathscr{M}$ an infinite set of formulas, then $\mathscr{M} \vdash_D F$ iff $\mathscr{M}' \vdash_D F$, for some finite subset $\mathscr{M}'$ of $\mathscr{M}$.

Another piece of notation:

We write $(w; \Sigma) \subset (w'; \Sigma')$ for the set $\{w < w', s_0 \subset s'_0, s_1 \subset s'_1, \ldots\}$; and $(w; \Sigma) \subseteq (w'; \Sigma')$ for $\{w \leqslant w', s_0 \subseteq s'_0, s_1 \subseteq s'_1, \ldots\}$.

Now for formulas and terms 'in' the variables w, Σ, Y (that is, containing no other variables), we define the notions 'universal formula', 'stable term', 'convergent term', in a way which obviously is relative to our choice of the sets W, Σ, Y.

DEFINITION 1. Let D be a calculus of Kind II, A formula $\mathscr{A}(w, \Sigma, Y)$ in the variables W, Σ, Y is called *universal in D* (via the relation constant R, and the term $\nu(w, \Sigma, Y)$ if:

(i) R is an NDS-relation in D;

(ii) $(w; \Sigma) \subset (w'; \Sigma') \vdash (\neg \mathscr{A}(w', \Sigma', Y) \rightarrow \langle \nu(w', \Sigma', Y), \nu(w, \Sigma, Y)\rangle \in R)$.

DEFINITION 2a. Let D be a calculus of Kind II. A term $t(w, \Sigma, Y)$ in the variables w, Σ, Y is called *stable in D* (via the formula $U(w, \Sigma, Y)$), if:

(i) $U(w, \Sigma, Y)$ is universal in D;

(ii) $(w; \Sigma) \subseteq (w'; \Sigma') \vdash_D U(w, \Sigma, Y) \to (t(w, \Sigma, Y) = t(w', \Sigma', Y))$.

DEFINITION 2b. Let D be a calculus of Kind II. A formula $\mathscr{A}(w, \Sigma, Y)$ in the variables w, Σ, Y is called *stable in D* (via the formula $U(w, \Sigma, Y)$), if:

(i) $U(w, \Sigma, Y)$ is universal in D;

(ii) $(w; \Sigma) \subseteq (w'; \Sigma') \vdash U(w, \Sigma, Y) \to (\mathscr{A}(w, \Sigma, Y) \leftrightarrow \mathscr{A}(w', \Sigma', Y))$.

DEFINITION 3. Let D be a calculus of Kind II. A term $\gamma(w, \Sigma, Y)$ in the variables w, Σ, Y is called *convergent in D* (via the predicate constant R, and the term $\nu(w, \Sigma, Y)$), if

(i) R is an NDS-relation in D;

(ii) $(w; \Sigma) \subseteq (w'; \Sigma') \vdash_D \gamma(w, \Sigma, Y) \subseteq \gamma(w', \Sigma', Y)$;

(iii) $(w; \Sigma) \subset (w'; \Sigma') \vdash_D \gamma(w, \Sigma, Y) \subset \gamma(w', \Sigma', Y) \vee \langle \nu(w', \Sigma', Y), \nu(w, \Sigma, Y) \rangle \in R$.

THEOREM 1. Let D be a calculus of Kind II. If $\mathscr{A}_1(w, \Sigma, Y)$ and $\mathscr{A}_2(w, \Sigma, Y)$ are two formulas in the variables w, Σ, Y, which are both universal in D, then so is their conjunction $\mathscr{A}_1(w, \Sigma, Y) \wedge \mathscr{A}_2(w, \Sigma, Y)$.

Proof: If the NDS-relations R_1 and R_2 are associated with the universality of the formulas $\mathscr{A}_1$ and $\mathscr{A}_2$, their sum $R_1 \oplus R_2$ can serve for the conjunction. The term $\nu(w, \Sigma, Y)$ to be associated with the new formula is also easy to construct.

THEOREM 2. Let D be a calculus of Kind II. If $\mathscr{A}_1$ and $\mathscr{A}_2$ are two formulas in the variables w, Σ, Y such that $\vdash_D \mathscr{A}_1 \to \mathscr{A}_2$, then if $\mathscr{A}_1$ is universal in D, then so is $\mathscr{A}_2$.

Proof: Obvious.

THEOREM 3. (Modus ponens preserves universality.)
Let D be a calculus of Kind II. If $\mathscr{A}$ and $\mathscr{B}$ are two formulas in the variables w, Σ, Y, and if $\mathscr{A}$ is universal in D, and also $\mathscr{A} \to \mathscr{B}$ is universal in D, then Ψ is universal in D.

Proof: From the two preceding theorems.

THEOREM 4. (Substitution, of certain kinds, preserves universality.)
Let D be a calculus of Kind II. Let $\mathscr{A}$ be a formula in the variables

w, Σ, Y, which is universal in D. Let t and γ be two terms in the variables w, Σ, Y, such that t is stable in D, and γ is convergent in D. Then

(a) If $\mathscr{A}'$ is the result of substituting t for one of the numerical parameters Y in $\mathscr{A}$, then $\mathscr{A}'$ is universal in D.

(b) If $\mathscr{A}''$ is the result of substituting γ for one of the sequence variables Σ in $\mathscr{A}$, then $\mathscr{A}''$ is universal in D.

THEOREM 5. (Substitution, of the proper kind, preserves stability and convergence.)
Let D be a calculus of Kind II. Let T, t, Γ, and γ be four terms in the variables w, Σ, Y, such that T and t are stable in D, and Γ and γ are convergent in D. Let $\mathscr{A}$ be a formula which is stable in D. Then

(a) The result of substituting t for one of the numerical parameters in T or in $\mathscr{A}$ will be stable in D.

(b) The result of substituting t for one of the numerical parameters in Γ will be convergent in D.

(c) The result of substituting γ for one of the sequence variables in T, or in $\mathscr{A}$, will be stable in D.

(d) The result of substituting γ for one of the sequence variables in Γ will be convergent in D.

THEOREM 6. (Bounded universal quantification preserves universality.)
Let D be a calculus of Kind II. We write $\bar{s}$ for $s_0, \ldots, s_{m-1}$, $\bar{y}$ for $y_0, \ldots, y_{k-1}$. We let u and z be two numerical parameters, distinct from $y_0, \ldots, y_{k-1}$.

Then, if $\mathscr{A}(w, \bar{s}, \bar{y}, u)$ is a formula in D in the variables shown, which is universal in D, then the formula $\forall u_{u<z}\mathscr{A}(w, \bar{s}, \bar{y}, u)$ is also universal in D.

THEOREM 7. (A connection between universality, and the notion of NDS-relation.)
Let D be a calculus of Kind II, R a binary predicate constant in D, s a sequence variable.

Then R is an NDS-relation in D iff the formula $\neg R$-desc(s) is universal in D.

Potential representing triples and quadruples. For calculi of Kind II, we can take over, word by word, the definitions given in Section II.9 for 'potential representing triples for type (m, k) (in D)', and 'potential representing quadruple (in D) for function (predicate, operation) constants of type (m, k)'.

The following theorem, which is easily verified, gives some connections between these notions, and the notions of universality, stability and convergence.

THEOREM 8. Let D be a calculus of Kind II. Then (writing $\bar{s}$ for $s_0, \ldots, s_{m-1}$, and $\bar{y}$ for $y_0, \ldots, y_{k-1}$):

(a) If $\langle Q, g, r\rangle$ is a potential representing triple in D for type (m, k), then the formula $r(\bar{s}, \bar{y}) \neq 0$ is a universal formula in D.

(b) When $\langle Q, g, r, B\rangle$ is a potential representing quadruple in D for function (predicate) constants of type (m, k), then $B(\bar{s}, \bar{y})$ is a stable term (formula) in D.

(c) When $\langle Q, g, r, B\rangle$ is a potential representing quadruple in D for operation constants of type (m, k), with $m \neq 0$, then $B(\bar{s}, \bar{y})$ is a convergent term in D.

(d) When $\langle Q, g, r, B\rangle$ is a potential representing quadruple in D for operation constants of type $(0, k)$, then $B(w, \bar{y})$ is a convergent term in D.

9. *Interpretations $\mathscr{C} \to D$, when $\mathscr{C}$ Is a Calculus of Kind I, D a Calculus of Kind II; the 'Translation' T_J Induced by an Interpretation*

When $\mathscr{C}$ and D are calculi of Kind I and Kind II, resp., then by an *interpretation* $\mathscr{C} \to D$ we will mean a translation J: $\mathscr{C}^0 \to D^0$, with the property

(i) For every constant A of $\mathscr{C}$, $J(\alpha_A)$ is an NDS-relation in D.

We call J a *strong interpretation* $\mathscr{C} \to D$, if it also satisfies the condition

(ii) For every R in $\mathscr{C}^0$, if R is an NDS-relation in $\mathscr{C}$, then $J(R)$ is an NDS-relation in D.

(We will prove below (Theorem 9, Cor.), that in fact every interpretation is a strong interpretation; but before we have proved that, we will keep the notions separate.)

We will now associate a 'translation', in the sense of a mapping of terms and formulas, with any given interpretation of this type. In defining this translation, we will again make use of the designated variable-sets W, Σ, Y introduced in the previous section. The definitions will be dependent on the choice of those sets.

We will also make the convenient assumption that the set of numerical variables of Kind I coincides with the set Y labelled 'numerical parameters' in $\mathscr{L}_{II}$. We also assume that the sequence $\phi_0, \phi_1, \ldots,$ constitutes all the function variables in $\mathscr{L}_I$.

DEFINITION. (Defining the translation T_J, for J a given interpretation $\mathscr{C} \to D$).
It is assumed that $\mathscr{C}$ and D are calculi of Kinds I and II, and J an interpretation $\mathscr{C} \to D$. When A is a constant of $\mathscr{C}$, we write V'_A for $J(V_A)$. The mapping T_J is now defined by recursion on the complexity of terms and formulas in $\mathscr{C}$:

(i) When y_i is a numerical variable in $\mathscr{L}_I$, $T_J(y_i) = y_i$.

(ii) When ϕ_i is the i:th function variable in $\mathscr{L}_I$, $T_J(\phi_i) = s_i$.

(iii) When A is a function, constant or predicate constant in $\mathscr{C}$ of type (m, k), or an operation constant in $\mathscr{C}$ of type (m, k), with $m \neq 0$, and if $\chi_0, \ldots, \chi_{m-1}$ are function terms in $\mathscr{C}$, and $t_0, \ldots, t_{k-1}$ numerical terms in $\mathscr{C}$, and their translations by T_J are, resp., $\chi'_o, \ldots, \chi'_{m-1}$, and $t'_o, \ldots, t'_{k-1}$, then the translation by T_J of the term (formula) $A(\chi_0, \ldots, \chi_{m-1}, t_0, \ldots, t_{k-1})$ is $V'_A(\chi'_o, \ldots, \chi'_{m-1}, t'_o, \ldots, t'_{k-1})$.

(iv) If Ψ is an operation constant in $\mathscr{C}$ of type $(0, k)$, if $t_0, \ldots, t_{k-1}$ are numerical terms in $\mathscr{C}$, and $t'_o, \ldots, t'_{k-1}$ their translations by T_J, then the translation, by T_J, of the term $\Psi(t_0, \ldots, t_{k-1})$ is $V'_\psi(w, t'_o, \ldots, t'_{k-1})$.

(v) If χ and t are, resp., a function term, and a numerical term in $\mathscr{C}$, and if χ' and t' are their translations by T_J, then the translation, by T_J, of $\{\chi\}(t)$ is $(\chi')_{t'}$.

(vi) If F and G are formulas in $\mathscr{C}$, and if F' and G' are their translations by T_J, then the translations, by T_J, of $\neg F$, $F \vee G$, $F \wedge G$, $F \to G$, $F \leftrightarrow G$, are, resp., $\neg F'$, $F' \vee G'$, $F' \wedge G'$, $F' \to G'$, and $F' \leftrightarrow G'$.

This concludes the definition of T_J.

It is immediate from the definition, that the translation T_J commutes with substitution in the following sense.

THEOREM 1. Let $\mathscr{C}$ and D be calculi of Kinds I and II, J an interpretation $\mathscr{C} \to D$, and T_J the associated translation. Then:

(a) If E is any term or formula in $\mathscr{C}$, t a numerical term in $\mathscr{C}$, F the result of substituting t for y_i in E, and E', t' and F' the translation of E, t, and F by T_J, then F' is the result of substituting t' for y_i in E'.

(b) If E is a term or formula in $\mathscr{C}$, χ a function term in $\mathscr{C}$, F the result of substituting χ for ϕ_i in E, and E', χ', F' are the translations of E, χ, F by T_J, then F' is the result of substituting χ' for s_i in E'.

It is also easy to show the following.

THEOREM 2. We assume that J is an interpretation $\mathscr{C} \to D$, with $\mathscr{C}$ of

Kind I, D of Kind II, and T_J the associated translation. Then:

(a) When t is a numerical term or a formula, in $\mathscr{C}$, its translation by T_J is a stable term, or stable formula, in D.

(b) When χ is a function term in $\mathscr{C}$, its translation by T_J is a convergent term in D.

Proof: We can prove (a) and (b) simultaneously by recursion on the length of terms in $\mathscr{C}$, using the results of the previous section (Theorems 9.5 and 9.6).

THEOREM 3. ('The translations of defining axioms are universal'.) We assume that J is an interpretation $\mathscr{C} \to D$, with $\mathscr{C}$ of Kind I, D of Kind II, and that T_J is the associated translation. We claim:

When A is any non-initial constant belonging to $\mathscr{C}$, then the translation, by T_J of any defining axiom of A is a universal formula in D.

Proof: The proof is done by cases, according to the form of the constant A. We will carry out the proof for two cases. The proofs for the remaining cases are similar, and can easily be reconstructed.

Case 1. A is an operation constant Ψ in $\mathscr{C}$ of type (m, k), with $m \neq 0$, and with defining axiom

$$\{\Psi(\phi, y)\}(z) = G(\phi, y, z).$$

The translation by T_J is then

$$(V'_\psi(\bar{s}, \bar{y}))_z = V'_G(\bar{s}, \bar{y}, z),$$

where $V'_\psi = J(V_\psi)$, and $V'_G = J(V_G)$.

We know from Theorem II, 10.4, and the definition of α, σ, ρ, V, that the canonical representing quadruples for Ψ and G stand in the relation $\Delta_7^{\mathscr{C}^0}{}_{,m,k}$. From this it follows, that the following is provable in $\mathscr{C}^0$:

$$\forall j_j \leqslant_z (\rho_G(\bar{s}, \bar{y}, j) \neq 0) \wedge \mathrm{lh}(\bar{s}) > z \to (V_\psi(\bar{s}, \bar{y}))_z = V_G(\bar{s}, \bar{y}, z)$$

Writing ρ'_G for $J(\rho_G)$, the following is then provable in D^0:

$$\forall j_j \leqslant_z (\rho'_G(\bar{s}, \bar{y}, j) \neq 0) \wedge \mathrm{lh}(\bar{s}) > z \to (V'_\psi(\bar{s}, \bar{y}))_z = V'_G(\bar{s}, \bar{y}, z)$$

The left hand side is easily seen to be universal in D. Therefore the right hand side is too.

Case 2. A is a function constant F in $\mathscr{C}$ of form $Fo^{(1)}(Q)$, so that the defining axioms are

$$\begin{cases} z = F(\phi) \rightarrow \langle \{\phi\}(z+1), \{\phi\}(z) \rangle \notin Q; \\ z < F(\phi) \rightarrow \langle \{\phi\}(z+1), \{\phi\}(z) \rangle \in Q. \end{cases}$$

The translation by T_J is:

$$\begin{cases} z = V_F'(s) \rightarrow \langle (s)_{z+1}, (s)_z \rangle \notin V_Q'; \\ z < V_F'(s) \rightarrow \langle (s)_{z+1}, (s)_z \rangle \in V_Q'; \end{cases}$$

where $V_F' = J(V_Q')$, $V_Q = J(V_Q)$.

Again, we know by Theorem II, 10.4, that the canonical representing quadruples for F and Q stand in the relation $\Delta_8^{\mathscr{C}^0}$; and from this we can infer that the following is provable in $\mathscr{C}^0$:

$$\rho_F(s) \neq 0 \rightarrow (z = V_F(s) \rightarrow \langle (s)_{z+1}, (s)_z \rangle \notin V_Q.$$
$$\rho_F(s) \neq 0 \rightarrow (z < V_F(s) \rightarrow \langle (s)_{z+1}, (s)_z \rangle \in V_Q$$

The translation, by J, of the left hand side is universal in D, and therefore also the right hand side.

THEOREM 4. (On the rule of induction.)
We assume again that J is an interpretation $\mathscr{C} \rightarrow D$, with $\mathscr{C}$ of Kind I, D of Kind II, and that T_J is the associated translation. We assume also that $A(0)$, $A(Sz)$, $A(y)$ are obtained by the obvious substitutions from $A(z)$.

Then we claim: If the translation, by T_J, of the formulas

(i) $A(0)$

and

(ii) $A(z) \rightarrow A(Sz)$

are universal in D, then the translation (by T_J) of

(iii) $A(y)$

is also universal in D.

Proof: If we write $B(z)$ for the translation of $A(z)$, then the translations of (i), (ii), and (iii) are respectively:

(i′) $B(0)$;

(ii′) $B(z) \rightarrow B(Sz)$;

(iii′) $B(y)$.

To say that (i′) and (ii′) are universal in D is to say that for some universal formulas U_1, $U_2(z)$, we can prove in D:

$$\begin{cases} U_1 \to B(0); \\ U_2(z) \to (B(z) \to B(S(z))); \end{cases}$$

When we can prove these two formulas in D, we can also prove:

$$U_1 \wedge \forall z_z \leqslant_y U_2(z) \to B(y)$$

Since the left hand side here is universal in D, the right hand side is too.
Q.E.D.

Using the results that we have already proved, it is easy to prove the following theorem, which deals with the remaining axioms and rules of the calculus $\mathscr{C}$.

THEOREM 5. We assume again that J is an interpretation $\mathscr{C} \to D$, with $\mathscr{C}$ of Kind I, D of Kind II, and that T_J is the associated translation. Then

(a) The translations, by T_J, of all the basic axioms of $\mathscr{C}$, are universal in D.

(b) When A and B are formulas in $\mathscr{C}$, where B is obtained from A by substitution, either of a numerical term for a numerical variable, or of a function term for a function variable, and A', B' are the translations, by T_J, of A and B, then if A' is universal in D, then B' is universal in D.

(c) If A_1, A_2, A_3 are formulas in $\mathscr{C}$, such that A_1 follows by modus ponens from A_2 and A_3, then, if A_1', A_2', A_3', are the translations of A_1, A_2, A_3 by T_J, then if A_2' and A_3' are universal in D, then so is A_1'.

What all this (Theorems 3–5) leads up to is clearly:

THEOREM 6. When J is an interpretation $\mathscr{C} \to D$, with $\mathscr{C}$ of Kind I, D of Kind II, and T_J is the associated translation, then the translation, by T_J, of any theorem of $\mathscr{C}$ is universal in D.

The following two theorems state some additional properties of the translation T_J. Both are easily verifiable.

With respect to the first theorem below (Theorem 7) it should be noticed, that when A is a strictly elementary formula (or term) of $\mathscr{C}$, it makes sense to speak of 'the translation of A into D by J^*', as well as by the translation T_J. It will follow from II, 10, 6 that:

THEOREM 7. (Under the same assumptions as above on $\mathscr{C}$, D, and J:) When A is a strictly elementary term of formula in $\mathscr{C}$, B its translation by T_J, B' its translation by J^*, then B and B' are equivalent in D.

THEOREM 8. (Same assumptions as above). When A is an elementary formula in $\mathscr{C}$, and B its translation by T_J, then if B is universal in D, it is also provable in D.

COROLLARY. When J, $\mathscr{C}$, D, T_J are as assumed above, then

(a) the translation, by T_J, of the elementary theorems in $\mathscr{C}$ are provable in D.

(b) the translations, by J^*, of the strictly elementary theorems in $\mathscr{C}$ are provable in D.

THEOREM 9. We assume again that J is an interpretation $\mathscr{C} \to D$, with $\mathscr{C}$ of Kind I, D of Kind II, and T_J the associated translation. We let R be an elementary predicate constant in $\mathscr{C}$, which is an NDS-relation in $\mathscr{C}$, i.e., for some constant F it is provable in $\mathscr{C}$:

$$\langle\{\phi\}(F(\phi) + 1), \{\phi\}(F(\phi)\rangle \notin R$$

We let R' be $J(V_R)$. Then, we claim, R' is an NDS-relation in D.

Proof: From the fact that the above formula for F and R is provable in $\mathscr{C}$, we get, by our main theorem on translation (Theorem 6), that the translation of the formula is universal in D. This means, that for some universal formula $U(s)$ we can prove in D:

$$(1) \qquad U(s) \to \langle (s)_{V'_F(s)+1}, (s)_{V'_F(s)}\rangle \notin R'$$

Then we can also prove in D:

$$(2) \qquad U(s) \wedge \rho'_F(s) \neq 0 \wedge \mathrm{lh}(s) \geqslant V'_F(s) + 2 \to \neg(R'\text{-desc}(s))$$

What we want to establish is that R' is an NDS-relation in D. By 8.7 we know that this is equivalent to establishing that the formula $\neg(R'$-desc$(s))$ is universal in D. Because of the provability of (2), (and since $U(s)$ is universal), it is sufficient, for our purposes, to establish the universality of the remaining conjunct of the left hand side of (2). We abbreviate this formula $L(s)$.

That means: we write $L(s)$ for the conjunction

$$\rho'_F(s) \neq 0 \wedge \mathrm{lh}(s) \geqslant V'_F(s) + 2.$$

To establish the universality of this formula, it is sufficient to find an NDS-relation Q in D, and a function g, for which it is provable:

$$\neg L(t) \wedge s \sqsubset t \to \langle g(t), g(s)\rangle \in Q.$$

For Q we take the relation $\omega + a'_F$, where $a'_F = J(a_F)$.

We define the function g by:

$$g(s) = \begin{cases} \langle \sigma'_F(s), 1 \rangle \text{ if } \rho'_F(s) = 0; \\ \langle (V'_F(s) + 2) - \mathrm{lh}(s), 0 \rangle \text{ if } \rho'_F(s) \neq 0. \end{cases}$$

It is easily verified that these constants Q and g have the desired properties.

COROLLARY. We make the same assumptions on $\mathscr{C}$, D, J, T_J as above. We claim that J is a *strong* interpretation, i.e.: whenever R in $\mathscr{C}^0$ is an NDS-relation in $\mathscr{C}$, then $J(R)$ is an NDS-relation in D.

Proof: Our assumption is that R (in $\mathscr{C}^0$) is an NDS-relation in $\mathscr{C}$. We observe, that the mapping J, which is assumed to be an interpretation $\mathscr{C} \to D$, will also be an interpretation $\mathscr{C}^+ \to D$. (This is easy to verify.) We can now apply the theorem to $\mathscr{C}^+$ and R. Since R is an NDS-relation in $\mathscr{C}^+$, we know by the theorem that $J(V_R)$ is an NDS-relation in D. But we know (II, 10, 6) that R and V_R are equivalent *in* $\mathscr{C}^0$. Therefore, $J(R)$ and $J(V_R)$ are equivalent in D, and $J(R)$ is an NDS-relation in D. Q.E.D.

10. The Calculus $\mathscr{C}^{\mathrm{Red}}$; ('the Calculus of Kind II Corresponding to the Given Calculus $\mathscr{C}$ of Kind I'); 'Translations and Interpretations' $D \to \mathscr{C}$, for D of Kind II, $\mathscr{C}$ of Kind I

Let D and $\mathscr{C}$ be calculi of Kinds II and I, resp., and let $\mathscr{C}^0$ be the elementary part of $\mathscr{C}$. Then by a *translation* B: $D \to \mathscr{C}$ we mean a mapping $D \to \mathscr{C}^0$, which satisfies the following conditions:

(1) Special constants in D are mapped on binary relation constants in $\mathscr{C}^0$, and the other constants in D are mapped on constants of the same type.

(2) The constants 0, S, Eq are mapped on themselves, ω is mapped on $<$.

(3) The induced mapping B^* on formulas carries theorems of D into theorems of $\mathscr{C}^0$.

(4) The images of the special constants of D are NDS-relations in $\mathscr{C}$.

Observation: The definition can be reworded, by replacing condition (3) by the more special:

(3′) For every constant K in D, the mapping B^* takes the defining axioms of K into theorems of $\mathscr{C}^0$.

The following lemma gives a criterion for the existence of a translation $D \to \mathscr{C}$.

THEOREM 1. Let $\mathscr{C}$ be a calculus of Kind I in standard form. Let M be a subset of $\mathscr{L}_{II}$, closed downwards (i.e., closed with respect to the predecessor relation). Let B be a mapping $M \to \mathscr{C}^0$, which satisfies the basic conditions (1) and (2) above, and also

(3_M) For every constant K of M, its defining axioms are carried, by B, into theorems of $\mathscr{C}^0$.

(4_M) When R is a special constant of M, then $B(R)$ is an NDS-relation in $\mathscr{C}$.

Then, for M' = the closure of M, B can be extended to a mapping $M' \to \mathscr{C}^0$, which satisfies conditions (1), (2), (3′) and (4), and which therefore is a translation $M' \to \mathscr{C}$.

Proof: We leave out the proof, which is rather obvious, and essentially depends on the fact that $\mathscr{C}^0$ is closed under operations corresponding to those in M'.

DEFINITION. When $\mathscr{C}$ is any calculus of Kind I, then by $\mathscr{C}^{\text{Red}}$ we mean the calculus of Kind II, which is the closure (under the obligatory operations of $\mathscr{L}_{II}$) of the set

$$\mathscr{C}^0 \cup \mathscr{S}$$

where $\mathscr{S}$ is the set of all constants $\text{Sp}(\alpha_A)$, for A in $\mathscr{C}$.

We get immediately from the definition:

THEOREM 2. When $\mathscr{C}$ is any calculus of Kind I, $\mathscr{C}$ is interpretable in $\mathscr{C}^{\text{Red}}$. (More specifically, the identity mapping will serve as an interpretation $\mathscr{C} \to \mathscr{C}^{\text{Red}}$.)

A more interesting result is:

THEOREM 3. When $\mathscr{C}$ is any calculus of Kind I, and D is $\mathscr{C}^{\text{Red}}$, then there is a translation $D \to \mathscr{C}$.

Proof: This follows easily from the lemma (Theorem 1) above. To apply the lemma, we only need to find a mapping on the set $\mathscr{C}^0 \cup \mathscr{S}$ which satisfies the conditions in the hypothesis of the lemma. The obvious mapping can be seen to work, that is, the mapping B defined by:

$$B(A) = A \text{ for } A \text{ in } \mathscr{C}^0;$$
$$B(\text{Sp}(\alpha_A)) = \alpha_A.$$

An important consequence is:

THEOREM 4. Any calculus $\mathscr{C}$ of Kind I is a conservative extension of its elementary part $\mathscr{C}^0$.

Proof: Assume that $\mathscr{C}$ is a calculus of Kind I in standard form. Let D

be $\mathscr{C}^{\text{Red}}$. Let $\mathscr{A}$ be a strictly elementary formula which is provable in $\mathscr{C}$. Let J be the identity mapping on $\mathscr{C}^0$.

We know, from what we have proved before (9.8, Corollary b),

(1) $\mathscr{A}$ is provable in D.

And now, since there is a translation, in our new sense, of D into $\mathscr{C}$, which is an extension of J, it follows:

(2) $\mathscr{A}$ is provable in $\mathscr{C}^0$.

We have then proved the theorem for calculi $\mathscr{C}$ in standard form. The more general claim follows immediately.

We will also here introduce the notion of *interpretation* $D \to \mathscr{C}$, for the case that D is a calculus of Kind II, and $\mathscr{C}$ a calculus of Kind I.

DEFINITION. Let $\mathscr{C}$ and D be calculi of Kinds I and II, resp. Let $\mathscr{C}^0$ and D^0 be their elementary parts. Then by an *interpretation* $D \to \mathscr{C}$ we will mean a translation J: $D^0 \to \mathscr{C}^0$, which satisfies the additional condition:

(1) When R is an NDS-relation in D, then $J(R)$ is an NDS-relation in $\mathscr{C}$.

It is easily seen that the difference between the two notions 'translation' $D \to \mathscr{C}$, and 'interpretation' $D \to \mathscr{C}$ is purely formal, and that practically anything that holds for one holds for the other. In particular:

THEOREM 5. When $\mathscr{C}$ and D are calculi of Kinds I and II, resp., then D is interpretable in $\mathscr{C}$ iff D is translatable in $\mathscr{C}$.

THEOREM 6. When $\mathscr{C}$ is any calculus of Kind I, $\mathscr{C}^{\text{Red}}$ is interpretable in $\mathscr{C}$.

11. The Calculus D^* (The Calculus of Kind I Corresponding to the Given Calculus D of Kind II)

DEFINITION. When D is a calculus of Kind I, we define D^* to be the calculus of Kind I, which is the closure (under the obligatory operations of $\mathscr{L}_I$) of the set

$$D^0 \cup \mathscr{S}$$

where $\mathscr{S}$ is the set of all constants $Fo^{(1)}(R)$, for R (in D^0) an NDS-relation in D.

It follows immediately from the definition, that

THEOREM 1. For any calculus D of Kind II, D is interpretable into D^* (via the identity mapping).

We now want to establish that interpretability also holds in the opposite direction. The following lemma gives a useful criterion for existence of interpretations.

THEOREM 2. Let D be a calculus of Kind II, and M a subset of $\mathscr{L}_I$, which is closed downwards, and let $\mathscr{C}$ be the closure of M under the obligatory operations of $\mathscr{L}_I$.

Then if for every constant A of M it holds:

(i) The constants α_A, σ_A, ρ_A, V_A of the representing quadruple belong to D, and

(ii) α_A is an NDS-relation in D,

then

(a): statements (i) and (ii) hold for every constant of $\mathscr{C}$, and

(b): $\mathscr{C}$ is interpretable in D.

Proof: The proof for (a) is by induction on the complexity of the constants of $\mathscr{C}$, and only requires verification that a calculus D (and the set of NDS-relations in D) has certain closure properties. (b) follows from (a), since the mapping V will give the required translation of elementary parts.

The following is the main result of this section.

THEOREM 3. For any calculus D of Kind II, D^* is interpretable in D.

Proof: By virtue of the lemma just proved (Theorem 2), it is sufficient to establish, that for every constant A in the set $D^0 \cup \mathscr{S}$ the constants α_A, σ_A, ρ_A, V_A belong to D, and that α_A is an NDS-relation in D.

For constants A in D^0, this is trivial. We have to verify our claim for constants A belonging to the set $\mathscr{S}$, i.e., for function constants F of form $Fo^{(1)}(R)$, with R (in D^0) an NDS-relation in D.

Looking at the rules for constructing the quadruples for constants of this form, we see that α_F is the constant $\Sigma(R)$. We know that this is an NDS-relation in D, when R is. (And we do know that R is an NDS-relation in D). The constants σ_F, ρ_F, and V_F are defined elementarily from the constants σ_R, ρ_R and V_R, and therefore belong to D. Q.E.D.

12. Summary of Results Concerning Interpretations

THEOREM 1. (Lemma on maximality of $\mathscr{C}^0$.)

When π is a translation $\mathscr{C}^0 \to \mathscr{C}$, C any elementary calculus, and t a translation $C \to \mathscr{C}$, then there is a translation s: $C \to \mathscr{C}^0$ which makes the diagram below commute, in the weak sense, that the different paths determine equivalent translations.

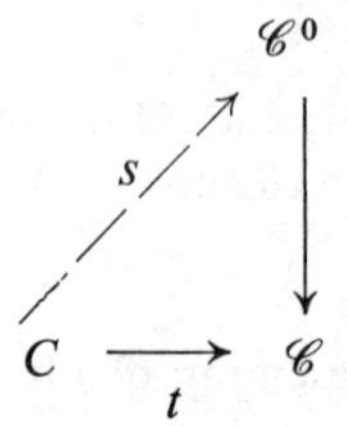

Proof: (a) We first treat the case where $\mathscr{C}$ is in standard form, and π is the identity mapping j. For every elementary constant K of $\mathscr{C}$ there is an equivalent constant $e(K)$ which is strictly elementary. We define s by the condition: $s(K) = e(t(K))$. It is immediate that the diagram commutes. From the fact that $\mathscr{C}$ is a conservative extension of $\mathscr{C}^0$, it follows, that s is actually a translation $C \to \mathscr{C}^0$.

(b) In the general case, when $\mathscr{C}$ is not necessarily in standard form, we get for a given translation t: $C \to \mathscr{C}$, an associated translation t': $C \to \mathscr{C}^+$. The translation s: $C \to \mathscr{C}^0$, constructed above, will be seen to have the desired properties.

DEFINITION. When K_1 and K_2 are any two calculi, of Kinds I or II (either both of the same kind, or of different kinds), and K_1^0 and K_2^0 are their elementary parts, then by an *interpretation* $K_1 \to K_2$ we mean a translation J: $K_1^0 \to K_2^0$, with the property: whenever a relation constant R of K_1^0 is an NDS-relation in K_1 then $J(R)$ is an NDS-relation in K_2.

It can be verified, that this definition is consistent with our previous definitions, i.e., in each particular case, for which we have previously defined 'interpretation', the old definition is equivalent to the new one.

The following is an immediate consequence of the definition.

THEOREM 2. When K_1, K_2, K_3 are any three calculi of Kinds I or II, if J_1 is an interpretation $K_1 \to K_2$, and J_2 an interpretation $K_2 \to K_3$, then the composite $J_2 \circ J_1$ is an interpretation $K_1 \to K_3$.

It turns out, that interpretability is also equivalent to some of the notions of translatability that we have introduced earlier. We now want to prove this.

The first result concerns theories of Kind I.

THEOREM 3. When $\mathscr{C}_1$ and $\mathscr{C}_2$ are two calculi of Kind I, then $\mathscr{C}_1$ is translatable into $\mathscr{C}_2$ iff $\mathscr{C}_1$ is interpretable in $\mathscr{C}_2$.

Proof: We have already proved the part that interpretability implies translatability (II, 11, 1). It remains to prove the converse implication.

We observe first: $\mathscr{C}_1$ is translatable in $\mathscr{C}_2$ iff $\mathscr{C}_1^+$ is translatable in $\mathscr{C}_2^+$; and similarly with interpretability. It is therefore sufficient for us to prove the theorem under the assumption that $\mathscr{C}_1$ and $\mathscr{C}_2$ are in standard form.

We assume then, that $\mathscr{C}_1$ and $\mathscr{C}_2$ are calculi in standard form, $\mathscr{C}_1^0$ and $\mathscr{C}_2^0$ their elementary parts, J_1 and J_2 the associated canonical translations, and T a translation $\mathscr{C}_1 \rightarrow \mathscr{C}_2$. By the lemma above (Theorem 1), we can find a translation s, which fits the diagram

$$\begin{array}{ccc} \mathscr{C}_1^0 & \xrightarrow{\ s\ } & \mathscr{C}_2^0 \\ \mathscr{J}_1 \downarrow & T & \downarrow \mathscr{J}_2 \\ \mathscr{C}_1 & \longrightarrow & \mathscr{C}_2 \end{array}$$

For every constant A of $\mathscr{C}_1^0$, $s(A)$ and $T(A)$ will be equivalent in $\mathscr{C}_2$. Since T evidently takes NDS-relations into NDS-relations, the same is true of s. So s is the interpretation we were looking for.

It also holds, and is easily verified, that

THEOREM 4. When D_1 and D_2 are two calculi of Kind II, then D_1 is translatable in D_2 if and only if D_1 is interpretable in D_2.

DEFINITION. Two calculi K_1 and K_2, of Kinds I or II, are called *equivalent* if each is interpretable in the other.

It follows by Theorem 3 and Theorem 4, that this definition is compatable with our previous, more special, definitions of equivalence between calculi.

We have already proved the following under different terminology

THEOREM 5. When $\mathscr{C}$ is a calculus of Kind I, and D a calculus of Kind II then $\mathscr{C}$ is equivalent to $\mathscr{C}^{\mathrm{Red}}$, and D is equivalent to D^*; and as a consequence, $(\mathscr{C}^{\mathrm{Red}})^*$ is equivalent to $\mathscr{C}$, and $(D^*)^{\mathrm{Red}}$ is equivalent to D.

Department of Philosophy,
University of Maryland

REFERENCES

Kleene, S. C., *Introduction to Metamathematics*, North-Holland, 1952.
Tait, W. W., 'Functionals defined by transfinite recursion', *J. of Symbolic Logic* **30** (1965), pp. 155–174.

D. VAN DALEN AND R. STATMAN

EQUALITY IN THE PRESENCE OF APARTNESS*

INTRODUCTION

The apartness relation was introduced by Brouwer, [1], [2], as a positive analogue of the inequality relation on the continuum. Subsequently Heyting introduced the notion of apartness axiomatically for a treatment of axiomatic geometry and algebra, [7], [8], [9]. In this paper we will take the axiomatic point of view and consider first-order intuitionistic theories of apartness and equality.

We will denote 'x lies apart from y' by $x \# y$, and $x \neq y$ will be an abbreviation for $\neg x = y$.

EQ, the theory of equality, has the (non-logical) axioms

$$\forall x(x = y)$$
$$\forall xyz(x = y \rightarrow (z = x \rightarrow y = z))$$

AP, the theory of apartness, has the axioms of **EQ**, plus

$$\forall xyz(x = y \rightarrow (x \# z \rightarrow y \# z))$$
$$\forall xy(\neg x \# y \leftrightarrow x = y)$$
$$\forall xy(x \# y \rightarrow y \# x)$$
$$\forall xyz(x \# y \rightarrow z \# x \vee z \# y)$$

It is well-known that **AP** is not conservative over **EQ**, as the equality in **AP** is *stable*, i.e.

$$\mathbf{AP} \vdash \forall xy(\neg\neg x = y \rightarrow x = y).$$

Bearing this in mind one could ask: is **AP** conservative over

$$\mathbf{EQ} + \forall xy(\neg\neg x = y \rightarrow x = y)?$$

This was answered negatively by the first author, when working on [4]. He observed that $\mathbf{AP} \vdash \neg\forall z(z \neq x \vee z \neq y) \rightarrow x = y$, but

$$\mathbf{EQ} \nvdash \forall xy(\neg x \neq y \rightarrow x = y) \rightarrow \forall xy(\neg\forall z(z \neq x \vee z \neq y) \rightarrow x = y)).$$

* The main results below were obtained in 1975 while the first author was a visiting fellow at Wolfson College, Oxford, and the second author a fellow of King's College, Cambridge.

J. Hintikka, I. Niiniluoto, and E. Saarinen (eds.), Essays on Mathematical and Philosophical Logic, 95–116.

So he conjectured that **AP** is conservative over

EQ + $\forall xy(\neg\forall z(z \neq x \vee z \neq y) \to x = y)$, bearing in mind that on the continuum apartness can be defined as $\forall z(z \neq x \vee z \neq y)$, (cf. [6], p. 71).

In the present paper the conjecture is refuted and the problem of axiomatizing the equality fragment of **AP** is satisfactorily solved. The clue to the solution is already contained in the previous conjecture.

The relation $\forall z(z \neq x \vee z \neq y)$ is a kind of inequality relation, which is in fact stronger than the ordinary inequality, hence the axiom $\neg\forall z(z \neq x \vee z \neq y) \to x = y$ is a stronger stability axiom. What we should do now is iterate the process:

define $x \neq_0 y := \neg x = y$
$x \neq_{n+1} y := \forall z(z \neq_n x \vee z \neq_n y)$.

It is immediate that **EQ** $\vdash x \neq_n y \to x \neq_m y$ for $n \geqslant m$.

Parallel to this sequence of inequalities we have a sequence of generalized stability axioms:

$$S_n := \forall xy(\neg x \neq_n y \to x = y).$$

It is a matter of routine to show that **AP** $\vdash S_n$ for all n (prove $x \# y \to x \neq_n y$ by induction on n).

As to be expected, the axioms S_n are increasing in strength:

EQ $\vdash S_n \to S_m$ for $n \geqslant m$.

Let us call **SEQ**^{n} the n-stable theory of equality **EQ** + S_n, and

SEQ$^\omega$ = **EQ** + $\{S_n | n \in \omega\}$,

the ω-stable theory of equality.

The main result of this paper is

AP is conservative over **SEQ**$^\omega$

or

SEQ$^\omega$ is the equality fragment of **AP**.

In Section 4 it is shown that if **AP** $\vdash A$, where A contains no apartness symbol, then **SEQ**$^n \vdash A$ for some n. For this purpose a normal form theorem for a natural deduction version of **AP** is derived.

In Section 5 it is shown that we actually need all the stability axioms, by proving **SEQ**$^n \neq$ **SEQ**$^{n+1}$ for all n. This result is obtained by means of Kripke models.

Finally Section 6 contains some miscellaneous results.

1. THE SYSTEMS NAP AND NEQ

We shall define a system **NAP** of natural (deduction) rules equivalent to **AP**. The logical rules of **NAP** are the rules of Gentzen's NJ (without $\neg$; we set $\neg B := B \to \text{人}$).

For a precise description see Prawitz [10], here we simply list the inference rules in their traditional form.

$$\wedge I)\quad \frac{A \quad B}{A \wedge B} \qquad\qquad \wedge E)\quad \frac{A_1 \wedge A_2}{A_i}\, i = 1, 2$$

$$\vee I)\quad \frac{A_i}{A_1 \vee A_2}\, i = 1,2 \qquad\qquad \vee E)\quad \frac{A_1 \vee A_2 \quad \begin{array}{c}[A_1]\\ B\end{array} \quad \begin{array}{c}[A_2]\\ B\end{array}}{B}$$

$$\to I)\quad \frac{\begin{array}{c}[A]\\ B\end{array}}{A \to B} \qquad\qquad \to E)\quad \frac{A \to B \quad A}{B}$$

$$\forall I)\quad \frac{A(a)}{\forall x A(x)} \qquad\qquad \forall E)\quad \frac{\forall x A(x)}{A(t)}$$

$$\exists I)\quad \frac{A(t)}{\exists x A(x)} \qquad\qquad \exists E)\quad \frac{\exists x A(x) \quad \begin{array}{c}[A(a)]\\ B\end{array}}{B}$$

$$\text{人})\quad \frac{\text{人}}{A}$$

The restrictions are as usual (cf. [9], p. 20).
Furthermore we have the axiom

$$\forall x(x = x)$$

and

$$=)\quad \frac{A(t_1)\ t_1 = t_2}{A(t_2)}$$

(where A is atomic, i.e., an equation or an apartness).

$$\#_1)\quad \frac{\begin{array}{c}[t_1 \# t_2]\\ \text{人}\end{array}}{t_1 = t_2} \qquad \#_2)\quad \frac{t_1 \# t_2\ t_1 = t_2}{\text{人}}$$

$$\#_3)\quad \frac{t_1 \# t_3 \quad \begin{array}{c}[t_1 \# t_2]\\ B\end{array} \quad \begin{array}{c}[t_2 \# t_3]\\ B\end{array}}{B}$$

NAP has the above rules, **NEQ** is obtained by leaving out $\#_1$, $\#_2$, $\#_3$. It is a routine exercise to show that **NAP(NEQ)** and **AP(EQ)** are equivalent.

We will denote derivations by $\mathscr{D}, \mathscr{D}_1, \mathscr{D}_2, \ldots,$

$$\begin{array}{l}\mathscr{D}\\ A\end{array}$$ is a derivation with conclusion A,

$$\begin{array}{l}[B]\\ \mathscr{D}\end{array}$$ is a derivation with B cancelled.

2. NAP-DERIVATIONS

Each rule of **NAP** can be written (not necessarily uniquely) in one of the following four ways

(1)
$$\begin{array}{c}[A]\\ \dfrac{B_1 \quad B_2}{C}\,R\end{array}$$

where $R \in \{\to I, \wedge I, \forall I, \vee I, \exists I, \#_1\}$, B_2 does not exist if $R \neq \wedge I$, and $[A]$ does not exist if $R \notin \{\to I, \#_1\}$.

(2) $$\dfrac{\curlywedge}{A}$$ (3) $$\dfrac{B_1\, B_2}{C}\,R$$

where $R \in \{\to E, \wedge E, \forall E, =\}$ and B_2 does not exist if $R \notin \{\to E, =\}$,

(4)
$$\dfrac{A_1 \quad \begin{array}{c}[B_1]\\ C\end{array} \quad \begin{array}{c}[B_2]\\ C\end{array}}{A_2}\,R$$

where $R \in \{\to E, \wedge E, \forall E, =, \exists E, \vee E, \#_3, \#_2\}$, $\begin{array}{c}[B_2]\\ C\end{array}$ does not exist if $R \notin \{\vee E, \#_3\}$, $\begin{array}{c}[B_1]\\ C\end{array}$ does not exist if $R \in \{\wedge E, \forall E\}$, and $[B_1]$ does not exist if $R \in \{\to E, =, \#_2\}$.

A *cut* is a combination of inferences of the form

$$\dfrac{\begin{array}{c}[A]\\ \dfrac{B_1\, B_2}{C_1}\,R_1\end{array} \quad \begin{array}{c}[F_1]\\ G\end{array} \quad \begin{array}{c}[F_2]\\ G\end{array}}{C_2}\,R_2\ ,$$

where $R_1 \in \{\to I, \wedge I, \forall I, \vee I; \exists I\}$ and $R_2 \in \{\to E, \wedge E, \forall E, \vee E, \exists E\}$, or of the form

$$\dfrac{\dfrac{A \quad \begin{matrix}[F_1]\\ G_1\end{matrix} \quad \begin{matrix}[F_2]\\ G_1\end{matrix}}{C_1}R_1 \quad \begin{matrix}[F_3]\\ G_2\end{matrix} \quad \begin{matrix}[F_4]\\ G_2\end{matrix}}{C_2}R_2$$

where $R_1 \in \{\exists E, \vee E, \#_3\}$ and $R_2 \in \{\to E, \wedge E, \forall E, \exists E, \vee E, =, \#_2, \#_3\}$, or of the form

$$\dfrac{\begin{matrix}\curlywedge\\ C_1\end{matrix} \quad \begin{matrix}[B_1]\\ C_2\end{matrix} \quad \begin{matrix}[B_2]\\ C_2\end{matrix}}{C_3}R$$

where $R \in \{\to E, \wedge E, \forall E, \exists E, \vee E, =, \#_2, \#_3\}$.

The *cut formula* of the cut is the formula C_1. Certain cuts are eliminated by the function *NEXT* defined as follows

$$\dfrac{\dfrac{\begin{matrix}[A]\\ \mathscr{D}_1\\ B\end{matrix}}{A \to B} \quad \begin{matrix}\mathscr{D}_2\\ A\end{matrix}}{B} \quad \xrightarrow{\text{NEXT}} \quad \begin{matrix}\mathscr{D}_2\\ [A]\\ \mathscr{D}_1\\ B\end{matrix}$$

$$\dfrac{\dfrac{\begin{matrix}\mathscr{D}_1\\ A_1\end{matrix} \quad \begin{matrix}\mathscr{D}_2\\ A_2\end{matrix}}{A_1 \wedge A_2}}{A_i} \quad \xrightarrow{\text{NEXT}} \quad \begin{matrix}\mathscr{D}_i\\ A_i\end{matrix}$$

$$\dfrac{\dfrac{\begin{matrix}\mathscr{D}\\ A_i\end{matrix}}{A_1 \vee A_2} \quad \begin{matrix}[A_1]\\ \mathscr{D}_1\\ B\end{matrix} \quad \begin{matrix}[A_2]\\ \mathscr{D}_2\\ B\end{matrix}}{B} \quad \xrightarrow{\text{NEXT}} \quad \begin{matrix}\mathscr{D}\\ [A_i]\\ \mathscr{D}_i\\ B\end{matrix}$$

$$\dfrac{\dfrac{\begin{matrix}\mathscr{D}(a)\\ A(a)\end{matrix}}{\forall x A(x)}}{A(t)} \quad \xrightarrow{\text{NEXT}} \quad \begin{matrix}\mathscr{D}(t)\\ A(t)\end{matrix}$$

$$\dfrac{\dfrac{\begin{array}{c}\mathcal{D}_1\\ A(t)\end{array}}{\exists x A(x)} \quad \begin{array}{c}[A(a)]\\ \mathcal{D}_2(a)\\ B\end{array}}{B} \quad \xrightarrow{\text{NEXT}} \quad \begin{array}{c}\mathcal{D}_1\\ [A(t)]\\ \mathcal{D}_2(t)\\ B\end{array}$$

(the other cuts will be eliminated according to the context).

The *cutrank* of a derivation is the maximum length of its cutformulae (if a derivation contains no cuts it has cutrank = 0).

The *height* of a derivation is the length of its left-most path.

The *depth* of a derivation is defined as follows:

$$\text{depth}(A) = 1$$

$$\text{depth}\left(\dfrac{\begin{array}{cc} [A] & \\ \mathcal{D}_1 & \mathcal{D}_2\\ B_1 & B_2\end{array}}{C}\right) = 1 + \max\{\text{depth}(\mathcal{D}_1), \text{depth}(\mathcal{D}_2)\}$$

$$\text{depth}\left(\dfrac{\begin{array}{c}\mathcal{D}\\ \curlywedge\end{array}}{A}\right) = 1 + \text{depth}(\mathcal{D})$$

$$\text{depth}\left(\dfrac{\begin{array}{ccc} & [B_1] & [B_2]\\ \mathcal{D} & \mathcal{D}_1 & \mathcal{D}_2\\ A & C_1 & C_1\end{array}}{C_2}\right) = 1 + \text{depth}(\mathcal{D}) + \max\{\text{depth}(\mathcal{D}_1), \text{depth}(\mathcal{D}_2)\}.$$

The *ordinal* of a derivation $\mathcal{D}$ is defined by $\text{ord}(\mathcal{D}) = \omega^2.\text{cutrank}(\mathcal{D}) + \omega.\text{depth}(\mathcal{D}) + \text{height}(\mathcal{D})$.

A derivation is *normal* if it contains no cuts.

3. NORMAL FORM THEOREM FOR NAP-DERIVATIONS

PROPOSITION: if cutrank $(\mathcal{D}) \neq 0$ then there is a $\mathcal{D}^*$ such that

(1) if $\mathcal{D}$ is a derivation of A from X then so is $\mathcal{D}^*$, and

(2) $\text{ord}(\mathcal{D}^*) < \text{ord}(\mathcal{D})$.

Proof: by induction on $\text{ord}(\mathcal{D})$

Basis: $\text{ord}(\mathcal{D}) = 0$ – then $\text{cutrank}(\mathcal{D}) = 0$.

Induction Step: cases according to the structure of the left-most path of $\mathscr{D}$

CASE 1 $\mathscr{D}$ =

$$
\dfrac{\dfrac{\dfrac{\begin{matrix}[A] & \\ \mathscr{D}_1 & \mathscr{D}_2 \\ B_1 & B_2\end{matrix}}{C_1}R \quad \begin{matrix}\mathscr{D}_3 \\ E_1\end{matrix}}{C_2} = \qquad \searrow \qquad \begin{matrix} & \mathscr{D}_{n+1} \\ C_{n-1} & E_{n-1}\end{matrix}}{C_n} =
$$

where $R \in \{\rightarrow I, \wedge I, \vee I, \forall I, \exists I, \#_1\}$, and $1 \leqslant n$ (when $n = 1$ there are no $=$-inferences)

then $\mathscr{D}^*$ =

$$
\dfrac{\dfrac{\dfrac{\begin{matrix}(A) & \\ \mathscr{D}_1^* & \mathscr{D}_2^* \\ B_1 & B_2\end{matrix}}{C_1}R \quad \begin{matrix}\mathscr{D}_3^* \\ E_1\end{matrix}}{C_2} = \qquad \searrow \qquad \begin{matrix} & \mathscr{D}_{n+1}^* \\ C_{n-1} & E_{n-1}\end{matrix}}{C_n} =
$$

where cutrank $(\mathscr{D}_i^*) = 0$ and (A) stands for all occurrences of A as an uncancelled assumption in $\mathscr{D}_1^*$.

CASE 2 $\mathscr{D} =$

$$\dfrac{\begin{matrix} & [B_1] & [B_2] \\ \mathscr{D}_1 & \mathscr{D}_2 & \mathscr{D}_3 \\ A_1 & C_1 & C_1 \end{matrix}}{A_2}\,R_1$$

$$\ddots$$

$$\dfrac{\begin{matrix} & [B_{2n-1}] & [B_{2n}] \\ & \mathscr{D}_{2n} & \mathscr{D}_{2n+1} \\ A_n & C_n & C_n \end{matrix}}{A_{n+1}}\,R_n$$

where $R_1 \in \{\exists E,\ \vee E,\ \#_3\}$, $R_i \in \{\rightarrow E,\ \wedge E,\ \forall E,\ =\}$ when $1 < i < n$, $R_n \in \{\rightarrow E, \wedge E, \forall E, =, \exists E, \vee E, \#_3, \#_2\}$ and $n \geqslant 2$

then $\mathscr{D}^* =$

$$\dfrac{\begin{matrix} \mathscr{D}_1 \\ A_1 \end{matrix} \quad \dfrac{\begin{matrix} [B_1] & [B_3] & [B_4] \\ \mathscr{D}_2 & \mathscr{D}_4 & \mathscr{D}_5 \\ A_2 & C_2 & C_2 \end{matrix}}{A_3}\,R_2 \quad \dfrac{\begin{matrix} [B_2] & [B_3] & [B_4] \\ \mathscr{D}_3 & \mathscr{D}_4 & \mathscr{D}_5 \\ A_2 & C_2 & C_2 \end{matrix}}{A_3}\,R_2}{A_3}\,R_1$$

$$\ddots$$

$$\dfrac{\begin{matrix} & [B_{2n-1}] & [B_{2n}] \\ & \mathscr{D}_{2n} & \mathscr{D}_{2n+1} \\ A_n & C_n & C_n \end{matrix}}{A_{n+1}}\,R_n$$

(Note that $\text{cutrank}(\mathscr{D}^*) \leqslant \text{cutrank}(\mathscr{D})$, $\text{depth}(\mathscr{D}^*) \leqslant \text{depth}(\mathscr{D})$, $\text{height}(\mathscr{D}^*) < \text{height}(\mathscr{D})$).

CASE 3 $\mathcal{D}$ =

$$\dfrac{\begin{array}{ccc}\mathcal{D}_1 & [B_1] & [B_2]\\ \curlywedge & \mathcal{D}_2 & \mathcal{D}_3\\ A_1 & C_1 & C_1\end{array}}{A_2}\,R_1$$

$$\searrow$$

$$\dfrac{\begin{array}{ccc} & [B_{2n-1}] & [B_{2n}]\\ & \mathcal{D}_{2n} & \mathcal{D}_{2n+1}\\ A_n & C_n & C_n\end{array}}{A_{n+1}}\,R_n$$

where $R_i \in \{\to E, \wedge E, \forall E, =\}$ when

$i < n$, $R_n \in \{\to E, \wedge E, \forall E, \exists E, \vee E, =, \#_3, \#_2\}$, and $0 \leqslant n$

then $\mathcal{D}^*$ =

$$\dfrac{\begin{array}{c}\mathcal{D}_1^*\\ \curlywedge\end{array}}{A_{n+1}}$$

where cutrank($\mathcal{D}_1^*$) < cutrank($\mathcal{D}$).

CASE 4 $\mathcal{D}$ =

$$\dfrac{\begin{array}{cc} & \mathcal{D}_1\\ A_1 & B_1\end{array}}{A_2}\,R_1$$

$$\searrow$$

$$\dfrac{\begin{array}{ccc} & [B_n] & [B_{n+1}]\\ & \mathcal{D}_n & \mathcal{D}_{n+1}\\ A_n & C & C\end{array}}{C}\,R_n$$

where $R_i \in \{\to E, \wedge E, \forall E, =\}$ when $i < n$, $R_n \in \{\to E, \wedge E, \forall E, =, \exists E, \vee E, \#_3, \#_2\}$, and $1 \leqslant n$

then $\mathscr{D}^* =$

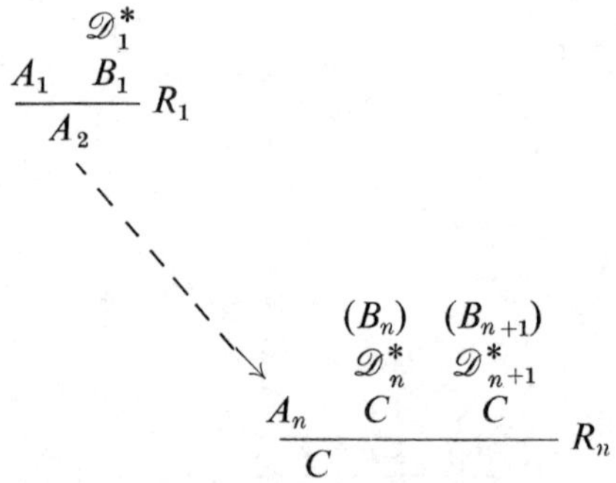

where $\text{cutrank}(\mathscr{D}_i^*) \leqslant \text{cutrank}(\mathscr{D}_i) \dot{-} 1$ and (B) is the set of all occurrences of B as an uncancelled assumption in $\mathscr{D}_i^*$.

CASE 5 $\mathscr{D} =$

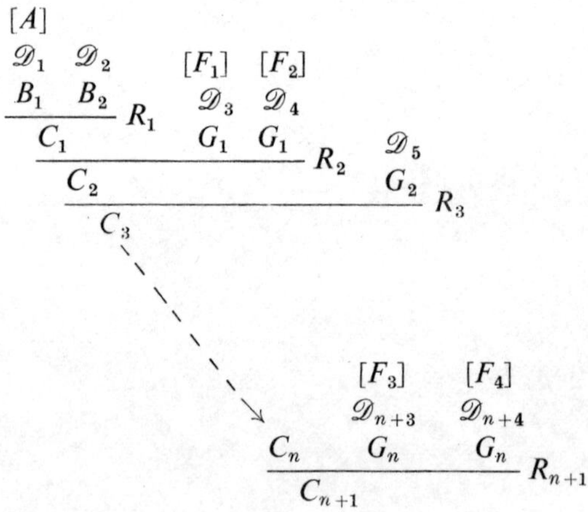

where $R_1 \in \{\to I, \wedge I, \vee I, \forall I, \exists I\}$, $R_2 \in \{\to E, \wedge E, \vee E, \forall E, \exists E\}$, $R_i \in \{\to E, \wedge E, \forall E, =\}$ when $2 < i \leqslant n$, $R_{n+1} \in \{\to E, \wedge E, \forall E, =, \vee E, \exists E, \#_3, \#_2\}$, and $1 \leqslant n$ when $R_1 \in \{\to I, \wedge I, \forall I\}$, and $1 = n$ otherwise.

then $D^* =$

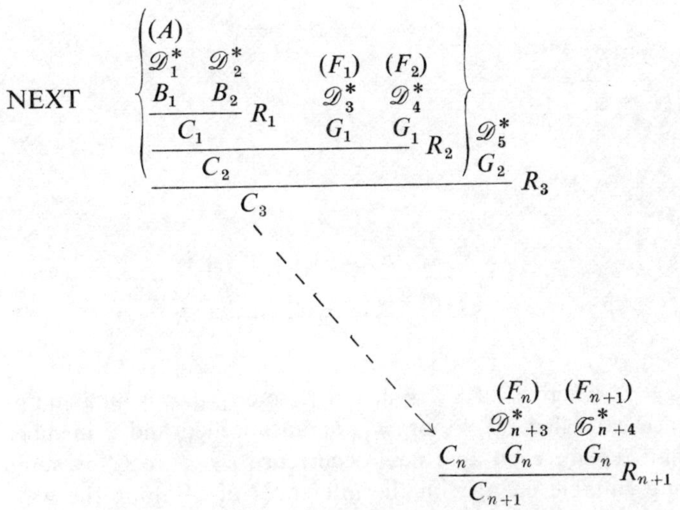

(Note that length(C_2) < length$(C_1) \leqslant$ cutrank$(\mathscr{D})$ when $1 < n$) where cutrank$(\mathscr{D}_i^*) \leqslant$ cutrank$(\mathscr{D}_i) \dot{-} 1$ and (H) is the set of all occurrences of H as an uncancelled assumption in $\mathscr{D}_i^*$.

COROLLARY. If $X \vdash_{\mathbf{AP}} A$ then there is a normal **NAP** derivation of A from X.

4. INTERPRETATION OF NORMAL NAP-DERIVATIONS AS NEQ-DERIVATIONS

Suppose that B is an equality formula, X is a set of equality formulae and Y is a set of apartnesses. Set $Y(n) = \{t_1 \neq_n t_2 : t_1 \# t_2 \in Y\}$

PROPOSITION: If $X \cup Y \vdash_{\mathbf{AP}} B$ then $X \cup Y(n) \vdash_{\mathbf{SEQ}^\omega} B$ for some n.

Proof: Let $\mathscr{D}$ be a normal **NAP**-derivation of B from $X \cup Y$. Observe that each apartness occurrence in $\mathscr{D}$ is contained in a sequence of inferences of the form

$$\dfrac{t_1 \# s_1 E_1}{t_2 \# s_2} =$$

$$\searrow$$

$$\dfrac{\dfrac{t_n \# s_n \quad E_n}{t_{n+1} \# s_{n+1}} = \quad \begin{matrix}[A_1] \\ B\end{matrix} \quad \begin{matrix}[A_2] \\ B\end{matrix}}{C} R$$

among the $t_j \# s_j$, where $R \in \{\#_2, \#_3\}$, $0 \leqslant n$, and $t_1 \# s_1$ is an assumption occurrence cancelled by $\#_3$ or $\#_1$, or uncancelled and a member of Y. We shall replace each apartness occurrence $t \# t'$ in $\mathscr{D}$ by some $t \neq_n t'$, making suitable changes in the inferences of $\mathscr{D}$ along the way. For convenience, we shall use the derived rules

$$\dfrac{B_1(t_1)\ t_1 = t_2}{B_1(t_2)} \qquad \dfrac{t_1 \neq_n t_2}{t_1 \neq_m t_2} \quad (m \leqslant n)$$

(B_1 an arbitrary formula)

in our transformations. We proceed by recursion over $\mathscr{D}$ to define a $\mathscr{D}^*$ which is an **NEQ**-derivation

CASE 1

$$\dfrac{\begin{matrix}\mathscr{D}_1 \\ t_1 \# s_1 \quad E_1\end{matrix}}{t_2 \# s_2} =$$

$$\searrow$$

$$\dfrac{\dfrac{\begin{matrix}\mathscr{D}_n \\ t_n \# s_n \quad E_n\end{matrix}}{t_{n+1} \# s_{n+1}} = \quad \begin{matrix}\mathscr{D}_{n+1} \\ t_{n+1} = s_{n+1}\end{matrix}}{\curlywedge} \quad \xrightarrow{\ *\ }$$

$$\dfrac{\overset{\mathcal{D}_1^*}{\neg(t_1 = s_1)\; E_1}}{\neg(t_2 = s_2)}=$$

$$\vdots$$

$$\dfrac{\dfrac{\overset{\mathcal{D}_n^*}{\neg(t_n = s_n)\; E_n}}{\neg(t_{n+1} = s_{n+1})}= \qquad \overset{\mathcal{D}_{n+1}^*}{t_{n+1} = s_{n+1}}}{\curlywedge}$$

CASE 2

$$\dfrac{\overset{\mathcal{D}_1}{t_1 \# s_1\; E_1}}{t_2 \# s_2}=$$

$$\vdots$$

$$\dfrac{\dfrac{\overset{\mathcal{D}_n}{t_n \# s_n\quad E_n}}{t_{n+1} \# s_{n+1}}= \quad \begin{matrix}[t_{n+1}\# r]\\ \mathcal{D}_{n+1}\\ C\end{matrix} \quad \begin{matrix}[r \# s_{n+1}]\\ \mathcal{D}_{n+2}\\ C\end{matrix}}{C}\#_3 \quad \xrightarrow{\;*\;}$$

$$\dfrac{\overset{\mathcal{D}_1^*}{t_1 \neq_{p+1} s_1\quad E_1}}{t_2 \neq_{p+1} s_2}=$$

$$\vdots$$

$$\dfrac{\dfrac{\dfrac{\overset{\mathcal{D}_n^*}{t_n \neq_{p+1} s_n\quad E_n}}{t_{n+1}\neq_{p+1}s_{n+1}}=}{t_{n+1}\neq_p \lor r \neq_p s_{n+1}}\forall E \quad \begin{matrix}\dfrac{[t_{n+1}\neq_p r]}{\{t_{n+1}\neq_{p(i)} r : 1\leqslant i\leqslant 1\}}\\ \mathcal{D}_{n+1}^*\\ C\end{matrix}\quad \begin{matrix}\dfrac{[r\neq_p s_{n+1}]}{\{r\neq_{q(i)} s_{n+1} : 1\leqslant i\leqslant m\}}\\ \mathcal{D}_{n+2}^*\\ C\end{matrix}}{C}$$

where $[t_{n+1} \# r] \overset{*}{\mapsto} \{t_{n+1} \neq_{p(i)} r : 1 \leqslant i \leqslant l\}$, $[r \# s_{n+1}] \overset{*}{\mapsto} \{r \neq_{q(i)} s_{n+1} : 1 \leqslant i \leqslant m\}$, and $p = \max\{p(i), q(j) : 1 \leqslant i \leqslant l,\ 1 \leqslant j \leqslant m\}$.
The sets $\{t_{n+1} \neq_{p(i)} r : 1 \leqslant i \leqslant l\}$, $\{r \neq_{q(i)} s_{n+1} : 1 \leqslant i \leqslant m\}$ are the assumptions obtained by the *-translation of the two right-hand derivations.

CASE 3

$$\begin{array}{c}[t \# s]\\ \mathcal{D}_1\\ \curlywedge \\ \hline t = s\end{array} \quad \overset{*}{\mapsto} \quad \dfrac{\dfrac{\dfrac{\forall x \forall y(\neg(x \neq_p y) \to x = y)}{\forall y(\neg(t \neq_p y) \to t = y)}}{\neg(t \neq_p s) \to t = s} \qquad \dfrac{\begin{array}{c}\dfrac{t \neq_p s}{\{t \neq_{p(i)} s : 1 \leqslant i \leqslant m\}}\\ \mathcal{D}_1^*\\ \curlywedge\end{array}}{\neg(t \neq_p s)}}{t = s}$$

where $[t \# s] \overset{*}{\mapsto} \{t \neq_{p(i)} s : 1 \leqslant i \leqslant m\}$, and $p = \max\{p(i) : 1 \leqslant i \leqslant m\}$.

6. KRIPKE MODELS FOR THE THEORIES OF n-STABLE EQUALITY

The result in Section 2 shows that $\mathbf{SEQ}^{\omega}$ axiomatizes the equality fragment of **AP**. It leaves open, however, the possibility that already some $\mathbf{SEQ}^n$ would suffice.

In order to show that the result is the best possible we will show by means of Kripke models that each $\mathbf{SEQ}^n$ is a proper sub-theory of $\mathbf{SEQ}^{n+1}$

We want to show $\mathbf{EQ} + S_n \nvdash S_{n+1}$ and we will do this by constructing for each n a Kripke model for **EQ** in which S_n holds, but S_{n+1} not. In the process of interpreting the stability axioms we have to break down the forcing clause $\alpha \Vdash \neg a \neq_n b$.

First we will treat the case $n = 1$, the general case then will be seen to be essentially the same, but more complicated.

Our main task will be to falsify S_2, the validity of S_1 will be ensured by adding a simple feature to the model.

Let us evaluate $\alpha \Vdash \neg a \neq_2 b$:

$$\begin{aligned}
&\alpha \Vdash \neg a \neq_2 b \Leftrightarrow \alpha \nVdash \neg \forall z(z \neq_1 a \vee z \neq_1 b) \Leftrightarrow\\
&\Leftrightarrow \textstyle\bigwedge \alpha_1 \geqslant \alpha \bigvee \alpha_2 \geqslant \alpha_1 \bigvee c \in \alpha_2\ \alpha_2 \nVdash c \neq_1 a \text{ and } \alpha_2 \nVdash c \neq_1 b\\
&\Leftrightarrow \textstyle\bigwedge \alpha_1 \geqslant \alpha \bigvee \alpha_2 \geqslant \alpha_1 \bigvee c \in \alpha_2\ \alpha_2 \nVdash \forall z(z \neq c \vee z \neq a) \text{ and}\\
&\quad \alpha_2 \nVdash \forall z(z \neq c \vee z \neq b)\\
&\Leftrightarrow \textstyle\bigwedge \alpha_1 \geqslant \alpha \bigvee \alpha_2 \geqslant \alpha_1 \bigvee c \in \alpha_2 [\bigvee \alpha_3 \geqslant \alpha_2 \bigvee d \in \alpha_3 (\bigvee \alpha_{3,0} \geqslant \alpha_3\\
&\quad \alpha_{3,0} \Vdash a = d \text{ and } \textstyle\bigvee \alpha_{3,1} \geqslant \alpha_3 \alpha_{3,1} \Vdash d = c) \text{ and}\\
&\quad \textstyle\bigvee \alpha_4 \geqslant \alpha_2 \bigvee e \in \alpha_4 (\bigvee \alpha_{4,0} \geqslant \alpha_4\ \alpha_{4,0} \Vdash e = c \text{ and}\\
&\quad \textstyle\bigvee \alpha_{4,1} \geqslant \alpha_4\ \alpha_{4,1} \Vdash e = b)]
\end{aligned}$$

('$c \in \alpha_2$' is used as a shorthand for 'c belongs to the domain of α_2', etc.)

We schematically represent this condition in Figure 1:

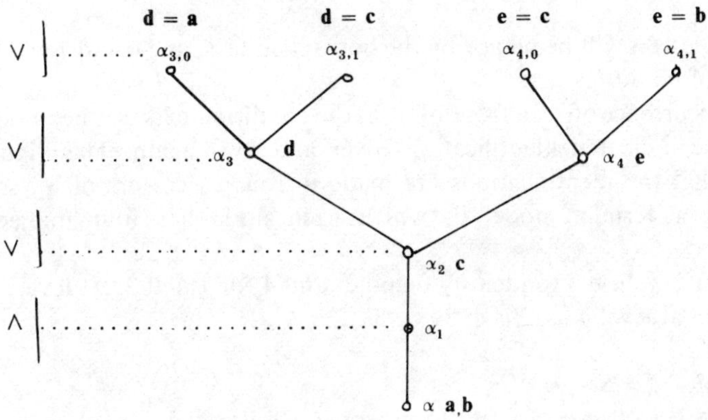

Fig. 1.

The quantifiers (in the metalanguage) indicate for the nodes and elements on the same horizontal line, how they are quantified in the condition. The various elements are denoted in bold-face at the node where they are introduced.

Finally, if the condition is obeyed then under S_2, $\alpha \Vdash a = b$. Figure 1 suggests what we have to do:

make S_1, trivially true by avoiding identifications $c = a$ in α_3 and $c = b$ in α_4 (so that we are never forced to identify a and b in α).

We falsify S_2 by making suitable identifications in the nodes $\alpha_{3,0}$, $\alpha_{3,1}$, $\alpha_{4,0}$, $\alpha_{4,1}$, as indicated above, without identifying a and b in α.

We now carry this out for a Kripke model.

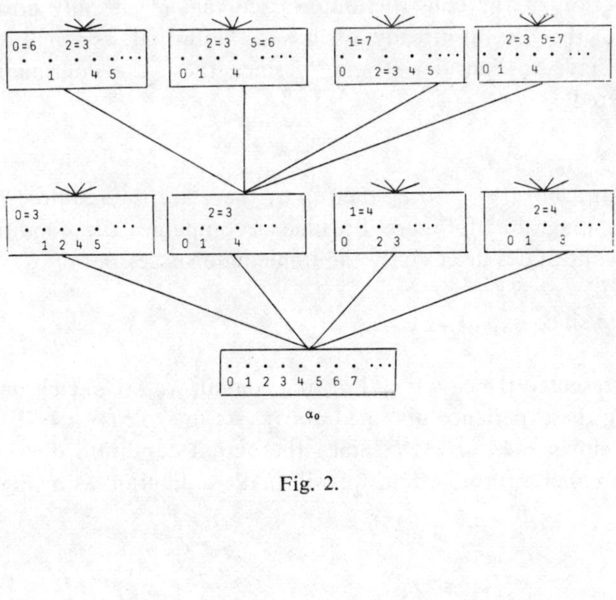

Fig. 2.

The role of α will be played by the bottom node α_0 and 0 and 1 will play the role of a and b.

The occurrence of $\wedge$ in front of α_1 in the condition makes it necessary to repeat the indicated identifications over and over again. The Figure 2 tells us that the identifications are made in four successors of α_1, so we construct a Kripke model in which each node has four immediate successors.

At level 2 we use 3 to identify 0 and 2, and 4 for 1 and 2. At level 3 use 5, 6, 7 and at level n use $3n-4$, $3n-3$, $3n-2$.

CLAIM 1. $\alpha_0 \Vdash S_1$

Suppose $\alpha \Vdash \neg\forall z(z \neq a \vee z \neq b)$ for some a and b, then

$$\bigwedge \alpha_1 \geqslant \alpha \bigvee \alpha_2 \geqslant \alpha_1 \bigvee c \in \alpha_2 [\bigvee \alpha_{2,0} \geqslant \alpha_2\ \alpha_{2,0} \Vdash a = c \text{ and } \bigvee \alpha_{2,1} \geqslant \alpha_2\ \alpha_{2,1} \Vdash b = c](*)$$

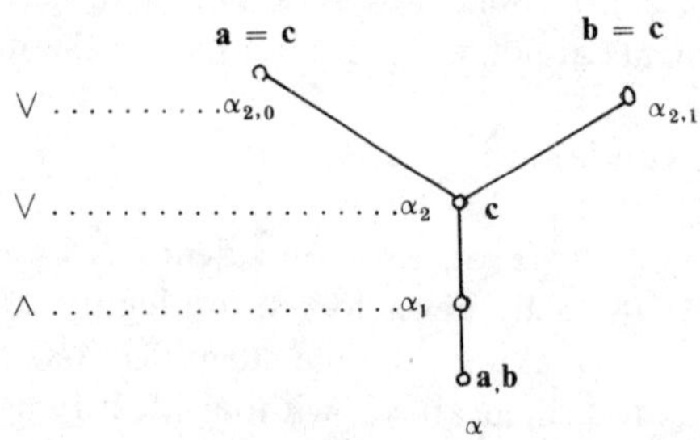

Fig. 3.

By inspection of the construction, we see that (*) is only possible (on account of the $\bigvee \alpha_1$) if already $\alpha \Vdash a = b$. Note that, as a matter of fact, we only have to consider 0 and 1, since these are continually being reconsidered.

CLAIM 2. $\alpha_0 \Vdash \neg 0 \neq_2 1$.

Choose an arbitrary $\alpha_1 \geqslant \alpha_0$, then in α_1 there are the required c, d, e to finish the diagram of Figure 1 which accompanies the condition. The four relevant nodes are exactly the immediate successors of α_1.

CLAIM 3. $\alpha_0 \nVdash S_2$.

Immediate since $\alpha_0 \nVdash 0 = 1$.

Having presented the case $n = 1$ in some detail, we will sketch the general case using the experience obtained above. As in the case $n = 2$, we want to break down $\alpha \Vdash \neg a \neq_n b$. Since the actual condition does not lend itself to a concise formulation, we will make a diagram as in Figure 1.

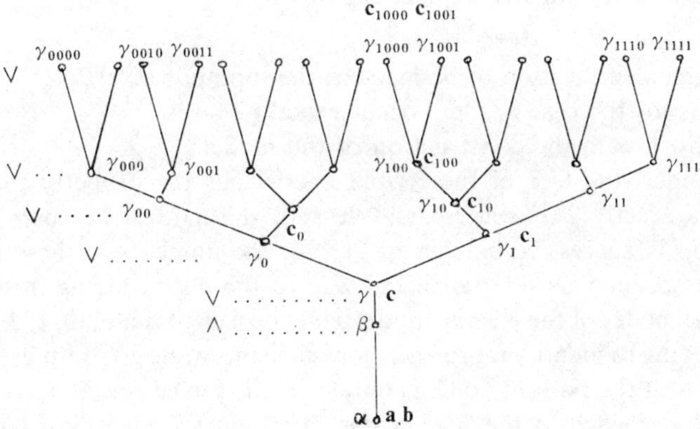

Fig. 4. The tree is precisely the parsing tree of the condition $\alpha \Vdash \neg\, a \neq_n b$.

For $n = 1$, we stop at γ_0 and γ_1. For $n = 2$, we stop at the γ_{ij}'s. The result is the tree of Figure 1.

In the general case we stop at the nodes with an index, which is a string of length n. On the left-hand side the quantifiers in the condition are indicated (note that there is only one $\wedge$).

For the construction of the model it is important what is going on at the top nodes.

We have denoted the nodes of the tree by γ's with indices. At each node γ_s also an element c_s is introduced, c_s is shown boldface in Figure 4. The pairs that should be identified in the top nodes are listed below for

$$\begin{array}{lllll} n = 1 & (c, a) & & (c, b) & \\ n = 2 & (c_0, c) & (c_0, a) & (c_1, c), & (c_1, b) \end{array}$$

In order to generalize this to all n, we replace the c's by their indices. (c has the empty sequence as index, denoted by –)
We now list $n = 1, 2, 3$.

$$\begin{array}{l} n = 1 \quad (-, a) \quad ; \quad (-, b) \\ n = 2 \quad (0, -), \quad (0, a) \quad ; \quad (1, -), \quad (1, b) \\ n = 3\, (00, 0), (00, -), (01, 0), (01, a); (10, 1), (10, -), (11, 1), (11, b) \end{array}$$

By now it is clear how to proceed from n to $n + 1$.

Let the list for n be given.

If (s, t) is at an odd place, replace the pair by

$$(s*\langle 0\rangle, s), (s*\langle 0\rangle, t)$$

If (s, t) is at an even place, replace the pair by

$$(s*\langle 1\rangle, s), (s*\langle 1\rangle, t).$$

Finally the identification to be made at the top node (s, t) is $c_s = c_t$. Note that for the case n, there occur exactly 2^n–1cs.

We now sketch the construction of the model.

The underlying tree of the Kripke model has the property that each node has exactly 2^{n+1} successors. We split the natural numbers greater than 1 up in successive batches of 2^{1+n}–1, the numbers in these batches B_k are associated in a certain fixed way to the c's occurring in the case $n + 1$. All nodes of the Kripke model have the fixed domain $\{0, 1, 2, 3, \ldots\}$. We are going to identify more and more elements while going up in the tree.

level 0 – at the bottom node α_0 only $n = n$ is forced for all n.

level 1 – in each of the 2^{n+1} nodes carry out the identification of the numbers of B_1 according to the identification of the c's in the condition (as indicated by the above instructions)

level m – in each of the 2^{n+1} successors of any node of level m–1 carry out the identification of the numbers of B_m(similar to level 1).

CLAIM 1. $\alpha_0 \Vdash \forall xy(\neg x \neq_n y \to x = y)$

It suffices to show that for any a, b and $\beta \geqslant \alpha$ the identifications as required by $\neg a \neq_n b$ are not going to be made. This follows immediately from the construction: the identifications are made among the 2^{n+1}–1 elements of the batches B_m, in such a way that the identifications of 2^n–1 elements of the smaller groups belonging to the case n do not occur in all those β. Hence if $\alpha \nVdash k = 1$, then $\alpha \nVdash \neg k \neq_n 1$ (again it suffices to consider 0 and 1). So S_n is trivially valid.

CLAIM 2. $\alpha_0 \Vdash \neg 0 \neq_{n+1} 1$.

Immediate from the construction.

CLAIM 3. $\alpha_0 \nVdash S_{n+1}$.

Immediate.

This finishes the proof that SEQ$^n \neq$ SEQ^{n+1} for all n.

The results of Sections 2 and 3 can be summed up as follows:

(i) **AP** is conservative over **SEQ**$^\omega$

(ii) **SEQ**$^\omega$ is a proper extension of each **SEQ**^{n}

(iii) The equality fragment of **AP** is not finitely axiomatizable.

7. COMMENTS

The following question arises immediately: Is in a model of **SEQ**$^\omega$, in view of the preceding results, the intersection of all the inequalities $\neq_n$ an apartness relation?

If so (or even in case of a weaker result: in each model of **SEQ**$^\omega$ there is some apartness relation), the proof theoretic part of the paper could be suppressed in favour of a simple model theoretic argument.

We will answer the question in the negative, following a suggestion of Smorynski, by constructing a model of **SEQ**$^\omega$ in which no apartness relation can be defined.

We first note, that given a node α such that $\alpha \Vdash a \neq_k b$, we can specify a set of successors such that $\alpha \nVdash a \neq_{k+1} b$, by introducing auxiliary elements in the domain of α, this can be seen by writing down the full definitions of $a \neq_k b$ and $a \neq_{k+1} b$ (cf. Section 6). More precisely, we will take care that in nodes α of length k $\alpha\, \sigma\, a \neq_{k+1} b$ in a certain systematic way, thereby automatically forcing $a \neq_k b$.

We can use the diagram of Figure 4 (leaving out the β) for $n = k + 1$. So at α of length k we introduce for each pair of elements in the domain of α $2^{k+1}-1$ auxiliary elements, which are identified in the 2^{k+1} immediate successors (per pair considered) in accordance to the diagram (cf. the instructions for identification). By this process we automatically ensure $\alpha \nVdash a \neq_k b$.

We show the first steps in Figure 5. The auxiliary elements are to the right of the dotted line.

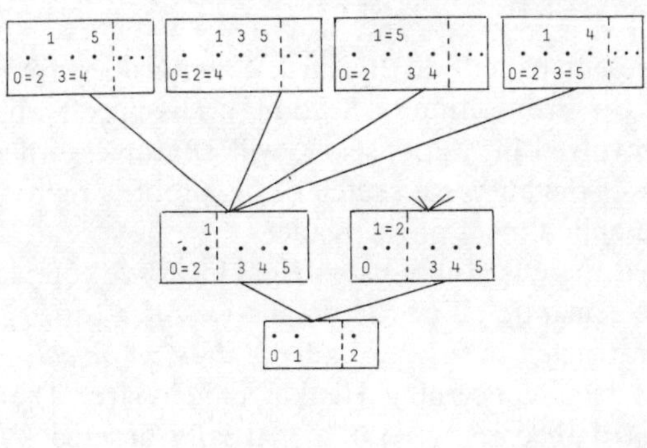

Fig. 5

So, as we go up in the tree, we force sharper and sharper inequalities.

CLAIM. $\Vdash \forall xy(\neg x \neq_n y \rightarrow x = y)$ for all n.

Suppose for some α and a, b in the domain of α $\alpha \Vdash \neg a \neq_n b$. Then, for no $\beta \geqslant \alpha$ $\beta \Vdash a \neq_n b$.

If $\text{lth}(\alpha) \geqslant n$, then, by the construction, $\alpha \Vdash a = b$.

If $\text{lth}(\alpha) < n$, then in each successor β of α of length n a and b must be identified. This can only be the case if already $\alpha \Vdash a = b$. This shows that we have a model of $\mathbf{SEQ}^{\omega}$.

The model does not carry an apartness relation, for suppose there are α, a and b such that $\alpha \Vdash a \# b$, then also $\alpha \Vdash a \neq_{n+1} b$ for $n = \text{lth}(\alpha)$. Contradiction, so $\alpha \nVdash a \# b$.

In particular $\alpha \nVdash 0 \# 1$ for all α, so the bottom node forces $\neg 0 \# 1$. But then, by the axioms of **AP**, it also forces $0 = 1$. Contradiction.

This shows that the obvious model theoretic approach to prove the main result of the paper fails. It also shows that what one would expect at a first glance, i.e. that the intersection of all the inequalities $\neq_n$ is an apartness relation, is incorrect. One might be tempted to attribute this to the fact that in this model there is no apartness relation at all. However we can make the following slight modification:

Extend the above model by adding maximal elements – each infinite branch is the maximal element of all its elements.

Let f be such a branch, then we put $f \Vdash a = b$ if $\bar{f}k \Vdash a = b$ for some k. Define $f \Vdash a \# b$ if $f \nVdash a = b$. Then the new model satisfies **AP**.

Now let $\alpha \Vdash a \neq_{\omega} b$ if $\alpha \Vdash a \neq_n b$ for all n. Clearly $\alpha \Vdash \neg 0 \neq_{\omega} 1$, but $\alpha \nVdash 0 = 1$ (for arbitrary α). So even in this model $\cap \neq_n$ is not an apartness relation.

After the present paper was finished, Smorynski surprised us with a short proof of the Proposition of Section 4. His proof, which avoids the proof theory involved in Sections 1–4, will appear elsewhere, [12]. With the permission of the author a sketch of his proof is reproduced here for the benefit and enlightment of the reader.

We recall that there is a number of Henkin's style completeness proofs for the Kripke semantics. For the details we refer to [11], 5.1.5–5.1.11. The model constructed in 5.1.9 is called a *Henkin model.*

The relevant facts concerning Henkin models are: The nodes of the tree are saturated theories Γ, Δ, . . ., partially ordered by inclusion. If $\Gamma \cup \{F_i | i \in \omega\}$ is consistent, then there is a node $\Delta \supseteq \Gamma$ such that F_0, F_1, F_2, . . . $\in \Delta$. The forcing relation, $\Vdash$, satisfies $\Delta \Vdash F \Leftrightarrow F \in \Delta$.

We now show that **AP** is conservative over $\mathbf{SEQ}^\omega$. Suppose $\mathbf{SEQ}^\omega \nvdash A$, for some equality sentence A. Let K be the Henkin model obtained by the construction of 5.1.9. Define $\Gamma \Vdash a \# b$ iff $\Gamma \Vdash a \neq_n b$ for all n. Claim: # is an apartness relation on K. The only interesting axiom is $\forall xy(\neg x \# y \to x = y)$. We show its validity by contradiction. Suppose $\Gamma_0 \Vdash \neg a \# b$ and $\Gamma_0 \nVdash a = b$ for some Γ_0. Since K is a model of $\mathbf{SEQ}^\omega$, $\Gamma_0 \nVdash a \neq_n b$ for all n. If $\Gamma_0 \cup \{a \neq_n b | n \in \omega\}$ were inconsistent, then $\Gamma_0 \cup \{a \neq_m b\} \vdash \curlywedge$ for some m, or $\Gamma_0 \vdash \neg a \neq_m b$. Contradiction.

So $\Gamma_0 \cup \{a \neq_n b | n \in \omega\}$ is consistent and therefore there exists a saturated $\Delta \supseteq \Gamma_0$ in K with $a \neq_n b \in \Delta$ for all n, i.e. $\Delta \Vdash a \neq b$. This contradicts $\Gamma_0 \Vdash \neg a \neq_m b$. Hence $\Gamma_0 \Vdash a = b$. Now it immediately follows that $\mathbf{AP} \nVdash A$.

From the preceding results we obtain.

PROPOSITION. There is no equality formula $A(x, y)$ such that for all equality formulae $B(x, y)$ $\vdash_{\mathbf{AP}} x \# y \to B(x, y) \Leftrightarrow \vdash_{\mathbf{AP}} A(x, y) \to B(x, y)$.

Proof: Suppose there is such an $A(x, y)$.
Then by the proposition of Section 4 $\vdash_{\mathbf{SEQ}^\omega} x \neq_n y \Leftrightarrow A(x, y)$ for some n.
Also $\vdash_{\mathbf{SEQ}^\omega} A(x, y) \to x \neq_m y$ for all m.
So $\vdash_{\mathbf{SEQ}^\omega} x \neq_n y \to x \neq_m y$ for all m.
However, thc Kripke model of Figure 5 is a model of $\mathbf{SEQ}^\omega$, so we produce contradiction by taking a node of length n and putting $m = n + 1$.

COROLLARY. There is no equality formula $A(x, y)$ such that

$$\vdash_{\mathbf{AP}} x \# y \leftrightarrow A(x, y).$$

The corollary illustrates the basic character of the apartness relation.

The natural deduction methods of Sections 2, 3, 4 can be extended to the theory of order. In particular one can establish the following result, [12], 3. Theorem 1; *The theory of linear order is conservative over the theory of apartness.*

The theory of linear order has axioms

$$x < y \wedge y < z \to x < z,$$
$$\neg(x < y) \wedge \neg(y < x) \to x = y,$$
$$x < y \to z < y \vee x < z.$$

and was called the theory of *pseudo-order* by Heyting, [9], p. 106. The proof is again based on a normal form theorem.*

The role of the various inequalities $\neq_n$ is rather opaque. For decidable species $\neq_0$ is an apartness relation (although, in general, by no means the only one, e.g. the universe of lawless sequences has two natural apartness relations, cf. [5], 7) and on $\mathbf{R}$ $\neq_1$ is the standard apartness relation. Question: are there non-trivial examples of species on which $\neq_n (n > 1)$ is an apartness relation?

Department of Mathematics and *Department of Philosophy,*
University of Utrecht
and
King's College, Cambridge University

BIBLIOGRAPHY

[1] L. E. J. Brouwer, 'Begründing der Mengenlehre unabhängig vom Logischen Satz Satz vom ausgeschlossenen Dritten. Zweiter Teil: Theorie der Punktmengen', *Koninklijke Nederlandse Akademie van Wetenschappen, Verhandelingen le sectie* **12** 7 (1919), pp. 3–33. Also in [3], pp. 191–221.

[2] L. E. J. Brouwer, 'Intuitionistische Zerlegung mathematischer Grundbegriffe', *Jahresberichte der Deutschen Mathematiker Verein* **33** (1923), pp. 251–256. Also in [3], pp. 275–280.

[3] L. E. J. Brouwer, *Collected Works*, Vol. I. ed. A. Heyting, Amsterdam 1975.

[4] D. van Dalen, 'Lectures on Intuitionism', *Cambridge Summer School in Mathematical Logic 1971*, Springer Lecture Notes, Vol. 337, 1973, pp. 1–94.

[5] D. van Dalen, 'An interpretation of intuitionistic analysis', *Annals of Mathematical Logic* **13** (1978), pp. 1–43.

[6] D. van Dalen and C. E. Gordon, 'Independence problems in sub-systems of intuitionistic arithmetic', *Indagationes Math.* **33** (1971), pp. 448–456.

[7] A. Heyting, *Intuïtionistische axiomatiek der projectieve meetkunde*, Diss., 1925.

[8] A. Heyting, 'Die Theorie der linearen Gleichungen in einer Zahlenspezies mit nicht kommutativer Multiplikation', *Math. Annalen* **98** (1927), pp. 465–490.

[9] A. Heyting, *Intuitionism*, Amsterdam, 1956.

[10] D. Prawitz, *Natural Deduction*, Stockholm, 1965.

[11] C. Smorynski, 'Applications of Kripke models', in [13].

[12] C. Smorynski, 'On Axiomatizing Fragments' (to appear).

[13] A. S. Troelstra, *Metamathematical Investigation of Intuitionistic Arithmetic and Analysis*, Berlin, 1973.

* *Added in proof:* The details have been provided by W. den Hartog, 'A proof theoretic study. The theory of pseudo-order as a conservative extension of the theory of apartness relation', Dept. of Math., Utrecht, preprint no. 77 (1978).

PART II

INFINITARY LANGUAGES

VEIKKO RANTALA

GAME-THEORETICAL SEMANTICS AND BACK-AND-FORTH

INTRODUCTION

We shall consider languages for which game-theoretical notions seem to form the most appropriate semantical tool and which form a natural syntactical framework for the back-and-forth technique.

The study of these languages gave rise to certain extended notions of partial isomorphism. However, the results obtained in this direction are not as strong as was originally anticipated, due to ignorance of certain properties of these languages.

These languages were introduced in [2] and they can be considered as generalizations of several well-known languages. The game-theoretical semantics for infinitary languages have been used, e.g., by Henkin and Keisler. In these cases, however, this technique has been applied only to specific kinds of strings of quantifiers, not to formulas in their entirety.

An informal introduction to the full game-theoretical approach can be found in [2]. What will be done here is to make this approach more explicit. We shall also state an exact definition of a specific kind of formulas which are generalizations of Hintikka's constituents for infinitary languages, present exact proofs for certain results in [2], and study various properties of these formulas. Clearly, closed infinitary constituents can be considered as counterparts to Scott sentences of $L_{\kappa\lambda}$ languages. It turns out that the structure of the constituents, even of the infinitary ones, is in a sense very "compact", although at first sight their structure looks complicated.

However, the way in which the formulas of these languages are defined gives rise also to certain complications. First, in contrast to the constituents, these languages allow also formulas with respect to which it is very difficult to decide whether or not one should count them as satisfiable in a model. Hence it seems rather difficult to find a definition of truth which in the best possible way would correspond to our intuition. Secondly, since the class of the formulas is not defined inductively, to prove that something holds for a class of formulas it is not possible to use the familiar method of induction on the complexity of a formula. It seems, however,

J. Hintikka, I. Niiniluoto, and E. Saarinen (eds.), Essays on Mathematical and Philosophical Logic, 119–151.

that when a formula is considered as a tree-like entity, this method can be replaced by induction on the length of a tree, or, in semantical cases, by induction on the length of a play of a game.

In this paper, we shall not deal with these problems. We shall state a definition of truth which corresponds to one in [2] and which works satisfactorily for a wide class of sentences, especially for the constituents. Also, instead of developing the mentioned kind of induction for an arbitrary class of formulas, we shall use another kind of induction for an arbitrary constituent. This is possible since these formulas themselves are defined inductively.

1. LANGUAGE

Let κ be a cardinal and α an ordinal. In [2], the language $N_{\kappa\alpha}$ was introduced as follows. Let L be a first-order language of an arbitrary similarity type τ.

DEFINITION 1.1. A formula of $N_{\kappa\alpha}$ is a 4-tuple

$$\phi = \langle T, \prec, g, S \rangle$$

where

(1) $\langle T, \prec \rangle$ is a tree of length (height) $\leqslant \alpha$ such that
  (i) through each element t of T there passes at least one branch with a maximal element,
  (ii) if the tree-rank of t, $t' \in T$ is a limit ordinal and t, t' have the same set of predecessors, then $t = t'$. It follows that T has the least element.

(2) $S = At \cup \{\neg, \vee, \wedge\} \cup (\{\exists, \forall\} \times V)$, where V is a set of variable and At is the set of all those atomic formulas of L whose variables are in V.

(3) g is a mapping from T into S such that
  (i) $g(t) \in At$ iff t is maximal,
  (ii) t has only one immediate successor iff $g(t) \in \{\neg\} \cup (\{\exists, \forall\} \times V)$,
  (iii) if $g(t) \in \{\vee, \wedge\}$, then t has fewer than κ immediate successors,
  (iv) for every $t \in T$, there are only finitely many predecessors t' of t such that $g(t') = \neg$.[1]

Let $\check{t} = \{t' \in T | t \preccurlyeq t'\}$. For each $t \in T$ there is a unique 4-tuple $\varphi' = \langle \check{t}, \prec', g', S \rangle$ such that $\prec' = \prec \restriction \check{t}$ and $g' = g \restriction \check{t}$. φ' is a formula of $N_{\kappa\alpha}$ since it satisfies the conditions (1)–(3) of Definition 1.1. φ' is the *sub-*

formula of φ *determined by* t. The *subformula-rank* of φ' is the tree-rank of t in T.

From a given 4-tuple satisfying (1)–(3) of Definition 1.1 we obtain the usual kind of notation for formulas as follows. Let

$$\varphi = \langle T, \prec, g, S\rangle$$

be a formula and Φ the set of its subformulas of rank 1, and t_0 the least element of T:

(a) If $g(t_0) = \bigvee$ or $\bigwedge$, φ can be denoted by $\bigvee\Phi$ or $\bigwedge\Phi$, respectively. In particular, if $\Phi = \{\varphi_1, \varphi_2\}$, φ can be denoted by $\varphi_1 \vee \varphi_2$ or $\varphi_1 \wedge \varphi_2$, respectively.

(b) If $\Phi = \{\varphi'\}$ and $g(t_0) = \neg$, $\langle\exists, x\rangle$, or $\langle\forall, x\rangle$, φ can be denoted by $\neg\varphi'$, $\exists x\varphi'$, or $\forall x\varphi'$, respectively.

(c) If $g(t_0) = \psi \in At$, φ can be denoted by ψ.

Conversely, from a formula written in the usual notation we can obtain corresponding entities $\langle T, \prec\rangle$, g, and S. Although there are several trees (and several sets S) corresponding to a given usual notation, these trees are isomorphic. This gives rise to the following definition.

DEFINITION 1.2. Formulas $\langle T, \prec, g, S\rangle$ and $\langle T', \prec', g', S'\rangle$ are *structurally similar* iff there is an isomorphism $f: \langle T, \prec\rangle \to \langle T', \prec'\rangle$ such that $g(t) = g(f(t))$.

It would be also possible to represent the formulas set-theoretically in a suitable class to obtain more uniqueness and preciseness. However, this will not be done here. At present, we do not know precisely how this representation will look like.

The free and bound variables of a formula can be defined as usual, likewise the operation of substitution. Our notation will be mostly standard.

If $\bar{x} = (x_\xi)_{\xi<\beta}$ is a sequence of distinct variables, by '$\varphi(\bar{x})$' we mean that φ has at most the $x_\xi (\xi < \beta)$ as its free variables.

Let $\varphi(\bar{x}) = \langle T, \prec, g, S\rangle$, where $S = At \cup \{\neg, \bigvee, \bigwedge\} \cup (\{\exists, \forall\} \times V)$, be a formula with the free variables $x_\xi(\xi < \beta)$ and $c_\xi(\xi < \beta)$ new individual constants. Let At' be At augmented by the new atomic formulas of the extended language. Then by '$\varphi(\bar{c})$' we mean the formula $\langle T, \prec, g', S'\rangle$ such that

$$S' = At' \cup \{\neg, \bigvee, \bigwedge\} \cup (\{\exists, \forall\} \times V)$$

and for every $t \in T$ such that $g(t) = \psi \in At$, if $x_{\xi_1}, \ldots, x_{\xi_k} \in \{x_\xi | \xi < \beta\}$ are the free variables occurring in ψ, then

$$g'(t) = \psi(c_{\xi_1}/x_{\xi_1}, \ldots, c_{\xi_k}/x_{\xi_k}).$$

Let λ be a cardinal. Every formula of $L_{\kappa\lambda}$ can be represented as a formula of $N_{\kappa\lambda}$ such that the length of each branch of the corresponding tree is a successor ordinal. Hence the formulas of $L_{\kappa\lambda}$ can be 'identified' with certain formulas of $N_{\kappa\lambda}$. The formulas of the game-quantifier language $L(\lambda)$, where the game-quantifier is well-ordered, can be identified with formulas of $N_{\infty\lambda^+}$.

By the notation '$N_{\infty\alpha}$' we naturally mean the language which includes the formulas of $N_{\kappa\alpha}$ for every cardinal κ. $N_{\infty\infty}$ is the language which includes the formulas of $N_{\infty\alpha}$ for every ordinal α.

If τ is the type of L, it is also the *type* of any of the languages defined. If all the members of τ are present in a formula φ, we shall often use the notation '$\varphi(\tau)$' for φ. Similarly, if we want to emphasize the type of a language, we shall write '$L(\tau)$', '$N_{\kappa\alpha}(\tau)$', etc.

2. SEMANTICAL GAME

Let $\varphi = \langle T, \prec, g, S \rangle$ be a sentence of $N_{\infty\infty}$ and $\mathfrak{M}$ a model for L. With φ and $\mathfrak{M}$ we associate a two-person game $G(\varphi, \mathfrak{M})$ as follows.

The *negation-rank* of a $t \in T$, $nr(t)$, is the number of all $t' \prec t$ such that $g(t') = \neg$. If $g(t) = \langle \exists, x \rangle (x \in V)$ or $g(t) = \bigvee$, and $nr(t)$ is even, we denote $g(t) = \langle \exists, x \rangle^+$, $\bigvee^+$, respectively. If $g(t) = \langle \forall, x \rangle$ or $g(t) = \bigwedge$, and $nr(t)$ is odd, we denote $g(t) = \langle \exists, x \rangle^-$, $\bigvee^-$, respectively. Similarly are defined $\langle \forall, x \rangle^\pm$, $\bigwedge^\pm$.

Let

$$\begin{aligned} T^\exists &= \{t \in T | g(t) = \langle \exists, x \rangle^\pm (x \in V)\} \\ T^\vee &= \{t \in T | g(t) = \textstyle\bigvee^\pm\} \\ T^\forall &= \{t \in T | g(t) = \langle \forall, x \rangle^\pm (x \in V)\} \\ T^\wedge &= \{t \in T | g(t) = \textstyle\bigwedge^\pm\} \\ T_I &= T^\exists \cup T^\vee \\ T_{II} &= T^\forall \cup T^\wedge. \end{aligned}$$

Let I_t be the set of all immediate $\prec$-successors of t. Let $\hat{t} = \{t' \in T_I \cup T_{II} | t' \prec t\}$. Given a $t \in T_I \cup T_{II}$, we denote by H^t the set of all functions $h \in {}^{\hat{t}}(T \cup M)$ (where M is the universe of $\mathfrak{M}$) such that

$h(t') \in I_{t'}$ if $t' \in T^{\vee} \cup T^{\wedge}$
$h(t') \in M$ if $t' \in T^{\exists} \cup T^{\forall}$
$h(t') \preccurlyeq t$ if $h(t') \in T$.

Let

$$H_I = \{\langle t, h\rangle | t \in T_I \;\&\; h \in H^t\}$$
$$H_{II} = \{\langle t, h\rangle | t \in T_{II} \;\&\; h \in H^t\}.$$

DEFINITION 2.1. $\Sigma(\varphi, \mathfrak{M})$ is the set of all functions $\sigma : H_I \to T \cup M$ such that $\sigma(t, h) \in I_t$ if $t \in T^{\vee}$ and $\sigma(t, h) \in M$ if $t \in T^{\exists}$. $\Pi(\varphi, \mathfrak{M})$ is the set of all functions $\pi : H_{II} \to T \cup M$ such that $\pi(t, h) \in I_t$ if $t \in T^{\wedge}$ and $\pi(t, h) \in M$ if $t \in T^{\forall}$. The members of $\Sigma(\varphi, \mathfrak{M})$ and $\Pi(\varphi, \mathfrak{M})$ are called *strategies* of $G(\varphi, \mathfrak{M})$.

LEMMA 2.1. Consider a pair of strategies $\sigma \in \Sigma(\varphi, \mathfrak{M})$, $\pi \in \Pi(\varphi, \mathfrak{M})$. Let $t_0 \prec \ldots \prec t_\mu \prec \ldots$ $(\mu < \alpha)$ be a chain in T and let $Z = \{\langle t_\mu, h_\mu, s_\mu\rangle | \mu < \alpha\}$ be a subset of $\sigma \cup \pi$ such that $h_0 \subset \ldots \subset h_\mu \subset \ldots$. If the set

$$I_\alpha = \{t \in T_I \cup T_{II} | \forall_{\mu<\alpha}(t_\mu \prec t) \;\&\; \neg\exists t' \in T_I \cup T_{II} \forall_{\mu<\alpha}(t_\mu \prec t' \prec t)\}$$

is not empty, then there is exactly one $\langle t, h, s\rangle \in \sigma \cup \pi$ such that $\bigcup_{\mu<\alpha} h_\mu \subseteq h$ and $t \in I_\alpha$.

Proof: Suppose first that α is a successor ordinal, $\alpha = \beta + 1$.

Let $t^{\exists} \in T_\beta \cup T^{\forall}$. Then, by the definition of a formula, I_α is a unit set, say $I_\alpha = \{t\}$. Consider the set

$$Z_0 = \{\langle t, h, s\rangle \in \sigma \cup \pi | h \in H^t \;\&\; s \in T \cup M\}.$$

Since $t_\beta \in T^{\exists} \cup T^{\forall}$, $s_\beta \in M$ and, by the definition of H^t, $H^t = \{h_\beta \cup \{\langle t_\beta, a\rangle\} | a \in M\}$. Thus there is exactly one $\langle t, h, s\rangle \in Z_0$ such that $h = h_\beta \cup \{\langle t_\beta, s_\beta\rangle\}$ since $\sigma \cup \pi$ is a function.

Let $t_\beta \in T^{\vee} \cup T^{\wedge}$ and $Z_0 = \{\langle t, h, s\rangle \in \sigma \cup \pi | t \in I_\alpha \;\&\; h \in H^t \;\&\; s \in T \cup M\}$. Again, $H^t = \{h_\beta \cup \{\langle t_\beta, t'\rangle\} | t' \in I_\alpha\}$. Since $t_\beta \in T^{\vee} \cup T^{\wedge}$ and $s_\beta \in I_\alpha$, there is exactly one $\langle t, h, s\rangle \in Z_0$ such that $h = h_\beta \cup \{\langle t_\beta, s_\beta\rangle\}$.

Suppose α is a limit ordinal. Then, by the definition of a formula, I_α is a unit set, $I_\alpha = \{t\}$ where $t \succcurlyeq \sup\{t_\mu | \mu < \alpha\}$. We claim that $\langle t, h, s\rangle \in \sigma \cup \pi$ for a unique $s \in T \cup M$, where $h = \bigcup_{\mu<\alpha} h_\mu$. Clearly $h \in {}^{\bar{t}}(T \cup M)$ since α is a limit ordinal. Thus $\langle t, h\rangle \in H_I \cup H_{II}$, say $\langle t, h\rangle \in H_I$. Since σ is a function from H_I into $T \cup M$, there is an $s \in T \cup M$ such that $\sigma(t, h) = s$ whence $\langle t, h, s\rangle \in \sigma \cup \pi$. s is uniquely determined since σ is a function and $H_I \cap H_{II} = \emptyset$. □

Given a subset Z of $\sigma \cup \pi$ as in Lemma 2.1, we denote the unique element mentioned in this Lemma by z_α.

Suppose now a subset Z of $\sigma \cup \pi$ is given as in Lemma 2.1 but $I_\alpha = \emptyset$. If $\alpha = \beta + 1$, then

(a) $s_\beta \in T$ and $\check{s}_\beta$ is a linearly ordered finite set with the greatest element t_α such that $g(t_\alpha) \in At$ and $g(t) = \neg$ for all the other $t \in \check{s}_\beta$.

If α is a limit ordinal, then either

(b) the set $\check{t}$ determined by $t = \sup \{t_\mu | \mu < \alpha\}$ is a linearly ordered finite set with the greatest element t_α as in (a), or

(c) there is no $t \in T$ such that $t_\mu \prec t$ for all $\mu < \alpha$.

Suppose $\rho \subset Z \cup T$ where Z is a subset of $\tau \cup \pi$ as in Lemma 2.1, $\sigma \in \Sigma(\varphi, \mathfrak{M})$, $\pi \in \Pi(\varphi, \mathfrak{M})$. We denote

$$Z_\rho = \begin{cases} \{z_\alpha\} \text{ if } I_\alpha \neq \emptyset \text{ and } \rho \cap T = \emptyset \\ \{t_\alpha\} \text{ if } I_\alpha = \emptyset \text{ and (a) or (b) holds, and } \rho \cap T = \emptyset \\ \emptyset \text{ if } I_\alpha = \emptyset \text{ and (c) holds, or if } \rho \cap T \neq \emptyset. \end{cases}$$

Let now t_0 be the least element of $T_I \cup T_{II}$. Then $\langle t_0, \emptyset, s_0 \rangle \in \sigma \cup \pi$ for some $s_0 \in T \cup M$ since $H^{t_0} = \{\emptyset\}$. Let a class $\rho_{(\sigma,\pi)}$ be defined as follows:

$$\begin{aligned} \rho^0_{(\sigma,\pi)} &= \{\langle t_0, \emptyset, s_0 \rangle\}, \\ \rho^{\alpha+1}_{(\sigma,\pi)} &= \rho^\alpha_{(\sigma,\pi)} \cup Z_{\rho^\alpha_{(\sigma,\pi)}} \\ \rho^\lambda_{(\sigma,\pi)} &= \cup_{\alpha<\lambda} \rho^\alpha_{(\sigma,\pi)} \text{ if } \lambda \text{ is a limit ordinal,} \\ \rho_{(\sigma,\pi)} &= \cup_{\alpha \in \mathrm{Ord}} \rho^\alpha_{(\sigma,\pi)} \end{aligned}$$

The following properties of $\rho_{(\sigma,\pi)}$ can be easily checked: There is a least ordinal α such that $\rho_{(\sigma,\pi)} = \cup_{\mu<\alpha} \rho^\mu_{(\sigma,\pi)}$. For each $z \in \rho_{(\sigma,\pi)}$ there is an ordinal $\alpha_z < \alpha$ such that z appears in $\rho^{\alpha_z}_{(\sigma,\pi)}$ for the first time, i.e., $z \in \rho^{\alpha_z}_{(\sigma,\pi)}$ but $z \notin \rho^\beta_{(\sigma,\pi)}$ for $\beta < \alpha_z$. Then $\rho_{(\sigma,\pi)}$ is well-ordered by the ordering $\prec_{(\sigma,\pi)}$ defined as follows: $z_1 \prec_{(\sigma,\pi)} z_2$ iff $\alpha_{z_1} < \alpha_{z_2}$.

An ordered pair $\langle \sigma, \pi \rangle$ is called a *play* of the game $G(\varphi, \mathfrak{M})$ if $\sigma \in \Sigma(\varphi, \mathfrak{M})$, $\pi \in \Pi(\varphi, \mathfrak{M})$. The *length* of $\langle \sigma, \pi \rangle$, $l(\sigma,\pi)$ = the order-type of $\rho_{(\sigma,\pi)} = OT(\rho_{(\sigma,\pi)}) = \alpha$. If $l(\sigma, \pi)$ is a successor ordinal $\beta + 1$, the greatest element of $\rho_{(\sigma,\pi)}$ is t_β and the other elements of $\rho_{(\sigma,\pi)}$ are of the form $\langle t, h, s \rangle \in \sigma \cup \pi$. If $l(\sigma, \pi)$ is a limit ordinal, $\rho_{(\sigma,\pi)} \subset \sigma \cup \pi$.

It follows from Lemma 2.1 that if $\langle t_1, h_1, s_1 \rangle \prec_{(\sigma,\pi)} \langle t_2, h_2, s_2 \rangle$, then $t_1 \prec t_2$ and $h_1 \subset h_2$. Thus, intuitively speaking, the play $\langle \sigma, \pi \rangle$ proceeds along a branch of T and the function h in a $\langle t, h, s \rangle$ includes the information about the choices made at the points of $T_I \cup T_{II}$ preceding t.

Consider now a play $\langle \sigma, \pi \rangle$ of a game $G(\varphi, \mathfrak{M})$. There is a unique maximal sequence

$$\langle t_0, h_0, a_0\rangle \prec_{(\sigma,\pi)} \ldots \prec_{(\sigma,\pi)} \langle t_\xi, h_\xi, a_\xi\rangle \prec_{(\sigma,\pi)} \ldots$$
$$(\xi < \mu < l(\sigma, \pi))$$

of elements of $\rho_{(\sigma,\pi)}$ such that for each ξ, $g(t_\xi) = \langle\exists, x_\xi\rangle$ or $g(t_\xi) = \langle\forall, x_\xi\rangle$ for some sequence $\bar{x} = (x_\xi)_{\xi<\mu}$ of variables, and $a_\xi \in M$. Let $s^{(\sigma,\pi)} = (a_\xi)_{\xi<\mu}$ and $\bar{c} = (c_\xi)_{\xi<\mu}$ where the c_ξ are new individual constants, distinct from each other, interpreted as the a_ξ, respectively.

Suppose $l(\sigma, \pi)$ is a successor ordinal. Let the greatest element of $\rho_{(\sigma,\pi)}$ be now denoted by $t_{(\sigma,\pi)}$, $g(t_{(\sigma,\pi)}) \in At$. Denote the expanded model $(\mathfrak{M}, a_\xi)_{\xi<\mu}$ by $(\mathfrak{M}, s^{(\sigma,\pi)})$ and by $\varphi_{(\sigma,\pi)}$ the atomic sentence of the extended language obtained when each x_ξ occurring in $g(t_{(\sigma,\pi)})$ is replaced by c_ξ.

Suppose $l(\sigma, \pi)$ is a limit ordinal. Given a $t \in \bigcup\bigcup \operatorname{dom}(\rho_{(\sigma,\pi)})$, let z^t be the unique element $\langle t', h, s\rangle$ of $\rho_{(\sigma,\pi)}$ such that $t = t'$, and let

$$\Sigma^t_{(\sigma,\pi)} = \{\sigma' \in \Sigma(\varphi, \mathfrak{M}) | \rho_{(\sigma',\pi)} \cap \rho_{(\sigma,\pi)} = \{z \in \rho_{(\sigma,\pi)} | z \prec_{(\sigma,\pi)} z^t\}\},$$
$$\Pi^t_{(\sigma,\pi)} = \{\pi' \in \Pi(\varphi, \mathfrak{M}) | \rho_{(\sigma,\pi')} \cap \rho_{(\sigma,\pi)} = \{z \in \rho_{(\sigma,\pi)} | z \prec_{(\sigma,\pi)} z^t\}\}$$

and

$$T^\wedge_{(\sigma,\pi)} = \{t \in \bigcup\bigcup \operatorname{dom}(\rho_{(\sigma,\pi)}) \cap T^\wedge | \forall\sigma' \in \Sigma^t_{(\sigma,\pi)} \operatorname{Lim}(l(\sigma',\pi))\},$$
$$T^\vee_{(\sigma,\pi)} = \{t \in \bigcup\bigcup \operatorname{dom}(\rho_{(\sigma,\pi)}) \cap T^\vee | \forall\pi' \in \Pi^t_{(\sigma,\pi)} \operatorname{Lim}(l(\sigma, \pi'))\}.$$

$T^\wedge_{(\sigma,\pi)}$ and $T^\vee_{(\sigma,\pi)}$ are well-ordered by $\prec$. Intuitively, when the play (σ, π) proceeds along a branch of T, these sets consist of those points of the branch from which onwards a player can 'force' any play corresponding to his original strategy to go to a limit play.

DEFINITION 2.2. The function $p : \Sigma(\varphi, \mathfrak{M}) \times \Pi(\varphi, \mathfrak{M}) \to \{-1, 0, 1\}$, defined as follows, is the *payoff function* of $G(\varphi, \mathfrak{M})$:
If $l(\sigma, \pi)$ is a successor ordinal, then

(1) $p(\sigma, \pi) = 1$ if $(\mathfrak{M}, s^{(\sigma,\pi)}) \models \varphi_{(\sigma,\pi)}$ and $nr(t_{(\sigma,\pi)})$ is even or $(\mathfrak{M}, s^{(\sigma,\pi)}) \not\models \varphi_{(\sigma,\pi)}$ and $nr(t_{(\sigma,\pi)})$ is odd,

(2) $p(\sigma, \pi) = -1$ if $(\mathfrak{M}, s^{(\sigma,\pi)}) \not\models \varphi_{(\sigma,\pi)}$ and $nr(t_{(\sigma,\pi)})$ is even or $(\mathfrak{M}, s^{(\sigma,\pi)}) \models \varphi_{(\sigma,\pi)}$ and $nr(t_{(\sigma,\pi)})$ is odd.

If $l(\sigma, \pi)$ is a limit ordinal, then

(3) $p(\sigma, \pi) = 1$ if $OT(T^\wedge_{(\sigma,\pi)}) > OT(T^\vee_{(\sigma,\pi)})$,
(4) $p(\sigma, \pi) = -1$ if $OT(T^\wedge_{(\sigma,\pi)}) < OT(T^\vee_{(\sigma,\pi)})$,
(5) $p(\sigma, \pi) = 0$ if $OT(T^\wedge_{(\sigma,\pi)}) = OT(T^\vee_{(\sigma,\pi)})$.

Conditions (3)–(5) are not needed if $\varphi \in L_{\infty\infty}$, $L(\infty)$. They are explications of the conditions stated in [2] for the limit case, in a more relevant form. (3)–(4) say, intuitively, that if one player has more possibilities to avoid a successor play than the other player, then the latter wins. There are different possible ways to define the payoff function for the case that the length of a play is a limit ordinal. But all the definitions chosen must agree in certain important and intuitively clear cases.[2]

DEFINITION 2.3. φ is *true in* $\mathfrak{M}$, $\mathfrak{M} \models \varphi$, iff there is a winning strategy σ in $\Sigma(\varphi, \mathfrak{M})$, that is, a strategy σ satisfying: For every $\pi \in \Pi(\varphi, \mathfrak{M})$, $p(\sigma, \pi) = 1$.

The *satisfaction* relation will be defined in an obvious way. Let $\bar{x} = (x_\xi)_{\xi<\beta}$ and $\bar{a} = (a_\xi)_{\xi<\gamma} \in {}^\gamma M$, where $\gamma \geqslant \beta$. Then

$$\mathfrak{M} \models \varphi(\bar{x})[\bar{a}]$$

means that

$$(\mathfrak{M}, \bar{a}) \models \varphi(\bar{c})$$

where $\bar{c} = (c_\xi)_{\xi<\beta}$ and the c_ξ, $\xi < \beta$, are new individual constants, distinct from each other, such that the interpretation of c_ξ in $\mathfrak{M}$ is a_ξ, for every $\xi < \beta$.

If there is no winning strategy in $\Sigma(\varphi, \mathfrak{M})$ in a game $G(\varphi, \mathfrak{M})$, it may happen that there is no winning strategy in $\Pi(\varphi, \mathfrak{M})$, either. Then we can consider our logic as a two-valued or three-valued logic. If we define φ to be false in $\mathfrak{M}$ iff there is no winning strategy in $\Sigma(\varphi, \mathfrak{M})$, we have a two-valued logic. If we define φ to be false iff there is a winning strategy in $\Pi(\varphi, \mathfrak{M})$, and undetermined if there is no winning strategy in $\Sigma(\varphi, \mathfrak{M})$ nor in $\Pi(\varphi, \mathfrak{M})$, we have the three-valued case. However, since there are sentences which intuitively seem to be neither true nor false, it is perhaps more natural to adopt the three-valued case. A simple example of such a sentence can be found in [2]. However, whichever possibility we adopt, it does not affect our later considerations. Moreover, as is pointed out in [2], the games associated with the constituents are always determinate, i.e., one of the sets of strategies includes a winning strategy.

Because of the possibility of indeterminateness, the implication and equivalence symbols ($\rightarrow$ and $\leftrightarrow$) will have only a semantical meaning here which is defined as usual.

LEMMA 2.2. If φ_1 and φ_2 are structurally similar, then $\models \varphi_1 \leftrightarrow \varphi_2$.

Proof. Let $\langle T_1, \prec_1\rangle$, $\langle T_2, \prec_2\rangle$ be the trees corresponding to φ_1, φ_2, respectively, and $f\colon T_1 \rightarrow T_2$ a mapping in question. We can suppose that φ_1 and φ_2 are sentences, without any loss of generality.

Suppose $\mathfrak{M}$ is a model such that $\mathfrak{M} \models \varphi_2$, and let $\sigma_2 \in \Sigma(\varphi_2, \mathfrak{M})$ be a winning strategy in $G(\varphi_2, \mathfrak{M})$. Then the strategy $\sigma_1 \in \Sigma(\varphi_1, \mathfrak{M})$ defined as follows is a winning strategy in $G(\varphi_1, \mathfrak{M})$:

$$\sigma_1(t, h) = \sigma_2(f(t), h'),$$

where $h' \in H^{f(t)}$ satisfying the following condition:

$$h'(f(t_0)) = \begin{cases} f(h(t_0)) \text{ if } h(t_0) \in T_1 \\ h(t_0) \text{ if } h(t_0) \in M. \end{cases}$$

The proof in the other direction is symmetrical. □

3. CONSTITUENTS

Let $z_\xi(\xi < \beta)$ be distinct variables and $\bar{z} = (z_\xi)_{\xi < \beta}$. Let $\Delta(\bar{z})$ be the set of all basic formulas of $L(\tau)$ (atomic formulas and their negations) which can be formed by using the constants of τ, the identity symbol, and the variables z_ξ, $\xi < \beta$. By $A(\bar{z})$, with possible subscripts, will be denoted a subset of $\Delta(\bar{z})$ which for each atomic formula $\varphi \in \Delta(\bar{z})$ contains either φ or $\neg\varphi$ (but not both). Let $A^+(\bar{z})$ $(A^-(\bar{z}))$ be the subset of the positive (negative) formulas of $A(\bar{z})$. If $\beta = 0$, we omit the parentheses; thus Δ is the set of all basic sentences of $L(\tau)$, etc.

The following fact is obvious.

LEMMA 3.1. Let $\bar{z}$ be as above. If $A_1(\bar{z})$ and $A_2(\bar{z})$ are jointly satisfiable, i.e., if there is a model $\mathfrak{M}$ and an $\bar{a} \in {}^\beta M$ such that $\mathfrak{M} \models \bigwedge A_1(\bar{z})[\bar{a}]$ and $\mathfrak{M} \models \bigwedge A_2(\bar{z})[a]$, then $\models \bigwedge A_1(\bar{z}) \leftrightarrow \bigwedge A_2(\bar{z})$.

Let $\varphi = \langle T, \prec, g, S \rangle$ be a formula of $N_{\infty\infty}(\tau)$. The *quantifier-rank* of a $t \in T$ is defined as follows:

$$\text{qr}(t) = OT\{t' \prec t \mid g(t') = \langle \exists, x \rangle \text{ or } g(t') = \langle \forall, x \rangle, x \in V\}.$$

The quantifier-rank of φ is

$$\text{qr}(\varphi) = \sup\{\text{qr}(t) \mid t \in T\}.$$

Let

$$T^{(\mu)} = \{t \in T \mid g(t) = \bigwedge \text{ and } \text{qr}(t) = \mu\},$$
$$\hat{t} = \{t' \in T \mid t' \prec t\}.$$

Let $\sim$ be the relation on $T^{(\mu)}$ defined as follows:

$t_1 \sim t_2$ iff t_1 and t_2 belong to the same level of T and there is a mapping $f\colon \hat{t}_1 \xrightarrow[\text{onto}]{1-1} \hat{t}_2$ such that: $g(t) = g(f(t))$, and $f(t) \prec f(t')$ iff $t \prec t'$.

Since $\sim$ is an equivalence relation, it gives rise to the partition $T^{(\mu)}/\sim$ of $T^{(\mu)}$.

Let U_μ be the μth level of T and $T|\mu = \cup_{\xi<\mu} U_\xi$. Let $\bar{z} = (z_\xi)_{\xi<\beta}$ be as above and let $x_0, \ldots, x_\mu, \ldots$ $(\mu < \alpha)$ be any list of variables which are different from each other and from the variables $z_\xi (\xi < \beta)$.

DEFINITION 3.1. A *constituent* $C^{(\alpha)}(\tau, \bar{z})$ *of depth* α *and of type* τ (and with the free variables z_ξ, $\xi < \beta$) is a formula $\langle T, \prec, g, S\rangle$ of $N_{\infty\infty}(\tau)$ of quantifier-rank α which satisfies the following conditions:

(1) If the tree-rank of a $t \in T$ is a limit ordinal, then $g(t) = \wedge$.

(2) Let $g(t) = \wedge$ and $\text{qr}(t) = \mu \leqslant \alpha$. Then there is a partition $\{I_t^\Delta, I_t^\forall, I_t^\exists\}$ of I_t such that

(i) there is an $A_t(\bar{z} \frown \bar{x}) \subset \Delta(\bar{z} \frown \bar{x})(\bar{x} = (x_\xi)_{\xi<\mu})$ and a mapping $f_t : A_t(\bar{z} \frown \bar{x}) \xrightarrow[\text{onto}]{1-1} I_t^\Delta$ such that

$$g(f_t(\varphi)) = \begin{cases} \varphi \text{ if } \varphi \in A_t^+(\bar{z} \frown \bar{x}) \\ \neg \text{ if } \varphi \in A_t^-(\bar{z} \frown \bar{x}), \end{cases}$$

(ii) if $\mu < \alpha$, then $|I_t^\forall| = 1$ and $g(t') = \langle \forall, x_\mu\rangle$ for the element $t' \in I_t^\forall$,

(iii) if $\mu < \alpha$, then $I_t^\exists \neq \emptyset$ and $g(t') = \langle \exists, x_\mu\rangle$ for all elements $t' \in I_t^\exists$,

(iv) if $\mu = \alpha$, then $I_t^\forall = I_t^\exists = \emptyset$.

(3) If $g(t_1) = g(t_2) = \wedge$ and $t_1 \prec t_2$, then $A_{t_1}(\bar{z} \frown \bar{x}) \subset A_{t_2}(\bar{z} \frown \bar{x})(\bar{x} = (x_\xi)_{\xi<\text{qr}(t_1)}, \bar{y} = (x_\xi)_{\xi<\text{qr}(t_2)})$.

(4) If $g(t) = \neg$, t_0 is the immediate predecessor of t, and $f_{t_0}(\neg\varphi) = t$, where φ is atomic, then $g(t') = \varphi$ for the $t' \in I_t$.

(5) If $g(t) = \langle \exists, x_\mu\rangle$ or $g(t) = \vee$, then $g(t') = \wedge$ for all $t' \in I_t$.

(6) If $g(t) = \langle \forall, x_\mu\rangle$, then $g(t') = \vee$ for the $t' \in I_t$.

(7) For every $\mu \leq \alpha$ and for every $S_1, S_2 \in T^{(\mu)}/\sim$, if $S_1 \neq S_2$, there is a mapping $h : S_1 \xrightarrow[\text{onto}]{1-1} S_2$ such that the subformulas determined by t and $h(t)$ are structurally similar.

(8) Let $\mu < \alpha$ be a limit ordinal. Then for every μ-branch B of $T|\mu$, there is a $t \in T$ of tree-rank μ such that $\hat{t} = B$.

By induction on $n \in \omega$ $(n \neq 0)$, using (6), (5), (2i), and (4), it can be shown that if α is a successor ordinal, $\alpha = \omega \cdot \lambda + n(n \neq 0)$, the height of T is $\omega \cdot \lambda + 3(n + 1)$ or $\omega \cdot \lambda + (3n + 2)$. If $\alpha \neq 0$ is a limit ordinal, the height of T is α. If $\alpha = 0$, the height is 3 or 2. Thus always $C^{(\alpha)}(\tau, \bar{z}) \in N_{\infty\alpha}{}^*(\tau)$, where

$$\alpha^* = \begin{cases} \alpha + (2n + 1) & \text{if } \alpha \text{ is an infinite successor ordinal} \\ \alpha & \text{if } \alpha \text{ is a limit ordinal} \neq 0 \\ \omega & \text{if } \alpha \text{ is finite.} \end{cases}$$

Given κ, α, if $|\tau|$ and $|\beta|$ are small enough, there are constituents $C^{(\alpha)}(\tau, \bar{z})$ which belong to $N_{\kappa\alpha^*}(\tau)$, that is, the cardinalities of the sets of the form $A_t(\bar{z} \frown \bar{x})$ are $< \kappa$.

Henceforth, we can omit the reference to the type of a constituent if no confusion will arise. Thus we may write $C^{(\alpha)}$ for $C^{(\alpha)}(\tau)$, etc.

We see that if $C^{(\alpha)}(\bar{z})$ is a constituent whose tree is $\langle T, \prec \rangle$, then its subformula determined by a $t_0 \in T^{(\mu)}$ is a constituent of depth $\leqslant \alpha$ and of the same type as $C^{(\alpha)}(\bar{z})$ (and with the free variables indicated), denoted by $C_{t_0}(\bar{z} \frown \bar{x})(\bar{x} = (x_\xi)_{\xi < \mu})$. If we denote

$$K_{t_0} = \{t \in T^{(\mu+1)} | t_0 \prec t\},$$

it follows from the condition (7) that this subconstituent can be written into the form

$$C_{t_0}(\bar{z} \frown \bar{x}) = \bigwedge A_{t_0}(\bar{z} \frown \bar{x}) \wedge \bigwedge_{t \in K_{t_0}} \exists x_\mu C_t(\bar{z} \frown \bar{x} \frown x_\mu) \wedge \forall x_\mu \bigvee_{t \in K_{t_0}} C_t(\bar{z} \frown \bar{x} \frown x_\mu).$$

This will be rewritten, more conveniently,

$$C_{t_0}(\bar{z}, \bar{x}) = \bigwedge A_{t_0}(\bar{z}, \bar{x}) \wedge \bigwedge_{t \in K_{t_0}} \exists x_\mu C_t(\bar{z}, \bar{x}, x_\mu) \wedge \forall x_\mu \bigvee_{t \in K_{t_0}} C_t(\bar{z}, \bar{x}, x_\mu).$$

We shall consider a method of proving a sentence to be true in a model, which is well-suited for the constituents.

Let φ be a sentence of $N_{\infty\infty}$ of quantifier-rank α and $\mathfrak{M}$ a model for L. Given a strategy $\sigma \in \Sigma(\varphi, \mathfrak{M})$, let $\{\Pi_\beta^\sigma(\varphi, \mathfrak{M}) | \beta \leqslant \alpha\}$ be a partition of $\Pi(\varphi, \mathfrak{M})$ such that

$$\Pi_\beta^\sigma(\varphi, \mathfrak{M} = \{\pi \in \Pi(\varphi, \mathfrak{M}) | OT(s^{(\sigma,\pi)}) = \beta\}.$$

Consider the following property of an ordinal β:

$(w)_\beta$ For every $\pi \in \Pi_\beta^\sigma(\varphi, \mathfrak{M})$, $p(\sigma, \pi) = 1$.

Then σ is a winning strategy iff $(w)_\beta$ holds for every $\beta \leqslant \alpha$.

Let now φ be a closed constituent $C^{(\alpha)}$ with the tree $\langle T, \prec\rangle$ and

$$\alpha^0 = \begin{cases} \alpha & \text{if } \alpha \text{ is a limit ordinal} \neq 0 \\ \alpha + 1 & \text{if } \alpha \text{ is a successor ordinal or } 0 \end{cases}$$

LEMMA 3.2. $\sigma \in \Sigma(C^{(\alpha)}; \mathfrak{M})$ is a winning strategy in $G(C^{(\alpha)}, \mathfrak{M})$ iff $(w)_\beta$ holds for every $\beta < \alpha^0$.

Proof. For every $\pi \in \Pi_\alpha^\sigma(C^{(\alpha)}, \mathfrak{M})$, where $\alpha \neq 0$ is a limit ordinal, $OT(T^\wedge_{(\sigma,\pi)} = \alpha > OT(T^\vee_{(\sigma,\pi)}) = 0$. □

We denote:

$$S_\beta^\exists = T^\exists \cap \{t \in T | \mathrm{qr}(t) = \beta\}$$
$$S_{\beta+1}^\forall = T^\forall \cap \{t \in T | \mathrm{qr}(t) = \beta + 1\}.$$

We shall mean by the H^β, $\beta \in \mathrm{Ord}$, the sets defined inductively as follows:

$$H^0 = \emptyset$$
$$H^{\beta+1} = H^\beta \cup \{\langle t, h\rangle \in H_I | t \in S_\beta^\exists \cup S_{\beta+1}^\forall\}$$
$$H^\lambda = \cup_{\beta<\lambda} H^\beta \text{ if } \mathrm{Lim}(\lambda).$$

Since the quantifier-rank of $C^{(\alpha)}$ is α, $H^{\alpha+1} = H^\alpha$; thus $H_I = H^\alpha$. If $\sigma^0 \subset \ldots \subset \sigma^\alpha$ is an increasing sequence of functions such that for every $\beta \leqslant \alpha$, $\sigma^\beta : H^\beta \to T \cup M$ and

$$(\sigma)_\beta \quad \begin{array}{l} \sigma^\beta(t, h) \in I_t \text{ if } t \in T^\forall \\ \sigma_\beta(t, h) \in M \text{ if } t \in T^\exists, \end{array}$$

it is clear that $\sigma^\alpha \in \Sigma(C^{(\alpha)}, \mathfrak{M})$.

In what follows, we shall also use the following notations: It follows from the structure of a constituent that given a $\sigma \in \Sigma(C^{(\alpha)}, \mathfrak{M})$, $\Pi_{\beta+1}^\sigma(C^{(\alpha)}, \mathfrak{M})$ can be partitioned into two sets

$$\Pi_{\beta+1}^{\sigma\exists}(C^{(\alpha)}, \mathfrak{M}) = \{\pi \in \Pi_{\beta+1}^\sigma(C^{(\alpha)}, \mathfrak{M}) | \text{there is a } \langle t, h, s\rangle \in p_{(\sigma,\pi)} \text{ such that } t \in T^{(\beta)} \ \&\ s \in T^\forall\}$$

$$\Pi_{\beta+1}^{\sigma\forall}(C^{(\alpha)}, \mathfrak{M}) = \{\pi \in \Pi_{\beta+1}^\sigma(C^{(\alpha)}, \mathfrak{M}) | \text{there is a } \langle t, h, s\rangle \in p_{(\sigma,\pi)} \text{ such that } t \in T^{(\beta)} \ \&\ s \in T^\forall\}.$$

Given a $t \in T_I \cup T_{II}$, $\mathrm{qr}(t) = \beta$, we say that a sequence $\bar{a} \in {}^\beta M$ *occurs in* an $h \in H^t$ if $\langle t_0, a_0\rangle, \ldots, \langle t_\mu, a_\mu\rangle \ldots \in h$ and $t_0 \prec \ldots \prec t_\mu \prec \ldots \prec t (\mu < \beta)$.

Given a $\sigma \in \Sigma(C^{(\alpha)}, \mathfrak{M})$ and a $\pi \in \Pi^{\sigma}_{\beta}(C^{(\alpha)}, \mathfrak{M})(\beta < \alpha^0)$, we shall denote by $\langle t^{\sigma}_{\pi}, h^{\sigma}_{\pi}, s^{\sigma}_{\pi}\rangle$ the unique element of $\rho_{(\sigma,\pi)}$ such that $t^{\sigma}_{\pi} \in T^{(\beta)}$ (hence $s^{\sigma}_{\pi} \in I^{\Delta}_{t_{\pi}\sigma}$). If σ is fixed and no confusion arises, we may omit the superscript σ and write simply $\langle t_{\pi}, h_{\pi}, s_{\pi}\rangle$.

LEMMA 3.3. Given a $\sigma \in \Sigma(C^{(\alpha)}, \mathfrak{M})$, $(w)_{\beta}$ holds iff for every $\pi \in \Pi^{\sigma}_{\beta}(C^{(\alpha)}, \mathfrak{M})$, $\mathfrak{M} \models \bigwedge A_{t\pi}(\bar{x})[s^{(\pi,\sigma)}](\bar{x} = (x_{\xi})_{\xi<\beta})$.

Proof. ($\Rightarrow$) Let $\pi \in \Pi^{\sigma}_{\beta}(C^{(\alpha)}, \mathfrak{M})$. For each $s \in I^{\Delta}_{t_{\pi}}$ there is a strategy $\pi_0 \in \Pi^{t_{\pi}}_{(\sigma\pi)} \cap \Pi^{\sigma}_{\beta}(C^{(\alpha)}, \mathfrak{M})$ such that $\langle t_{\pi_0}, h_{\pi_0}, s_{\pi_0}\rangle = \langle t_{\pi}, h_{\pi}, s\rangle$ from which the conclusion follows, since $p(\sigma, \pi_0) = 1$ and $s^{(\sigma,\pi_0)} = s^{(\sigma,\pi)}$ for all such strategies π_0. ($\Leftarrow$) Let $\pi \in \Pi^{\sigma}_{\beta}(C^{(\alpha)}, \mathfrak{M})$. Since $s_{\pi} \in I^{\Delta}_{t_{\pi}}$, either $g(t_{(\sigma,\pi)}) \in A_{t\pi}(\bar{x})$ or $\neg g(t_{(\sigma,\pi)}) \in A_{t\pi}(\bar{x})$, by Definition 3.1, (1i), whence $p(\sigma, \pi) = 1$, by the assumption and Definition 2.2, (1). □

THEOREM 3.4. $\sigma \in \Sigma(C^{(\alpha)}, \mathfrak{M})$ is a winning strategy iff for every $\beta < \alpha^0$ and for every $\pi \in \Pi^{\sigma}_{\beta}(C^{(\alpha)}, \mathfrak{M})$, $\mathfrak{M} \models A_{t\pi}(\bar{x})[s^{(\sigma,\pi)}]$.

Proof. Lemmas 3.2 and 3.3. □

LEMMA 3.5. Let $\mathfrak{M} \models C^{(\alpha)}$ and let $\sigma \in \Sigma(C^{(\alpha)}, \mathfrak{M})$ be a winning strategy in $G(C^{(\alpha)}, \mathfrak{M})$. Then for every $\beta < \alpha^0$ and every $\pi \in \Pi^{\sigma}_{\beta}(C^{(\alpha)}, \mathfrak{M})$, $(\mathfrak{M}, s^{(\sigma,\pi)}) \models C_{t_{\pi}}(\bar{d})$, where the $d_{\xi}(\xi < \beta)$ are new constants and $\bar{d} = (d_{\xi})_{\xi<\beta}$.

Proof. Let $\beta < \alpha^0$ and $\pi \in \Pi^{\sigma}_{\beta}(C^{(\alpha)}, \mathfrak{M})$ be given. $\langle T, \prec\rangle$ is supposed to be the tree corresponding to $C^{(\alpha)}$. Recall $\langle t_{\pi}, h_{\pi}, s_{\pi}\rangle$ is the unique element of $\rho_{(\sigma,\pi)}$ such that $t_{\pi} \in T^{(\beta)}$.

Define:

$$\langle T^*, \prec^*\rangle = \langle \check{t}_{\pi}, \prec\restriction\check{t}_{\pi}\rangle, \text{ the subtree determined by } t_{\pi},$$
$$T^*_I = T_I \cap T^*,\ T^*_{II} = T_{II} \cap T^*,$$
$$\overset{*}{t} = \{t' \in T^*_I \cup T^*_{II} | t' \prec^* t\},$$
$$H^{*t} = \{h\restriction\overset{*}{t} | h \in H^t\},$$
$$H^*_I = \{\langle t, h^*\rangle | t \in T^*_I \ \&\ h^* \in H^{*t}\},$$
$$H^*_{II} = \{\langle t, h^*\rangle | t \in T^*_{II} \ \&\ h^* \in H^{*t}\}.$$

Given a $t \in T^*_I \cup T^*_{II}$ and an $h^* \in H^{*t}$, let $^*h = h_{\pi} \cup h^*$. It is clear that $^*h \in H^t$, since $\mathring{t} = \mathring{t}_{\pi} \cup \overset{*}{t}$. Define

$$\sigma^* = \{\langle t, h^*, s\rangle | \langle t, h^*\rangle \in H^*_I \ \&\ s = \sigma(t, {}^*h)\}.$$

Since σ is a function, $\sigma^*\colon H^*_I \to T^* \cup M$, and since $\sigma \in \Sigma(C^{(\alpha)}, \mathfrak{M})$, it follows that

$\sigma^*(t, h^*) \in I_t$ if $t \in T^\forall \cap T^*$,
$\sigma^*(t, h^*) \in M$ if $t \in T^\exists \cap T^*$,

whence $\sigma^* \in \Sigma(C_{t_\pi}(\bar{d}), (\mathfrak{M}, s^{(\sigma,\pi)}))$.

Given any $\pi_0^* \in \Pi(C_{t_\pi}(\bar{d}), (\mathfrak{M}, s^{(\sigma,\pi)}))$, define

$$\pi_0 = \{\langle t, h, s\rangle | \langle t, h\rangle \in H_{II} \;\&\; s = \pi_0^*(t, h\restriction \overset{*}{t}) \text{ if } t \in T_{II}^* \;\&\; s = \pi(t, h) \text{ if } t \notin T_{II}^*\}.$$

$\pi_0 : H_{II} \to T \cup M$, since π_0^* and π are functions. π_0 is a strategy in $\Pi(C^{(\alpha)}, \mathfrak{M})$, since π_0^* and π are strategies in $\Pi(C_{t_\pi}(\bar{d}), (\mathfrak{M}, s^{(\sigma,\pi)}))$ and $\Pi(C^{(\alpha)}, \mathfrak{M})$, respectively. Let now $\langle t, h^*\rangle \in H_{II}^*$. By the definition of $^*h$, $h^* = {}^*h\restriction t$, $^*h \in H^t$, whence $\pi_0^*(t, h^*) = \pi_0^*(t, {}^*h\restriction t)$. By the definition of π_0, $\pi_0^*(t, {}^*h\restriction \overset{*}{t}) = \pi_0(t, {}^*h)$. Thus $\pi_0^*(t, h^*) = \pi_0(t, {}^*h)$ for all $(t, h^*) \in H_{II}^*$.

Let $\pi_0^* \in \Pi_\mu^{\sigma*}(C_{t_\pi}(\bar{d}), (\mathfrak{M}, s^{(\sigma,\pi)}))$ (hence $OT(s^{(\sigma*,\pi*)}) = \mu$), where $\mu \subset \alpha_1^0$, α_1 = the depth of $C_{t_\pi}(\bar{d})$.

Consider the set $\rho' = \rho_{(\sigma,\pi)} - \{\langle t_\pi, h_\pi, s_\pi\rangle, t_{(\sigma,\pi)}\}$. We claim that $\rho' \subset \rho_{(\sigma,\pi_0)}$. Let

$$z_0' \prec_{(\sigma,\pi)} \ldots \prec_{(\sigma,\pi)} z_\xi' \prec_{(\sigma,\pi)} \ldots (\xi < \gamma_1 = OT(\rho_{(\sigma,\pi)}) - 2)$$

be the well-ordering of ρ' (cf. Section 2), where $z_\xi' = \langle t_\xi', h_\xi', s_\xi'\rangle$, t_0' is the least element of $T(t_0' \in T^\wedge)$, and $h_0' = \emptyset$.

We shall prove the claim by induction on $\xi < \gamma_1^0$:

1. Let $\xi = 0$. Since $\pi_0(t_0', \emptyset) = \pi(t_0', \emptyset) = s_0'$, $z_0' = \langle t_0', \emptyset, s_0'\rangle$ is in $\sigma \cup \pi_0$, hence by the definition of $\rho_{(\sigma,\pi_0)}$, $z_0' \in \rho_{(\sigma,\pi_0)}$.

2. Let $\xi < \gamma_1^0$ and suppose that $z_\nu' \in \rho_{(\sigma,\pi_0)}$ for all $\nu < \xi$. Since $z_\xi' \in \rho_{(\sigma,\pi)}$, $t_\xi' \in I_\xi$, thus $I_\xi \neq \emptyset$ and $\cup_{\mu<\xi} h_\nu' \subseteq h_\xi'$ (cf. Lemma 2.1). Thus by induction hypothesis, Lemma 2.1, and the definition of $\rho_{(\sigma,\pi_0)}$, $\langle t_\xi', h_\xi', s\rangle \in \rho_{(\sigma,\pi_0)}$ for some $s \in T \cup M$. If $t_\xi' \in T_I$, $s = \sigma(t_\xi', h_\xi')$. Since $\langle t_\xi', h_\xi', s_\xi'\rangle \in \rho_{(\sigma,\pi)}$, $s_\xi' = \sigma(t_\xi', h_\xi')$. If $t' \in T_{II}$, $s = \pi_0(t_\xi', h_\xi') = \pi(t_\xi', h_\xi')$. Since $\langle t_\xi', h_\xi', s_\xi'\rangle \in \rho_{(\sigma,\pi)}$, $s_\xi' = \pi(t_\xi', h_\xi')$. Thus in any case, $s = s_\xi'$, whence $z_\xi' \in \rho_{(\sigma,\pi_0)}$. This proves the claim.

Let $\langle t_{\pi^*_0}, h_{\pi^*_0}, s_{\pi^*_0}\rangle$ be the unique element of $\rho_{(\sigma*,\pi^*_0)}$ such that the quantifier-rank of $t_{\pi^*_0}$ in T^* is μ and $g(t_{\pi^*_0}) = \wedge$. Consider the set $\{\langle t, h^*, s\rangle \in \rho_{(\sigma*,\pi^*_0)} | t_\pi \leqslant^* t \leqslant^* t_{\pi^*_0}\}$ and its well-ordering

$$z_0^* \prec_{(\sigma*,\pi^*_0)} \ldots \prec_{(\sigma*,\pi^*_0)} z_{\gamma_2}^* (\gamma_2 = OT(\rho_{(\sigma*,\pi^*_0)}) - 1),$$

where $z_\xi^* = \langle t_\xi, h_\xi^*, s_\xi\rangle (\xi \leqslant \gamma_2)$, $t_0 = t_\pi$, $h_0^* = \emptyset$, and $t_{\gamma_2} = t_{\pi^*_0}$, $h_{\gamma_2} = h_{\pi^*_0}$. We claim that for every $\xi \leqslant \gamma_2$, $z_\xi = \langle t_\xi, {}^*h_\xi, s_\xi\rangle$ belongs to $\rho_{(\sigma,\pi_0)}$.

We shall prove the claim by induction on $\xi \leqslant \gamma_2$:

1. Let $\xi = 0$. $z_0^* = \langle t_\pi, \emptyset, s_0 \rangle$ and $z_0 = \langle t_\pi, h_\pi, s_0 \rangle$. Since $\rho' \subset \rho_{(\sigma,\pi_0)}$ and $t_\pi \in I_{\gamma_1}$, there is an $s \in T \cup M$ such that $\langle t_\pi, h_\pi, s \rangle \in \rho_{(\sigma,\pi_0)}$, by Lemma 2.1 and the definition of $\rho_{(\sigma,\pi_0)}$. But $s = \pi_0(t_\pi, h_\pi) = \pi_0^*(t_\pi, \emptyset) = s_0$, whence $z_0 \in \rho_{(\sigma,\pi_0)}$.

2. Suppose $z_\nu \in \rho_{(\sigma,\pi_0)}$ for all $\nu < \xi$. Again, since $z_\xi^* \in \rho_{(\sigma*,\pi^*_0)}$, $t_\xi \in I_\xi$, whence $I_\xi \neq \emptyset$; and $h_0^* \subset \ldots \subset h_\nu^* \subset \ldots (\nu < \xi)$.
Hence by Lemma 2.1 and the definition of $\rho_{(\sigma*,\pi^*_0)}$:

(i) If ξ is a successor ordinal, $h_\xi^* = h_{\xi-1}^* \cup \{\langle t_{\xi-1}, s_{\xi-1}\rangle\}$. Thus $^*h_\xi = h_\pi \cup h_\xi^* = (h_\pi \cup h_{\xi-1}^*) \cup \{\langle t_{\xi-1}, s_{\xi-1}\rangle\} = {}^*h_{\xi-1} \cup \{\langle t_{\xi-1}, s_{\xi-1}\rangle\}$.

(ii) If ξ is a limit ordinal, $h_\xi^* = \bigcup_{\nu<\xi} h_\nu^*$. Hence

$$^*h_\xi = h_\pi \cup h_\xi^* = \bigcup_{\nu<\xi}(h_\pi \cup h_\nu^*) = \bigcup_{\nu<\xi} {}^*h_\nu.$$

Thus $^*h_\xi \subseteq \bigcup_{\nu<\xi} {}^*h_\nu$, whence by induction hypothesis, Lemma 2.1, and the definition of $\rho_{(\sigma,\pi_0)}$, $\langle t_\xi, {}^*h_\xi, s\rangle \in \rho_{(\sigma,\pi_0)}$ for some $s \in T \cup M$. If $t_\xi \in T_I$, $s = \sigma(t_\xi, {}^*h_\xi) = \sigma^*(t_\xi, h_\xi^*) = s_\xi$, whence $\langle t_\xi, {}^*h_\xi, s_\xi\rangle \in \rho_{(\sigma,\pi_0)}$. If $t \in T_{II}$, $s = \pi_0(t_\xi, {}^*h_\xi) = \pi_0^*(t_\xi, h_\xi^*) = s_\xi$, whence $\langle t_\xi, {}^*h_\xi, s_\xi\rangle \in \rho_{(\sigma,\pi_0)}$. This, completes the proof of the claim.

It follows from these claims that $\pi_0 \in \Pi^\sigma_{\beta+\mu}(C^{(\alpha,)}\ \mathfrak{M})$, $s^{(\sigma,\pi_0)} = s^{(\sigma,\pi)} \frown s^{(\sigma*,\pi_0^*)}$ and $t_{\pi_0} = t_{\pi^*_0}$.

We can show now that $\sigma^* \in \Sigma(C_{t\pi}(\bar{d}), (\mathfrak{M}, s^{(\sigma,\pi)}))$ is a winning strategy. By Theorem 3.4, it is sufficient to show that for every $\mu < \alpha_1^0$ and for every $\pi_0^* \in \Pi^{\sigma*}_\mu(C_{t_\pi}(\bar{d}), (\mathfrak{M}, s^{(\sigma,\pi)}))$, $(\mathfrak{M}, s^{(\sigma,\pi)}) \models \bigwedge A_{t_{\pi^*_0}}(\bar{d}, \bar{y})[s^{(\sigma*,\pi_0^*)}]$, where $\bar{y} = (x_{\beta+\xi})_{\xi<\mu}$. So, let $\mu < \alpha_1^0$ and $\pi_0^* \in \Pi^{\sigma*}_\mu(C_{t_\pi}(\bar{d}), (\mathfrak{M}, s^{(\sigma,\pi)})$. Consider the corresponding strategy $\pi_0 \in \Pi^\sigma_{\beta+\mu}(C^{(\alpha)}, \mathfrak{M})$. Since $\sigma \in \Sigma(C^{(\alpha)}, \mathfrak{M})$ is a winning strategy, then by Theorem 3.4,

$$\mathfrak{M} \models \bigwedge A_{t_{\pi_0}}(\bar{x} \frown \bar{y})[s^{(\sigma,\pi_0)}],$$

hence

$$\mathfrak{M} \models \bigwedge A_{t_{\pi_0}}(\bar{x} \frown \bar{y})[s^{(\sigma),\pi} \frown s^{(\sigma*,\pi_0^*}]$$

which implies

$$(\mathfrak{M}, s^{(\sigma,\pi)}) \models \bigwedge A_{t_{\pi_0}}(\bar{d}, \bar{y})[s^{(\sigma*,\pi_0^*)}].$$

But $t_{\pi_0} = t_{\pi^*_0}$, whence

$$(\mathfrak{M}, s^{(\sigma,\pi)}) \models \bigwedge A_{t_{\pi^*_0}}(\bar{d}, \bar{y})[s^{(\sigma*,\pi^*_0)}]. \quad \square$$

The following two corollaries of Lemma 3.5 show that the structure of the constituents can be considered as a syntactical counterpart to an extended back-and-forth property.

COROLLARY 3.6. Suppose $\mathfrak{M} \models C^{(\alpha)}$. Then for every $\beta < \alpha^0$ and every $\bar{a} \in {}^{\beta}M$, there is a subconstituent $C(\bar{x})(\bar{x} = (x_\xi)_{\xi<\beta})$ of $C^{(\alpha)}$ such that $\mathfrak{M} \models C(\bar{x})[\bar{a}]$.

Proof: Let $\sigma \in \Sigma(C^{(\alpha)}, \mathfrak{M})$ be a winning strategy and $\bar{a} = (a_\xi)_{\xi<\beta}$. Then there is a $\pi \in \Pi^\sigma_\beta(C^{(\alpha)}, \mathfrak{M})$ such that $\bar{a} = s^{(\sigma,\pi)}$. In fact, it is easy to check that this equality holds for every $\pi \in \Pi^\sigma_\beta(C^{(\alpha)}, \mathfrak{M})$ which satisfies the following condition:
For every $\mu < \beta$ and every $\langle t, h\rangle \in H_{II}$ such that $\mathrm{qr}(t) = \mu$ and $(a_\xi)_{\xi<\mu}$ occurs in h, $\pi(t, h) \in T^\forall$ if $t \in T^\wedge$ and $\pi(t, h) = a_\mu$ if $t \in T^\vee$.

Then, take $C_{t_\pi}(\bar{x})$ for $C(\bar{x})$. $\mathfrak{M} \models C_{t_\pi}(\bar{x})[s^{(\sigma,\pi)}]$, by Lemma 3.5; and $\bar{a} = s^{(\sigma,\pi)}$. □

COROLLARY 3.7. Suppose $\mathfrak{M} \models C^{(\alpha)}$. Then for every subconstituent of $C^{(\alpha)}$ of the form $C(\bar{x})(\bar{x} = (x_\xi)_{\xi<\beta}, \beta<\alpha^0)$, there is an $\bar{a} \in {}^{\beta}M$ such that $\mathfrak{M} \models C(\bar{x})[\bar{a}]$.

Proof: By (7) of Definition 3.1, there is a subconstituent $C'(\bar{x})$ of $C^{(\alpha)}$, determined by $t \in T^{(\beta)}$, say, such that it is structurally similar with $C(\bar{x})$ and for every $t' \in \hat{t}$, $t' \in T^\wedge$ or $t' \in T^\exists$. Let $\sigma \in \Sigma(C^{(\alpha)}, \mathfrak{M})$ be a winning strategy. Then there is a $\pi \in \Pi^\sigma_\beta(C^{(\alpha)}, \mathfrak{M})$ such that $t = t_\pi$. This equality holds for every strategy $\pi \in \Pi^\sigma_\beta(C^{(\alpha)}, \mathfrak{M})$ which satisfies the following: For every $\mu < \beta$, if $t' \in T^{(\mu)}$, then $\pi(t', h) \in T^\exists$ for all $h \in H^{t'}$.

Since σ is a winning strategy, $\mathfrak{M} \models C'(\bar{x})[s^{(\sigma,\pi)}]$, by Lemma 3.5. Since $C(\bar{x})$ is structurally similar with $C'(\bar{x})$, $\mathfrak{M} \models C(\bar{x})[s^{(\sigma,\pi)}]$, by Lemma 2.2. Thus, take $\bar{a} = s^{(\sigma,\pi)}$. □

THEOREM 3.8. For every $\alpha \leqslant \omega$, if $\mathfrak{N} \models C_1^{(\alpha)}$ and $\mathfrak{N} \models C_2^{(\alpha)}$, where $C_1^{(\alpha)}$ and $C_2^{(\alpha)}$ are of the same type, then $\models C_1^{(\alpha)} \leftrightarrow C_2^{(\alpha)}$.

Proof: It is sufficient to consider the case $\alpha = \omega$.
Suppose $\mathfrak{M} \models C_2^{(\omega)}$. We have to show that $\mathfrak{M} \models C_1^{(\omega)}$. Let $C_1^{(\omega)} = \langle T, \prec, g, S\rangle$ and $C_2^{(\omega)} = \langle T', \prec', g', S'\rangle$. We define a winning strategy $\sigma \in \Sigma(C_1^{(\omega)}, \mathfrak{M})$ by defining functions

$$\sigma^\beta : H^\beta \to T \cup M(\beta \leqslant \omega)$$

satisfying $(\sigma)_\beta$.

1. Suppose $\sigma^0, \ldots, \sigma^\beta$ have been defined, $\beta < \omega$.

(i) First we have to define $\sigma^{\beta+1}(t, h)$ for all $\langle t, h\rangle \in H_I$ such that $t \in S^\exists_\beta$, $h \in H^t$.

Let $t \in S^\exists_\beta$ be given and let an $\bar{a} \in {}^{\beta}M$ occur in an $h \in H^t$. Let t_0 be the immediate $\prec$-predecessor of t, $t_0 \in T^\wedge$, and t_1 the immediate successor of

$t, t_1 \in T^\wedge$. Suppose the following condition holds for $\langle t, h\rangle$:

$(*)_{th}$ There is a $\bar{b} \in {}^\beta N$ and a $t_0' \in T'^{(\beta)} = \{t \in T' | \mathrm{qr}(t) = \beta$ & $g'(t) = \wedge\}$ such that $\mathfrak{N} \models C_{t_0}(\bar{x})[\bar{b}]$, $\mathfrak{N} \models C_{t_0}'(\bar{x})[\bar{b}]$, and $\mathfrak{M} \models C_{t_0}'(\bar{x})[\bar{a}](\bar{x} = (x_\xi)_{\xi<\beta})$.

Since $\mathfrak{N} \models C_{t_0}(\bar{x})[\bar{b}]$, there is a $b_\beta \in N$ such that $\mathfrak{N} \models C_{t_1}(\bar{x}, x_\beta)[\bar{b} \frown b_\beta]$, by Corollary 3.7. Since $\mathfrak{N} \models C_{t_0}'(\bar{x})[\bar{b}]$, there is a $t_1' \in T'^{(\beta+1)}$ such that $t_0' \prec' t_1'$ and $\mathfrak{N} \models C_{t_1}'(\bar{x}, x_\beta)[\bar{b} \frown b_\beta]$, by Corollary 3.6. Since $\mathfrak{M} \models C_{t_0}'(\bar{x})[\bar{a}]$, there is an $a_\beta \in M$ such that $\mathfrak{M} \models C_{t_1}'(\bar{x}, x_\beta)[\bar{a} \frown a_\beta]$, by Corollary 3.7. Let a_h be a fixed element of M such that $\mathfrak{M} \models C_{t_1}'(\bar{x}, x_\beta)[\bar{a} \frown a_h]$. Define now:

$$\sigma^{\beta+1}(t, h) = \begin{cases} a_h \text{ if } (*)_{th} \text{ holds} \\ \text{an arbitrary, fixed element of } M \text{ if } (*)_{th} \text{ does} \\ \text{not hold.} \end{cases}$$

(ii) Next we have to define $\sigma^{\beta+1}(t, h)$ for all $\langle t, h\rangle \in H_I$ such that $t \in S_{\beta+1}^\vee$, $h \in H^t$.

Let $t \in S_{\beta+1}^\vee$ be given and let an $\bar{a} \in {}^{\beta+1}M$ occur in $h \in H^t$. Let t_0 be the element of T for which $t_0 \prec t$, $t_0 \in T^{(\beta)}$. Suppose the following condition holds for $\langle t, h\rangle$:

$(**)_{th}$ There is a $t_0' \in T'^{(\beta)}$ and a $\bar{b} \in {}^\beta N$ such that $\mathfrak{M} \models C_{t_0}'(\bar{x})[\bar{a}\restriction\beta]$, $\mathfrak{N} \models C_{t_0}'(\bar{x})[\bar{b}]$, and $\mathfrak{N} \models C_{t_0}(\bar{x})[\bar{b}](\bar{x} = (x_\xi)_{\xi<\beta})$.

Since $\mathfrak{M} \models C_{t_0}'(\bar{x})[\bar{a}\restriction\beta]$, there is a $t_1' \in T'^{(\beta+1)}$ such that $t_0' \prec' t_1'$ and $\mathfrak{M} \models C_{t_1}'(\bar{x}, x_\beta)[\bar{a}]$, by Corollary 3.6. Since $\mathfrak{N} \models C_{t_0}'(\bar{x})[\bar{b}]$, there is a $b_\beta \in N$ such that $\mathfrak{N} \models C_{t_1}'(\bar{x}, x_\beta)[\bar{b} \frown b_\beta]$, by Corollary 3.7. Since $\mathfrak{N} \models C_{t_0}(\bar{x})[\bar{b}]$, there is a $t_1 \in I_t$ such that $\mathfrak{N} \models C_{t_1}(\bar{x}, x_\beta)[\bar{b} \frown b_\beta]$, by Corollary 3.6. Let t_h be a fixed element of I_t such that $\mathfrak{N} \models C_{th}(\bar{x}, x_\beta)[\bar{b} \frown b_\beta]$. We define:

$$\sigma^{\beta+1}(t,h) = \begin{cases} t_h \text{ if } (**)_{th} \text{ holds} \\ \text{and arbitrary, fixed element of } I_t \text{ if } (**)_{th} \text{ does} \\ \text{not hold.} \end{cases}$$

(iii) If $\langle t, h\rangle \in H^\mu$, $\mu \leqslant \beta$, let $\sigma^{\beta+1} = \sigma^\mu(t, h)$. Clearly $(\sigma)_{\beta+1}$ holds.

2. Let $\sigma^\omega = \cup_{\beta<\omega}\sigma^\beta$. Then $(\sigma)_\omega$ holds, whence $\sigma = \sigma^\omega \in \Sigma(C_1^{(\omega)}, \mathfrak{M})$. Next we prove that the following holds for every $\beta < \omega$:

$(+)_\beta$ For every $\pi \in \Pi_\beta^\sigma(C_1^{(\omega)}, \mathfrak{M})$, there is a $\bar{b} \in {}^\beta N$ and a $t_0' \subset T'^{(\beta)}$ such that $\mathfrak{N} \vdash C_{t_\pi}(x)[\bar{b}]$, $\mathfrak{N} \models C_{t_0}'(\bar{x})[\bar{b}]$, and $\mathfrak{M} \models C_{t_0}'(\bar{x})[s^{(\sigma,\pi)}]$ (here $t_\pi = t_\pi^\sigma$).

The claim will be proved by induction on β.

1. The case $\beta = 0$ is obvious, by the assumptions; $\bar{b} = \emptyset$, $t_0' =$ the least element of T', and $t_\pi =$ the least element of T.

2. Let $\gamma = \beta + 1$, and suppose $(+)_\beta$ holds.

(a) Suppose $\pi \in \Pi_\gamma^{\sigma\exists}(C_1^{(\omega)}, \mathfrak{M})$. Let t_0 be the element of T such that $t_0 \prec t_\pi$, $t_0 \in T^{(\beta)}$. For any $\pi^* \in \Pi_{(\sigma,\pi)}^{\mathring{t}_0} \frown \Pi_\beta^\sigma(C_1^{(\omega)}, \mathfrak{M})$, $\langle t_0, h_\pi \upharpoonright \mathring{t}_0, s\rangle = \langle t_{\pi^*}, h_{\pi^*}, s_{\pi^*}\rangle$ for some $s \in I_{t_0}^{\Delta}$. Thus by induction hypothesis, there is a $\bar{b} \in {}^\beta N$ and a $t_0' \in T'^{(\beta)}$ such that $\mathfrak{N} \vDash C_{t_{\pi*}}(\bar{x})[\bar{b}]$, $\mathfrak{N} \vDash C_{t_0}'(\bar{x})[\bar{b}]$, and $\mathfrak{M} \vDash C_{t_0}'(\bar{x})[s^{(\sigma,\pi^*)}]$. Thus the condition $(*)_{th}$ holds, where $t_0 \prec t \prec t_\pi (t \in S_\beta^\exists)$, $h = h_{\pi*} \cup \{\langle t_0, t\rangle\} \in H^t$, and $s^{(\sigma,\pi^*)}$ occurs in h. By 1(i), above, there is a $b_\beta \in N$ and a $t_1' \in T'^{(\gamma)}$ such that

$$\mathfrak{N} \vDash C_{t_\pi}(\bar{x}, x_\beta)[\bar{b} \frown b_\beta],\ \mathfrak{N} \vDash C_{t_1}'(\bar{x}, x_\beta)[\bar{b} \frown b_\beta],$$

and $\mathfrak{M} \vDash C_{t_1}'(\bar{x}, x_\beta)[s^{(\sigma,\pi^*)} \frown \sigma^{\beta+1}(t, h)]$.

But $\sigma^{\beta+1}(t, h) = \sigma(t, h)$, whence $s^{(\sigma,\pi^*)} \frown \sigma^{\beta+1}(t, h) = s^{(\sigma,\pi^*)} \frown \sigma(t, h) = s^{(\sigma,\pi)}$, by the definition of $s^{(\sigma,\pi)}$. Thus $\mathfrak{M} \vDash C_{t_1}'(\bar{x}, x_\beta)[s^{(\sigma,\pi)}]$, whence $\bar{b} \frown b_\beta$ and t_1' are the required elements of ${}^\gamma N$ and $T'^{(\gamma)}$, respectively.

(b) Suppose $\pi \in \Pi_\gamma^{\sigma\forall}(C_1^{(\omega)}, \mathfrak{M})$. Let t_0 be the element of $T^{(\beta)}$ such that $t_0 \prec t_\pi$. Let $s^{(\sigma,\pi)} = (a_\xi)_{\xi \leqslant \beta}$. Consider again a $\pi^* \in \Pi_{(\sigma,\pi)}^{\mathring{t}_0} \cap \Pi_\beta^\sigma(C_1^{(\omega)}, \mathfrak{M})$ such that $\langle t_0, h_\pi \upharpoonright \mathring{t}_0, s\rangle = \langle t_{\pi*}, h_{\pi*}, s_{\pi*}\rangle$ for a $s \in I_{t_0}^{\Delta}$. By induction hypothesis, there is a $\bar{b} \in {}^\beta N$ and a $t_0' \in T'^{(\beta)}$ such that $\mathfrak{N} \vDash C_{t_{\pi*}}(\bar{x})[\bar{b}]$, $\mathfrak{N} \vDash C_{t_0}'(\bar{x})[\bar{b}]$, and $\mathfrak{M} \vDash C_{t_0}'(\bar{x})[s^{(\sigma,\pi^*)}]$. Thus the condition $(**)_{th}$ holds, where t is the immediate predecessor of $t_\pi (t \in S_{\beta+1}^\forall)$ and $h = h_{\pi*} \cup \{\langle t_0, t_1\rangle, \langle t_1, a_\beta\rangle\}$ (where $t_1 \in T^\forall$), since $s^{(\sigma,\pi)}$ occurs in h and $s^{(\sigma,\pi^*)} = s^{(\sigma,\pi)} \upharpoonright \beta$. Then by 1(ii), above, there is a $b_\beta \in N$ and a $t_1' \in T'^{(\gamma)}$ such that $\mathfrak{M} \vDash C_{t_1}'(\bar{x}, x_\beta)[s^{(\sigma,\pi)}]$, $\mathfrak{N} \vDash C_{t_1}'(\bar{x}, x_\beta)[\bar{b} \frown b_\beta]$, and $\mathfrak{N} \vDash C_{th}(\bar{x}, x_\beta)[\bar{b} \frown b_\beta]$, where $t_h = \sigma^{\beta+1}(t, h) = \sigma(t, h) = t_\pi$. Hence $\bar{b} \frown b_\beta$ and t_1' are the required elements of ${}^\gamma N$ and $T'^{(\gamma)}$, respectively.

Thus $(+)_\beta$ holds for all $\beta < \omega$.

It follows that σ is a winning strategy. By Theorem 3.4, it is sufficient to show that for all $\beta < \omega$ and all $\pi \in \Pi_\beta^\sigma(C_1^{(\omega)}, \mathfrak{M})$, $\mathfrak{M} \vDash \bigwedge A_{t_\pi}(\bar{x})[s^{(\sigma,\pi)}]$.

So, let $\beta < \omega$, $\pi \in \Pi_\beta^\sigma(C_1^{(\omega)}, \mathfrak{M})$, and $\bar{b}$, t_0' elements of ${}^\beta N$, $T'^{(\beta)}$, respectively, mentioned in $(+)_\beta$. Since $\mathfrak{N} \vDash C_{t_\pi}(\bar{x})[\bar{b}]$ and $\mathfrak{N} \vDash C_{t_0}'(\bar{x})[\bar{b}]$, it follows that $\mathfrak{N} \vDash \bigwedge A_{t_\pi}(\bar{x})[\bar{b}]$ and $\mathfrak{N} \vDash \bigwedge A_{t_0}'(\bar{x})[\bar{b}]$. Thus by Lemma 3.1, $\vDash \bigwedge A_{t_\pi}(\bar{x}) \leftrightarrow \bigwedge A_{t_0}'(\bar{x})$. Since $\mathfrak{M} \vDash C_{t_0}'(\bar{x})[s^{(\sigma,\pi)}]$, it follows that $\mathfrak{M} \vDash \bigwedge A_{t_0}'(\bar{x})[s^{(\sigma,\pi)}]$, whence $\mathfrak{M} \vDash \bigwedge A_{t_\pi}(\bar{x})[s^{(\sigma,\pi)}]$. □

It is not known, at present, whether Theorem 3.8 holds for $\alpha > \omega$. It seems that what would be needed to prove $(+)_\beta$ for a limit ordinal β

is the following 'extension' property of an arbitrary constituent: Let $\langle T, \prec \rangle$ be the tree corresponding to given constituent $C^{(\alpha)}$, $\alpha > \omega$, and $\mathfrak{M}$ a model. If $t_0 \prec \ldots \prec t_\xi \prec \ldots \prec t_\beta (\beta < \alpha^0)$ is any increasing sequence of elements of T such that $t_\xi \in T^{(\xi)}$ and β is a limit ordinal, and if $\bar{a}_0 \subseteq \ldots \subseteq \bar{a}_\xi \subseteq \ldots \subseteq \bar{a}_\beta$ is any increasing sequence such that $\bar{a}_\xi \in {}^\xi M$ and

$$\mathfrak{M} \models C_{t_\xi}(\bar{x}_\xi)[\bar{a}_\xi](\bar{x}_\xi = (x_\mu)_{\mu<\xi}) \text{ for all } \xi < \beta,$$

then

$$\mathfrak{M} \models C_{t_\beta}(\bar{x}_\beta)[\bar{a}_\beta](\bar{x}_\beta = (x_\mu)_{\mu<\beta}).$$

Although it has been conjectured that this property holds, there is no proof for it, at present, neither a counterexample.

4. A CONSTRUCTION

Let $\mathfrak{M}$ be a model of type τ and $\Delta_{\mathfrak{M}}$ the diagram of $\mathfrak{M}$. It is supposed that for each member of M a new individual constant is introduced such that they are different from the constants of τ. Let $c_\xi(\xi < \delta)$ be the individual constants of $L(\tau)$ and $K_{\mathfrak{M}}$ the set of the new constants introduced. For simplicity, we use the new constants autonymously, that is, the name of an $a \in M$ is $a \in K_{\mathfrak{M}}$. Let $\Delta^+_{\mathfrak{M}}$ be the positive diagram of $\mathfrak{M}$, i.e., the set of all atomic sentences of $\Delta_{\mathfrak{M}}$ which hold in $(\mathfrak{M}, a)_{a \in M}$, and $\Delta^-_{\mathfrak{M}}$ the negative diagram of $\mathfrak{M}$, i.e., the set of the negated sentences of $\Delta_{\mathfrak{M}}$. If $\bar{a} = (a_\xi)_{\xi<\mu}$ is any sequence of individual constants of $K_{\mathfrak{M}}$ or $L(\tau)$, we now denote by $A_{\mathfrak{M}}(\bar{a})$ the set of all such members of $\Delta_{\mathfrak{M}}$ which have been formed by using the a_ξ.

THEOREM 4.1. For every ordinal α and every model $\mathfrak{M}$ of type τ there is a constituent $C^{(\alpha)}_{\mathfrak{M}}$ of type τ in $N_{\infty\alpha^*}(\tau)$ which is true in $\mathfrak{M}$.

Proof: For brevity, we shall denote $\langle \exists, x \rangle$, $\langle \forall, x \rangle$ by $\exists x$, $\forall x$, respectively, where $x \in \{x_\mu | \mu < \alpha\} \subseteq V$, V an appropriate set of variables. Let

$$\Delta_0 = \{\bar{c}\} \cup \Delta_{\mathfrak{M}} \cup \{\vee, \wedge\} \cup (K_{\mathfrak{M}} \times \{\wedge\}) \cup (K_{\mathfrak{M}} \times \{\exists x_\mu | \mu < \alpha\}) \cup \{\forall x_\mu | \mu < \alpha\}$$

where $\bar{c} = (c_\xi)_{\xi<\delta}$. Consider the tree $\langle T', \prec' \rangle$ where $T' = \cup_{\beta<\lambda_0} {}^\beta\Delta_0$, λ_0 is the least limit ordinal $\geq \alpha$, and $\prec'$ is the partial ordering on T' defined by inclusion: $t_1 \prec' t_2$ iff t_2 extends t_1.

From $\langle T', \prec' \rangle$ we construct its subtree $\langle T, \prec \rangle$, $T \subset T'$, $\prec\, = \,\prec' \restriction T$, by induction on levels U_β of T, for all $\beta < \alpha$ if α is a limit ordinal, and all $\alpha \leqslant \omega \cdot \lambda + 3(n + 1)$ if α is a successor ordinal, $\alpha = \omega \cdot \lambda + n (n \neq 0)$. Let $(\prec)_\beta$ be the following property of ordinals:

> $(\prec)_\beta$ For every ordinals ξ, μ and every $t \in T | \beta$ such that $\xi < \mu < \beta$, $t \in U_\xi$, and t is not of the form $t_0 \frown \varphi$, where $\varphi \in \Delta_{\mathfrak{M}}$, there is an $s \in U_\mu$ such that $t \prec s$.

We say that an $\bar{a} = (a_\xi)_{\xi < \mu} \in {}^\mu K_{\mathfrak{M}}$ *occurs in* a $t \in T$ if $\bar{a}$ is the maximal sequence such that

$$\langle \nu_0, \langle a_0, q_0 \rangle\rangle, \ldots, \langle \nu_\xi, \langle a_\xi, q_\xi \rangle\rangle, \ldots \in t \; (\xi < \mu),$$

where $q_\xi = \exists x_\xi$ or $q_\xi = \wedge$ for every $\xi < \mu$, and $\nu_0 < \ldots < \nu_\xi \ldots$

The tree $\langle T, \prec \rangle$ will be constructed as follows. We shall define first the levels U_0 and U_1.

1. $U_0 = \{\langle \bar{c}, \wedge \rangle\}$.

2. If $\alpha > 0$, $U_1 = \{\langle \bar{c}, \wedge \rangle \frown \varphi | \varphi \in A_{\mathfrak{M}}(\bar{c})\}$
$\cup \{\langle \bar{c}, \wedge \rangle \frown \langle a, \exists x_0 \rangle | a \in K_{\mathfrak{M}}\} \cup \{\langle \bar{c}, \wedge \rangle \frown \forall x_0\}$.

If $\alpha = 0$, $U_1 = \{\langle \bar{c}, \wedge \rangle \frown \varphi | \varphi \in A_{\mathfrak{M}}(\bar{c})\}$.

It is clear that $(\prec)_\xi$ holds for $\xi \leqslant 2$.

3. Suppose $\beta = \gamma + 1$ with γ a successor ordinal, and $U_0, \ldots, U_\gamma$ have been constructed so that $(\prec)_\xi$ holds for all $\xi \leqslant \beta$, and the last member of a $t \in U_\gamma$ is some of the following forms: φ (where $\varphi \in \Delta_{\mathfrak{M}}$), $\wedge$, $\vee$, $\langle a, \wedge \rangle$ (where $a \in K_{\mathfrak{M}}$), $\langle a, \exists x_\mu \rangle$ (for some $\mu < \beta$), $\forall x_\mu$ (for some $\mu < \beta$). For each $t \in U_\gamma$ we shall define the set I_t of its immediate $\prec$-successors as follows. I_t will be a subset of the set of all immediate $\prec'$-successors of t in T'.

(i) If $t = t_0 \frown \varphi$, where $\varphi \in \Delta^+_{\mathfrak{M}}$, $I_t = \emptyset$.

(ii) If $t = t_0 \frown \neg\varphi$, where $\neg\varphi \in \Delta^-_{\mathfrak{M}}$, $I_t = \{t \frown \varphi\}$.

(iii) If $t = t_0 \frown \wedge$ and an $\bar{a} = (a_\xi)_{\xi < \mu+1} \in {}^{\mu+1}K_{\mathfrak{M}}$ occurs in t_0, then:

(a) if $\mu + 1 < \alpha$,

$I_t = \{t \frown \varphi | \varphi \in A_{\mathfrak{M}}(\bar{c} \frown \bar{a})\} \cup \{t \frown \langle a, \exists x_{\mu+1} \rangle | a \in K_{\mathfrak{M}}\}$
$\cup \{t \frown \forall x_{\mu+1}\}$,

(b) if $\mu + 1 = \alpha$,

$I_t = \{t \frown \varphi | \varphi \in A_{\mathfrak{M}}(\bar{c} \frown \bar{a})\}$.

(iv) If $t = t_0 \frown \vee$, $I_t = \{t \frown \langle a, \vee \rangle | a \in K_{\mathfrak{M}}\}$.

(v) If $t = t_0 \frown \langle a_\mu, \wedge \rangle$, where $a_\mu \in K_{\mathfrak{M}}$, and an $\bar{a} = (a_\xi)_{\xi < \mu+1} \in {}^{\mu+1}K_{\mathfrak{M}}$ occurs in t (that is, $\bar{a} \restriction \mu$ occurs in t_0), then I_t is defined as in (iii).

(vi) If $t = t_0 \frown \langle a, \exists x_\mu \rangle$, where $a \in K_{\mathfrak{M}}$, then $I_t = \{t \frown \bigvee\}$.
(vii) If $t = t_0 \frown \forall x_\mu$, then $I_t = \{t \frown \bigvee\}$.
We let $\cup_{t \in U_\gamma} I_t$ be the βth level U_β of T.
It is obvious that $(\prec)_{\beta+1}$ holds.

4. Suppose $\beta \neq 0$ is a limit ordinal and $T|\beta$ has been constructed so that for every $\xi < \beta$, $(\prec)_\xi$ holds. Then $(\prec)_\beta$ holds since $T|\beta = \cup_{\xi<\beta} U_\xi$. It follows that for every $t \in T|\beta$ such that t is not of the form $t_0 \frown \varphi$, $\varphi \in \Delta_{\mathfrak{M}}$, there is an β-branch of $T|\beta$ extending t. For suppose that such a t is given, $t \in U_\xi$, $\xi < \beta$. By $(\prec)_\beta$, there is a sequence

$$\xi < \xi_0 < \ldots < \xi_\mu < \ldots (\mu < \gamma \leqslant \beta)$$

of ordinals such that $\cup_{\mu<\gamma} \xi_\mu = \beta$, and a sequence

$$t \prec t_0 \prec \ldots \prec t_\mu \prec \ldots (\mu < \gamma)$$

of elements of $T|\beta$ such that $t_\mu \in U_\mu$. Thus we can define:

$$U_\beta = \{B | B \text{ is a } \beta\text{-branch of } T|\beta\}.$$

It is obvious that $(\prec)_{\beta+1}$ holds and in every $t \in U_\beta$ a sequence $\bar{a} \in {}^\beta K_{\mathfrak{M}}$ occurs.

5. Let $\beta = \gamma + 1$ where $\gamma \neq 0$ is a limit ordinal. Suppose $(\prec)_\xi$ holds for all $\xi \leqslant \beta$. For each $t \in U_\gamma$ define I_t as follows:
If an $\bar{a} = (a_\xi)_{\xi<\gamma} \in {}^\gamma K_{\mathfrak{M}}$ occurs in t, then

$$I_t = \{t \frown \varphi | \varphi \in A_{\mathfrak{M}}(\bar{c} \frown \bar{a})\} \cup \{t \frown \langle a, \exists x_\gamma \rangle | a \in K_{\mathfrak{M}}\} \\ \cup \{t \frown \forall x_\gamma\}.$$

Let $U_\beta = \cup_{t \in U_\gamma} I_t$.
It is again obvious that $(\prec)_{\beta+1}$ holds.

If α is a successor ordinal, $\alpha = \omega\cdot\lambda + n (n \neq 0)$, it can be seen by looking at all the different kinds of extensions any element of $U_{\omega\cdot\lambda}$ can have according to rules 3,5, that, by 3(iii), (b), the construction comes to an end at the level $U_{\omega\cdot\lambda+(3n+3)}$ or at the level $U_{\omega\cdot\lambda+(3n+2)}$ (depending on whether in case 3(iii), (b), $A_{\mathfrak{M}}(\bar{c} \frown \bar{a})$ includes negated atomic sentences or not). If α is a limit ordinal, it is clear that T will be of length α. Thus, in any case we can define:

$$T = \cup_{\beta<(\alpha^*)^0} U_\beta.$$

Next, we define a mapping $g : T \to S$ as follows:

(a) If $t = t_0 \frown \varphi$, where $\varphi \in \Delta^+_{\mathfrak{M}}$, and a sequence $\bar{a} = (a_\xi)_{\xi<\mu} \in {}^\mu K_{\mathfrak{M}}$

occurs in t, then $g(t) = \varphi(x_0/a_0, \ldots, x_\xi/a_\xi, \ldots)$ $(\xi < \mu)$. Here the substitution x_ξ/a_ξ is considered 'vacuous' if a_ξ does not appear in φ.

(b) $g(t_0 \frown \neg\varphi) = \neg(\neg\varphi \in \Delta^{-}_{\mathfrak{M}})$.

(c) $g(t_0 \frown \bigwedge) = \bigwedge$.

(d) $g(t_0 \frown \bigvee) = \bigvee$.

(e) $g(t_0 \frown \langle a, \bigwedge\rangle) = \bigwedge$ $(a \in K_{\mathfrak{M}})$.

(f) $g(t_0 \frown \langle a, \exists x_\mu\rangle) = \langle\exists, x_\mu\rangle(a \in K_{\mathfrak{M}}, 0 \leqslant \mu < \alpha)$.

(g) $g(t_0 \frown \forall x_\mu) = \langle\forall, x_\mu\rangle(0 \leqslant \mu < \alpha)$.

(h) If $t \in U_\beta$ where β is a limit ordinal, $g(t) = \bigwedge$.

It is a routine matter, although tedious, to show that the conditions of Definition 3.1 are satisfied, so that $\langle T, \prec, g, S\rangle$ really is a constituent of depth α and of type τ in $N_{\infty\alpha^*(\tau)}$. So, it will not be done here. We shall only indicate how to verify the most difficult condition (7) of Definition 3.1. First, it is a straightforward consequence of the construction of T that given a sequence $\bar{a} \in {}^\beta K_{\mathfrak{M}}(\beta < \alpha)$, there is in each $S \in T^{(\beta)}/\sim$ a unique t such that $\bar{a}$ occurs in t. Conversely, given a $t \in T^{(\beta)}$, there is a unique $\bar{a} \in {}^\beta K_{\mathfrak{M}}$ which occurs in t. Now, a mapping h needed to verify the structural similarity of the subformulas determined by a $t \in S_1$, $f(t) \in S_2$ of (7) is defined as follows. Let T_1, T_2 be the subtrees of T determined by $t, f(t)$, respectively. If the tree-rank of a $t_1 \in T_1$ is a successor ordinal, then

> $h(t_1)$ = the unique member t_2 of T_2 such that the same sequence occurs in t_1 and t_2 and the last members of t_1 and t_2 are the same.

If the tree-rank of a $t_1 \in T_1$ is a limit ordinal, then

> $h(t_1)$ = the unique member t_2 of T_2 such that the same sequence occurs in t_1 and t_2.

We shall prove that the constructed constituent $C^{(\alpha)}_{\mathfrak{M}} = \langle T, \prec, g, S\rangle$ is true in $\mathfrak{M}$.

We define functions $\sigma^\beta : H^\beta \to T \cup M(\beta \leqslant \alpha)$ satisfying the conditions $(\sigma)_\beta$.

1. Suppose $\sigma^0, \ldots, \sigma^\beta$ have been defined $(\beta < \alpha)$.

(i) First we have to define $\sigma^{\beta+1}(t, h)$ for all $\langle t, h\rangle \in H_I$ such that $t \in S^{\exists}_\beta$, $h \in H^t$. Let $t \in S^{\exists}_\beta$. If $t_0 \in T^\wedge$ is the immediate $\prec$-predecessor of t, then $t = t_0 \frown \langle a_t, \exists x_\beta\rangle$ for a unique $a_t \in K_{\mathfrak{M}}$. Then we define:

$$\sigma^{\beta+1}(t, h) = a_t(a_t \in M) \text{ for all } h \in H^t.$$

(ii) Next, we define $\sigma^{\beta+1}(t, h)$ for all $\langle t, h\rangle \in H_I$ such that $t \in S^{\forall}_{\beta+1}$, $h \in H^t$. Let $t \in S^{\forall}_{\beta+1}$. If t_0 is the immediate $\prec$-predecessor of t, then $t_0 \in T^{\forall}$. Thus given $\langle t, h\rangle$ as mentioned, the last element of h is $\langle t_0, a_{th}\rangle$ for a unique $a_{th} \in M$. Define:

$$\sigma^{\beta+1}(t, h) = t \frown \langle a_{th}, \wedge\rangle (a_{th} \in K_{\mathfrak{M}}).$$

Let

$$\sigma^{\beta+1} = \sigma^{\beta} \cup \{\langle t, h, s\rangle | t \in S^{\exists}_{\beta} \;\&\; h \in H^t \;\&\; s = a_t\}$$
$$\cup \{\langle t, h, s\rangle | t \in S^{\forall}_{\beta+1} \;\&\; h \in H^t \;\&\; s = t \frown \langle a_{th}, \wedge\rangle\}.$$

Since $(\sigma)_{\mu}$ holds for all $\mu \leqslant \beta$, so also $(\sigma)_{\beta+1}$ holds.

2. Suppose β is a limit ordinal ($\beta \leqslant \alpha$) and $\sigma^0, \ldots, \sigma^{\mu}, \ldots (\mu < \beta)$ have been defined so that for every $\mu < \beta$, $(\sigma)_{\mu}$ holds. Then define:

$\sigma^{\beta} = \bigcup_{\mu<\beta}\sigma^{\mu}$. Since $H^{\beta} = \bigcup_{\mu<\beta}H^{\mu}$, it is clear that also $(\sigma)_{\beta}$ holds.

3. Let $\sigma = \sigma^{\alpha}$. Then $\sigma \in \Sigma(C^{(\alpha)}_{\mathfrak{M}}, \mathfrak{M})$.

To show that σ is a winning strategy, we shall first prove that the following condition holds for all $\gamma < \alpha^0$:

$(+)_{\gamma}$ For every $\pi \in \Pi^{\sigma}_{\gamma}(C^{(\alpha)}_{\mathfrak{M}}, \mathfrak{M})$, $s^{(\sigma,\pi)}$ occurs in t_{π}.

The claim will be proved by induction on γ. Let $\gamma < \alpha^0$ and suppose $(+)_{\mu}$ holds for all $\mu < \gamma$.

1. Suppose $\gamma = \beta + 1$.

(a) Let $\pi \in \Pi^{\sigma\exists}_{\beta+1}(C^{(\alpha)}_{\mathfrak{M}}, \mathfrak{M})$. Since $s^{(\sigma,\pi)} = (a_{\xi})_{\xi<\beta+1}$ occurs in h_{π}, it follows that $s^{(\sigma,\pi)} \restriction \beta$ occurs in $h_{\pi} \restriction \mathring{t}_0$, where $t_0 \in T^{(\beta)}$, $t_0 \prec t_{\pi}$. But for any $\pi^* \in \Pi^{t_0}_{(\sigma,\pi)} \cap \Pi^{\sigma}_{\beta}(C^{(\alpha)}_{\mathfrak{M}}, \mathfrak{M})$, $\langle t_0, h_{\pi} \restriction t_0, s\rangle = \langle t_{\pi*}, h_{\pi*}, s_{\pi*}\rangle$ for some $s \in I^{\Delta}_{t_0}$. Thus by induction hypothesis, $s^{(\sigma,\beta)} \restriction \beta$ occurs in $t_{\pi}* = t_0$. Let t be the element of T such that $t \in S^{\exists}_{\beta} \cap I_{t_0}$ and $t_0 \prec t \prec t_{\pi}$.

By the definition of $\rho_{(\sigma,\pi)}$ and of $s^{(\sigma,\pi)}$, $h_{\pi} = h' \cup \{\langle t, a_{\beta}\rangle\}$ where $h' = h_{\pi} \restriction \mathring{t}_0 \cup \{\langle t_0, t\rangle\}$, and $\langle t, h', a_{\beta}\rangle \in \rho_{(\sigma,\pi)}$, whence $\sigma(t, h') = a_{\beta}$. But by the definition of $\sigma^{\beta+1}$ (1(i), above), $\sigma^{\beta+1}(t, h') = a_t$, whence $\sigma(t, h') = a_t$, and $a_{\beta} = a_t$.

Thus, because $s^{(\sigma,\pi)} \restriction \beta$ occurs in t_0 and $t = t_0 \frown \langle a_{\beta}, \exists x_{\beta}\rangle$, also $s^{(\sigma,\pi)}$ occurs in t, whence it occurs in t_{π}.

(b) Let $\pi \in \Pi^{\sigma\forall}_{\beta+1}(C^{(\alpha)}_{\mathfrak{M}}, \mathfrak{M})$. As in (a), it can be shown that $s^{(\sigma,\beta)} \restriction \beta$ occurs in the immediate predecessor t of $t_{\pi} (t \in S^{\forall}_{\beta+1})$.

By the definition of $\rho_{(\sigma,\pi)}$, $\langle t_0, h_0, a_{\beta}\rangle \in \rho_{(\sigma,\pi)}$ where $t_0 \in T^{\forall}$ is the immediate predecessor of t and $h_0 \in H^{t_0}$. Thus the last element of h in

$\langle t, h, t_\pi \rangle \in \rho_{(\sigma,\pi)}$ is $\langle t_0, a_\beta \rangle$, whence by the definition of $\sigma^{\beta+1}$ (1(ii), above), $\sigma^{\beta+1}(t, h) = t \frown \langle a_\beta, \wedge \rangle$, hence $\sigma(t, h) = t \frown \langle a_\beta, \wedge \rangle$. On the other hand, $\sigma(t, h) = t_\pi$, whence $t_\pi = t \frown \langle a_\beta, \wedge \rangle$. Thus $s^{(\sigma,\pi)}$ occurs in t_π.

2. Suppose γ is a limit ordinal, $\gamma < \alpha^0$, and $\pi \in \Pi^\sigma_\gamma(C^{(\alpha)}_{\mathfrak{M}}, \mathfrak{M})$. Consider the sequence

$$t_0 \prec \ldots \prec t_\xi \prec \ldots \prec t_\pi (\xi < \gamma)$$

such that for each $\xi < \gamma$, $t_\xi \in T^{(\xi)}$, and $\langle t_\xi, h_\xi, s_\xi \rangle \in \rho_{(\sigma,\pi)}$ with the relevant h_ξ, s_ξ. Since $s^{(\sigma,\pi)}$ occurs in h_π, $s^{(\sigma,\pi)} \upharpoonright \xi$ occurs in $h_\xi = h_\pi \upharpoonright \mathring{t}_\xi$. But, as noticed before, $\langle t_\xi, h_\xi, s \rangle = \langle t_{\pi*}, h_{\pi*}, s_{\pi*} \rangle$ for some $\pi^* \in \Pi^\sigma_\xi(C^{(\alpha)}_{\mathfrak{M}}, \mathfrak{M})$ if $s \in I^\Delta_{t_\xi}$. Thus by induction hypothesis, $s^{(\sigma,\pi)} \upharpoonright \xi$ occurs in t_ξ.

On the other hand, since the t_ξ are increasing sequences of elements of Δ_0 and γ is a limit ordinal, $t_\pi = \bigcup_{\xi<\gamma} t_\xi$, whence $s^{(\sigma,\pi)}$ occurs in t_π.

Thus $(+)_\gamma$ holds for every $\gamma < \alpha^0$.

We can show now that σ is a winning strategy. By Theorem 3.4, it is sufficient to show that for all $\beta < \alpha^0$ and all $\pi \in \Pi^\sigma_\beta(C^{(\alpha)}_{\mathfrak{M}}, \mathfrak{M})$, $\mathfrak{M} \models \bigwedge A_{t_\pi}(\bar{c}, \bar{x})[s^{(\sigma,\pi)}]$.

Let $\beta < \alpha^0$ and $\pi \in \Pi^\sigma_\beta(C^{(\alpha)}_{\mathfrak{M}}, \mathfrak{M})$. By the previous claim, $s^{(\sigma,\pi)}$ occurs in t_π. Let $s^{(\sigma,\pi)} = (a_\xi)_{\xi<\beta}$.

If t_π is of the form $t_0 \frown \wedge$, then by the case 3(iii) of the construction of T, the case (a) of the definition of g (see above), and by the fact that $s^{(\sigma,\pi)}$ occurs in t_0, $\mathfrak{M} \models A_{t_\pi}(\bar{c}, \bar{x})[s^{(\sigma,\pi)}]$.

If t_π is of the form $t_0 \frown \langle a, \wedge \rangle (a \in K_{\mathfrak{M}})$, then $a = a_{\beta-1}$, since $s^{(\sigma,\pi)}$ occurs in t_π. Then by the case (v) of the construction of T and the case (a) of the definition of g, the conclusion follows.

If β is a limit ordinal, then by the case (5) of the construction of T and (a) again, and by the fact that $s^{(\sigma,\pi)}$ occurs in t_π, the conclusion follows. □

Let the power λ of $\mathfrak{M}$ be infinite. If $\alpha \leqslant \lambda$ and the cardinality of τ is μ, then $C^{(\alpha)}_{\mathfrak{M}} \in N_{\kappa^+\alpha^*}(\tau)$, where $\kappa = \max\{\lambda, \mu\}$.

Suppose the constructed constituent $C^{(\alpha)}_{\mathfrak{M}} \in N_{\kappa\alpha^*}(\tau)$. Then it is possible that there are constituents of the same depth and type which belong to languages $N_{\eta\alpha^*}(\tau)$ with $\eta < \kappa$ and which are true in $\mathfrak{M}$. But at least in the case $\alpha \leqslant \omega$ they are equivalent with $C^{(\alpha)}_{\mathfrak{M}}$ by Theorem 3.8:

THEOREM 4.2. For every $\alpha \leqslant \omega$, $C^{(\alpha)}_{\mathfrak{M}}$ is the unique constituent of depth α and type τ, up to logical equivalence, which is true in $\mathfrak{M}$.

THEOREM 4.3. $C^{(\alpha)}_{\mathfrak{M}}$ has no models of power $< \min\{\alpha, |M|\}$.

Proof: Let $\mathfrak{N}$ be a model for $L(\tau)$ of power $\kappa < \min\{\alpha, |M|\}$ and $\bar{b} = (b_\beta)_{\beta<\kappa}$ an enumeration of N. Then a strategy π satisfying the following conditions is a winning strategy in $\Pi(C_{\mathfrak{M}}^{(\alpha)}, \mathfrak{N})$:

(i) For every $t \in T^\wedge$ and every $h \in H^t$, if $\mathrm{qr}(t) = \beta < \kappa$, then $\pi(t, h) = t \frown \forall x_\beta$.

(ii) For every $t \in T^\forall$ and every $h \in H^t$, if $\mathrm{qr}(t) = \beta < \kappa$, then $\pi(t, h) = b_\beta$.

(iii) If $t \in T^{(\kappa)}$ and $\bar{b}$ occurs in an $h \in H^t$, then $\pi(t, h) = t'$ such that $g(t') = \langle \exists, x_\kappa \rangle$ and a sequence $\bar{a} = (a_\xi)_{\xi \leqslant \kappa} \in {}^{\kappa+1}M$ occurs in t' for which $a_\xi \neq a_\eta$ if $\xi \neq \eta (\xi, \eta \leqslant \kappa)$.

(iv) If $t \in T^{(\kappa+1)}$ and $\bar{b} \frown b_\beta$ occurs in an $h \in H^t$, then $\pi(t, h) = t \frown \neg(a_\kappa = a_\beta)$. □

5. ON EXTENDED NOTIONS OF PARTIAL ISOMORPHISM

We shall define two generalized notions of partial isomorphism which most naturally seem to correspond to the structure of the constituents and to the game-theoretical semantics. We shall also extend slightly the notions of element type and λ-saturated model to obtain an approximation theorem, analogous to Theorem 3.4 in [4].

Let $x_0, \ldots, x_\mu, \ldots$ ($\mu \in \mathrm{Ord}$) be a fixed ordering of the variables of L, $\bar{x} = (x_\mu)_{\mu<\beta}$, and $\bar{a} \in {}^\beta M$, $\bar{b} \in {}^\beta N$. We denote

$$(\mathfrak{M}, \bar{a}) \equiv (\mathfrak{N}, \bar{b})(\Delta(\bar{x}))$$

iff for every $\varphi \in \Delta(\bar{x})$, $\mathfrak{M} \models \varphi[\bar{a}]$ iff $\mathfrak{N} \models \varphi[\bar{b}]$, where each $x_\mu (\mu < \beta)$ is interpreted as a_μ, b_μ in $\mathfrak{M}$ and $\mathfrak{N}$, respectively. If β is a limit ordinal, then $(\mathfrak{M}, \bar{a}) \equiv (\mathfrak{N}, \bar{b})(\Delta(\bar{x}))$ iff for every $\mu < \beta$, $(\mathfrak{M}, \bar{a}\restriction\mu) \equiv (\mathfrak{N}, \bar{b}\restriction\mu)(\Delta(\bar{x}\restriction\mu))$, by the definition.

DEFINITION 5.1. $\mathfrak{M} \approx_\alpha \mathfrak{N}$ iff there is a sequence $F_0 \subseteq \ldots \subseteq F_\beta \subseteq \ldots$ $(\beta < \alpha^0)$ of sets of mappings such that

(i) Each $f \in F_\beta$ is a mapping from range($\bar{a}$) onto range($\bar{b}$) for some $\bar{a} \in {}^\beta M$, $\bar{b} \in {}^\beta N$, such that $f(a_\mu) = b_\mu$ for all $\mu < \beta$ and $(\mathfrak{M}, \bar{a}) \equiv (\mathfrak{N}, \bar{b})(\Delta(\bar{x}))$.

(ii) If $f \in F_\beta$, $\beta < \alpha$, and $a' \in M$, there is a $g \in F_{\beta+1}$ such that $f \subseteq g$ and $a' \in \mathrm{dom}(g)$.

(iii) If $f \in F_\beta$, $\beta < \alpha$, and $b' \in N$, there is a $g \in F_{\beta+1}$ such that $f \subseteq g$ and $b' \in \mathrm{range}(g)$.

(iv) If $f_0 \subseteq \ldots \subseteq f_\xi \subseteq \ldots$ ($\xi < \beta < \alpha$), where $f_0 \in F_0, \ldots$, $f_\xi \in F_\xi, \ldots$, then there is an $f \in F_\beta$ which extends every f_ξ.

In this definition, the 'empty' mapping is considered as being the unique member of F_0 such that $\mathfrak{M} \equiv \mathfrak{N}(\Delta)$.

THEOREM 5.1. If $\mathfrak{M} \approx_\alpha \mathfrak{N}$, then $\mathfrak{N} \vDash C^{(\alpha)}_{\mathfrak{M}}$.

Proof: Let $C^{(\alpha)}_{\mathfrak{M}} = \langle T, \prec, g, S \rangle$. We shall define a winning strategy $\sigma \in \Sigma(C^{(\alpha)}_{\mathfrak{M}}, \mathfrak{N})$ in the game $G(C^{(\alpha)}_{\mathfrak{M}}, \mathfrak{N})$. Notice that all the relevant notions defined in Sections 2–3 will now be related, in an obvious way, to $\mathfrak{N}$ (instead of $\mathfrak{M}$, as in Sections 2–3). We can also assume that the bound variables of $C^{(\alpha)}_{\mathfrak{M}}$ are $x_0, \ldots, x_\beta, \ldots$ ($\beta < \alpha$).

We shall define functions $\sigma^\beta : H^\beta \to T \cup N$ ($\beta \leqslant \alpha$) satisfying $(\sigma)_\beta$.

1. Suppose $\sigma^0, \ldots, \sigma^\beta$ have been defined.

(i) First define $\sigma^{\beta+1}(t, h)$ for all $(t, h) \in H_I$ such that $t \in S^{\exists}_\beta$ and $h \in H^t$. Let $\bar{a} = (a_\xi)_{\xi \leqslant \beta} \in {}^{\beta+1}M$ occur in a $t \in S^{\exists}_\beta$ and $\bar{b} \in {}^\beta N$ occur in an $h \in H^t$.

Suppose there is an $f \in F_\beta$ from range($\bar{a} \restriction \beta$) onto range($\bar{b}$). Then there is a $g \in F_{\beta+1}$ extending f such that $a_\beta \in$ dom(g). Fix such a g and define

$$\sigma^{\beta+1}(t, h) = g(a_\beta).$$

If there is no such f, let

$$\sigma^{\beta+1}(t, h) = \text{an arbitrary, fixed element of } N.$$

(ii) We define $\sigma^{\beta+1}(t, h)$ for all $(t, h) \in H_I$ such that $t \in S^{\forall}_{\beta+1}$ and $h \in H^t$. Let $\bar{a} = (a_\xi)_{\xi < \beta} \in {}^\beta M$ occur in a $t \in S^{\forall}_{\beta+1}$ and $\bar{b} = (b_\xi)_{\xi \leqslant \beta} \in {}^{\beta+1}N$ occur in an $h \in H^t$.

Suppose there is an $f \in F_\beta$ from range($\bar{a}$) onto range($\bar{b} \restriction \beta$). Then there is a $g \in F_{\beta+1}$ extending f such that $b_\beta \in$ range(g). Fix such a g and define

$$\sigma^{\beta+1}(t, h) = t \frown \langle g^{-1}(b_\beta), \wedge \rangle.$$

If there is no $f \in F_\beta$ from range($\bar{a}$) onto range($\bar{b} \restriction \beta$), let

$$\sigma^{\beta+1}(t, h) = \text{an arbitrary, fixed element of } I_t.$$

(iii) If $\langle t, h \rangle \in H^\beta$, let $\sigma^{\beta+1}(t, h) = \sigma^\beta(t, h)$.

Clearly $(\sigma)_{\beta+1}$ holds.

2. $\sigma^\beta = \cup_{\xi < \beta} \sigma^\xi$ if β is a limit ordinal. Then $(\sigma)_\beta$ holds if $(\sigma)_\xi$ holds for all $\xi < \beta$.

To show that $\sigma = \sigma^\alpha \in \Sigma(C^{(\alpha)}_{\mathfrak{M}}, \mathfrak{N})$ is a winning strategy, we prove that the following condition holds for all $\gamma < \alpha^0$:

$(+)_\gamma$ For every $\pi \in \Pi^\sigma_\gamma(C^{(\alpha)}_{\mathfrak{M}}, \mathfrak{N})$, if $\bar{a} \in {}^\gamma M$ occurs in t_π, there is an $f \in F_\gamma$ from range($\bar{a}$) onto range($s^{(\sigma,\pi)}$)

The claim will be proved by induction on $\gamma < \alpha^0$. Suppose $(+)_\xi$ holds for all $\xi < \gamma$.

1. Let $\gamma = \beta + 1$.

(a) Suppose $\pi \in \Pi_\gamma^{\sigma\exists}(C_{\mathfrak{M}}^{(\alpha)}, \mathfrak{N})$. Since $\bar{a} = (a_\xi)_{\xi<\gamma}$ occurs in t_π, then $\bar{a}\restriction\beta$ occurs in t_0 where $t_0 \in T^{(\beta)}$, $t_0 < t_\pi$. But for any $\pi^* \in \Pi_{(\sigma,\pi)}^{t_0} \cap \Pi_\beta^\sigma(C_{\mathfrak{M}}^{(\alpha)}, \mathfrak{N})$, $\langle t_0, h_\pi \restriction \mathring{t}_0, s\rangle = \langle t_{\pi^*}, h_{\pi^*}, s_{\pi^*}\rangle$ for some $s \in I_{t_0 0}^{\Delta}$. Thus by induction hypothesis, there is an $f \in F_\beta$ from range$(\bar{a}\restriction\beta)$ onto range$(s^{(\sigma,\pi^*)})$. Since $\bar{a}$ occurs in t where $t_0 \prec t \prec t_\pi$ and $s^{(\sigma,\pi^*)}$ occurs in h_{π^*}, $\sigma^{\beta+1}(t, h') = g(a_\beta)$ for a fixed $g \in F_\gamma$, $f \subseteq g$, where $h' = h_{\pi^*} \cup \{\langle t_0, t\rangle\} \in H^t$. Thus $\sigma(t, h') = g(a_\beta)$ and $s^{(\sigma,\pi)} = s^{(\sigma,\pi^*)} \frown g(a_\beta)$, whence g is a needed mapping.

(b) Suppose $\pi \in \Pi_\gamma^{\sigma\forall}(C_{\mathfrak{M}}^{(\alpha)}, \mathfrak{N})$ and $s^{(\sigma,\pi)} = (b_\xi)_{\xi<\gamma}$. Since $\bar{a}$ occurs in t_π, then $\bar{a}\restriction\beta$ occurs in t_0 where $t_0 \in T^{(\beta)}$, $t_0 \prec t_\pi$. Again, $t_0 = t_{\pi^*}$ for a $\pi^* \in \Pi_{(\sigma,\pi)}^{t_0} \cap \Pi_\beta^\sigma(C_{\mathfrak{M}}^{(\alpha)}, \mathfrak{N})$. Thus by induction hypothesis, there is an $f \in F_\beta$ from range$(\bar{a}\restriction\beta)$ onto range$(s^{(\sigma,\pi^*)})$. Since $\bar{a}\restriction\beta$ occurs in the immediate predecessor t of $t_\pi (t \in T^\vee)$ and $s^{(\sigma,\pi)}$ occurs in $h' = h_{\pi^*} \cup \{\langle t_0, t'\rangle, \langle t', b_\beta\rangle\} \in H^t$, where $t_0 \prec t' \prec t$, $t' \in T^\forall$, $\sigma(t, h') = \sigma^{\beta+1}(t, h') = t \frown \langle g^{-1}(b_\beta), \wedge\rangle$ for a fixed $g \in F_\gamma, f \subseteq g$. Thus g is a needed mapping.

2. Let γ be a limit ordinal. Let

$$t_0 \prec \ldots \prec t_\xi \prec \ldots \prec t_\pi$$

be the sequence where, for each $\xi < \gamma$, $t_\xi \in T^{(\xi)}$ and $\langle t_\xi, h_\xi, s_\xi\rangle \in \rho_{(\sigma,\pi)}$ with the relevant h_ξ, s_ξ. Since $s^{(\sigma,\pi)}$ occurs in h_π, then $s^{(\sigma,\pi)}\restriction\xi$ occurs in $h_\xi = h_\pi \restriction \mathring{t}_\xi$. But, again, $t_\xi = t_{\pi^*}$ for a $\pi^* \in \Pi_\xi^\sigma(C_{\mathfrak{M}}^{(\alpha)}, \mathfrak{N})$. Thus $s^{(\sigma,\pi)}\restriction\xi = s^{(\sigma,\pi^*)}$. If $\bar{a} \in {}^\gamma M$ occurs in t_π, then $\bar{a}\restriction\xi$ occurs in $t_\xi = t_{\pi^*}$. Hence by induction hypothesis, there is an $f_\xi \in F_\xi$ from range$(\bar{a}\restriction\xi)$ onto range$(s^{(\sigma,\pi)}\restriction\xi)$. If $\xi_0 < \xi_1 < \gamma$, then $\bar{a}\restriction\xi_1$ extends $\bar{a}\restriction\xi_0$ and $s^{(\sigma,\pi)}\restriction\xi_1$ extends $s^{(\sigma,\pi)}|\xi_0$, whence $f_{\xi_0} \subseteq f_{\xi_1}$. By (iv), there is an $f \in F_\gamma$ which extends every f_ξ, $\xi < \beta$. Clearly, f is a needed mapping.

Now it follows that σ is a winning strategy. Let $\beta < \alpha^0$ and $\pi \in \Pi_\beta^\sigma(C_{\mathfrak{M}}^{(\alpha)}, \mathfrak{N})$. If $\bar{a} \in {}^\beta M$ occurs in t_π, $\mathfrak{M} \models \bigwedge A_{t_\pi}(\bar{x})[\bar{a}]$. By $(+)_\beta$, there is an $f \in F_\beta$ from range$(\bar{a})$ onto range$(s^{(\sigma,\pi)})$, whence $\mathfrak{N} \models \bigwedge A_{t_\pi}(\bar{x})[s^{(\sigma,\pi)}]$, by the definition of F_β. □

It is not known, at present, whether the implication of Theorem 5.1 can be reversed. The reason seems to be the same as was mentioned after the proof of Theorem 3.8.

Let $\mathfrak{M}$ be a model for L and α an ordinal.

DEFINITION 5.2. An *α-type* Γ *in* $Th(\mathfrak{M})$ is the set of all $L_{\omega\omega}$ formulas with their free variables chosen from $\{x_\xi | \xi < \alpha\}$ which are satisfied in $\mathfrak{M}$ by a sequence $\bar{a} = (a_\xi)_{\xi<\alpha} \in {}^\alpha M$ when x_ξ is interpreted as a_ξ, for all $\xi < \alpha$. Γ is the *type of* $\bar{a}$.

DEFINITION 5.3. $\mathfrak{M}$ is α-*saturated* iff for every $\bar{a} \in {}^\beta M$ such that $\beta < \alpha$ and for every $(\beta + 1)$-type Γ in $Th(\mathfrak{M})$ which includes the type of $\bar{a}$ there is an $a \in M$ such that Γ is the type of $\bar{a} \frown a$.

If α is a cardinal λ, this notion obviously coincides with the usual notion of λ-saturated model.

Let $C^{(\alpha)} = \langle T, \prec, g, S\rangle$ (with the bound variables x_ξ, $\xi < \alpha$) be a constituent of depth α and β an ordinal $< \alpha$. Let T' be the smallest subset of T such that

(i) $T^{(\beta)} \subset T'$,
(ii) for every $t \in T^{(\beta)}$, $\hat{t} \subset T'$,
(iii) for every $t \in T^{(\beta)}$, $I_t^\Delta \subset T'$,
(iv) if $t' \in I_t^\Delta$ for some $t \in T^{(\beta)}$ and $g(t') = \neg$, then $I_{t'} \subset T'$.

Clearly $\langle T', \prec \upharpoonright T'\rangle$ is a tree and $\langle T', \prec \upharpoonright T', g \upharpoonright T', S\rangle$ is a constituent of depth β. This constituent is called the *approximation of* $C^{(\alpha)}$ *of depth* β and denoted by $C^{(\alpha)}|\beta$.

LEMMA 5.2. $\models C^{(\alpha)} \to C^{(\alpha)}|\beta$ for all $\beta < \alpha$.

Proof: Theorem 3.4. □

The converse of this implication does not hold in general, by Theorem 4.3.

THEOREM 5.3. Let α be an infinite ordinal. If $\mathfrak{M}$ and $\mathfrak{N}$ are α-saturated and $\mathfrak{N} \models C_{\mathfrak{M}}^{(\alpha)}|n$ for every $n \in \omega$, then $\mathfrak{N} \models C_{\mathfrak{M}}^{(\alpha)}$.

Proof: Define for every $\beta < \alpha$:

$$F_\beta = \{f : \text{range}(\bar{a}) \to \text{range}(\bar{b}) | \bar{a} \in {}^\beta M \ \&\ \bar{b} \in {}^\beta N \ \&\ (\mathfrak{M}, \bar{a}) \equiv_{\omega\omega} (\mathfrak{N}, \bar{b}) \ \&\ f(a_\xi) = b_\xi \text{ for all } \xi < \beta\}.$$

By Lemma 5.2, the last assumption implies that $\mathfrak{M} \equiv_{\omega\omega} \mathfrak{N}$. Thus it is obvious that the $F_\beta (\beta < \alpha)$ satisfy (i)–(iv) of Definition 5.1, whence $\mathfrak{M} \approx_\alpha \mathfrak{N}$, and by Theorem 5.1, $\mathfrak{N} \models C_{\mathfrak{M}}^{(\alpha)}$. □

In [3] it is proved that a weaker extension property than (iv) holds for the constituents We proceed to show that constituents have a stronger back-and-forth property than (ii)–(iii) of Definition 5.1.

DEFINITION 5.4. $\mathfrak{M} \sim_\alpha \mathfrak{N}$ iff there is a sequence $F_0 \subseteq \ldots \subseteq F_\beta \subseteq \ldots$ $(\beta < \alpha^0)$ of sets of mappings satisfying (i) of Definition 5.1 and

(ii) If $f \in F_\beta$ and $\bar{a} \in {}^\gamma M$, $\beta + \gamma < \alpha^0$, there is a $g \in F_{\beta+\gamma}$ such that $f \subseteq g$ and range$(\bar{a}) \subseteq$ dom(g).

(iii) If $f \in F_\beta$ and $\bar{b} \in {}^\gamma N$, $\beta + \gamma < \alpha^0$, there is a $g \in F_{\beta+\gamma}$ such that $f \subseteq g$ and range$(\bar{b}) \subseteq$ range(g).

THEOREM 5.4. If $\mathfrak{N} \models C^{(\alpha)}_{\mathfrak{M}}$, then $\mathfrak{M} \sim_\alpha \mathfrak{N}$.

Proof: Let $C^{(\alpha)}_{\mathfrak{M}} = \langle T, \prec, g, S\rangle$. It follows from the construction of $C^{(\alpha)}_{\mathfrak{M}}$ that for every $\beta < \alpha^0$ and every $\bar{a} \in {}^\beta M$, there is a $t \in T^{(\beta)}$ such that $\bar{a}$ occurs in t and the subconstituent $C_t(\bar{x})(\bar{x} = (x_\xi)_{\xi<\beta})$ determined by t is satisfied by $\bar{a}$ in $\mathfrak{M}$. This unique (up to structural similarity) subconstituent of $C^{(\alpha)}_{\mathfrak{M}}$ determined by $\bar{a}$ is now denoted by $C_{\bar{a}}(\bar{x})$. Thus $C^{(\alpha)}_{\mathfrak{M}} = C_\phi$.

For each $\beta < \alpha^0$, define:

$$F_\beta = \{f : \text{range}(\bar{a}) \to \text{range}(\bar{b}) | \bar{a} \in {}^\beta M \;\&\; \bar{b} \in {}^\beta N \;\&\; \mathfrak{N} \models C_{\bar{a}}(\bar{x})[\bar{b}] \;\&\; f(a_\xi) = b_\xi \text{ for all } \xi < \beta\}.$$

It is obvious that $F_0 \subseteq \ldots \subseteq F_\beta \subseteq \ldots (\beta < \alpha^0)$. We show that this sequence satisfies the conditions (i)–(iii) of Definition 5.4.

Let an $f \in F_\beta$ such that f : range$(\bar{a}) \to$ range$(\bar{b})$ be denoted by $f_{(\bar{a},\bar{b})}$.

(i) Let $\beta < \alpha^0$ and $f_{(\bar{a},\bar{b})} \in F_\beta$. Since $\mathfrak{N} \models C_{\bar{a}}(\bar{x})[\bar{b}]$, it follows from the definition of a constituent that $(\mathfrak{M}, \bar{a}) \equiv (\mathfrak{N}, \bar{b})(\Delta(\bar{x}))$.

(ii) Let $f_{(\bar{a},\bar{b})} \in F_\beta$ and $\bar{a}_0 \in {}^\gamma M$, $\beta + \gamma < \alpha^0$. Since $\mathfrak{N} \models C_{\bar{a}}(\bar{x})[\bar{b}]$, or $(\mathfrak{N}, \bar{b}) \models C_{\bar{a}}(\bar{d})$ (where $\bar{d} = (d_\xi)_{\xi<\beta}$, the d_ξ $(\xi < \beta)$ are new constants) and $C_{\bar{a}\frown\bar{a}_0}(\bar{d},\bar{x}_0)$ $(\bar{x}_0 = (x_\xi)_{\beta\leq\xi<\beta+\gamma})$ is a subconstituent of $C_{\bar{a}}(\bar{d})$, there is by Corollary 3.7, a sequence $\bar{b}_0 \in {}^\gamma N$ such that

$$(\mathfrak{N},\bar{b}) \models C_{\bar{a}\frown\bar{a}_0}(\bar{d}, \bar{x}_0)[\bar{b}_0],$$

whence

$$\mathfrak{N} \models C_{\bar{a}\frown\bar{a}_0}(\bar{x}, \bar{x}_0)[\bar{b} \frown \bar{b}_0].$$

Thus there is a mapping $f_{(a\frown\bar{a}_0\bar{b}\frown\bar{b}_0)} \in F_{\beta+\gamma}$, by the definition of $F_{\beta+\gamma}$. Clearly $f_{(\bar{a}\frown\bar{a}_0,\bar{b}\frown\bar{b}_0)}$ is an extension of $f_{(\bar{a},\bar{b})}$ such that range$(\bar{a}_0)$ is included. in its domain.

(iii): The proof is similar but we use Corollary 3.6. □

If λ is an infinite cardinal, it is obvious that $\mathfrak{M} \approx_\lambda \mathfrak{N}$ iff $\mathfrak{M} \cong^{p,e}_\lambda \mathfrak{N}$, where $\cong^{p,e}_\lambda$ is the extended notion of partial isomorphism defined in [1] (see also [5]). Similarly, if λ is an infinite cardinal, then $\mathfrak{M} \sim_\lambda \mathfrak{N}$ iff $\mathfrak{M} \cong_\lambda^p \mathfrak{N}$ (cf. [1], [5]).

Thus, since $\mathfrak{M} \cong_\lambda^p \mathfrak{N}$ iff $\mathfrak{M} \equiv_{\infty\lambda} \mathfrak{N}$, it follows from Theorem 5.4:

COROLLARY 5.5. For all infinite cardinals λ and all ordinals $\alpha \geqslant \lambda$, if $\mathfrak{N} \models C_{\mathfrak{M}}^{(\alpha)}$, then $\mathfrak{M} \equiv_{\infty\lambda} \mathfrak{N}$.

Combining the proofs of Theorems 5.1 and 5.4, we see that

$$\mathfrak{N} \models C_{\mathfrak{M}}^{(\omega)} \text{ iff } \mathfrak{M} \sim_\omega \mathfrak{N}.$$

Hence

$$\mathfrak{N} \models C_{\mathfrak{M}}^{(\omega)} \text{ iff } \mathfrak{M} \cong_p \mathfrak{N}.$$

Since

$$\mathfrak{N} \models C_{\mathfrak{M}}^{(\omega)} \text{ iff } \mathfrak{M} \equiv \mathfrak{N}(N_{\infty\omega})$$

(cf. [3]), we have

COROLLARY 5.6. $\mathfrak{M} \equiv \mathfrak{N}(N_{\infty\omega})$ iff $\mathfrak{M} \equiv_{\infty\omega} \mathfrak{N}$.

THEOREM 5.7. Let $\mathfrak{M}$ and $\mathfrak{N}$ be models for L of power κ. If $\mathfrak{N} \models C_{\mathfrak{M}}^{(\kappa}$ then $\mathfrak{M} \cong \mathfrak{N}$. Hence, if $\mathfrak{M} \equiv \mathfrak{N}(N_{\infty\kappa})$, then $\mathfrak{M} \cong \mathfrak{N}$.

Proof: Let $\bar{a}' = (a'_\alpha)_{\alpha<\kappa}$, $\bar{b}' = (b'_\alpha)_{\alpha<\kappa}$ be enumerations of M, N, respectively. We shall construct enumerations $\bar{a} = (a_\alpha)_{\alpha<\kappa}$ and $\bar{b} = (b_\alpha)_{\alpha<\kappa}$ of M and N such that $(\mathfrak{M}, \bar{a}) \equiv (\mathfrak{N}, \bar{b})(\Delta(\bar{x}))$, where $\bar{x} = (x_\alpha)_{\alpha<\kappa}$, the $x_\alpha (\alpha < \kappa)$ are the variables of $C_{\mathfrak{M}}^{(\kappa)}$.

An ordinal $\alpha = \omega \cdot \lambda + n$ is called even or odd according to whether n is even or odd. As in the proof of Theorem 5.1, all the relevant notions defined in Sections 2–3 will again be related to $\mathfrak{N}$.

Let $\sigma \in \Sigma(C_{\mathfrak{M}}^{(\kappa)}, \mathfrak{N})$ be a winning strategy in the game $G(C_{\mathfrak{M}}^{(\kappa)}, \mathfrak{N})$. Consider a strategy $\pi \in \Pi_\kappa^\sigma(C_{\mathfrak{M}}^{(\kappa)}, \mathfrak{N})$ which satisfies the following conditions:

(1) Suppose $\langle t, h \rangle \in H_{II}$ such that
 (i) $t \in T^{(\alpha)}$ and α is even,
 (ii) $\bar{a}_0 \in {}^\alpha M$ occurs in t,
 (iii) $\bar{b}_0 \in {}^\alpha N$ occurs in h.

Let a_α be the first member of $\bar{a}'$ which is different from the members of $\bar{a}_0$. There is a $t' \in I_t^\exists$ such that $\bar{a}_0 \frown a_\alpha$ occurs in t', by the construction of $C_{\mathfrak{M}}^{(\alpha)}$ (Theorem 4.1). Then

$$\pi(t, h) = t'.$$

(2) Suppose $\langle t, h\rangle \in H_{II}$ such that

$$t \in T^{(\alpha)} \text{ and } \alpha \text{ is odd.}$$

Let $I_t^{\forall} = \{t'\}$. Then for all $h \in H^t$,

$$\pi(t, h) = t'.$$

(3) Suppose $\langle t, h\rangle \in H_{II}$ such that
(i) $t \in T^{\forall}$ and $\text{qr}(t) = \alpha$ is odd,
(ii) $\bar{a}_0 \in {}^{\alpha}M$ occurs in t,
(iii) $\bar{b}_0 \in {}^{\alpha}N$ occurs in h.
Let b_α be the first member of $\bar{b}'$, which is different from the members of $\bar{b}_0$. Then

$$\pi(t, h) = b_\alpha.$$

Let $t_0 \prec \ldots \prec t_\alpha \prec \ldots (\alpha < \kappa)$ be the chain of T such that $t_\alpha \in {}^{\alpha}T$ and $\langle t_\alpha, h_\alpha, s_\alpha\rangle \in \rho_{(\sigma,\pi)}$ for the relevant h_α, s_α. Let $\bar{b} = s^{(\sigma,\pi)}$ and let $\bar{a} \in {}^{\kappa}M$ be the sequence such that $\bar{a}\restriction\alpha$ occurs in t_α, for every $\alpha < \kappa$. It follows from the back-and-forth property of π that $\bar{a}$ is an enumeration of M and $\bar{b}$ is an enumeration of N. Since σ is a winning strategy, then for all $\alpha < \kappa$,

$$\mathfrak{N} \models A_{t_\alpha}(\bar{x}_0)[\bar{b}_0]$$

where $\bar{b}_0$ is the sequence in ${}^{\alpha}N$ which occurs in h_α and $\bar{x}_0 = (x_\xi)_{\xi<\alpha}$. This, follows from Theorem 3.4 and from the fact that $\langle t_\alpha, h_\alpha, s\rangle = \langle t_{\pi^*}, h_{\pi^*}, s_{\pi^*}\rangle$ for an $s \in I_{t_\alpha}^{A}$ and for a $\pi^* \in \Pi_\alpha^\sigma(C_{\mathfrak{M}}^{(\kappa)}, \mathfrak{N})$, and $\bar{b}_0 = s^{(\sigma,\pi^*)}$. On the other hand, if $\bar{a}_0 \in {}^{\alpha}M$ occurs in t_α,

$$\mathfrak{M} \models A_{t_\alpha}(\bar{x}_0)[\bar{a}_0].$$

Thus

$$(\mathfrak{M}, \bar{a}) \equiv (\mathfrak{N}, \bar{b})(\Delta(\bar{x}))$$

whence the mapping $f : M \to N$ such that $f(a_\alpha) = b_\alpha$ is an isomorphism. □

This theorem shows that the language $N_{\infty\kappa}$ is essentially stronger (with respect to elementary equivalence) than $L_{\infty\kappa}$ when κ is a regular uncountable cardinal, although in the case $\kappa = \omega$ this does not hold (and possibly not in the case $cf(\kappa) = \omega$, $\kappa > \omega$).

This theorem implies also that Theorem 5.4 can not be reversed.

COROLLARY 5.8. Let $\mathfrak{M}$ be a model for L of power κ and $X \subseteq M$. Then the following conditions are equivalent.

(i) X is left fixed in every automorphism of $\mathfrak{M}$.

(ii) There is a formula $\varphi(x)$ of $N_{\infty\kappa}$ such that

$$(\mathfrak{M}, X) \models \forall x(Px \leftrightarrow \varphi(x)).$$

Proof: It is sufficient to show that not-(ii) implies not-(i). Since (ii) does not hold, there are elements $a \in M$ and $a' \in M$ such that $a \in X$, $a' \notin X$, and $\mathfrak{M} \models C_a(x_0)[a']$, where $C_a(x_0)$ is the subconstituent of $C^{(\kappa)}_{\mathfrak{M}}$ determined by a (cf. the proof of Theorem 5.4). For otherwise $(\mathfrak{M}, X) \models \forall x(Px \leftrightarrow \bigvee_{a \in X} C_a(x))$.

Let c be a new individual constant. Since $\mathfrak{M} \models C_a(x_0)[a']$, also $(\mathfrak{M}, a') \models C_\alpha(c)$. But by the construction in Theorem 4.1, $C_a(c)$ is $C^{(\kappa)}_{(\mathfrak{M},a)}$. Hence by Theorem 5.7, $(\mathfrak{M}, a) \cong {}^f(\mathfrak{M}, a')$ where $f(a) = a'$. Thus $a \in X$ but $f(a) \notin X$, whence (i) does not hold. □

In particular, if $|L| \leqslant \kappa$, then $\varphi(x) \in N_{\kappa^+\kappa}$ and we obtain the straightforward generalization (of the analogue for $N_{\omega_1\omega}$) of Scott's Definability Theorem for $L_{\omega_1\omega}$ for the larger cardinalities.

Department of Philosophy,
University of Helsinki
(*Academy of Finland*)

NOTES

[1] I am indebted to Maaret Karttunen for this necessary addendum.

[2] The question of the best possible definition of truth is still under discussion. In [6], the whole question will be evaluated by means of a reduction to game quantifier languages.

REFERENCES

[1] M. A. Dickmann: *Large Infinitary Languages*, North-Holland, Amsterdam, 1975.
[2] J. Hintikka and V. Rantala: 'A New Approach to Infinitary Languages', *Annals of Mathematical Logic* **10**, no. 1, 1976.
[3] M. Karttunen: 'Infinitary Languages $N_{\infty\lambda}$ and Generalized Partial Isomorphisms', this volume, pp. 152–168.

[4] H. J. Keisler: 'Formulas with Linearly Ordered Quantifiers', in J. Barwise (ed.), *The Syntax and Semantics of Infinitary Languages*, Lecture Notes in Mathematics, vol. 72, Springer-Verlag, 1968.
[5] D. W. Kueker: 'Back-And-Forth Arguments and Infinitary Logics', in D. W. Kueker (ed.), *Infinitary Logic: In Memoriam Carol Karp*, Lecture Notes in Mathematics, vol. 492, Springer-Verlag, 1975.
[6] J. Oikkonen: 'Second-Order Definability, Game Quantifiers and Related Expressions', *Commentationes Physico-Mathematicae* **48** (1978), 39–101.

MAARET KARTTUNEN

INFINITARY LANGUAGES $N_{\infty\lambda}$ AND GENERALIZED PARTIAL ISOMORPHISMS

INTRODUCTION

In [3] Hintikka and Rantala define a new hierarchy of infinitary languages (the so-called $N_{\kappa\lambda}$-languages). In our paper we consider the relationship between these languages and some generalized partial isomorphisms. With these generalizations we can give an algebraic characterization to the elementary equivalence of $N_{\infty\lambda}$.

In [1] Dickmann defines one kind of extension of partial isomorphisms ($\cong_{\lambda}^{p,e}$, see p. 313; see also [5], p. 33, where a similar definition is given). The main point of this definition is that we can take unions over chains of length less than λ, i.e. if $I: \mathscr{A} \cong_{\lambda}^{p,e} \mathscr{B}$ and $F \subseteq I$ is totally ordered by $\subseteq$ and $\bar{\bar{F}} < \lambda$, then there is a $g \in I$ which extends all the $f \in F$. Unfortunately any connections between $\cong_{\lambda}^{p,e}$ and infinitary languages have not been found.

We cannot get the equivalence between $\cong_{\lambda}^{p,e}$ and $\equiv_{N_{\infty\lambda}}$, either, but instead of $\cong_{\lambda}^{p,e}$ we can define a little weaker notion of a generalized partial isomorphism, which in many respects resembles that of $\cong_{\lambda}^{p,e}$, and this generalization gives the desired equivalence.

1. LANGUAGES $N_{\kappa\lambda}$

The syntax of $N_{\kappa\lambda}$

The following definition of languages $N_{\kappa\lambda}$ is very much the same as in [3]. Only some slight changes are made.

In what follows L is a (finitary) first-order language, V the set of its variables and AF the set of its atomic formulas.

A formula ϕ of $N_{\kappa\lambda}$ ($\kappa, \lambda \in \text{Card}$) is a quadruple

$$\phi = \langle T, <_T, g, S \rangle$$

where

(i) $\langle T, <_T \rangle$ is a tree of length $\leqslant \lambda$ such that

(a) through each element $t \in T$ there passes at least one branch with a maximal element,

J. Hintikka, I. Niiniluoto, and E. Saarinen (eds.), Essays on Mathematical and Philosophical Logic, 153–168.

(b) if the rank of $t, t' \in T$ is a limit ordinal and t, t' have the same set of predecessors, then $t = t'$,

(ii) $S = AF \cup \{\sim, \bigvee, \bigwedge\} \cup (\{\exists, \forall\} \times V)$,

(iii) g is a mapping from T into S satisfying the following conditions:

(a) $g(t) \in AF$ iff t is maximal,

(b) t has only one immediate successor iff $g(t) \in \{\sim\} \cup (\{\exists, \forall\} \times V)$,

(c) if $g(t) \in \{\bigvee, \bigwedge\}$, t has fewer than κ successors,

(iv) for each $t \in T$ there are only finitely many $t' <_T t$ *s.t.* $g(t') = \sim$.

The only difference between this definition and that in [3] is the fact (iv); without this demand there are sentences that are equivalent with their negations.

Our main interest is in the class $N_{\infty\lambda} = \bigcup_{\kappa \in \mathrm{Card}} N_{\kappa\lambda}$.

Some definitions

We give here some notions about the tree-structure of formulas in $N_{\kappa\lambda}$. In what follows we also use the terminology and definitions concerning trees found in Drake's book [2], see Chapter 7, Section 1.

Let ϕ and ψ be formulas in $N_{\kappa\lambda}$.

Tree$_\phi$: the corresponding tree-structure $\underline{T} = \langle T, <_T \rangle$ of ϕ, (we usually identify Tree$_\phi$ and ϕ)

root (ϕ): the root of Tree$_\phi$,

$\phi \leqslant \psi$: ψ is a subformula of ϕ (which, for example, means that if Tree$_\phi$ $= \langle T, <_T \rangle$ and Tree$_\psi = \langle U, <_U \rangle$ then $U = \{t \in T | t' \leqslant_T t\}$ for some $t' \in T$).

Let t be a node in Tree$_\phi$ and let $\phi \leqslant \psi$.

$qr_\phi(t) = OT\langle\{t' \in \mathrm{Tree}_\phi | t' <_T t, g(t') = \exists x \text{ or } \forall x\}, <_T\rangle$,

$qr_\phi(\psi) = qr_\phi(\mathrm{root}(\psi))$,

$nr_\phi(t) = OT\langle\{t' \in \mathrm{Tree}_\phi | t' <_T t, g(t') = \sim\}, <_T\rangle$,

$nr_\phi(\psi) = nr_\phi(\mathrm{root}(\psi))$.

Usually we denote $nr(t)$, $nr(\psi)$, $qr(t)$ and $qr(\psi)$ if ϕ is clear from the context.

If $g(t) = \exists x$ or $\vee$ and $nr_\phi(t)$ is even then we denote $g(t) = \exists^+ x, \vee^+$; if $g(t) = \forall x$ or $\wedge$ and $nr_\phi(t)$ is odd, then we denote $g(t) = \exists^- x, \vee^-$. Similarly we get $\forall^\pm x$ and $\wedge^\pm$.

Beside these definitions our set-theoretical and model-theoretical notation is standard.

In what follows $\alpha, \beta, \gamma, \delta, \eta$ are intended to be ordinals and κ, λ infinite cardinals.

The definition of satisfaction

Let ϕ be a formula in $N_{\kappa\lambda}$ and let $\mathbf{x} = \langle x_\eta | \eta < \alpha \rangle$, $\alpha < \lambda$, contain its free variables. Let $\mathscr{A}$ be a model for L and $\mathbf{a} \in {}^\alpha A$ elements from its domain.

We associate with ϕ, $\mathscr{A}$ and $\mathbf{a}$ a semantical game $G(\phi, \mathscr{A}, \mathbf{a})$. (Here we mainly follow the definition in [3].)

There are two players, Myself and Nature. In a play of that game we go along one branch of Tree_ϕ proceeding from root(ϕ) to the top of that branch, and when we meet quantifiers, we pick up elements from the domain of $\mathscr{A}$. The play proceeds in the following way:

Suppose that we have gained a node $t \in \text{Tree}_\phi$.

If $g(t) = \vee^\pm$, I choose one of the immediate successors of t;
if $g(t) = \wedge^\pm$, Nature makes a similar choice;
if $g(t) = \exists^\pm x$, I pick up an element from A;
if $g(t) = \forall^\pm x$, Nature makes a similar move;
if $g(t) = \sim$, we just jump over and proceed with its immediate successor (and remember that the negation-rank increases);
if $g(t) \in AF$, the play ends.

If the order-type of those nodes we have gained is a limit ordinal and the branch does not end yet, we take the supremum of those nodes and go on playing with it. (Remember that the supremum is unique.)

Let Σ be my strategies and Π the set of Nature's strategies in G. Let $\langle \sigma, \pi \rangle$ be the play of G determined by the strategies $\sigma \in \Sigma$, $\pi \in \Pi$. As above there is a branch $D(\sigma, \pi)$ associated with $\langle \sigma, \pi \rangle$ in an obvious way. Let $l(\sigma, \pi)$ be the length of that branch.

We define the pay-off function

$$p : \Sigma \times \Pi \to \{-1, 0, 1\}$$

in the following way:

Let $\langle \sigma, \pi \rangle \in \Sigma \times \Pi$.

If $l(\sigma, \pi)$ is a successor ordinal, then there is an atomic formula θ at the top of $D(\sigma, \pi)$ and during the play $\langle \sigma, \pi \rangle$ we have chosen elements $\mathbf{b} = \langle b_\eta | \eta < \beta \rangle \in {}^\beta A$ where $\beta = qr(\theta)$.

In these cases:

$$p(\sigma, \pi) = \begin{cases} +1 \text{ if } \mathscr{A} \models \theta[\mathbf{a} ^\wedge \mathbf{b}] \text{ and } nr(\theta) \text{ is even,} \\ \quad \text{or } \mathscr{A} \not\models \theta[\mathbf{a} ^\wedge \mathbf{b}] \text{ and } nr(\theta) \text{ is odd;} \\ -1 \text{ if } \mathscr{A} \models \theta[\mathbf{a} ^\wedge \mathbf{b}] \text{ and } nr(\theta) \text{ is odd,} \\ \quad \text{or } \mathscr{A} \not\models \theta[\mathbf{a} ^\wedge \mathbf{b}] \text{ and } nr(\theta) \text{ is even.} \end{cases}$$

(Here $^\wedge$ is the catenation sign.)

We have many possibilities to define $p(\sigma, \pi)$ when $l(\sigma, \pi)$ is a limit ordinal but unfortunately all these definitions have turned out to be somewhat unsatisfactory (in the sense that there are sentences which intuitively are true in a model but the pay-off function says something else about the truthness of these sentences). See, for example, the definition in [3].

Intuitively speaking, we can define a 'bad choice' corresponding to disjunctions and conjunctions. We put $p(\sigma, \pi)$ to be $+1$ if Nature has made 'bad choices' more times than I have, -1 if the same happens to me, and 0 if we have made equally many 'bad choices'. (The number of these choices must be understood as ordinal numbers.) We have many ways to define these 'bad choices'. An ordinary one is to compare the complexity of the subformulas of which the disjunctions and conjunctions are formed. However, we do not give any of these definitions here because our results are very independent of the details in them.

My strategy $\sigma \in \Sigma$ is a winning one iff $p(\sigma, \pi) = +1$ for each $\pi \in \Pi$ and a strategy $\pi \in \Pi$ is a winning one for Nature iff $p(\sigma, \pi) = -1$ for every $\sigma \in \Sigma$.

We say that $\mathbf{a}$ makes ϕ true in $\mathscr{A}$, $\mathscr{A} \models \phi[\mathbf{a}]$, iff I have a winning strategy in the game $G(\phi, \mathscr{A}, \mathbf{a})$ and $\mathbf{a}$ makes ϕ false in $\mathscr{A}$ iff Nature has a winning one in G. (So: $\mathbf{a}$ makes ϕ false in $\mathscr{A}$ iff $\mathscr{A} \models \sim\phi[\mathbf{a}]$.) If ϕ is a sentence, then ϕ is true in $\mathscr{A}$ iff $\mathscr{A} \models \phi[\emptyset]$ and ϕ is false in $\mathscr{A}$ iff $\mathscr{A} \models \sim\phi[\emptyset]$.

DEFINITION 1.1. Let ϕ and ψ be formulas in $N_{\kappa\lambda}$ such that $\phi(\mathbf{x}) \leqslant \psi(\mathbf{x} \wedge \mathbf{y})$.
Let $\mathscr{A}$ be a model and $\mathbf{a}$ be elements in A s.t. $\mathscr{A} \models \phi(\mathbf{x})[\mathbf{a}]$.
We write

$$\mathscr{A} \models_{\sigma} \psi(\mathbf{x}^{\wedge}\mathbf{y})[\mathbf{a}^{\wedge}\mathbf{b}]$$

when I have followed my winning strategy σ and Nature one of his strategies in a play of the game $G(\phi, \mathscr{A}, \mathbf{a})$, and in that play we have gained ψ and chosen elements $\mathbf{b}$ from A by following our strategies.
(We sometimes write $\models$ instead of $\models_{\sigma}$ when σ is inessential or clear from the context.)

It is reasonable to require that the pay-off function has been defined in such a way that if $\phi \leqslant \psi$, $\mathscr{A} \models \phi[\mathbf{a}]$ and $\mathscr{A} \models_{\sigma} \psi[\mathbf{a} \wedge \mathbf{b}]$ then $\mathscr{A} \models \psi[\mathbf{a} \wedge \mathbf{b}]$ when $nr_{\phi}(\psi)$ is even and $\mathscr{A} \models \sim\psi[\mathbf{a} \wedge \mathbf{b}]$ when $nr_{\phi}(\psi)$ is odd.

If the length $l(\sigma, \pi)$ is a successor ordinal, this demand is clearly satisfied, In limit cases we simply have to define $p(\sigma, \pi)$ to be $+1$ if there is an ordinal $\alpha < l(\sigma, \pi)$ s.t. after the level α I do not make any 'bad choices' but for each ordinal β, $\alpha \leqslant \beta < l(\sigma, \pi)$, there are 'bad choices' made by Nature after the level β. Similarly we define $p(\sigma, \pi)$ to be -1 when I am in trouble. Finally, $p(\sigma, \pi) = 0$ if neither of these cases happen. In this way we can easily modify the definition in [3].

A reduced form of a formula

Let $\phi = \langle T, <_T, g, S\rangle$ be a formula in $N_{\kappa\lambda}$. *A reduced form of* ϕ,

$$\phi^r = \langle T^r, <_{T^r}, g^r, S\rangle$$

is a formula in $N_{\kappa\lambda}$ which we get from ϕ in the following way:

(i) $S^r = S$,

(ii) $\langle T^r, <_{T^r}\rangle$: At first we omit all those nodes $t \in T$ for which $g(t) = \sim$. Call this tree $\mathscr{T}' = \langle T', <_{T'}\rangle$. Then for each $t \in T'$ s.t. $g(t) \in AF$ and $nr_\phi(t)$ is odd we add an extra node $t\sim$ to precede t immediately. This is how you get $\langle T^r, <_{T^r}\rangle$.

(iii) For each $t \in T^r$ we define:

$$g^r(t) = \begin{cases} \sim \text{ if } t = t\sim, \\ \exists x \text{ if } g(t) = \exists^{\pm} x, \\ \forall x \text{ if } g(t) = \forall^{\pm} x, \\ \vee \text{ if } g(t) = \vee^{\pm}, \\ \wedge \text{ if } g(t) = \wedge^{\pm}, \\ \theta \text{ if } g(t) = \theta \text{ and } \theta \in AF. \end{cases}$$

In other words, we get ϕ^r from ϕ by pushing all the negation signs appearing in Tree_ϕ towards the top of that tree as far as they go and changing $\exists x$, $\forall x$, $\vee$ and $\wedge$ when needed. (Remember that for each $t \in \text{Tree}_\phi$ there are only finitely many negation signs preceding t.)

LEMMA 1.2. For each $\mathscr{A}$, $\phi(\mathbf{x})$ (where $\mathbf{x} = \langle x\eta \mid \eta < \alpha\rangle$) and $\mathbf{c} \in {}^{\alpha}A$:

$$\mathscr{A} \models \phi[\mathbf{c}] \text{ iff } \mathscr{A} \models \phi^r[\mathbf{c}].$$

Proof: trivial. □

2. CONSTITUENTS IN $N_{\infty\lambda}$

Our constituents are modified versions of those in [3].

From now on we often identify nodes t and their values $g(t)$.

Let $\mathscr{A}$ be a model for L and $\mathbf{a} \in {}^{\alpha}A$, $\alpha < \lambda$. We construct a formula $C_{\langle \mathscr{A}, \mathbf{a} \rangle}$ of $N_{\infty\lambda}$, the free variables of which are $\mathbf{x} = \langle x_\eta | \eta < \alpha \rangle$, called *a constituent for* $\langle \mathscr{A}, \mathbf{a} \rangle$. (If $\mathbf{a} = \emptyset$, then $C_{\langle \mathscr{A}, \mathbf{a} \rangle} = C_{\mathscr{A}}$ which is a constituent for $\mathscr{A}$.)

First we make a tree $C^{\exists}_{\langle \mathscr{A}, \mathbf{a} \rangle}$:
Let $\bar{\bar{A}} = \kappa$ and $h : \kappa \to A$ be a bijection.

Step 0: Make a tree $\wedge \bigwedge (\pm)\, \theta(\mathbf{x})$ consisting of all those atomic formulas or their negations the free variables of which are among $\mathbf{x}$ and for which $\mathscr{A} \models (\pm)\, \theta(\mathbf{x})\, [\mathbf{a}]$. Call this tree T_0. Define: $f_0 = \emptyset$.

Step η: Suppose that for all $\gamma < \eta$ you have defined T_γ and a function
$(\eta < \lambda)$ $f_\gamma : \{t \in T_\gamma | t = \exists x^{\beta}_{\alpha+\delta},\ \beta < \kappa,\ \delta < \gamma\} \to A$,
and if $\gamma_1 \leqslant \gamma_2$ then T_{γ_1} is an initial subtree of T_{γ_2} (i.e. $T_{\gamma_1} \subseteq T_{\gamma_2}$ and if $t \in T_{\gamma_1} \cap T_{\gamma_2}$ then for each $t' <_{T_{\gamma_2}} t$: $t' <_{T_{\gamma_1}} t$), $f_{\gamma_1} \subseteq f_{\gamma_2}$ and for each $\gamma < \eta$ when in T_γ there is a subtree ST of a form

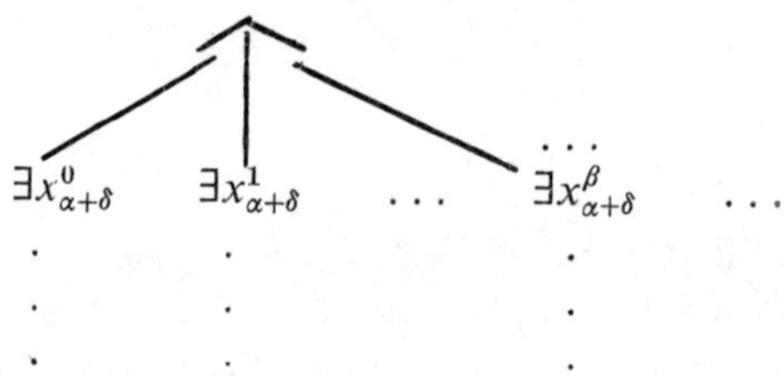

then $f'_\gamma = f_\gamma \restriction \{t \in ST | t = \exists x^{\beta}_{\alpha+\delta},\ \beta < \kappa\}$ is a bijection from the set $\{t \in ST | t = \exists x^{\beta}_{\alpha+\delta},\ \beta < \kappa\}$ to A and $f'_\gamma(\exists x^{\beta}_{\alpha+\delta}) = h(\beta)$.

$\lim(\eta)$: Put $T'_\eta = \sup\limits_{\gamma < \eta} T_\gamma$.

After each branch D in T'_η s.t. length$(D) = \eta$ you add a tree of a form $\wedge \bigwedge (\pm)\, \theta(\mathbf{x})$ where $\mathbf{x} = \langle x_\gamma | \gamma < \alpha + \eta \rangle$ and the formulas $(\pm)\, \theta(\mathbf{x})$ are all those atomic formulas (or their negations) with free variables among $\mathbf{x}$ s.t. $\mathscr{A} \models (\pm)\, \theta(\mathbf{x})\, [\mathbf{a}^\frown\mathbf{b}]$ where $\mathbf{b} = \langle b_\gamma | b_\gamma = f_{\gamma+1}(\exists x_{\alpha+\gamma}),\ \exists x_{\alpha+\gamma} \in D,\ \gamma < \eta \rangle$. This is how you get T_η. Put $f_\eta = \bigcup\limits_{\gamma < \eta} f_\gamma$.

succ(η): $\eta = \gamma + 1$.
You change each subtree $ST = \wedge \bigwedge (\pm)\, \theta(\mathbf{x})$ of T_γ where $q_r(\bigwedge) = \gamma$ in the following way:
In the place of ST you put a tree ST':

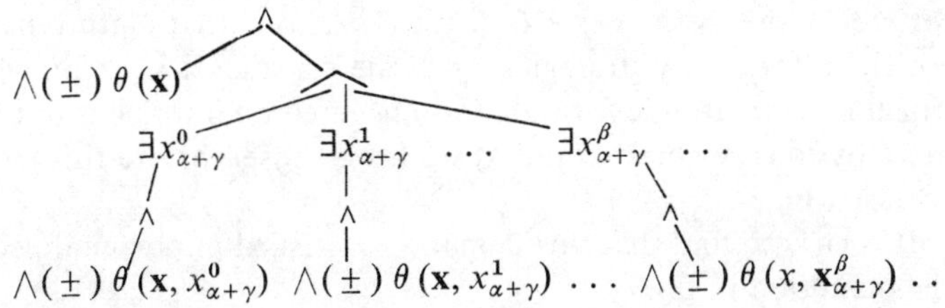

where $\wedge(\pm)\,\theta(\mathbf{x}, x^\beta_{\alpha+\gamma})$ consists of all those atomic formulas or their negations s.t. $\mathscr{A} \models (\pm)\,\theta(\mathbf{x}, x^\beta_{\alpha+\gamma})[\mathbf{c}, h(\beta)]$. (You get $\mathbf{c}$ by means of the branch at the top of which ST' is; $\mathbf{c} = \mathbf{a}\,^\wedge\,\mathbf{b}$, $\mathbf{b} = \langle b_\delta | \delta < \gamma \rangle$.) Put $f_\eta = f_\gamma \cup \{\langle \exists x^\beta_{\alpha+\gamma}, h(\beta)\rangle | \beta < \kappa\}$.

$C^{\exists}_{\langle \mathscr{A}, \mathbf{a}\rangle} = \sup\limits_{\eta<\lambda} T_\eta.$

Now you make $C_{\langle \mathscr{A}, \mathbf{a}\rangle}$ by adding the '$\forall$-parts':

$C^0 = C^{\exists}_{\mathscr{A}, \langle \mathbf{a}\rangle}$.

Suppose that you have made C^γ for each $\gamma < \eta$. Moreover, if $\gamma_1 \leqslant \gamma_2$, then C^{γ_1} is an initial subtree of C^{γ_2}.

If $\lim(\eta)$, then $C^\eta = \sup\limits_{\gamma<\eta} C^\gamma$.

If $\text{succ}(\eta)$, $\eta = \gamma + 1$, then you change each subtree ST of C^γ where ST is of a form $\wedge(\pm)\,\theta(\mathbf{x}) \wedge \bigwedge\limits_{\beta<\kappa} \exists x^\beta_{\alpha+\gamma} ST_\beta(\mathbf{x}, x^\beta_{\alpha+\gamma})$ and $qr(ST) = \gamma$ to a form

$$\wedge(\pm)\,\theta(\mathbf{x}) \wedge \bigwedge_{\beta<\kappa} \exists x_{\alpha+\gamma}\, ST_\beta(\mathbf{x}, x_{\alpha+\gamma}) \wedge \forall x_{\alpha+\gamma} \bigvee_{\beta<\kappa} ST_\beta(\mathbf{x}, x_{\alpha+\gamma}).$$

$C_{\langle \mathscr{A}, \mathbf{a}\rangle} = \sup\limits_{\eta<\lambda} C^\eta.$

Now it is easy to show: $\mathscr{A} \models C_{\langle \mathscr{A}, \mathbf{a}\rangle}[\mathbf{a}]$.

In other words, a constituent for $\langle \mathscr{A}, \mathbf{a}\rangle$ is of a form

$$C_{\langle \mathscr{A}, \mathbf{a}\rangle}(\mathbf{x}) = \wedge(\pm)\,\theta(\mathbf{x}) \wedge \bigwedge_{b\in A} \exists y C_{\langle \mathscr{A}, \mathbf{a}, b\rangle}(\mathbf{x}, y) \wedge \forall y \bigvee_{b\in A} C_{\langle \mathscr{A}, \mathbf{a}, b\rangle}(\mathbf{x}, y)$$

where $C_{\langle \mathscr{A}, \mathbf{a}, b\rangle}$:s are constituents for $\langle \mathscr{A}, \mathbf{a}, b\rangle$ and each of these are of depth λ.

Notice that a constituent in $N_{\infty\lambda}$ is of depth λ and every branch D in the constituent s.t. length(D) is a limit ordinal has a length λ.

A reasonable demand

The pay-off function must be defined in such a way that the only possibility for me to lose in the game $G(C_{\langle \mathscr{A}, \mathbf{a} \rangle}, \mathscr{A}, \mathbf{a})$ is that Nature has a strategy $\pi \in \Pi$ s.t. for all my strategies $\sigma \in \Sigma$ there is a level $\delta_{\langle \sigma, \pi \rangle}$ in which Nature can choose an atomic formula (or a negated one) which is not (is) satisfied in $\mathscr{A}$ by those elements (and $\mathbf{a}$) we have chosen before this level. In other cases I win.

It is easy to find that this demand is satisfied in our modified version of the definition in [3].

3. GENERALIZED PARTIAL ISOMORPHISMS

In this section we first define some algebraic and metamathematical relations between models. Then in our main theorem we consider the relationship between these relations, constituents and the elementary equivalence of $N_{\infty\lambda}$. Some of our definitions are not new; we have taken them here to show the strength or weakness of the others.

In what follows λ is a regular cardinal and the models we are dealing with are for the language L.

DEFINITION 3.1. Models $\mathscr{A}$ and $\mathscr{B}$ are partially isomorphic with a strong λ-extension property,

$$I: \mathscr{A} \cong_{s}^{\lambda e} \mathscr{B}$$

if there is a non-empty family I of functions such that

(1) every $f \in I$ is an isomorphism from a substructure of $\mathscr{A}$ to a substructure of $\mathscr{B}$.

(2) for every $f \in I$ and $a \in A$ there is a $g \in I$ s.t. $f \subseteq g$ and $a \in \mathrm{dom}(g)$, for every $f \in I$ and $b \in B$ there is a $g \in I$ s.t. $f \subseteq g$ and $b \in rng(g)$,

(3) if $F \subseteq I$ is totally ordered by $\subseteq$ and $\bar{\bar{F}} < \lambda$, then there is a $g \in I$ s.t. for each $f \in F: f \subseteq g$.

(See, for example, [1], p. 313: $\cong_{\lambda}^{p,e}$; also p. 316.)

DEFINITION 3.2. Models $\mathscr{A}$ and $\mathscr{B}$ are partially isomorphic with a weak λ-extension property,

$$I: \mathscr{A} \cong_{w}^{\lambda e} \mathscr{B},$$

if there is a non-empty set U, a binary relation $\leqslant_U$ on U, a non-empty family I of functions and a mapping $g: U \to I$ s.t.

(0) $\langle U, <_U \rangle$ is a tree,

(1) as (1) in the definition 3.1,

(2) for every $u \in U$ and $a \in A$ there is a $v \in U$ s.t. $u \leqslant_U v$ and $a \in$

dom($g(v)$),
for every $u \in U$ and $b \in B$ there is a $v \in U$ s.t. $u \leqslant_U v$ and $b \in rng(g(v))$,

(3) if $u \leqslant_U v$ then $g(u) \subseteq g(v)$,

(4) if $V \subseteq U$ is totally ordered by $\leqslant_U$ and $V < \lambda$, then there is a $v \in U$ s.t. for each $u \in V$: $u \leqslant_U v$.

DEFINITION 3.3. Models $\mathscr{A}$ and $\mathscr{B}$ are λ-partially isomorphic,

$$I: \mathscr{A} \cong {}^{\lambda}\mathscr{B}$$

if there is a non-empty family I of functions such that

(1) as (1) in the definition 3.1,

(2) for every $f \in I$ and $X \subseteq A$ s.t. $\bar{\bar{X}} < \lambda$ there is a $g \in I$ s.t. $f \subseteq g$ and $X \subseteq \mathrm{dom}(g)$,
for every $f \in I$ and $Y \subseteq B$ s.t. $\bar{\bar{Y}} < \lambda$ there is a $g \in I$ s.t. $f \subseteq g$ and $Y \subseteq rng(g)$.

Clearly: $\mathscr{A} \cong {}_{s}^{\lambda e}\mathscr{B} \Rightarrow \mathscr{A} \cong {}_{w}^{\lambda e}\mathscr{B} \Rightarrow \mathscr{A} \cong {}^{\lambda}\mathscr{B}$ and if $\bar{\bar{A}} \leqslant \lambda, \bar{\bar{B}} \leqslant \lambda, \mathscr{A} \cong {}_{w}^{\lambda e}\mathscr{B}$ then $\mathscr{A} \cong \mathscr{B}$.

DEFINITION 3.4. Let $\mathscr{L}$ be a language.
Models $\mathscr{A}$ and $\mathscr{B}$ are $\mathscr{L}$-equivalent with a λ-extension back-and-forth property,

$$\mathscr{A} \equiv {}_{\mathscr{L}}^{\lambda}\mathscr{B},$$

if

(1) $\mathscr{A} \equiv {}_{\mathscr{L}}\mathscr{B}$ (i.e. $\mathscr{A}$ and $\mathscr{B}$ are elementarily equivalent in $\mathscr{L}$),

(2) for each $\alpha < \lambda$, $\mathbf{a} \in {}^{\alpha}A$, $\mathbf{b} \in {}^{\alpha}B$, if $(\mathscr{A}, \mathbf{a}) \equiv {}_{\mathscr{L}}(\mathscr{B}, \mathbf{b})$ then

(a) $\forall a \in A\, \exists b \in B$ s.t. $(\mathscr{A}, \mathbf{a}, a) \equiv {}_{\mathscr{L}}(\mathscr{B}, \mathbf{b}, b)$,

(b) $\forall b \in B\, \exists a \in A$ s.t. $(\mathscr{A}, \mathbf{a}, a) \equiv {}_{\mathscr{L}}(\mathscr{B}, \mathbf{b}, b)$,

(3) for each $\alpha < \lambda$ s.t. $\lim(\alpha)$, $\alpha \neq 0$, $\mathbf{a} \in {}^{\alpha}A$, $\mathbf{b} \in {}^{\alpha}B$, if $(\mathscr{A}, \mathbf{a}\restriction\beta) \equiv {}_{\mathscr{L}}(\mathscr{B}, \mathbf{b}\restriction\beta)$ for each $\beta < \alpha$, then $(\mathscr{A}, \mathbf{a}) \equiv {}_{\mathscr{L}}(\mathscr{B}, \mathbf{b})$.

If $\mathscr{L} = L_{\omega\omega}$, then (3) is always true and we shall see that if $\mathscr{L} = N_{\infty\lambda}$ then (2) always holds.
Now we give our main theorem:

THEOREM 3.5. Let $\mathscr{A}$ be a model and $C_{\mathscr{A}}$ a constituent for $\mathscr{A}$ in $N_{\infty\lambda}$. Then for every model $\mathscr{B}$ (of the same similarity type as $\mathscr{A}$) the following holds:

$$\mathscr{A} \equiv {}_{L_{\omega\omega}}^{\lambda}\mathscr{B} \Rightarrow \mathscr{A} \equiv {}_{N_{\infty\lambda}}^{\lambda}\mathscr{B} \Rightarrow \mathscr{A} \cong {}_{s}^{\lambda e}\mathscr{B} \Rightarrow$$
$$\mathscr{A} \cong {}_{w}^{\lambda e}\mathscr{B} \Leftrightarrow \mathscr{A} \equiv {}_{N_{\omega\lambda}}\mathscr{B} \Leftrightarrow \mathscr{B} \models C_{\mathscr{A}} \Rightarrow \mathscr{A} \cong {}^{\lambda}\mathscr{B}.$$

Proof: The implications

$$\mathscr{A} \cong_{s}^{\lambda e} \mathscr{B} \Rightarrow \mathscr{A} \cong_{w}^{\lambda e} \mathscr{B} \Rightarrow \mathscr{A} \cong^{\lambda} \mathscr{B} \text{ and } \mathscr{A} \equiv_{N_{\infty\lambda}} \mathscr{B} \Rightarrow \mathscr{B} \models C_{\mathscr{A}}$$

are clear.

<u>I: $\mathscr{A} \cong_{w}^{\lambda e} \mathscr{B} \Rightarrow \mathscr{A} \equiv_{N_{\infty\lambda}} \mathscr{B}$:</u>

Let $\phi \in N_{\infty\lambda}$ be a sentence s.t. $\mathscr{A} \models \phi$, and ϕ^r the reduced form of ϕ.

Let σ be my winning strategy in the game $G(\phi^r, \mathscr{A}, \emptyset)$ and $\langle U, <_U \rangle$, g as in the definition 3.2.

We give a winning strategy for me in the game $G(\phi^r, \mathscr{B}, \emptyset)$ by induction (we only give an intuitive idea how to define it corresponding to one play of that game):

Let $\alpha < \lambda$ and suppose that for each $\eta < \alpha$ the following holds:

($*_\eta$): we have found a subformula $\psi_\eta \geqslant \phi^r$ s.t. ψ_η is in the level η in ϕ^r (i.e. its root is in that level) and $qr_{\phi^r}(\psi_\eta) = \tau_\eta$, and we have found $\mathbf{a}_\eta \in {}^{\tau_\eta}A$, $\mathbf{b}_\eta \in {}^{\tau_\eta}B$ and $u_\eta \in U$ s.t. for each $\beta < \tau_\eta$: $g(u_\eta)(a_\beta) = b_\beta$, and $\mathscr{A} \models_\sigma \psi_\eta[\mathbf{a}_\eta]$,

and if $\gamma \leqslant \eta$ then $u_\gamma \leqslant_U u_\eta$, $\psi_\gamma \leqslant \psi_\eta$ and $\mathbf{a}_\gamma = \mathbf{a}_\eta \restriction \tau_\gamma$, $\mathbf{b}_\gamma = \mathbf{b}_\eta \restriction \tau_\gamma$.

We have to prove that the above requirements hold for α.

lim(α): If $\alpha = 0$ take $\psi_0 = \phi^r$, $\tau_0 = 0$, $\mathbf{a}_0 = \emptyset$, $\mathbf{b}_0 = \emptyset$, u_0 any $u \in U$.

Let $\alpha \neq 0$.

If the nodes $\{t_\eta = \text{root}(\psi_\eta) | \eta < \alpha\}$ form a branch (i.e. a maximal chain) in ϕ^r, then the play ends; otherwise there is a node $t \in \text{Tree}_{\phi^r}$ and a formula $\psi \geqslant \psi_\eta$, $\eta < \alpha$, s.t. $\text{root}(\psi) = t = \sup_{\eta < \alpha} t_\eta$.

Put $\psi_\alpha = \psi$, $\mathbf{a}_\alpha = \sup_{\alpha < \eta} \mathbf{a}_\eta$, $\mathbf{b}_\alpha = \sup_{\alpha < \eta} \mathbf{b}_\eta$, $\tau_\alpha = \sup_{\eta < \alpha} \tau_\eta$ and $u_\alpha = v \in U$ where $u_\eta \leqslant_U v$ for each $\eta < \alpha$. (There is such a $v \in U$ because $V = \{u_\eta | \eta < \alpha\}$ is totally ordered by $\leqslant_U$ and $\bar{\bar{V}} < \lambda$.)

Now: $\psi_\eta \leqslant \psi_\alpha$ for each $\eta < \alpha$, $qr(\psi_\alpha) = \tau_\alpha$ and $g(u_\alpha)(a_\beta) = b_\beta$ for each $\beta < \tau_\alpha$. Also we know that $\mathscr{A} \models_\sigma \psi_\alpha[\mathbf{a}_\alpha]$.

Clearly all the other requirements we need are satisfied.

succ(α): (1) $\psi_\eta = \bigvee_{j \in J} \psi_j$.
($\alpha = \eta + 1$)

Because $\mathscr{A} \models_\sigma \psi_\eta[\mathbf{a}_\eta]$ then there is a $j_0 \in J$ s.t. $\mathscr{A} \models_\sigma \psi_{j_0}[\mathbf{a}_\eta]$. I take this j_0 to be the value of my strategy corresponding to $\langle \mathscr{B}, \psi_\eta, \mathbf{b}_\eta \rangle$ (i.e. in the situation where we have gained ψ_n and chosen $\mathbf{b}_\eta$ by following our strategies).

(2) $\psi_\eta = \bigwedge_{j \in J} \psi_j$
Let Nature choose $j_0 \in J$.
Because $\mathscr{A} \models {}_\sigma\psi_\eta[\mathbf{a}_\eta]$ we know that also $\mathscr{A} \models {}_\sigma\psi_{j_0}[\mathbf{a}_\eta]$. So, put him choose j_0 also in the play corresponding to $\langle \mathscr{A}, \psi_\eta, \mathbf{a}_\eta \rangle$.
In these cases $\psi_\alpha = \psi_{j_0}$, $u_\alpha = u_\eta$, $\mathbf{a}_\alpha = \mathbf{a}_\eta$, $\mathbf{b}_\alpha = \mathbf{b}_\eta$, $\tau_\alpha = \tau_\eta$, $\mathscr{A} \models {}_\sigma\psi_\alpha[\mathbf{a}_\alpha]$.

(3) $\psi_\eta = \exists x \psi(x)$.
Put $\psi_\alpha = \psi$ and $\tau_\alpha = \tau_\eta + 1$.
Since $\mathscr{A} \models {}_\sigma\psi_\eta[\mathbf{a}_\eta]$, there is $a\tau_\eta \in$A s.t.

$$\mathscr{A} \models {}_\sigma\psi_\alpha[\mathbf{a}_\eta, a\tau_\eta].$$

Let $\mathbf{a}_\alpha = \mathbf{a}_\eta \cup \{\langle \tau_\eta, a\tau_\eta \rangle\}$.
Now there is a $u \in U$ s.t. $u_\eta \leqslant_U u$ and $a\tau_\eta \in$dom$(g(u))$.
Put $u_\alpha = u$.
Let $b_{\tau_\eta} = g(u_\alpha)(a_{\tau_\eta})$ and put $\mathbf{b}_\alpha = \mathbf{b}_\eta \cup \{\langle \tau_\eta, b_{\tau_\eta} \rangle\}$.
Now $g(u_\alpha)(a_\beta) = b_\beta$ for each $\beta < \tau_\alpha$.
I choose b_{τ_η} to be the value of my strategy corresponding to the situation $\langle \mathscr{B}, \psi_\eta, \mathbf{b}_\eta \rangle$.
(4) $\psi_\eta = \forall x \psi(x)$.
Put $\psi_\alpha = \psi$ and $\tau_\alpha = \tau_\eta + 1$.
Let Nature choose $b_{\tau_\eta} \in B$.
There is a $u \in U$ s.t. $u_\eta \leqslant_U u$ and $b_{\tau_\eta} \in rng(g(u))$.
Let $a_{\tau_\eta} = (g(u))^{-1}(b_{\tau_\eta})$.
Define u_α, $\mathbf{a}_\alpha$ and $\mathbf{b}_\alpha$ as in (3).
Put Nature choose this a_{τ_η} in the play corresponding to $\langle \mathscr{A}, \psi_\eta, \mathbf{a}_\eta \rangle$. Now $\mathscr{A} \models {}_\sigma\psi_\alpha[\mathbf{a}_\alpha]$.
(2) and (4) are the places where the definition is branching corresponding to the choices Nature makes, i.e. to Nature's different strategies.
(5) Let ψ_η be an atomic formula or a negated one.
Since $(*_\eta)$ holds, we know that $\mathscr{A} \models {}_\sigma\psi_\eta[\mathbf{a}_\eta]$.
It follows that $\mathscr{A} \vDash \psi_\eta[\mathbf{a}_\eta]$. But we also know that $\mathscr{B} \vDash \psi_\eta[\mathbf{b}_\eta]$ because $g(u_\eta)(a_\beta) = b_\beta$ for each $\beta < \tau_\eta$. So we have succeeded.

In the same way we can prove that if $\mathscr{B} \vDash \phi$ then $\mathscr{A} \vDash \phi$.

$\mathscr{B} \vDash C_{\mathscr{A}} \Rightarrow \mathscr{A} \cong {}^{\lambda}_{w}{}^{e}\mathscr{B}$:

For simplicity, suppose that we have only predicate symbols in our language.

Let σ_1 be my winning strategy in $G(\mathscr{A}, C_{\mathscr{A}}, \emptyset)$ and σ_2 in the game $G(\mathscr{B}, C_{\mathscr{A}}, \emptyset)$.

Let $\mathscr{C} = \{C | C \text{ is a constituent}, C_{\mathscr{A}} \leqslant C\}$.

We shall now construct I, $\underline{U}$ and g by induction.

Let $\alpha < \lambda$ and suppose that for each $\eta < \alpha$ the following holds:

$(*_\eta)$: We have constructed a non-empty tree $\underline{U}_\eta = \langle U_\eta, <_\eta \rangle$, a non-empty family I_η of functions and a mapping $g_\eta : U_\eta \to I_\eta$ s.t.

(1) every $f \in I_\eta$ is an isomorphism from a substructure of $\mathscr{A}$ to a substructure of $\mathscr{B}$,

(2) if $u \in U_\eta$ s.t. u has at least one $<_\eta$-successor in $\underline{U}_\eta$ then:

– $\forall a \in A \exists v \in U_\eta : u \leqslant_\eta v$ and $a \in \text{dom}(g_\eta(v))$,

– $\forall b \in B \exists v \in U_\eta : u \leqslant_\eta v$ and $b \in rng(g_\eta(v))$,

(3) if $u \leqslant_\eta v$ then $g_\eta(u) \subseteq g_\eta(v)$,

(4) if D is a branch in $\underline{U}_\eta$ then length$(D) = \eta + 1$.

Moreover, we have a mapping $h_\eta : U_\eta \to \mathscr{C} \times \bigcup_{\beta<\lambda} {}^\beta A \times \bigcup_{\beta<\lambda} {}^\beta B$ which satisfies the following conditions:

(a) for each $u \in U_\eta$:

if $h_\eta(u) = \langle C(\mathbf{x}), \mathbf{a}, \mathbf{b} \rangle$ and rank$(u) = OT\langle \{v \in U_\eta | v <_\eta u\}, <_\eta \rangle = \gamma$ then $qr_{C\mathscr{A}}(C) = \gamma$, $\mathbf{x} = \langle x_\beta | \beta < \gamma \rangle$, $\mathbf{a} = \langle a_\beta | \beta < \gamma \rangle$, $\mathbf{b} = \langle b_\beta | \beta < \gamma \rangle$, $\mathscr{A} \models_{\sigma_1} C(\mathbf{x})[\mathbf{a}]$, $\mathscr{B} \models_{\sigma_2} C(\mathbf{x})[\mathbf{b}]$, $\text{dom}(g_\eta(u)) = \{a_\beta | \beta < \gamma\}$ and $g_\eta(u)(a_\beta) = b_\beta$ for each $\beta < \gamma$,

(b) if $u, v \in U_\eta$, $u \leqslant_\eta v$, $h_\eta(u) = \langle C_u, \mathbf{a}_u, \mathbf{b}_u \rangle$, $h_\eta(v) = \langle C_v, \mathbf{a}_v, \mathbf{b}_v \rangle$ and rank$(u) = \gamma$ then $C_u \leqslant C_v$, $\mathbf{a}_u = \mathbf{a}_v \restriction \gamma$, $\mathbf{b}_u = \mathbf{b}_v \restriction \gamma$,

and if $\gamma \leqslant \eta$ then $\underline{U}_\gamma$ is an initial subtree of $\underline{U}_\eta$, $I_\gamma \subseteq I_\eta$, $g_\gamma = g_\eta \restriction U_\gamma$, $h_\gamma = h_\eta \restriction U_\gamma$.

We have to prove that the above requirements hold for α.

lim(α): If $\alpha = 0$ put $U_0 = \{0\}$, $<_0 = \emptyset$, $I_0 = \{\emptyset\}$, $g_0 = \{\langle 0, \emptyset \rangle\}$, $h_0 = \{\langle 0, \langle C_{\mathscr{A}}, \emptyset, \emptyset \rangle \rangle\}$.

Let $\alpha \neq 0$.

Let $\underline{U}' = \langle U', <' \rangle = \bigcup_{\eta<\alpha} \underline{U}_\eta$, $I' = \bigcup_{\eta<\alpha} I_\eta$, $g' = \bigcup_{\eta<\alpha} g_\eta$, $h' = \bigcup_{\eta<\alpha} h_\eta$

(our hypothesis ensures that g' and h' are functions, $g' : U' \to I'$, $h' : U' \to \mathscr{C} \times \bigcup_{\beta<\lambda} {}^\beta A \times \bigcup_{\beta<\lambda} {}^\beta B$).

$\underline{U}'$ is a tree where the length of each branch is α.

Let $\mathscr{D} = \{D | \; D \text{ is a branch in } \underline{U}'\}$.

For each $D \in \mathscr{D}$ take a new element d_D and define:
$U_\alpha = U' \cup \{d_D | D \in \mathscr{D}\}$, $<_\alpha = <' \cup \{\langle u, d_D \rangle | D \in \mathscr{D}, u \in D\}$
$\underline{U}_\alpha = \langle U_\alpha, <_\alpha \rangle$.
Take any $D \in \mathscr{D}$. For each $u \in D$ there are C_u, $\mathbf{a}_u$ and $\mathbf{b}_u$ s.t. $h'(u) = \langle C_u, \mathbf{a}_u, \mathbf{b}_u \rangle$. Since $u \leqslant' v$ implies $C_u \leqslant C_v$, we know that there is a constituent $C_D \in \mathscr{C}$ s.t.
$C_D \geqslant C_u$ for each $u \in U'$, $qr_{C\mathscr{A}}(C_D) = \alpha$ and $C_D = C_D(\mathbf{x})$ where $\mathbf{x} = \langle x_\beta | \beta < \alpha \rangle$.
Let $\mathbf{a}_D = \sup_{u \in D} \mathbf{a}_u$, $\mathbf{b}_D = \sup_{u \in D} \mathbf{b}_u$. Now $\mathrm{dom}(\mathbf{a}_D) = \mathrm{dom}(\mathbf{b}_D) = \alpha$ and $\mathscr{A} \models_{\sigma_1} C_D(\mathbf{x})[\mathbf{a}_D]$, $\mathscr{B} \models_{\sigma_2} C_D(\mathbf{x})[\mathbf{b}_D]$.
Let $h_\alpha = h' \cup \{\langle d_D, \langle C_D, \mathbf{a}_D, \mathbf{b}_D \rangle \rangle | D \in \mathscr{D}\}$.

The induction hypothesis and the definition of g' ensure that if we have $f_D = \bigcup_{u \in D} g'(u)$ for each $D \in \mathscr{D}$, then f_D is an isomorphism from a substructure of $\mathscr{A}$ to a substructure of $\mathscr{B}$ and if $\mathbf{a}_D = \langle a_\beta | \beta < \alpha \rangle$, $\mathbf{b}_D = \langle b_\beta | \beta < \alpha \rangle$ then $\mathrm{dom}(f_D) = \{a_\beta | \beta < \alpha\}$ and $f_D(a_\beta) = b_\beta$ for each $\beta < \alpha$.
Put $I_\alpha = I' \cup \{f_D | D \in \mathscr{D}\}$ and $g_\alpha = g' \cup \{\langle d_D, f_D \rangle | D \in \mathscr{D}\}$.

It is now easy to show that $\underline{U}_\alpha$, I_α, g_α and h_α satisfy the desired conditions.

succ(α): Let $\mathscr{D} = \{D | D$ is a branch in $\underline{U}_\eta\}$.
($\alpha = \eta + 1$) For each $D \in \mathscr{D}$ take a $u \in D$ s.t. $\mathrm{rank}(u) = \eta$. Now there are $C(\mathbf{x})$, $\mathbf{a}_D$, $\mathbf{b}_D$ s.t. $h_\eta(u) = \langle C(\mathbf{x}), \mathbf{a}_D, \mathbf{b}_D \rangle$, $\mathscr{A} \models_{\sigma_1} C(\mathbf{x})[\mathbf{a}_D]$, $\mathscr{B} \models_{\sigma_2} C(\mathbf{x})[\mathbf{b}_D]$ and $C(\mathbf{x})$ is of the form $C(\mathbf{x}) = \bigwedge_{j \in J} (\pm)\theta_j(\mathbf{x}) \wedge \bigwedge_{k \in K} \exists x C_k(\mathbf{x}, x) \wedge \forall x \bigvee_{k \in K} C_k(\mathbf{x}, x)$.

Take any $a \in A$. There is a $k \in K$ s.t. $\mathscr{A} \models_{\sigma_1} C_k[\mathbf{a}, {}_D\, a]$.
Now there is a $b \in B$ s.t. $\mathscr{B} \models_{\sigma_2} C_k[\mathbf{b}, {}_D\, b]$.
Put $f_D^a = g_\eta(u) \cup \{\langle a, b \rangle\}$ and $C_D^a = C_k$. Clearly f_D^a is an isomorphism from a substructure of $\mathscr{A}$ to a substructure of $\mathscr{B}$, since C_η gives us the information needed in showing this.
Put $\mathbf{a}_D^a = \mathbf{a}_D \cup \{\langle \eta, a \rangle\}$, $\mathbf{b}_D^a = \mathbf{b}_D \cup \{\langle \eta, b \rangle\}$.
Similarly you can define f_D^b, C_D^b, $\mathbf{a}_D^b$ and $\mathbf{b}_D^b$ for each $b \in B$.
Let $E = \{d_D^c | c \in A \cup B, D \in \mathscr{D}\}$ be a set so that $E \cap U_\eta = \emptyset$. (You can suppose that $A \cap B = \emptyset$.) Now define:

$I_\alpha = I_\eta \cup \{f_D^c | c \in A \cup B, D \in \mathscr{D}\}$,

$U_\alpha = U_\eta \cup E,$
$<_\alpha = <_\eta \cup \{\langle u, d^c_D\rangle | c \in A \cup B,\ D \in \mathcal{D},\ u \in D\},$
$g_\alpha = g_\eta \cup \{\langle d^c_D, f^c_D\rangle | c \in A \cup B,\ D \in \mathcal{D}\},$
$h_\alpha = h_\eta \cup \{\langle d^c_D, \langle C^c_D, \mathbf{a}^c_D, \mathbf{b}^c_D\rangle\rangle | c \in A \cup B,\ D \in \mathcal{D}\}.$
Now it is easy to check that our requirements are satisfied.

Finally put $\underline{U} = \bigcup_{\alpha<\lambda} \underline{U}_\alpha$, $I = \bigcup_{\alpha<\lambda} I_\alpha$ and $g = \bigcup_{\alpha<\lambda} g_\alpha$. Clearly $I : \mathcal{A} \cong^{\lambda e}_{w} \mathcal{B}$.

$\underline{\mathcal{A} \equiv^{\lambda}_{L\omega\omega} \mathcal{B} \Rightarrow \mathcal{A} \equiv^{\lambda}_{N\infty\lambda} \mathcal{B}:}$

First we prove that $\mathcal{A} \equiv^{\lambda}_{L\omega\omega} \mathcal{B} \Rightarrow \mathcal{A} \cong^{\lambda e}_{s} \mathcal{B}$. (For simplicity, suppose that there are only predicate symbols in L.)

Define: $I = \{f |\ f$ is an isomorphism from a substructure of $\mathcal{A}$ to a substructure of $\mathcal{B}$ s.t. $\overline{\mathrm{dom}(f)} < \lambda$ and if $\mathrm{dom}(f) = \{a_\beta | \beta < \eta\}$ and $b_\beta = f(a_\beta)$ for all $\beta < \eta$ then $(\mathcal{A}, \mathbf{a}) \equiv_{L\omega\omega} (\mathcal{B}, \mathbf{b})\}$.

(1) of $\equiv^{\lambda}_{L\omega\omega} \Rightarrow \emptyset \in I$, (2) of $\equiv^{\lambda}_{L\omega\omega} \Rightarrow$ back-and-forth,
(3) of $\equiv^{\lambda}_{L\omega\omega} \Rightarrow$ the strong extension property.
So:

$$\mathcal{A} \cong^{\lambda e}_{s} \mathcal{B}.$$

(1) This implies that $\mathcal{A} \equiv_{N\infty\lambda} \mathcal{B}$.
(2) Back-and-forth: always true (use constituents).
(3) λ-extension property:
Let $\alpha < \lambda$, $\mathbf{a} \in {}^\alpha A$, $\mathbf{b} \in {}^\alpha B$ s.t. for each $\beta < \alpha$.

$$(\mathcal{A}, \mathbf{a}\restriction\beta) \equiv_{N\infty\lambda} (\mathcal{B}, \mathbf{b}\restriction\beta).$$

Now $\forall\beta < \alpha$:$(\mathcal{A}, \mathbf{a}\restriction\beta) \equiv_{L\omega\omega} (\mathcal{B}, \mathbf{b}\restriction\beta)$ and therefore $(\mathcal{A}, \mathbf{a}) \equiv_{L\omega\omega} (\mathcal{B}, \mathbf{b})$.
This and the fact that $\mathcal{A} \equiv^{\lambda}_{L\omega\omega} \mathcal{B}$ imply that $(\mathcal{A}, \mathbf{a}) \equiv^{\lambda}_{L\omega\omega} (\mathcal{B}, \mathbf{b})$.
As above we get: $(\mathcal{A}, \mathbf{a}) \equiv_{N\infty\lambda} (\mathcal{B}, \mathbf{b})$.
So: $\mathcal{A} \equiv^{\lambda}_{N\infty\lambda} \mathcal{B}$.
$\underline{\mathcal{A} \equiv^{\lambda}_{N\infty\lambda} \mathcal{B} \Rightarrow \mathcal{A} \cong^{\lambda e}_{s} \mathcal{B}:}$

This is a similar proof to the one for

$$\mathcal{A} \equiv^{\lambda}_{L\omega\omega} \mathcal{B} \Rightarrow \mathcal{A} \cong^{\lambda e}_{s} \mathcal{B}.$$

(See the previous proof! Change only "$(\mathcal{A}, \mathbf{a}) \equiv_{L\omega\omega} (\mathcal{B}, \mathbf{b})$" to a form: "$(\mathcal{A}, \mathbf{a}) \equiv_{N\infty\lambda} (\mathcal{B}, \mathbf{b})$" in the definition of I.)
The theorem is now proved. □

As we mentioned before, we can define the satisfaction in several ways (i.e. the pay-off function in those cases when there is a limit play in the game). However, the previous proof works no matter which of these definitions we take to be our basic one. In proving $\mathscr{A} \cong_w^{\lambda e} \mathscr{B} \Rightarrow (\mathscr{A} \models \phi^r \Rightarrow \mathscr{B} \models \phi^r)$ when there is a limit play $\langle \sigma', \pi' \rangle$ in $G(\phi^r, \mathscr{B}, \emptyset)$, where σ' is the strategy we gave to me in that proof, then there is a similar play $\langle \sigma, \pi \rangle$ in $G(\phi^r, \mathscr{A}, \emptyset)$ where σ is my winning strategy. Every time we have to choose a successor of a disjunction or a conjunction in $\langle \sigma', \pi' \rangle$ we choose similarly in $\langle \sigma, \pi \rangle$. So we make the same 'bad choices' in $\langle \sigma', \pi' \rangle$ as we make in $\langle \sigma, \pi \rangle$. Therefore $p(\sigma', \pi') = 1$ since $p(\sigma, \pi) = 1$.

It is easy to see that the theorem can be proved if our languages are arbitrarily deep provided that in each formula ϕ the quantifier-rank is $< \lambda$ for every node $t \in \text{Tree}_\phi$.

COROLLARY 3.6. Let $\mathscr{A}$ and $\mathscr{B}$ be λ-saturated models s.t. $\mathscr{A} \equiv_{L_{\omega\omega}} \mathscr{B}$. Then $\mathscr{A} \equiv_{N_{\infty\lambda}} \mathscr{B}$.

Proof: $\mathscr{A} \equiv_{L_{\omega\omega}}^{\lambda} \mathscr{B}$. □

COROLLARY 3.7. Let $\mathscr{A}$ be λ-homogeneous and $\alpha < \lambda$, $\mathbf{a} \in {}^{\alpha}A$, $\mathbf{b} \in {}^{\alpha}A$ s.t.

$$(\mathscr{A}, \mathbf{a}) \equiv_{L_{\omega\omega}} (\mathscr{A}, \mathbf{b}).$$

Then $(\mathscr{A}, \mathbf{a}) \equiv_{N_{\infty\lambda}} (\mathscr{A}, \mathbf{b})$.

Proof: $(\mathscr{A}, \mathbf{a}) \equiv^{\lambda}{}_{L_{\omega\omega}} (\mathscr{A}, \mathbf{b})$. □

COROLLARY 3.8. Let $\mathscr{A}$ and $\mathscr{B}$ be models s.t. $\bar{\bar{A}} \leqslant \lambda$, $\bar{\bar{B}} \leqslant \lambda$, $\mathscr{A} \equiv_{N_{\infty\lambda}} \mathscr{B}$. Then $\mathscr{A} \cong \mathscr{B}$.

Proof: $\mathscr{A} \cong_w^{\lambda e} \mathscr{B}$. □

COROLLARY 3.9. If $\lambda = \omega$, we get:

$$\mathscr{A} \equiv_{N_{\infty\omega}} \mathscr{B} \Leftrightarrow \mathscr{A} \equiv_{L_{\infty\omega}} \mathscr{B}\ (\Leftrightarrow \mathscr{A} \equiv_{L(\omega)} \mathscr{B}) \Leftrightarrow \mathscr{B} \models C_{\mathscr{A}} \Leftrightarrow \mathscr{B} \models S_{\mathscr{A}},$$

where $S_{\mathscr{A}}$ is the Scott-sentence of $\mathscr{A}$. ($L(\omega)$ is one of the languages defined in [4].)

Proof: Now $I : \mathscr{A} \cong_s^{\omega e} \mathscr{B} \Leftrightarrow I : \mathscr{A} \cong_w^{\omega e} \mathscr{B} \Leftrightarrow I : \mathscr{A} \cong_p \mathscr{B}$ (where $\mathscr{A} \cong_p \mathscr{B}$ means that $\mathscr{A}$ and $\mathscr{B}$ are partially isomorphic). □

Department of Computer Science,
University of Helsinki

REFERENCES

[1] Dickmann, M. A., *Large Infinitary Languages*, North-Holland, Amsterdam, 1975.

[2] Drake, F. R., *Set Theory. An Introduction to Large Cardinals*, North-Holland, Amsterdam, 1973.

[3] Hintikka, J. and Rantala, V., 'A new approach to infinitary languages', *Annals of Mathematical Logic* **10** (1976), 95–115.

[4] Keisler, H. J., 'Formulas with linearly ordered quantifiers', in *The Syntax and Semantics of Infinitary Languages*, ed. by J. Barwise, Lecture Notes in Mathematics **72**, Springer, Berlin, 1968, pp. 96–130.

[5] Kueker, D. W., 'Back-and-forth arguments and infinitary logics', in *Infinitary Logic: In Memoriam Carol Karp*, ed. by D. W. Kueker, Lecture Notes in Mathematics **492**, Springer, Berlin, 1975, pp. 17–71.

PART III

SET THEORY AND MODEL THEORY

SOLOMON FEFERMAN

GENERALIZING SET-THEORETICAL MODEL THEORY AND AN ANALOGUE THEORY ON ADMISSIBLE SETS*

1. INTRODUCTION

By *set-theoretical model theory* I mean the kind of model theory for finitary first order predicate calculus $\mathscr{L}_{\omega,\omega}$ exemplified by the book by Chang and Keisler [C, K]. As we know, parts of this (such as the completeness theorem) can be carried out in a completely elementary framework. The distinctively set-theoretical parts of [C, K] use the theory of transfinite ordinals and cardinals, the axiom of choice and, on occasion, the continuum hypothesis and its generalizations.

By *model theory on admissible sets* I mean here specifically a development due to Cutland [C2] which presents *analogues* for various standard notions and theorems of model theory in the following setting: one takes any admissible set $(M, \in)$ which has a $\Sigma_1^{(M)}$ global well-ordering and considers structures $\mathfrak{A} = (A, \ldots)$ with $A \subseteq M$ for which the satisfaction relation on $\mathfrak{A}$ is $\Sigma_1^{(M)}$. In the case that M is locally countable (i.e., where each element of M is countable in M), we get analogues of classical results in which (i) the elements of M correspond to the countable sets and (ii) the $\Sigma_1^{(M)}$ subsets of M correspond to the sets of cardinality $\leqslant \aleph_1$. The special case $M = L_{\omega_1}$ (ω_1 = least non-recursive ordinal) had been previously treated in [C1], where for $\mathfrak{A}$ coded in the natural numbers, (i) HYP corresponds to countable and (ii) Π_1^1 corresponds to cardinality $\leqslant \aleph_1$. Among results obtained in [C1] are analogues of existence and uniqueness of $\aleph_1$-saturated structures, the Ehrenfeucht-Mostowski theorem, the Vaught two-cardinal theorem, and results connected with categoricity in uncountable powers.

Cutland's work is not accounted for by the kind of generalization of model theory to admissible sets developed by Barwise [B]. In the latter one uses languages $\mathscr{L}_M$ rather than $\mathscr{L}_{\omega,\omega}$, and membership in M plays instead a role analogous to being finite, with $\Sigma_1^{(M)}$ subsets of M corresponding to the (recursively) enumerable sets.

To explain the remarkable success of Cutland's analogue treatment and to see to what extent such work can be carried further, one should set up a theoretical framework for which set theory provides one interpreta-

J. Hintikka, I. Niiniluoto, and E. Saarinen (eds.), Essays on Mathematical and Philosophical Logic, 171–195.

tion and admissible sets another, and in which one can prove general statements of model theory. These should directly generalize the theorems of, say, [C, K] on the one hand and of [C2] on the other. At the same time, other interpretations would extend the range of analogues which can be obtained.

A means to do this was suggested in [F], p. 125. Two *theories of operations and classes* T_0 and T_1 were presented there. It was outlined how T_0 could serve to generalize those parts of classical mathematics which have constructive counterparts and/or recursive analogues. Similarly it was indicated how T_1 could be used to generalize those parts of classical mathematics which have (so-called) hyperarithmetic analogues. This paper constitutes a detailed explanation of how the latter may be carried out for set-theoretical and admissible model theory. It has proved more convenient (and perhaps perspicuous) for this purpose to introduce a new theory T_Ω of *operations, classes, and ordinals*; T_Ω can be interpreted in T_1. To admit direct recursion-theoretic ("effective") interpretations, the theories T_0, T_1, and T_Ω are taken to be *non-extensional.* As is familiar this is no real loss, since extensionality is entirely dispensable by the systematic use of classes equipped with equivalence relations (the way mathematics used to be done). The intuitive interpretation of the theory T_Ω is that operations are entities given by rules of definition and that classes are entities given by membership-determining properties (not necessarily decidable, i.e., not necessarily having characteristic functions). In T_Ω, ordinals may be thought of as being provided by unique names, while in T_1 they were treated more generally without uniqueness.

The axioms of T_Ω are set out in Section 2. Of special interest are the added statements of (i) EOC, which expresses that every ordinal is countable and (ii) CH, which expresses the continuum hypothesis. It is shown in Section 3 how to obtain a variety of models $\mathfrak{M}^*$ of T_Ω, among them certain ones built from set-theoretical and well-ordered admissible models $\mathfrak{M} = (M, \in)$. In the former, Ω may be interpreted as the ordinals below any regular cardinal to satisfy T_Ω, and as $\aleph_1$ when EOC is also to be satisfied. In the latter, Ω may be interpreted as $o(M)$ for T_Ω, which further satisfies EOC when M is locally countable.

Section 4 presents the fundamental machinery for the mathematics of operations, classes, and ordinals in T_Ω. This leads directly in Section 5 to closure conditions on denumerable and Ω-enumerable classes. One general classical result that fails is that if B is denumerable and A is a subclass of B, then A is also denumerable; in fact, one can even get a counter-example with A being Ω-enumerable (5.3). In the admissible interpretation, this is simply an example of a $\Sigma_1^{(M)}$ subset of ω which is not in M. As it turns out, one can get a useful substitute for transfer of denumerability to subclasses, namely when both A and $B - A$ are Ω-enumerable (5.4). Basic model theory is carried out in T_Ω + EOC, where the hypothesis EOC is merely added for simplicity. The main new notion introduced is that of an *Ω-valuated structure*, i.e., an $\mathfrak{A}$ for which the satisfaction relation $\text{Sat}_\mathfrak{A}$ is Ω-enumerable. In the set-theoretical interpretation, every Ω-enumerated structure is automatically Ω-valuated. In the admissible interpretation, these are the same as the M-structures in Cutland's sense. All the further work is done with Ω-valuated structures. This proceeds through the Elementary Chain Theorem and the downward Löwenheim-Skolem theorem in Section 6.

The topic of *Ω-saturated structures* is chosen (in Section 7) as the principal one to illustrate our means of generalizing both set-theoretical and (Cutland's) admissible model theory. For the purpose of obtaining existence of such the continuum hypothesis CH (as expressed in T_Ω) is added. In the models $\mathfrak{M}^*$ built for set-theoretical models $\mathfrak{M}$ of ZFC we have CH true just in case it is true in its usual sense in $\mathfrak{M}$. On the other hand, as it happens, CH is *always* true in the $\mathfrak{M}^*$ built from locally countable globally Σ_1-well-ordered $\mathfrak{M}$. Section 8 sketches a generalization of the basic properties of ultrapowers and their saturation under the hypotheses T_Ω + EOC + CH (no analogue was given for this in [C2]). The paper concludes with some final comments and suggestions for further work.

$\aleph_1$-saturated models have been used in model theory as a *tool* to derive results such as interpolation theorems, preservation theorems, completeness of certain theories, etc. It is becoming clear from the work of Barwise and Schlipf [B, S] that countable models having the more refined recursive saturation property serve the same purpose. On the other hand, $\aleph_1$-saturated and related models still play an important role in uncountable model theory, and continue to have an interest in their own right. They were chosen here primarily to reveal the methodology of the present kind of generalization, independently of their eventual status in the subject.

2. THE THEORY T_Ω OF OPERATIONS, CLASSES, AND ORDINALS

The first part of the theory T_Ω (in 2.3 below) is essentially the same as the theory T_0 of [F], without the axiom scheme V of inductive generation, but with classical logic. The axioms specifically for ordinals (in 2.5) are new. Familiarity with [F] is not presumed but would be helpful. Certain minor syntactical and notational changes are made here.

2.1. *Syntax of* T_Ω

Individual (*or general*) *variables*: $a, b, c, \ldots, x, y, z$.
Individual constants: a series of constants $\underline{k}, \underline{s}, \ldots$ introduced in the axioms (letters for constants are distinguished from variables by underlining).
Class variables: $A, B, C, \ldots, X, Y, Z$.
Class constants: Ω, $<$.
Atomic formulas: (i) equations between variables or constants of either sort, and (ii) formulas of the form $\mathrm{App}(r, s, t)$ and $t \in W$, where r, s, t are individual variables or constants and W is a class variable or constant.
Connectives, quantifiers: $\neg, \vee, \wedge, \Rightarrow, \forall, \exists$.
Formulas: generated from atomic formulas by the connectives and the quantifiers applied to either sort of variable.
Elementary second order formulas: formulas which contain no equations containing class variables or constants and no bound class variables.

2.2. *Notation and abbreviations*

(a) Application terms are built up in an extension of this syntax as follows: each term of T_Ω is an application term; if t_1, t_2 are application terms, then so also is t_1t_2. We read $t_1t_2 \ldots t_n$ by associating to the left, $(\ldots(t_1t_2)\ldots)t_n$. The formula $t \simeq y$ is explained as a formula of T_Ω by:

(i) $t \simeq y$ is $t = y$ when t is a term of T_Ω.

(ii) $t_1t_2 \simeq y$ is $\exists z_1, z_2[t_1 \simeq z_1 \wedge t_2 \simeq z_2 \wedge \mathrm{App}(z_1, z_2, y)]$.

Then we write

(iii) $t_1 \simeq t_2$ for $\forall y[t_1 \simeq y \Leftrightarrow t_2 \simeq y]$.

(iv) $t\downarrow$ for $\exists y(t \simeq y)$.

(v) $t \in W$ for $\exists y(t \simeq y \wedge y \in W)$.

(vi) $t_1 = t_2$ is written for $t_1 \simeq t_2$ in case both $t_1\downarrow$ and $t_2\downarrow$ have been proved, and $t_1 \neq t_2$ for $\neg(t_1 \simeq t_2)$ under the same conditions.

(b) *Pairing, n-tupling, etc.* Among the constants is one denoted $\underline{p}$. We write (t_1, t_2) for $\underline{p}t_1t_2$, and $(t_1, \ldots, t_n, t_{n+1})$ for $((t_1, \ldots, t_n), t_{n+1})$.

(c) *Abstraction.* Among the constants are some denoted $\underline{c}_k(k < \omega)$. Given any elementary second order Φ and list of variables $\mathbf{v}$ of the form $x, y_1, \ldots, y_n, Y_1, \ldots, Y_m$ with $n \geqslant 0$, $m \geqslant 0$ and which includes all the variables of Φ we write $\{x|\Phi(x, y_1, \ldots, y_n, Y_1, \ldots, Y_m\}$ or $\{x|\Phi\}$ for

$\underset{\sim}{c}_k(y_1, \ldots, y_n, Y_1, \ldots, Y_m)$ where k is the pair of Gödel numbers $\ulcorner\Phi\urcorner$, $\ulcorner\mathbf{v}\urcorner$.

(d) *Restricted quantifier notation*. $\forall x \in X\Phi$ is written for $\forall x[x \in X \Rightarrow \Phi]$, similarly $\forall x_1, x_2 \in X\Phi$, $\exists x \in X\Phi$, $\exists x_1, x_2 \in X\Phi$, etc.

2.3. *The axioms of T_Ω for operations and classes*

The logic is that of classical 2-sorted predicate calculus with equality (allowing equality between sorts). The axioms are as follows:

I. (i) $xy \simeq z_1 \wedge xy \simeq z_2 \Rightarrow z_1 = z_2$.
   (ii) $\forall X \exists x(X = x)$.

II. (i) (Constant functions). $(\underset{\sim}{k}xy\downarrow) \wedge \underset{\sim}{k}xy = x$.
   (ii) (Substitution). $(\underset{\sim}{s}xy)\downarrow \wedge \underset{\sim}{s}xyz \simeq xz(yz)$.
   (iii) (Definition by cases). $(\underset{\sim}{d}xyab)\downarrow \wedge (a = b \Rightarrow \underset{\sim}{d}xyab = x) \wedge (a \neq b \Rightarrow \underset{\sim}{d}xyab = y)$.
   (iv) (Pairing, projections). $(\underset{\sim}{p}x_1x_2)\downarrow \wedge (\underset{\sim}{p}_1z)\downarrow \wedge (\underset{\sim}{p}_2z)\downarrow \wedge \underset{\sim}{p}_i(x_1, x_2) = x_i$.

III. (*Elementary comprehension scheme*). For each elementary second order $\Phi(x, y_1, \ldots, y_n, Y_1, \ldots, Y_m)$,

$$z \in \{x \mid \Phi(x, \ldots)\} \Leftrightarrow \Phi(z, \ldots).$$

IV. (*Join*). $\forall x \in A \exists Y(fx \simeq Y) \Rightarrow (\underset{\sim}{j}Af)\downarrow \wedge$

$$\forall z\{z \in (\underset{\sim}{j}Af) \Leftrightarrow \exists x \in A \exists y \exists Y[z = (x, y) \wedge fx \simeq Y \wedge y \in Y]\}.$$

Remark. Pairing and the projections could be defined using $\underset{\sim}{d}$, simply taking $\underset{\sim}{d}$ for $\underset{\sim}{p}$ and $\underset{\sim}{p}_1z \simeq z00$, $\bar{p}_2z \simeq z01$ (where 0, 1 are defined below).

2.4. *Further notation and abbreviation*

Before completing the axiom list of T_Ω, it is useful to make the following abbreviations:

(i) We write

$$f : A \to B \text{ for } \forall x\{x \in A \Rightarrow fx \in B\}.$$
$$f : A \underset{\text{onto}}{\to} B \text{ for } (f : A \to B) \wedge \forall y \in B \exists x \in A(fx = y).$$
$$f : A \overset{1-1}{\to} B \text{ for } (f : A \to B) \wedge \forall x_1, x_2 \in A[x_1 \neq x_2 \Rightarrow fx_1 \neq fx_2].$$
$$B^A \text{ or } (A \to B) \text{ for } \{f \mid f : A \to B\}.$$
$$f \equiv_{(A \to B)} g \text{ for } f, g \in B^A \wedge \forall x \in A[fx = gx].$$

When A, B are given we write $\equiv$ for $\equiv_{(A \to B)}$.

(ii) Various classes A are considered with 'equality relations' $\equiv_A$ on them, i.e., equivalence relations $E_A \subseteq A^2$ on A. Because our system is non-extensional, we cannot in general form quotients $A/\equiv_A$; however, the preceding notation can be extended to these situations by writing:

$$f:(A/\equiv_A) \to (B/\equiv_B) \text{ for } f: A \to B \wedge \forall x_1, x_2 \in A$$
$$[x_1 \equiv_A x_2 \Rightarrow fx_1 \equiv_B fx_2].$$
$$f:(A/\equiv_A) \underset{\text{onto}}{\to} (B/\equiv_B) \text{ for } f:(A/\equiv_A) \to (B/\equiv_B) \wedge$$
$$\forall y \in B \exists x \in A (fx \equiv_B y).$$

We treat similarly $f:(A/\equiv_A) \overset{1-1}{\to} (B/\equiv_B)$. Mixed notations such as $f: A \underset{\text{onto}}{\to} (B/\equiv_B)$ are understood as $f:(A/=) \underset{\text{onto}}{\to} (B/\equiv_B)$, i.e., where $\equiv_A$ is taken as $\{(x, y)|x, y \in A \wedge x = y\}$.

(iii) Lower case Greek letters $\alpha, \beta, \gamma, \ldots, \xi, \eta, \zeta$ are used as variables to range (by convention) over the elements of Ω. We write $\alpha < \beta$ for $(\alpha, \beta) \in (<)$, $\alpha \leqslant \beta$ for $\alpha < \beta \vee \alpha = \beta$, $[0, \alpha)$ for $\{\xi | \xi < \alpha\}$ and $[0, \alpha]$ for $\{\xi | \xi \leqslant \alpha\}$. We write $\forall \xi < \alpha \Phi$ for $\forall \xi (\xi < \alpha \Rightarrow \Phi)$, and similarly for $\exists \xi < \alpha < \Phi$. Finally, we identify α with $[0, \alpha)$ in mapping situations, e.g.,

$$f: \alpha \to \Omega \text{ is written for } f: [0, \alpha) \to \Omega,$$
$$f: \alpha \to \beta \text{ is written for } f: [0, \alpha) \to [0, \beta), \text{ etc.}$$

It is assumed we have a constant $\underline{s}_1$ representing successor. We write α' for $\underline{s}_1 \alpha$.

(iv) We write $\mathrm{Lim}(\alpha)$ for $\alpha \neq 0 \wedge \forall \xi (\xi < \alpha \Rightarrow \xi' < \alpha)$.

2.5. *The axioms for* Ω *and* $<$

These use additional constants 0, ω, $\underline{s}_1$, $\underline{l}$, $\underline{q}$, and $\underline{c}$.

V. (i) (*Linear ordering.*) $\alpha \leq \alpha \wedge (\alpha \leq \beta \vee \beta \leq \alpha) \wedge$
$(\alpha \leq \beta \wedge \beta \leq \gamma \Rightarrow \alpha \leq \gamma) \wedge (\alpha \leq \beta \wedge \beta \leq \alpha \Rightarrow \alpha = \beta)$.

(ii) (*Least element.*) $0 \in \Omega \vee \forall \alpha (0 \leq \alpha)$.

(iii) (*Successor.*) $\alpha' \in \Omega \wedge \alpha < \alpha' \wedge \forall \xi (\alpha < \xi \Rightarrow \alpha' \leq \xi)$.

(iv) (ω) $\mathrm{Lim}(\omega) \wedge \forall \alpha [\mathrm{Lim}(\alpha) \Rightarrow \omega \leq \alpha]$.

(v) (*Transfinite induction.*) For each formula $\Phi(\alpha, \ldots)$,

$$\forall \alpha [\forall \xi < \alpha \Phi(\xi, \ldots) \Rightarrow \Phi(\alpha, \ldots)] \Rightarrow \forall \alpha \Phi(\alpha, \ldots).$$

(vi) (*Reflection.*) $(f: \alpha \to \Omega) \Rightarrow \exists \beta (f: \alpha \to \beta)$.

(vii) (*Decidability of membership in initial segments.*)

$$(\underline{l}\alpha x \simeq 0 \vee \underline{l}\alpha x \simeq 1) \wedge (\alpha \underline{l} x = 0 \Leftrightarrow x < \alpha).$$

(viii) (*Computability of bounded quantification.*)

$$\forall z [\underline{q} \alpha f \simeq z \Rightarrow z = 0 \vee z = 1]$$

$\wedge[\underline{q}\alpha f \simeq 0 \Leftrightarrow \forall \xi < \alpha(f\xi \simeq 0)]$
$\wedge[\underline{q}\alpha f \simeq 1 \Leftrightarrow \exists \xi < \alpha(f\xi \neq 0)]$.

(ix) (*Selection.*) $[f\alpha \simeq 0 \Rightarrow (\underline{c}f) \in \Omega \wedge f(\underline{c}f) \simeq 0]$.

2.6. *Three further axioms*

These use additional constants $\underline{e}$, $\underline{g}$, and $\underline{g}^*$:

EOC (*Every ordinal countable*)

$$\forall \alpha(\underline{e}\alpha : \omega \underset{\text{onto}}{\rightarrow} [0, \alpha]).$$

CH (*Continuum hypothesis*) $\underline{g} : \Omega \underset{\text{onto}}{\rightarrow} (\omega^\omega/\equiv)$.

GCH* (*A consequence of the generalized continuum hypothesis*)

$$\forall \alpha[\underline{g}^*\alpha : \Omega \underset{\text{onto}}{\rightarrow} (\alpha^\alpha/\equiv)].$$

Remarks

(i) In the presence of EOC is it sufficient to take (viii) only for ω, i.e., computability of quantification over ω. Furthermore, GCH* follows from CH in that case.

(ii) The theory $T_1^{(N)}$ of [F], p. 123 is obviously interpreted in T_Ω by matching N with $[0, \omega)$.

2.7. *Denumerable and Ω-enumerable classes*

The following definition will be used in explaining features of the models $\mathfrak{M}^*$ of T_Ω to be described in the next section.

DEFINITION. (i) A is said to be *C-enumerable* (mod $\equiv_A$) if A is empty or $\exists f[f : C \underset{\text{onto}}{\rightarrow} A/\equiv_A]$. (ii) A is said to be *denumerable* (mod $\equiv_A$) if it is ω-enumerable (mod $\equiv_A$).

When A is taken with the identity relation for $\equiv_A$ we drop 'mod $\equiv_A$'.

Under the hypothesis EOC the classes which are Ω-enumerable but not denumerable correspond to the sets of cardinality ω_1 in set theory. The exact relationship is established in 3.2 of the next section.

3. SOME MODELS OF T_Ω

The reader is advised to look first at the statements of theorems at the end of 3.2, 3.3.

3.1. *The general method of forming models*

It is shown in [F], pp. 104–107 how to associate with any model $\mathfrak{M} = (M, \text{App}, k, s, d, p, p_1, p_1, p_2)$ of axioms I(i) and II, a collection $Cl \subseteq M$ and a relation $\in \subseteq M \times Cl$ which satisfies the remaining axioms I(ii), III, and IV of 2.3.

Following Friedman [Fr], p. 115, we shall call any such $\mathfrak{M}$ an *enumerative system*; this is another form of the abstract Wagner-Strong axioms for enumerative recursion theory on a set M with a given pairing operator $P : M^2 \to M$ and projections $P_i : M \to M$ where P, P_1, P_2 are to be 'recursive' on M. All known generalizations of recursion theory to such situations which set up analogues of the relation $\{e\}(x) \simeq y$ (from ordinary recursion theory) satisfy these axioms. These do not include all generalizations of recursion theory, such as those which set up analogues of the relation $W(e, x) \Leftrightarrow x \in W_e$ (where W_e is an enumeration of the r.e. sets); to pass to an enumeration of graphs of functions we need an 'effective' method of uniformizing generalized r.e. relations – which usually requires some sort of effective selection operators on M.

We can always generate an enumerative system to include any given P, P_1, P_2. In fact, given any collection $\mathscr{F}_n$ of partial functions of n arguments in M to M, including such P, P_1, P_2 and where the cardinality of each $\mathscr{F}_n$ is at most that of M, we can set up an enumerative system such that for each $F \in \mathscr{F}_n$ there is an f with

(i) $\qquad fx_1 \ldots x_{n-1}\downarrow$ and $fx_1 \ldots x_{n-1}x_n \simeq F(x_1, \ldots, x_n)$.

One simply assigns a code to each $F \in \mathscr{F}_n$ and picks codes k, s, d, P, P_1, P_2 and then generates $\simeq$ inductively to satisfy (i) and the axioms II. {For example, we may take $k = 0$, $kx \simeq (0, x)$ and $(0, x)y \simeq x$ so $kxy\downarrow$ and $kxy = x$. For s we may take $s = 1$, $sx \simeq (1, x)$, $(1, x)y \simeq (1, x, y)$ and $(1, x, y)z \simeq xz(yz)$; in other words,

$$yz \simeq w \wedge xz \simeq u \wedge uw \simeq v \Rightarrow (1, x, y)z \simeq v.\}$$

The preceding shows how to get a great variety of models of T_Ω(I–IV). To get models of all of T_Ω we turn to more special situations.

3.2. *Starting with set-theoretic models*

Let $(M, \in_M)$ be any model of ZFC. Take pairing P and projections P_i on M as usual. Let $\mathscr{F}_1$ be the collection of partial functions whose graph is a set in M, together with P_1, P_2 and the successor operation $x \mapsto x'$. Let

$\mathscr{F}_2$ consist of P together with the characteristic function of $\in_M$. Let $\mathscr{F}_3$ consist of the function Sep with $\text{Sep}(a, b, \Phi) = \{x \in_M a | (M, \in_M) \models \Phi(x, b)\}$. We generate an enumerative system from $\mathscr{F}_1$, $\mathscr{F}_2$, and $\mathscr{F}_3$ as just indicated, but also with the following inductive clauses for a suitable code q:

$$\text{(i)} \qquad \forall x \in_M a(fx \simeq 0) \Rightarrow qfa \simeq 0$$
$$\forall x \in_M a \exists y \neq 0(fx \simeq y) \Rightarrow qfz \simeq 1.$$

Now the transfinite inductive definition of $fx \simeq y$ is set-theoretically definable; it only needs the ordinals of M for its stages. (To make sense of this when M is not standard, we work formally within ZFC instead.) There is a set-theoretical relation $\text{Der}(d, f, x, y)$ which expresses that the set d codes a derivation tree of the relation $fz \simeq y$, so that $fz \simeq y \Leftrightarrow \exists d$ $\text{Der}(d, f, x, y)$. Since $\simeq$ is definable we may in this case take Cl to consist of all codes of formulas $\Phi(x, a_1, \ldots, a_n)$ in the language of $(M, \in_M)$ and with parameters a_i in M. The result is a model $\mathfrak{M}^*$ of T_Ω(I–IV) with the following features

(ii) *If $h \in M$ is a set of ordered pairs which is a function in M, then there is $\hat{h} \in M$ with $\hat{h}x \simeq y \Leftrightarrow (x, y) \in h \Leftrightarrow h(x) = y$.*

(iii) *If $a \in M$, then for each f there is a function f_a in M with $f_a(x) = y \Leftrightarrow x \in_M a \wedge fx \simeq y$.*

(iv) *If $a \in M$, then there is a class a^* such that $x \in_M a \Leftrightarrow x \in a^*$.*

(v) *If A is a class, then for each $a \in M$ there is a (unique) set $b \in M$ with $x \in_M b \Leftrightarrow x \in_M a \wedge x \in A$.*

Now to satisfy axiom V, take Ω to be $\{x | x \in_M \kappa\}$ where κ is any regular uncountable cardinal in M, and $<$ to be the $<$-relation on κ. $\alpha, \beta, \gamma, \ldots, \xi, \eta, \zeta$ thus range over the predecessors of κ. By construction, V(i)–(ix) are directly satisfied. For the selection axiom (x), we use the minimum function μ on subsets of κ which lies in M, i.e., $\mu : \mathscr{P}_M(\kappa) \to \kappa$, such that for $a \subseteq \kappa$ with $a \neq 0$, $\mu(a) \in_M a$. Then we take $cf \simeq \mu(\text{Sep}(\kappa, f, \Phi))$ where $\Phi(x, f)$ is $\text{Ord}(x) \wedge \exists d \, \text{Der}(d, f, x, 0)$, i.e., $cf \simeq \mu(\{\alpha | f\alpha \simeq 0\})$. Thus all of T_Ω is now seen to be true in this model.

Let A be any class with $\equiv_A$ an equivalence relation on A. We may use Scott's method (taking equivalence classes to be sets of elements of least rank) to actually identify $A/\equiv_A$ with a class. Then

(vi) *A is denumerable* (mod $\equiv_A$) *in the sense of $\mathfrak{M}^*$ iff $A/\equiv_A$ is denumerable in M, and*

(vii) *A is Ω-enumerable* (mod $\equiv_A$) *in the sense of $\mathfrak{M}^*$ iff $A/\equiv_A$ is of cardinality $\leq \kappa$ in M.*

In particular, for each α the class $(\alpha^\alpha/\equiv)$ is in 1–1 correspondence with the set of all functions from α to α in M. In conclusion, we have the following:

THEOREM. *With each model* $\mathfrak{M}$ *of* ZFC *and regular uncountable* κ *in* $\mathfrak{M}$ *is associated a model* $\mathfrak{M}^*$ *of* T_E *in which* Ω *is the class of ordinals* $< \kappa$, *with* $\mathfrak{M}^*$ *satisfying* (ii)–(vii) *above, and further*:

(viii) CH *holds in* $\mathfrak{M}^*$ *just in case the usual* CH *holds in* $\mathfrak{M}$,

(ix) GCH* *holds in* $\mathfrak{M}^*$ *if* GCH *holds below* κ *in* $\mathfrak{M}$, *and*

(x) EOC *holds in* $\mathfrak{M}^*$ *just in case* $\kappa = \aleph_1(\mathfrak{M})$.

3.3 *Starting with admissible models*

Let $(M, \in)$ be a pure admissible set with $o(M) < \omega$ (cf. [B]) satisfying the following condition called 'effective global well-ordering' [in C2] and being 'recursively listed' in [B]: (Σ_1–GWO). There is a function $\nu : o(M) \underset{\text{onto}}{\overset{1-1}{\rightarrow}} M$ definable in Σ_1-form (allowing parameters).

As we know, (Σ_1-GWO) holds in $M = L_\alpha$ for every admissible α.

Under this hypothesis, the uniformization theorem for M – r.e. (i.e., $\Sigma_1{}^{(M)}$) relations holds and we get an enumeration $f (f \in M)$ of the M-partial recursive (i.e., $\Sigma_1{}^{(M)}$) functions. This leads us directly to a model $\mathfrak{M}^*$ of T_Ω (I–IV), taking $fx \simeq y$ to hold when the fth function at x has value y. Now we interpret Ω to be the class of all ordinals of M and $<$ to be the $\in$ relation on Ω. The axioms V(i)–(ix) are directly verified. To establish selection, let $\psi(f, x, y, z)$ be a Δ_0-formula such that $fx \simeq y \Leftrightarrow \exists z \psi(f, x, y, z)$. Then $\exists \alpha (f\alpha \simeq 0) \Leftrightarrow \exists x, z[\text{Ord}(x) \wedge \psi(f, x, 0, z)]$.
We take $cf \simeq x \Leftrightarrow \exists \beta \exists z \in TC(\nu(\beta))(\nu(\beta) = (x, z) \wedge \text{Ord}(x) \wedge \psi(f, x, 0, z)) \wedge$ $\wedge \forall \gamma < \beta \neg \exists z \in TC(\nu(\beta))(\nu(\beta) = (x, z) \wedge \text{Ord}(x) \wedge \psi(f, x, 0, z))$.
This determines c as a Σ_1-definable function.

We can associate with each class A and equivalence relation $\equiv_A$ on A, a subclass of A acting as $A/\equiv_A$; ν is used to select a unique member from each equivalence class. Then we have:

(i) *A is denumerable* (mod $\equiv_A$) *in the sense of* $\mathfrak{M}^*$ *iff* $A/\equiv_A$ *is countable in* M, *and*

(ii) *A is Ω-enumerable* (mod $\equiv_A$) *in the sense of* $\mathfrak{M}^*$ *iff* $A/\equiv_A$ *is* M – r.e.

Given any α, we have $f \in \alpha^\alpha \Leftrightarrow \forall \xi < \alpha \exists \eta < \alpha (f\xi \simeq \eta)$. Thus α^α is M – r.e. and so α^α and *a fortiori*, $(\alpha^\alpha/\equiv)$ is Ω-enumerable. To conclude, we have the following:

THEOREM. *With each pure admissible set* $(M, \in)$ *satisfying* (Σ_1-GWO) *and with* $o(M) > \omega$ *is associated a model* $\mathfrak{M}^*$ *of* T_Ω *in which* (i) *and* (ii) *above hold, and further*:

(iii) GCH* *holds in* $\mathfrak{M}^*$, *and*

(iv) EOC *holds in* $\mathfrak{M}^*$ *just in case* M *is locally countable* (i.e. *every element of* M *is countable in* M).

To see (iv), suppose EOC. By Σ_1-GWO, every set in M is listed by an ordinal, hence is countable. Conversely, if M is locally countable, then every ordinal α is countable, so $\exists\beta(\nu(\beta) : \omega \xrightarrow[\text{onto}]{1-1} \alpha)$. We can then select β uniformly as an M-recursive function of α and hence get $e\alpha : \omega \xrightarrow[\text{onto}]{1-1} \alpha$.

The paradigms for (iv) are of course, $(L_{\omega_1}, \in)$ where ω_1 is the least non-recursive ordinal and $(L_{\aleph_1(L)}, \in)$.

4. MATHEMATICS OF ORDINALS AND ENUMERABILITY IN T_Ω

We review consequences of axioms I–IV in 4.1, 4.2. (These were proved in [F] but can also be proved easily by the reader.) We then take up ordinal arithmetic in order to obtain a pairing function on ordinals.

4.1. *Closure conditions on operations*

The following are proved in [F], pp. 95–96:

(i) *Abstraction.* With each application term t and variable x is associated a new application term t^* with variables among those of t different from x, for which we can prove

$$(t^*\downarrow) \wedge \forall x[t^*x \simeq t].$$

We write $\lambda x \cdot t$ or $\lambda x \cdot t[x]$ for t^*, so $(\lambda xt[x])y \simeq t[y]$. (That $\lambda x \cdot t[x]$ is always defined whether or not any $t[x]$ is defined makes sense if we think of $\lambda x \cdot t[x]$ as a well-determined rule.)

(ii) *Recursion.* There is a closed term r such that the following is provable for any n:

$$(rf\downarrow) \wedge \forall x_1, \ldots, x_n[(rf)x_1, \ldots, x_n \simeq f(rf)x_1, \ldots, x_n].$$

4.2. *Closure conditions on classes*

These are established in [F] pp. 96–97.

(i) *Finitary class operations.* Each of the following is defined in the form

$\{x|\Phi(x, a_1, \ldots, a_n, A_1, \ldots, A_m\}$ for suitable elementary 2nd order Φ:

$$V = \{x|x = x\},\ \Lambda = \{x|x \neq x\},\ \{a_1, a_2\},\ \{a_1\},\ -A_1,$$
$$A_1 \cap A_2,\ A_1 \times A_2,\ A_2^{A_1}.$$

We take $A^1 = A$, $A^{n+1} = A^n \times A$.

(ii) *Infinitary class operations*. Each of the following is proved to exist under the hypothesis $\forall x \in A \exists Y(fx \simeq Y)$, where we write B_x for fx:

$$\bigcup_{x \in A} B_x,\ \bigcap_{x \in A} B_x,\ \sum_{x \in A} B_x,\ \prod_{x \in A} B_x.$$

Here $\coprod_{x \in A} B_x$ is the *disjoint sum* of $\langle B_x \rangle_{x \in A}$, which we can take simply as jAf from the join axiom. Then

$$\prod_{x \in A} B_x = \{y|\forall x \in A(yx \in B_x)\} = \{y|\forall x \in A \exists u(yx \simeq u \wedge (x, u) \in jAf)\}.$$

4.3. *Operations on ordinals and ordinal functions*

We now make use of all of the axioms of T_Ω.

(i) *Successor*. $\alpha' = \beta' \Rightarrow \alpha = \beta$.

(ii) *Predecessor*. We shall obtain this by selection, applied to $f\alpha$ where $f\alpha\beta \simeq 0 \Leftrightarrow \beta' = \alpha$. For this use definition by cases, giving $f = \lambda\alpha\cdot\lambda\beta\cdot\underline{d}01\alpha'\beta$, then take $\underline{pd} = \lambda\alpha\cdot\underline{c}(f\alpha)$. Thus $\exists\beta(\beta' = \alpha) \Rightarrow (\underline{pd}(\alpha))' = \alpha$. This holds exactly when $\alpha \neq 0$ and $\neg\,\mathrm{Lim}(\alpha)$, for then $\exists\xi(\xi < \alpha \wedge \alpha \leq \xi')$.

(iii) *Least element principle*. $\exists\alpha\Phi(\alpha) \Rightarrow \exists\alpha\{\Phi(\alpha) \wedge \forall\xi < \alpha \neg\,\Phi(\xi)\}$. This is the contrapositive of transfinite induction, by classical logic. Further, α in the conclusion is unique.

(iv) *Supremum*. By reflection and the least element principle, if $(f : \alpha \to \Omega)$ then $\exists!\beta[f : \alpha \to \beta \wedge \forall_\gamma < \beta \neg(f : \alpha \to \gamma)]$. We write $\beta = \sup_{\xi<\alpha}(f\xi)$. We shall show that β can be obtained uniformly from f and α. First we show how to obtain m with

$$(\dagger) \qquad mf\alpha\beta \simeq 0 \Leftrightarrow \forall\xi < \alpha\exists\eta < \beta(f\xi \simeq \eta) \wedge \forall\gamma < \beta\exists\xi < \alpha\, \exists\eta < \beta(f\xi \simeq \eta \wedge \gamma \leq \eta).$$

Using definition by cases, we associate f^0 with f so that $f^0xy \simeq 0 \Leftrightarrow fx \simeq y$, by $f^0 = \lambda x\lambda y\cdot\underline{d}01(fx)y$. (This and the function f_0 with $f_0yx \simeq 0 \Leftrightarrow fx \simeq y$ will be used repeatedly in the following.) The bounded quantifiers will be taken care of by axiom V(ix). By that we have $\underline{q}h\alpha \simeq 0 \Leftrightarrow \forall\xi < \alpha\cdot(h\xi \simeq 0)$, and $\underline{q}h\alpha \simeq 1 \Leftrightarrow \exists\xi < \alpha(h\xi \neq 0)$. Note that we need not have $\forall\xi < \alpha(h\xi\downarrow)$ in the latter case. Let $\bar{h}x \simeq \underline{d}10(hx)0$, so when $hx\downarrow$ we have $\bar{h}x \simeq 1$ if $hx \simeq 0$, $hx \simeq 0$ if $\bar{h}x \neq 0$. Thus $\underline{q}\bar{h}\alpha \simeq 1 \Leftrightarrow \exists\xi < \alpha(h\xi \simeq 0)$. We now put these together:

$$\overline{\overline{q(f^0\xi)}}\beta \simeq 0 \Leftrightarrow \exists\eta < \beta(f\xi \simeq \eta)$$

and with $h = \lambda\beta\lambda\xi\ \overline{\overline{q(f^0\xi)}}\beta$,

$$q(h\beta)\alpha \simeq 0 \Leftrightarrow \forall\xi < \alpha(h\beta\xi \simeq 0) \Leftrightarrow \forall\xi < \alpha\exists\eta < \beta(f\xi \simeq \eta).$$

This argument has been carried out in detail to show manipulations with bounded quantifiers. To treat the second conjunct on the r.h.s. of (†), we proceed similarly, using also decidability of membership in initial segments, so $\gamma \leq \eta \Leftrightarrow l\gamma\eta' \simeq 0$. The propositional truth functions such as $\wedge$ are accounted for by suitable operations, so we can assemble all these to obtain m satisfying (†). Finally $\sup_{\xi<\alpha} f\xi \simeq c(mf\alpha)$.

(v) *Decidability of limit numbers.* $\text{Lim}(\alpha) \Leftrightarrow \forall\xi < \alpha(\xi' < \alpha)$, so we have an operation $\underline{L} : \Omega \to \{0, 1\}$ with $\underline{L}\alpha \simeq 0 \Leftrightarrow \text{Lim}(\alpha)$.

(vi) *Recursion on the ordinals.* From any α_0, g we can find f uniformly satisfying

$$\text{(Rec)}\qquad f0 \simeq \alpha_0,\ f\alpha' \simeq g\alpha(f\alpha),\ \text{Lim}(\alpha) \Leftrightarrow f\alpha \simeq \sup_{\xi<\alpha}(f\xi).$$

This is obtained by recursion, using definition by cases. First take

$$ahx \simeq \begin{cases} \alpha_0 \text{ if } x \simeq 0 \\ g(\underline{pd}x)(h(\underline{pd}x)) \text{ if } x \simeq (\underline{pd}x)' \\ \sup_{\xi<x}(h\xi) \text{ if } \underline{L}x \simeq 0. \end{cases}$$

Then take f to be (ra) where r is given by 4.1 (ii), so $f\alpha \simeq af\alpha$ and f satisfies (Rec). It also follows that if $(g\alpha\beta)\downarrow$for all α, β then $(f\alpha)\downarrow$ for all α.

(vii) *Ordinal arithmetic.* Using (vi) we can define $\beta + \alpha$, $\beta.\alpha$, β^α, etc. as usual, and establish their familiar properties. Further we can show for any normal $f : \Omega \to \Omega$ that $\alpha \leq f\alpha$ and if $\alpha \geq f0$ then $\exists!\xi(f\xi \leq \alpha < f\xi')$. This permits us to establish the division algorithm, representation to base α for any $\alpha > 1$, etc.

(viii) *Pairing on ordinals.* Any ordinal $\alpha \neq 0$ is uniquely representable in the form $\alpha = 2^\xi\cdot(2\eta + 1)$, with $\xi < \alpha$, $\eta < \alpha$. We are thus led to operations π, π_1, π_2 with

$$\pi(\xi, \eta) \simeq \begin{cases} pd(2^\xi(2\eta + 1)) \text{ if } \xi, \eta < \omega \\ 2^\xi(2\eta + 1) \text{ otherwise,} \end{cases}$$
$$\pi_1(\pi(\xi, \eta)) = \xi, \pi_2(\pi(\xi, \eta)) = \eta.$$

Thus $\pi : \Omega^2 \to \Omega$ is a pairing function on ordinals, with projections π_1, π_2. The map $\pi^{-1}\alpha \simeq (\pi_1\alpha, \pi_2\alpha)$ satisfies:

$$\pi^{-1} : \Omega \underset{\text{onto}}{\to} \Omega^2.$$

5. CLOSURE CONDITIONS ON DENUMERABLE AND Ω-ENUMERABLE CLASSES

These closure conditions are formulated only for classes with the identity relation. (For example, that A denumerable and B denumerable implies $A \times B$ denumerable.) This is done for simplicity and also because for extensional operations on classes the results transfer easily to classes with equivalence relations. (For example, that A denumerable (mod $\equiv A$) and B denumerable (mod $\equiv B$) implies $A \times B$ denumerable (mod $\equiv A \times B$).) All theorems are in T_Ω.

5.1. *Closure under Boolean and Cartesian operations*

We say that C *is pairable* if there are operations b, b_1, b_2 such that $b : C^2 \underset{\text{onto}}{\to} C$, $b_i : C \to C$ and $b_i(b(x_1, x_2)) = x_i$ for $x_1, x_2 \in C$. Then we have $b^{-1} : C \underset{\text{onto}}{\to} C^2$ given by $b^{-1}z = (b_1z, b_2z)$. As we have seen, Ω is pairable and, of course, this also holds for ω (i.e. for $[0, \omega)$). Define $A^{(\omega)}$ to be the class of $x : \omega \to A$ which are eventually constant, i.e. $A^{(\omega)} = \{x | x : \omega \to A \wedge \exists n < \omega \forall m < \omega(m \geq n \Rightarrow x_m = x_n)\}$. The following is easily proved.

THEOREM. *Suppose C is pairable.*

(i) *If A, B are C-enumerable, then so also are $A \cup B$, $A \cap B$, $A \times B$ and A^n for each $n \in \omega$.*

(ii) *If A is C-enumerable and ω is C-enumerable then $A^{(\omega)}$ is C-enumerable.*

(iii) *If B is C-enumerable and for each $x \in B$, $fx : C \underset{\text{onto}}{\to} A_x$ then $\bigcup_{x \in B} A_x$ is C-enumerable.*

Note, for the applications below, that ω is trivially Ω-enumerable by taking $f : \Omega \underset{\text{onto}}{\to} \omega$ to be the identity on ω and 0 for $x \geq \omega$.

5.2. *Uniformization and choice for Ω-enumerable classes*

THEOREM. *Suppose R is* Ω*-enumerable where* $R \subseteq A \times B$.

(i) $\exists f \forall x\{\exists y \cdot (x, y) \in R \Rightarrow (x, fx) \in R\}$.

(ii) *If, in addition, B is* Ω*-enumerable, then*

$$\exists f \forall x\{\exists y \in B \cdot (x, y) \in R \Rightarrow fx \in B \wedge (x, fx) \in R\}$$

(iii) *Under the same hypothesis as* (ii) *we have*

$$\forall x \in A \exists y \in B \cdot (x, y) \in R \Rightarrow \exists f : A \to B \forall x \in A \cdot (x, fx) \in R.$$

Proof: For (i), let $g : \Omega \underset{\text{onto}}{\to} R$, so $(x, y) \in R \Leftrightarrow \exists \alpha \cdot g\alpha \simeq (x, y)$, and $\exists y \cdot (x, y) \in R \Leftrightarrow \exists \alpha \cdot p_1(g\alpha) \simeq x \Leftrightarrow \exists \alpha \cdot h\alpha \simeq x \Leftrightarrow \exists \alpha \cdot h_0 x\alpha \simeq 0$ where h = $\lambda z \cdot p_1(gz)$. (Recall that $h_0 xz \simeq 0 \Leftrightarrow hz \simeq x$.) Now we can apply selection on Ω: let $f_1 = \lambda x \cdot c(h_0 x)$, so that $\exists \alpha \cdot h_0 x\alpha \simeq 0 \Rightarrow f_1 x \in \Omega \wedge h_0 x(f_1 x) \simeq 0$. Finally we take $f = \lambda x \cdot p_2(f_1 x)$.

(ii) follows by applying (i) to the relation $R_1 = (\mathscr{D}(R) \times B) \cap R$, where $\mathscr{D}(R) = \{x | \exists y \cdot (x, y) \in R\}$. Note that $p_1 : R \underset{\text{onto}}{\to} \mathscr{D}(R)$, so under the hypothesis that R and B are Ω-enumerable, so also is $\mathscr{D}(R)$ and thence R_1.

(iii) is an immediate corollary.

5.3. *A counter-example to denumerability of sub-denumerable classes*

In set theory, if B is C-enumerable and $A \subseteq B$ then A is C-enumerable. This is not a theorem of T_Ω. For a counter-example we refer to the model $\mathfrak{M}^*$ described in 3.3. when $\mathfrak{M} = (L_{\omega_1}, \in)$ with ω_1 the least non-recursive ordinal. Consider the subclasses of ω in $\mathfrak{M}^*$. As we know from 3.3 (i), (ii) and [B], we have for $A \subseteq \omega$ that:

(i) A is denumerable iff A is hyperarithmetic and

(ii) A is Ω-enumerable iff A is Π^1_1.

Thus any Π^1_1–Δ^1_1 class A serves to give a counter-example to the above for $B = \omega$ and $C = \Omega$.

5.4. *A sufficient condition for transfer of enumerability to subclasses*

THEOREM. *Suppose* $A \subseteq B$ *and that both A and* $B - A$ *are* Ω*-enumerable. Then A is B-enumerable.*

Proof: Suppose $g : \Omega \underset{\text{onto}}{\to} A$ and $h : \Omega \underset{\text{onto}}{\to} (B - A)$.

As before, we take $g_0 = \lambda y \lambda x \cdot d01(gx)y$ so that $g_0 yx\downarrow \Leftrightarrow gx\downarrow$ and $g_0 yx = 0 \Leftrightarrow gx = y$. Thus $g_0 y\xi$ is defined for all y and ξ. Let h_0 be obtained from h in the same way. For each $y \in B$, $\exists \xi(g\xi = y) \vee \exists \xi(h\xi = y)$ so $\exists \xi(g_0 y\xi = 0 \vee h_0 y\xi = 0)$ and, finally $\exists \xi \cdot (g_0 y\xi) \cdot (h_0 y\xi) = 0$. The

function $fy\xi = (g_0y\xi)\cdot(h_0y\xi)$ is defined for all y and ξ and $fy\xi = \Leftrightarrow 0\ g_0y\xi = 0 \vee h_0y\xi = 0$.
We apply selection to obtain

$$\exists\xi(fy\xi = 0) \Rightarrow c(fy) \in \Omega \wedge fy(c(fy)) = 0.$$

Furthermore $\forall y \in B\exists\xi(fy\xi = 0)$. Hence $y \in B \Rightarrow [y \in A \Leftrightarrow g(c(fy)) = y]$. To complete the argument, suppose that A is non-empty and fix some $a \in A$. We may define $g^* : B \underset{\text{onto}}{\rightarrow} A$ by $g^* = \lambda y\cdot dya(g(cfy))y$, which means for $y \in B$ that $g^*y = y$ if $g(c(fy)) = y$, i.e. if $y \in A$, and $g^*y = a$ otherwise.

This result will play an important role in Section 6.7.

5.5. *A sufficient condition for extensionalization*

THEOREM. *Suppose that A, E and $A^2 -\ E$ are Ω-enumerable, where E is an equivalence relation on A. Then we can find $A_1 \subseteq A$ such that:*

(i) A_1 is Ω-enumerable.
(ii) $\forall x \in A\exists!y \in A_1[(x, y) \in E]$.

Proof: Let $g : \Omega \underset{\text{onto}}{\rightarrow} A$ enumerate A, and $h_1 : \Omega \underset{\text{onto}}{\rightarrow} E, h_2 : \Omega \underset{\text{onto}}{\rightarrow} A^2 - E$. Given ξ and $x = g\xi$, there is a least η with $(x, g\eta) \in E$. We shall show that η is a function f_1 of ξ. We have

$$\forall\xi\exists\eta\cdot\{(g\xi, g\eta) \in E \wedge \forall\zeta < \eta((g\xi, g\zeta) \notin E)\}, \text{ so}$$
$$\forall\xi\exists\eta\cdot\{\exists\mu(h_1\mu = (g\xi, g\eta) \wedge \forall\zeta < \eta\exists\nu(h_2\nu = (g\xi, g\zeta))\}.$$

The inner quantifiers $\exists\mu$, $\exists\nu$ may be eliminated using the selection operator c. The condition within braces is then built up from equations by conjunction and bounded quantification, so is equivalent to an equation $f\xi\eta = 0$. Finally we may take $f_1\xi \simeq c(f\xi)$. We take $A_1 = \{f_1\xi | \xi \in \Omega\}$. To show that A_1 is Ω-enumerable, define $f_2\xi = g(f_1\xi)$ if $f_1\xi = \xi$ and $f\xi = g0$ otherwise, so $f : \Omega \underset{\text{onto}}{\rightarrow} A_1$.

Remark. Under the given hypothesis, A_1 may serve as A/E.

5.6. *One-one enumeration*

The following is easily proved by the usual recursive argument.

THEOREM. *If A is Ω-enumerable then either there exists $f : \Omega \underset{\text{onto}}{\overset{1-1}{\rightarrow}} A$ or there exists α and $f : \alpha \underset{\text{onto}}{\overset{1-1}{\rightarrow}} A$.*

6. BASIC MODEL THEORY IN T_Ω + EOC

We now adjoin the hypothesis EOC of uniform countability of the initial segments of the ordinals. This is done both for simplicity and because EOC is satisfied in the paradigm applications, namely: in the models $\mathfrak{M}^*$ built from set-theoretical models $\mathfrak{M}$ in which $\Omega = \aleph_1^{(\mathfrak{M})}$ (3.2) and in the models $\mathfrak{M}^*$ built from admissible $\mathfrak{M}$ satisfying local countability, particularly $\mathfrak{M} = (L_{\omega_1}, \in)$ where $\Omega = \omega_1$. All of the results of this section can be established in more general form in T_Ω. In this and Section 7.8 we follow Chang, Keisler [C, K] in notation, notions and proofs of theorems as far as possible.

6.1. *Languages and structures*

Let $\mathscr{L}$ be a language for the single-sorted 1st order predicate calculus with equality given by any finite set of relation, function and constant symbols. We shall also consider expansions $\mathscr{L}_I$ of $\mathscr{L}$ by arbitrary classes $\{c_i | i \in I\}$ of distinct constant symbols. If the symbols for $\mathscr{L}$ other than the identity are $P_0, \ldots, P_n, F_0, \ldots, F_m, c_0, \ldots, c_q$, a model for $\mathscr{L}$ is a structure of the form $\mathscr{A} = (A, R_0, \ldots, R_n, f_0, \ldots, f_m, z_0, \ldots, z_q)$ where $A, R_0, \ldots, R_n$ are classes with $R_j \subseteq A^{k_j}$(k_j = number of arguments of P_j), and $f_j : A^{l_j} \to A$(l_j the number of arguments of F_j), and finally each $z_j \in A$. Unless otherwise indicated, equality is interpreted as the identity on A. To indicate a more general interpretation by a congruence relation E_A or $\equiv_A$ with respect to $\mathscr{A}$, we write $(\mathscr{A}, \equiv_A)$. If $a : I \to A$ and $a_i = ai$ for each $i \in I$ then $(\mathscr{A}, a_i)_{i \in I}$ is a model for $\mathscr{L}_I$.

DEFINITION. $\mathscr{A} = (A, R_0, \ldots, R_n, f_0, \ldots, f_m, z_0, \ldots, z_q)$ is called *C-enumerable* if A and each of its relations R_i are C-enumerable. Equivalently, $\mathscr{A}$ is C-enumerable if the disjoint sum $(A + R_0 + \ldots + R_n)$ is C-enumerable. By a *C-enumeration of* $\mathscr{A}$ we mean an $f : C \xrightarrow[\text{onto}]{} (A + R_0 + \ldots + R_n)$; we write $f : C \xrightarrow[\text{onto}]{} \mathscr{A}$ in this case. These notions are extended to $(\mathscr{A}, E_A)$ and $(\mathscr{A}, a_i)_{i \in I}$ by including E_A, resp. I in the disjoint sum.

6.2. *Satisfaction*

Let $\mathrm{Sat}_{\mathscr{A}}^{(k)}$ be the class of all (φ, x, q) where $\mathscr{A} \models \varphi[x_1, \ldots, x_q]$ and φ is a formula with at most $v_1, \ldots, v_q$ free, $x = (x_1, \ldots, x_q) \in A^q$ and φ is of logical degree $\leq k$. $\mathrm{Sat}_{\mathscr{A}}^{(k)}$ is a subclass of $Fm_{\mathscr{L}} \times \bigcup_{q<\omega} A^q \times \omega$. It is easy to see that there is an operation g such that for each k, $g(\mathrm{Sat}_{\mathscr{A}}^{(k)}) = \mathrm{Sat}_{\mathscr{A}}^{(k+1)}$. Hence we can define f by recursion so that $fk = \mathrm{Sat}_{\mathscr{A}}^{(k)}$ for each k. It follows that $\mathrm{Sat}_{\mathscr{A}}^{(k)} = \bigcup_{k<\omega} \mathrm{Sat}_{\mathscr{A}}^{(k)}$ is also a class. Moreover, all of this is uniform in $\mathscr{A}$, i.e. we have an operation *Sat* such that for each $\mathscr{A}$, $Sat\ \mathscr{A} \simeq \mathrm{Sat}_{\mathscr{A}}$. Finally, all of this extends directly and uniformly to structures $(\mathscr{A}, \equiv_A)$ and $(\mathscr{A}, a_i)_{i\in I}$.

6.3. *Denumerability of satisfaction for denumerable structures*

THEOREM. *If* $\mathscr{A}$ *is denumerable so also is* $\mathrm{Sat}_{\mathscr{A}}$ *(uniformly in any* ω*-enumeration of* $\mathscr{A}$*).*

Proof: Suppose $g : \omega \xrightarrow[\text{onto}]{} (A + R_0 + \ldots + R_n)$ is an ω-enumeration of $\mathscr{A}$. We may reduce the work to the case that $A = \omega$ and $0,'$ appear in $\mathscr{A}$, and that there are no repetitions in the enumerations of the R_i. Then there is a closed term $0^{(n)}$ for each individual and we need only deal with satisfaction for sentences, i.e. with the theory $\mathrm{Th}(\mathscr{A})$ of $\mathscr{A}$. First it is shown that $\mathrm{Th}_{\mathscr{A}}$ has a characteristic function, i.e. an $S_{\mathscr{A}} : \mathrm{Th}(\mathscr{A}) \to \{0,1\}$ with $S_{\mathscr{A}}\varphi = 0 \Leftrightarrow \varphi \in \mathrm{Th}(\mathscr{A})$ whenever $\varphi \in Stc_{\mathscr{L}}$. This is defined by recursion on φ with $S_{\mathscr{A}}(\neg\varphi) = 1 - S_{\mathscr{A}}\varphi$, $S_{\mathscr{A}}(\varphi \vee \psi) \simeq S_{\mathscr{A}}\varphi \cdot S_{\mathscr{A}}\psi$ and, finally,

$$S_{\mathscr{A}}(\forall v_i \varphi) \simeq \begin{cases} 0 \text{ if } \forall n < \omega(S_{\mathscr{A}}(\varphi(0^{(n)}/v_i)) \simeq 0) \\ 1 \text{ otherwise.} \end{cases}$$

Using the operator for bounded quantification this recursion determines $S_{\mathscr{A}}$ as an operation. The ω-enumeration of $\mathrm{Th}_{\mathscr{A}}$ is easily obtained from $S_{\mathscr{A}}$.

6.4. Ω-*valuated structures*

The following is the crucial notion for the further work.

DEFINITION. $\mathscr{A} = (A, \ldots)$ is Ω-*valuated* if $\mathscr{A}$ is Ω-enumerable and $\mathrm{Sat}_{\mathscr{A}}$ is Ω-enumerable.

It is not in general true that if $\mathscr{A}$ is Ω-enumerable then $\mathscr{A}$ is Ω-valuated. For a counter-example in $\mathfrak{M}^*$ built from the admissible $\mathfrak{M} = (L_{\omega_1}, \in)$, let $\mathscr{A} = (A, R)$ where A is Σ_1 and R is a Σ_1 subclass of A whose complement to A is not Σ_1. If $\mathrm{Sat}_{\mathscr{A}}$ were Ω-enumerable, we could enumerate $\{x | x \in A \wedge \neg R(x)\}$, which is a contradiction.

On the other hand, it *is* true in all the set-theoretical models described in 3.2 that if $\mathscr{A}$ is Ω-enumerable then $\mathscr{A}$ is Ω-valuated. Thus, *any theorem in our system which applies to* Ω*-valuated structures generalizes a* *corresponding theorem from set-theoretical model theory for $\aleph_1$-enumerable structures.*

Finally, if $\mathfrak{M}^*$ is built from admissible $\mathfrak{M} = (M, \in)$ as in 3.3, then $\mathscr{A}$ is Ω-valuated just in case $\mathrm{Sat}_{\mathscr{A}}$ is $\Sigma_1^{(M)}$. This is the same as Cutland's notion of M-structure in [C2] (p. 270). Cutland calls $\mathscr{A}$ an M^*-structure if $\mathscr{A} \notin M$. In case of locally countable M, the structures in M are the same as the countable $\mathscr{A}$ (in $\mathfrak{M}^*$). So then $\mathscr{A}$ is an M^*-structure in Cutland's sense just in case it is Ω-valuated but not countable (in $\mathfrak{M}^*$). *Cutland's theorems for M-structures are special cases of our theorems for Ω-valuated structures.*

Note. In the case of $M = L_{\omega_1}$ and $\mathscr{A}$ coded up in ω, $\mathscr{A}$ is an M-structure if $\mathscr{A}$ is Π_1^1-*covered* in the sense of [C1].

6.5. *Extensionalizing Ω-valuated $(\mathscr{A}, \equiv_A)$*

The notion of being Ω-valuated of course makes sense more generally for structures $(\mathscr{A}, E_A)$ where E_A is a congruence relation $\equiv_A$ on $\mathscr{A}$. But then both E_A and $A^2 - E_A$ must be Ω-enumerable. Hence we can apply the theorem of 5.5 to obtain Ω-enumerable $A_1 \subseteq A$ with $\forall x \in A \exists ! y \in A_1 (x \equiv_A y)$. Thus we may identify $\mathscr{A}/\equiv_A$ with the substructure $\mathscr{A}_1 = \mathscr{A} \upharpoonright A_1$ of $\mathscr{A}$. Furthermore, $\mathrm{Sat}_{\mathscr{A}_1}$ is the restriction of $\mathrm{Sat}_{\mathscr{A}}$ to sequences from A_1, so $\mathscr{A}_1$ is also Ω-valuated.

6.6. *The completeness theorem*

We use consistency in its syntactic sense.

THEOREM. *For each Ω-enumerable I and consistent set of sentences Γ in $\mathscr{L}_I$ we can find (uniformly from Γ, I) an Ω-valuated model $(\mathscr{A}, a_i)_{i \in I}$ of Γ which further is denumerable if I is denumerable.*

Proof: By the usual Gödel-Henkin completeness argument. We adjoin Ω new constant symbols $d_\xi (\xi < \Omega)$ to $\mathscr{L}_I$, or just countably many symbols $d_n (n < \omega)$ when I is countable. We write $\{c_\xi | \xi \in \Omega\}$ for the entire class of constants used. Then enumerate all existential sentences $\exists v_{i_\xi} \psi_\xi (v_{i_\xi})$ of $\mathscr{L}_\Omega$ and a sequence of distinct constants c_{η_ξ} which do not appear in any ψ_ζ for $\zeta \leq \xi$. We pass to the class $\Gamma \cup \{\exists v_{i_\xi} \psi_\xi (v_{i_\xi}) \rightarrow \psi_\xi (c_{\eta_\xi}) | \xi < \Omega\}$, which is consistent, and thence by transfinite recursion to a complete and consistent Ω-enumerated extension Γ^+ for which the constants c_ξ forms a class of witnesses. The model $(\mathscr{A}, \equiv_A)$ is defined as usual, taking $A = \Omega$, $(\xi \equiv_A \eta) \Leftrightarrow (c_\xi \equiv c_\eta) \in \Gamma^+$, and so on for the relations and operations in such a way that for any formula $\varphi(v_1, \ldots, v_q)$ and any $\xi_1, \ldots, \xi_q$ we have

$(\mathscr{A}, \equiv_A) \models \varphi[\xi_1, \ldots, \xi_q] \Leftrightarrow \varphi(c_{\xi_1}, \ldots, c_{\xi_q}) \in \Gamma^+$. This shows directly that $(\mathscr{A}, \equiv_A)$ is Ω-valuated. We may finally pass to the model $(\mathscr{A}/\equiv_A)$ of Γ by the extensionalizing procedure just described in 6.5. Of course if I is denumerable, everything in sight may also be kept denumerable.

6.7. *Elementary substructures and chains*

The notions of *elementary substructure* $\mathscr{A} \prec \mathscr{B}$ and of *elementary chain* $\mathscr{A}_0 \prec \ldots \prec \mathscr{A}_\xi \prec \ldots (\xi < \mathscr{A}$ or $\xi < \Omega)$ are explained as usual. We can prove the Tarski-Vaught theorem for any such elementary chain without hypotheses on enumerability. However we also want a stronger form for Ω-valuated structures.

DEFINITION. Let C be $[0, \alpha)$ or Ω and $f\xi = \mathscr{A}_\xi$ for each $\xi \in C$. We say that the sequence $\langle \mathscr{A}_\xi \rangle_{\xi \in C}$ *is uniformly* Ω-*valuated* if there are g, h such that for each $\xi \in C$, $g\xi : \Omega \underset{\text{onto}}{\rightarrow} \mathscr{A}_\xi$ and $h\xi : \Omega \underset{\text{onto}}{\rightarrow} \mathrm{Sat}_{\mathscr{A}_\xi}$.

The following is then easily proved.

THEOREM. *Suppose* $\mathscr{A}_0 \prec \ldots \prec \mathscr{A}_\xi \prec \ldots (\xi \in C)$ *where* $\langle \mathscr{A}_\xi \rangle_{\xi \in C}$ *is uniformly* Ω-*valuated. Then* $\mathscr{A} = \bigcup_{\xi \in C} \mathscr{A}_\xi$ *is* Ω-*valuated and is an elementary extension of each* $\mathscr{A}_\eta$ *for* $\eta \in C$.

COROLLARY. *If* $\mathscr{A}_0 \prec \ldots \prec \mathscr{A}_\xi \prec \ldots (\xi \in \Omega)$ *where* $\langle \mathscr{A}_\xi \rangle_{\xi < \Omega}$ *is uniformly denumerable then* $\bigcup_{\xi \in \Omega} \mathscr{A}_\xi$ *is* Ω-*valuated.*

6.8. *Downward Löwenheim-Skolem for* Ω-*valuated structures*

THEOREM. *Suppose* $\mathscr{A}$ *is* Ω-*valuated. With each denumerable* $X \subseteq A$ *can be associated uniformly (in any enumeration of* X*) a denumerable* B *with* $X \subseteq B \subseteq A$ *and* $\mathscr{A} \restriction B \prec \mathscr{A}$.

Proof: Fix $x_0 \in A$. For each $\varphi(v_0, v_1, \ldots, v_q)$ and $x = (x_1, \ldots, x_q) \in A_q$

$$\exists y\{\mathscr{A} \models \varphi[y, x_1, \ldots, x_q] \vee \mathscr{A} \models (\neg \exists v_0 \varphi(v_0, v_1, \ldots, v_q))[x_1, \ldots, x_q] \wedge y = x_0\}.$$

The relation within brackets is Ω-enumerable, so by the Ω-choice theorem of 5.2 (iii) we can find f which is total on $Fm \times \omega \times \bigcup_{q < \omega} A_q$ such that $f\varphi q x \simeq y$ is a solution of the above. In other words $f\varphi q$ is a Skolem-function in $\mathscr{A}$ for $\varphi(v_0, v_1, \ldots, v_q)$. Let $B =$ the closure of X under all these functions. From any $g : \omega \underset{\text{onto}}{\rightarrow} X$ may be obtained $h : \omega \underset{\text{onto}}{\rightarrow} B$.

COROLLARY. *$\mathscr{A}$ is Ω-valuated iff there exists a sequence $\langle \mathscr{A}_\xi \rangle_{\xi < \Omega}$ of uniformly denumerable substructures of $\mathscr{A}$ with*

(i) $\mathscr{A}_0 \prec \ldots \prec \mathscr{A}_\xi \prec \ldots \prec \mathscr{A}$, and

(ii) $\mathscr{A} = \bigcup_{\xi \in \Omega} \mathscr{A}_\xi$.

Proof: Given an enumeration of $\mathscr{A}$, iterate the L–S theorem to build up an elementary chain which exhausts $\mathscr{A}$. The converse was proved at the end of 6.7.

6.9. *Theories of denumerable substructures*

THEOREM. *Suppose $\mathscr{A}$ is Ω-valuated, and X is denumerable, $X \subseteq A$ Then* Th$(\mathscr{A}, x)_{x \in X}$ *is denumerable, uniformly in (any ω-enumeration of) X.*

Proof: Let C be the class of closed formulas in $\mathscr{L}_\omega$, i.e. in $\mathscr{L}$ augmented by constants $c_0, \ldots, c_n, \ldots (n < \omega)$. Given $g : \omega \xrightarrow[\text{onto}]{} X$, we denote Th$(\mathscr{A}, gn)_{n < \omega}$ by T_g. Then $T_g \subseteq C$ is Ω-enumerable. Also C–$T_g = \{\varphi | \varphi$ is closed in $\mathscr{L}_\omega$ and $(\mathscr{A}, gn)_{n < \omega} \models \neg\varphi\}$ is Ω-enumerable. C itself is denumerable. Hence by the transfer of enumeration to subclasses (5.4), T_g is denumerable.

7. Ω-SATURATED MODELS IN T_Ω + EOC(+CH)

The new hypothesis CH enters only in the existence Theorem 7.3.

7.1. DEFINITION

$\mathscr{A}$ is called *Ω-saturated* if for each denumerable $X \subseteq A$ and each denumerable class $\{\varphi_n(v_0) | n < \omega\}$ of formulas in $\mathscr{L}_X$ we have

$$(*) \qquad \forall n \exists y \in A \cdot (\mathscr{A}, x)_{x \in X} \models \bigwedge_{i \le n} \varphi_i[y] \Rightarrow \exists y \in A \forall n \cdot (\mathscr{A}, x)_{x \in X} \models \varphi_n[y].$$

LEMMA. *If $\mathscr{A}$ is Ω-valuated and Ω-saturated, then the y in the conclusion of (*) can be found uniformly as a function of any ω-enumeration of X and of $\lambda n \cdot \varphi_n$.*

Proof: As shown at the end of Section 6, Th$(\mathscr{A}, x, y)_{x \in X}$ is denumerable uniformly in y (and X). Let $hy : \omega \xrightarrow[\text{onto}]{} \text{Th}(\mathscr{A}, x, y)_{x \in X}$. For any $f : \omega \to Fm$ with $fn = \varphi_n(v_0)$, we can define

$$gfy \simeq \begin{cases} 0 \text{ if } \forall n (\mathscr{A}, x)_{x \in X} \models \varphi_n[y] \\ 1 \text{ otherwise,} \end{cases}$$

for $(\mathscr{A}, x)_{x\in X} \models \varphi_n[y] \Leftrightarrow (\mathscr{A}, x, y)_{x\in X} \models \varphi_n(c) \Leftrightarrow \exists m(hym \simeq \varphi_n(c))$, fixing c as the constant for y; thus we are only using bounded quantification to obtain g. This allows us to Ω-enumerate those y in A satisfying the conclusion of (*). A particular y can be chosen by uniformization of Ω-enumerable relations.

7.2. *Uniqueness theorem for Ω-saturated models*

THEOREM. *If $\mathscr{A}$, $\mathscr{B}$ are both Ω-saturated and Ω-valuated and $\mathscr{A} \equiv \mathscr{B}$ then $\mathscr{A} \simeq \mathscr{B}$.*

Proof: By the usual back-and-forth argument, as in [C, K], pp. 220–221. Let $a : \Omega \xrightarrow[\text{onto}]{} A$, $b : \Omega \xrightarrow[\text{onto}]{} B$. We define $\bar{a} : \Omega \rightarrow A$, $\bar{b} : \Omega \rightarrow B$ by transfinite recursion in such a way that for each α,

$$(\dagger) \qquad (\mathscr{A}, \bar{a}_\xi)_{\xi<\alpha} \equiv (\mathscr{B}, \bar{b}_\xi)_{\xi<\alpha}$$

and each $a_\eta = \bar{a}_{2\eta}$ and $b_\eta = \bar{b}_{2\eta+1}$. Given (†), if $\alpha = 2\beta$ we put $\bar{a}_\alpha = a_\beta$ and let $T_\alpha = \mathrm{Th}(\mathscr{A}, \bar{a}_\xi)_{\xi\leq\alpha}$. This is denumerable and so we can produce a sequence $\varphi_n(v_0)$ of formulas from $\mathscr{L}_\alpha$ (i.e. $\mathscr{L}$ with constants c_ξ for $\xi < \alpha$) such that $\varphi_n(c_\alpha)$ enumerates T_α when c_α is interpreted as $\bar{a}_\alpha$. By (†), $\forall n \exists y \in B \cdot (\mathscr{B}, \bar{b}_\xi)_{\xi<\alpha} \models \varphi_n[y]$. Then using the preceding lemma, we can choose $\bar{b}_\alpha$ (uniformly as a function of the $\bar{b}_\xi(\xi < \alpha)$) such that $\forall n \cdot (\mathscr{B}, \bar{b}_\xi)_{\xi<\alpha} \models \varphi_n[\bar{b}_\alpha]$. If $\alpha = 2\beta + 1$ we take $\bar{b}_\alpha = b_\beta$ and define $\bar{a}_\alpha$ in a way symmetric to the preceding case. When the process is completed, we have $\bar{a} : \Omega \xrightarrow[\text{onto}]{} A$, $\bar{b} : \Omega \xrightarrow[\text{onto}]{} B$ and $(\mathscr{A}, \bar{a}_\xi)_{\xi\in\Omega} \equiv (\mathscr{B}, \bar{b}_\xi)_{\xi\in\Omega}$. The isomorphism $f : \mathscr{A} \simeq \mathscr{B}$ may be defined by $fa \simeq \bar{b}_\xi$ where ξ is the least ordinal with $a = \bar{a}_\xi$.

7.3. *Existence theorem for Ω-saturated models*

THEOREM (CH). *If $\mathscr{A}_0$ is any denumerable structure we can find an Ω-saturated, Ω-valuated $\mathscr{A}$ with $\mathscr{A}_0 \prec \mathscr{A}$.*

Proof: Again this proceeds by the standard argument, as in [C, K], pp. 215–216. We make use of the following.

LEMMA. *With each denumerable $\mathscr{A}_0$ and denumerable $X \subseteq A_0$ and sequence $\{\varphi_n(v_0) | n < \omega\}$ in $\mathscr{L}_X$ such that $\forall n \exists y \in A_0(\mathscr{A}, x)_{x\in X} \models \varphi_n[y]$, may be associated denumerable $\mathscr{A}_1$ and $y \in A_1$ with $\mathscr{A}_0 \prec \mathscr{A}_1$ and $\forall n.(\mathscr{A}_1, x)_{x\in X} \models \varphi_n[y]$.*

Proof of the lemma: By the completeness theorem applied to $\mathrm{Th}(\mathscr{A}_0, a)_{a\in A_0} \cup \{\varphi_n(c) | n < \omega\}$ where c is a new constant. □

To prove the theorem we define a sequence $\mathscr{A}_\alpha$ and $f\alpha$ by transfinite recursion in such a way that

(i) $f\alpha : \omega \underset{\text{onto}}{\rightarrow} A_\alpha$,

(ii) $\beta < \alpha \Rightarrow \mathscr{A}_\beta \prec \mathscr{A}_\alpha$, and

(iii) every 1-type that is finitely realizable in $(\mathscr{A}_\beta, x)_{x \in X}$ with X denumerable is realized in some $\mathscr{A}_\alpha$ with $\alpha > \beta$.

To carry this out, we use the fact that at each stage α satisfying (i) we can Ω-enumerate, up to $\equiv$, all functions $g : \omega \rightarrow A_\alpha$, in other words all denumerable subclasses of A_α. This is where CH is required. Let $g_{\alpha,\beta}$ range over this enumeration and $X_{\alpha,\beta} =$ range of $g_{\alpha,\beta}$. We can then Ω-enumerate all sequences $\{\varphi_n(v_0) | n < \omega\}$ which are finitely realizable in $(\mathscr{A}_\alpha, X_{\alpha,\beta})$. Using a tripling function (α, β, γ) on ordinals and the lemma we may choose $\mathscr{A}_{(\alpha,\beta,\gamma)+1} \succ \mathscr{A}_{(\alpha,\beta,\gamma)}$ to realize the γ-th such type.

8. ULTRAPOWERS AND Ω-SATURATED MODELS IN T_Ω + EOC + CH

The ideas are indicated very sketchily. Note that CH is assumed throughout.

8.1. *Existence of ultrafilters in* $S_\Omega(\omega)$

Let $S_\Omega(\omega)$ be the collection of denumerable subsets of ω, or of all $f : \omega \rightarrow \{0, 1\}$. By CH, these can be Ω-enumerated, up to $\equiv$.

THEOREM. *Any denumerable* $S \subseteq S_\Omega(\omega)$ *with the finite intersection property has an* Ω*-enumerated ultrafilter extension* D.

Proof: By the standard argument, we Ω-enumerate a subcollection D of $S_\Omega(\omega) = \{X_\alpha | \alpha < \Omega\}$, putting X_α in D at stage α if X_α together with the preceding choices has the f.i.p., otherwise $\omega - X_\alpha$.

COROLLARY. *There exist* Ω*-enumerable non-principal ultrafilters in* $S(_\Omega\omega)$.

8.2. *Ultrapowers*

Given an ultrafilter $D \subseteq S_\Omega(\omega)$ and any $\mathscr{A}$, the ultrapower $\mathscr{A}^\omega/D$ with the equivalence relation $\equiv$ is defined as usual.

THEOREM. *If* $\mathscr{A}$ *is* Ω*-valuated and* D *is an* Ω*-enumerated ultrafilter in* $S_\Omega(\omega)$, *then* $(\mathscr{A}^\omega/D, \equiv)$ *is* Ω*-valuated and* $\mathscr{A} \prec \mathscr{A}^\omega/D$.

Proof: The usual ultrapower theorem

$$\text{(UP)} \qquad (\mathscr{A}^{\omega}/D, \equiv) \models \varphi[f_1, \ldots, f_n] \Leftrightarrow \{i \mid \mathscr{A} \models \varphi[f_1(i), \ldots, f_n(i)]\} \in D,$$

is proved by induction on φ. Uniformization of Ω-enumerable classes allows us to pass from φ to $\exists v_0 \varphi$. CH gives us an Ω-enumeration of A^{ω}. Then the r.h.s. of (UP) gives us an enumeration of the satisfaction relation for $(\mathscr{A}^{\omega}/D, \equiv)$, in other words an Ω-valuation. We may extensionalize $(\mathscr{A}^{\omega}/D, \equiv)$ to get the usual ultrapower.

8.3. *Saturation of ultrapowers*

THEOREM. *If $\mathscr{A}$ is Ω-valuated and D is Ω-enumerable and non-principal then $\mathscr{A}^{\omega}/D$ is Ω-saturated.*

Proof: Just as in [C, K] pp. 305–306.

9. CONCLUSION

9.1. *Comparison with Cutland's work*

It is clear from 6.4 that by following the pattern of arguments in Sections 6–8, all of Cutland's work in [C2] for locally countable admissible $(M, \in)$ may be generalized by a theorem of T_{Ω} + EOC + CH, giving at the same time a generalization of a classical result of model theory. Various of the results in [C2] are proved without the hypothesis of local countability. I presume that all of these can be generalized in T_{Ω} + GCH*, though I have not checked the details.

9.2. *Directions for further work*

Besides verifying the preceding, we wish to see to what extent this program can be carried out in model theory and in other parts of set-theoretical mathematics as well. This raises two questions:

(i) What are the obstacles to the program? In other words, what theorems of set-theoretical model theory fail to have a generalization in T_{Ω} + EOC($\pm$CH)?

(ii) What is an appropriate theoretical framework to generalize the use of higher cardinals $\aleph_{\alpha}$ by 'effective' properties of higher number classes Ω_{α}?

Department of Mathematics,
Stanford University

REFERENCES

* Research supported in part by NSF grant MCS76-07163.

[B] J. Barwise, *Admissible Sets and Structures*, Springer, Berlin, 1975.

[B,S] J. Barwise and J. Schlipf, 'An introduction to recursively saturated and resplendent models', *J. Symbolic Logic* **41** (1976), 531–536.

[C,K] C. C. Chang and H. J. Keisler, *Model Theory*, North-Holland, Amsterdam, 1973.

[C1] N. Cutland, 'Π_1^1-models and Π_1^1-categoricity', pp. 42–63 in *Conference in Mathematical Logic – London '70*, Springer Lecture Notes **255**

[C2] N. Cutland, 'Model theory on admissible sets', *Annals Math. Logic* **5** (1973), 257–289.

[F] S. Feferman, 'A language and axioms for explicit mathematics', pp. 87–139 in *Algebra and Logic*, Springer Lecture Notes **450**

[Fr] H. Friedman, 'Axiomatic recursive function theory', pp. 113–137 in *Logic Colloquium '69*, North-Holland, Amsterdam, 1971.

JUHA OIKKONEN

HIERARCHIES OF MODEL THEORETIC DEFINABILITY – AN APPROACH TO SECOND ORDER LOGICS

ABSTRACT. We define hierarchies of model theoretic definability by iterating PC- and RPC-definitions and their duals. These hierarchies can be considered generalized second order logics. We show that the model theoretic RPC-hierarchy defined in terms of $\mathscr{L}_{\infty\omega}(\mathrm{WF})$ essentially coincides with the Lévy hierarchy of classes. These results can be applied to get new information about $\mathscr{L}^{\mathrm{II}}$.

0. INTRODUCTION

Among the formal languages studied in mathematical logic there is a famous triple: first order predicate calculus, second order predicate calculus, and the theory of types. What we call second order logic here, is second order logic with relation variables on strong models. Predicate calculus has a beautiful and useful model theory, which is the subject of intensive research work even today. On the other hand, the well known limitations of predicate calculus have led to the study of many stronger logics like infinitary languages and logics with generalized quantifiers, which are closely connected to predicate calculus.

The ideas of the theory of types have perhaps found a more fruitful realization in the study of the cumulative hierarchies of axiomatic set theory. But the ordinary hierarchies of set theory are built using only the empty set. The models of predicate calculus and second order logic are ordinary mathematical relational structures. So especially the sentences of second order logic describe mathematical structures and their relations. Hence to have a full analogue with predicate calculus and second order logic, we must allow cumulative hierarchies with any mathematical structure $\mathfrak{M}$ as the set of urelements which are used in building sets. The theory of admissible sets with urelements in Barwise [1] is a major contribution to this direction.

So the model theories of first order logic and the theory of types have found rich modern forms. The situation with second order logic $\mathscr{L}^{\mathrm{II}}$ is different: There is hardly any general model theory of $\mathscr{L}^{\mathrm{II}}$. For instance

J. Hintikka, I. Niiniluoto, and E. Saarinen (eds.), Essays on Mathematical and Philosophical Logic, 197–225.

the ordinary completeness theorem requires non-standard models, which are not allowed here. On the other hand it is known that the study of second order definability on specific structures, especially the natural and real numbers in recursion theory and in descriptive set theory, leads to an interesting theory. It is also known that the general properties of second order logic are strongly tied to metatheory: it is for instance easy to write a second order sentence, which has a model if and only if the continuum hypothesis holds.

One way of approaching second order logic is to consider its fragments and other logics which are weaker than $\mathscr{L}^{\mathrm{II}}$ or some of its analogues. Certain strengthenings of predicate calculus via generalized quantifiers have proven to be connected to second order logic. So the various strengthenings of predicate calculus can be viewed as approximations of second order logic from below. Krynicki [11] compares, for example, second order logic with logics with so called Henkin quantifiers.

The ordinary second order logic is obtained from predicate calculus by adding set and relation quantifiers. We propose some more general logics: our approach to second order logics is to study second order definability in general and to look in what cases an interesting theory arises. To study second order logics from a general point of view like this, it is reasonable to consider all second order quafitifications needed in model theory, and to apply them to different logics.

It is well known that the so called PC- and RPC-classes have an important role in model theory. In a definition of a PC-class one uses the model theoretic quantification 'there exists an expansion . . .'. The RPC-classes are defined in a similar way by the quantification 'there exists an extented expansion . . .' Both of these quantifications are of a second order character. (Barwise [2] calls the RPC-classes the Σ_1^1-classes.) In the case of predicate calculus the PC-quantification coincides with a finite block of quantifiers of the form

$$\exists R_0 \ldots R_{n-1} F_0 \ldots F_{m-1} x_0 \ldots x_{k-1}$$

where the R's are relation variables, and the F's are function variables. So every PC-class of predicate calculus can be defined by a Σ_1^1-sentence of $\mathscr{L}^{\mathrm{II}}$. Allowing function variables does not strengthen the logic, because functions can be replaced by relations. Hence the PC-quantification is certainly a second order quantification. The RPC-quantification is stronger, but it is in many cases closely connected to the PC-quantification, as we shall see below. So it also is a natural object of study, when one is interested in second order definability. From our point of view it is natural to proceed one step further, and to consider also the RPC-quantification a second order quantification. We define here several second order logics in terms of the second order hierarchies of model theoretic definability where we use alternating strings of PC- and RPC-quantifications and their duals. The classes thus obtained are called the Σ_n^{PC}-, Π_n^{PC}-, Δ_n^{PC}-, Σ_n^{RPC}- Π_n^{RPC}- and Δ_n^{RPC}-classes of the logic used in the definition. So the Σ_1^{PC}- and Σ_1^{RPC}-classes are just the PC- and RPC-classes. Our interest in second order logics lies mainly in the behaviour of the PC- and RPC-hierarchies. So it is reasonable from our point of view to study these hierarchies defined using any logic. If the logic used is a first order logic like predicate calculus, then there is even more reason to consider the PC- and RPC-hierarchies second order logics. Hence if the absolute logics of Barwise [1] are good first order logics, then the PC- and RPC-hierarchies, or at least the former ones, of absolute logics are good second order logics. All the specific logics discussed below besides $\mathscr{L}^{\text{II}}$ and $\mathscr{L}^{\text{III}}$ will be absolute.

This general approach to second order definability has many advantages: it has proven that several local results of descriptive set theory have global correspondences in the theory of the lower levels of the PC-hierarchy of $\mathscr{L}_{\omega_1\omega}$, when the considerations are restricted to countable structures. Vaught [16] is a recent contribution to this area. We give another example: one of the most important problems connected to second order definability is to find a characterization of second order Δ_1-definability, which does not mention second order quantifications. In the case of recursion theory, where only the structure $\mathscr{N}$ of the natural numbers is studied, this was the problem to characterize the Δ_1^1-relations of $\mathscr{N}$, or in other words those structures $(\mathscr{N}, \mathrm{R})$ which belong to some Δ_1^{PC}-class of the language of first order arithmetic expanded with one relation symbol. The solution of this problem was given by Kleene's theory of hyperarithmetical relations. (We restricted ourselves above to numerical relations for notational simplicity.) In the general model theoretic case the same problem is that of giving a syntactical description to the Δ-extensions of different logics. This problem has been solved in several special cases, but there are still many important open questions about the Δ-extensions. (See Barwise [2] and Makowsky, Shelah and Stavi [13].) These local and global forms of the problem of second order Δ_1-definability are similar also in that they both are closely connected to separation or interpolation properties: the final step in the analysis of the Δ_1^1-relations is the Lusin-Addison separation theorem for hyperarithmetical relations. On the other hand, the Δ-

extensions were first defined for finding extensions of given logics, which have certain interpolation properties.

We noticed above that ordinary second order logic is very strongly connected to underlying metatheory. So it is natural to ask what kind of connections second order definability and set theory can have. One gets even more reason to this kind of research, when one remembers that set theory is the modern heir of type theory. In [1] Barwise showed that a class of $\mathscr{L}$-structures, which is closed under isomorphism, is definable by a Δ_1^{KP}-formula if and only if it is the class of all models of some sentence of $\mathscr{L}_{\omega_1 \omega}$. Here we must suppose that $\mathscr{L}$ together with the other parameters is an element of HC, the collection of hereditarily countable sets. This result gives rise to a general study of the relations between model theoretic and set theoretic definability. We study here how the PC- and RPC-hierarchies of $\mathscr{L}_{\infty\omega}$(WF), the logic obtained from $\mathscr{L}_{\infty\omega}$ by adding new atomic sentences of the form '*E* is well-founded', where *E* is a binary relation symbol, are connected to the Lévy hierarchy of classes. We show for instance that a class of $\mathscr{L}$-structures closed under isomorphism is Σ_1 if and only if it is a Σ_1^{RPC}-class of $\mathscr{L}_{\infty\omega}$(WF). Analogous results hold also about the first levels of the PC-hierarchy of $\mathscr{L}_{\infty\omega}$(WF).

It is easy to see that any Σ_n^{PC}- or Π_n^{PC}-class of $\mathscr{L}_{\infty\omega}$(WF) or any other absolute logic is Δ_2. This result follows from the fact that the PC-quantifications and their duals can be restricted to certain power sets. So the above results about the first levels of the PC-hierarchy of $\mathscr{L}_{\infty\omega}$(WF) cannot be lifted to arbitary levels. The absolute logics are, very loosely speaking, defined by requiring that the model classes are Δ_1. So the model classes of first order logics can be thought to be Δ_1. (See the discussion in Barwise [1].) Hence the model classes of $\mathscr{L}^{\text{II}}$ are Δ_2. This promising analogy between the order of a logic and the subscript on the Δ_n does not go further. For example if $\mathscr{L}^{\text{III}}$ is the third order analogue of $\mathscr{L}^{\text{II}}$, then also the model classes of $\mathscr{L}^{\text{III}}$ are Δ_2. The same holds also for higher order logics.

In our main theorem it is shown that also the higher levels of the RPC-hierarchy of $\mathscr{L}_{\infty\omega}$(WF) essentially coincide with the respective level of the Lévy hierarchy of classes. This result shows a deep difference between PC- and RPC-quantifications. But the above remarks about higher order logics show that this difference does not compel us to think that the RPC-quantification cannot be of second order. On the other hand our results show that the RPC-quantification is a neat approximation from above of the PC-quantification and second order logics using it. As a special case our results yield information about $\mathscr{L}^{\text{II}}$ and its RPC-analogue. Some of these applications are remarkable from our point of view in that their proofs require considering the RPC-hierarchies of infinitary languages.

1. PRELIMINARIES

Our metatheory is ZFC and our set theoretic notation is standard. The sequences $\langle a, b\rangle$, $\langle a, b, c\rangle, \ldots$ are considered functions and (a, b), $(a, b, c), \ldots$ are as usual ordered pairs, triples etc. The formulas of the language of set theory are denoted by $\mathscr{P}, \mathscr{Q}, \ldots$ The necessary set theoretic background can be found in Drake [7] and in Barwise [3]. The reader should be familiar with the elements of finitary and infinitary model theory. Chang-Keisler [5] and Keisler [10] can be used to fill the gaps.

An *alphabet* is a set of non-logical symbols – i.e. constant, predicate and function symbols. The equality sign is taken to be a logical symbol. Constant symbols are of the form (0, a); n-ary predicate symbols are of the form $(1, n, \mathrm{a})$; n-ary function symbols are of the form $(2, n, \mathrm{a})$. Alphabets are denoted by $\mathscr{K}, \mathscr{L}, \ldots$

Let $\mathscr{L}$ be an alphabet. *$\mathscr{L}$-structures* are conceived as pairs $\mathfrak{M} = (M, f_{\mathfrak{M}})$ where M is the domain of $\mathfrak{M}$ and $f_{\mathfrak{M}}$ is the interpretation of $\mathscr{L}$ in $\mathfrak{M}$. $\mathscr{L}$-structures are denoted by $\mathfrak{M}, \mathfrak{N}, \ldots$ The interpretation $f_{\mathfrak{M}}(s)$ of a symbol $s \in \mathscr{L}$ is often denoted by $s^{\mathfrak{M}}$. Capital letters $M, N, \ldots$ always denote the domains of the corresponding structures. For example M_0 is the domain of $\mathfrak{M}_0$. If $\mathfrak{M}$ is an $\mathscr{L}$-structure and $\mathscr{K} \subseteq \mathscr{L}$ then $\mathfrak{M}\restriction\mathscr{K}$ denotes the structure $(M, f_{\mathfrak{M}}\restriction\mathscr{K})$ which is called the *$\mathscr{K}$-restriction* of $\mathfrak{M}$. $\mathfrak{M}$ is called an *expansion* of $\mathfrak{M}\restriction\mathscr{K}$ to $\mathscr{L}$. In addition let $U \in \mathscr{L} - \mathscr{K}$ be a unary predicate symbol. Now $(\mathfrak{M}\restriction\mathscr{K})^U$ is the substructure of $\mathfrak{M}\restriction\mathscr{K}$ which is generated by $U^{\mathfrak{M}}$. The structure $(\mathfrak{M}\restriction\mathscr{K})^U$ is often called the $\mathscr{K}$, *U-restriction* of $\mathfrak{M}$.

Let $\mathscr{L}$ be an alphabet. An *$\mathscr{L}$-logic* is a pair $(\mathbf{A}, \mathbf{R})$ of two classes where

(i) the elements of $\mathbf{R}$ are pairs $(\mathfrak{M}, a)$ where $\mathfrak{M}$ is an $\mathscr{L}$-structure and $a \in \mathbf{A}$,

(ii) if $(\mathfrak{M}, a) \in \mathbf{R}$ and $\mathfrak{M} \simeq \mathfrak{N}$ then $(\mathfrak{N}, a) \in \mathbf{R}$.

Requirement (ii) is called the *isomorphism condition*. The elements of $\mathbf{A}$ are called *sentences* and $\mathbf{R}$ is called the *satisfaction relation*. Sentences are usually denoted by $\varphi, \psi, \ldots$ and sets of sentences by $\Phi, \Psi, \ldots$ We write $\mathfrak{M}\mathbf{R}\varphi$ for $(\mathfrak{M}, \varphi) \in \mathbf{R}$. We use the notation $(\mathscr{L}^*, \models^*_{\mathscr{L}})$, $(\mathscr{L}^\dagger, \models^\dagger_{\mathscr{L}}), \ldots$ or just

$(\mathscr{L}^*, \models^*)$, $(\mathscr{L}^\dagger, \models^\dagger)$, ... for $\mathscr{L}$-logics. $\mathscr{L}$-logics are often called just *logics*.

We say that an operation * which assigns to each alphabet $\mathscr{L}$ an $\mathscr{L}$-logic $(\mathscr{L}^*, \models^*_{\mathscr{L}})$ and which satisfies the following for all alphabets $\mathscr{L}$ and $\mathscr{K}$ s.t. $\mathscr{L} \subseteq \mathscr{K}$

(i) $\mathscr{L}^* \subseteq \mathscr{K}^*$,

(ii) for all $\mathscr{K}$-structures $\mathfrak{M}$ and all $\varphi \in \mathscr{L}^*$
$\mathfrak{M} \models^*_{\mathscr{K}} \varphi$ and $(\mathfrak{M} \upharpoonright \mathscr{L}) \models^*_{\mathscr{L}} \varphi$
are equivalent,

is a *system of logics*. Systems of logics are denoted by *, †, ... Condition (ii) implies that we may omit the subscript $\mathscr{L}$ from $\models^*_{\mathscr{L}}$ when * is a system of logics.

We say that a logic $(\mathscr{L}^*, \models^*)$ is *absolute* if there are Σ_1-formulas $\mathscr{R}(x, y, z)$ and $\mathscr{S}(x, y, z, u)$ and a Π_1-formula $\mathscr{P}(x, y, z, u)$ of the language of set theory and a set a s.t.

(i) $\mathscr{L}^* = \{\varphi | \mathscr{R}(\varphi, \mathscr{L}, a)\}$,

(ii) $\models^* = \{(\mathfrak{M}, \varphi) | \mathscr{S}(\mathfrak{M}, \varphi, \mathscr{L}, a)\}$,

(iii) $\forall xy\, (\mathscr{R}(y, \mathscr{L}, a) \rightarrow (\mathscr{S}(x, y, \mathscr{L}, a) \leftrightarrow \mathscr{P}(x, y, \mathscr{L}, a)))$.

This definition makes weaker requirements than that of Barwise [1]. The difference is that Barwise's definition suits well for proving theorems about our metatheory whereas our weaker requirements are enough for studying one universe of set theory; if $a \in H(\omega_1)$ then the results of [1] remain unchanged. The notion of an absolute logic is meant to be some kind of formal version of that of a first order logic, and it is discussed in Barwise [1]. We say that a system of logics * is *absolute* if there are $\mathscr{R}(x, y, z)$, $\mathscr{S}(x, y, z, u)$ and $\mathscr{P}(x, y, z, u)$ and a parameter a s.t. the above claims are satisfied for all alphabets $\mathscr{L}$.

We have so far omitted all examples. Now that we have finished our list of definitions about abstract logics it is time to see how our familiar logics fit to this general framework. First we recall that predicate calculus $\mathscr{L}_{\omega\omega}$ and its extensions like $\mathscr{L}_{\infty\omega}$ are often considered in a gödelized form where sentences are sets. For example $\neg\varphi$ might be defined to be $(\neg, \varphi)$. In this sense $\mathscr{L}_{\omega\omega}$ and $\mathscr{L}_{\infty\omega}$ fit our abstract framework well: They are both absolute systems of logics which are closed under negation and conjunction. (See Barwise [1].) An important example of an absolute system of logics (without parameters) for us is $\mathscr{L}_{\infty G}$ which is obtained from $\mathscr{L}_{\infty\omega}$ by adding the following rule of formation: If $\varphi_n(x_0, \ldots,$

$x_{k_n-1}, y_0, \ldots, y_{l-1})$ is for each $n < \omega$ a formula of $\mathscr{L}_{\infty G}$ with all free variables displayed then

$$\forall x_0 \exists x_1 \forall x_2 \ldots \bigvee_{n<\omega} \varphi_n(x_0, \ldots, x_{n-1}, y_0, \ldots, y_{m-1})$$

and

$$\forall x_0 \exists x_1 \forall x_2 \ldots \bigwedge_{n<\omega} \varphi_n(x_0, \ldots, x_{n-1}, y_0, \ldots, y_{m-1})$$

are formulas of $\mathscr{L}_{\infty G}$ whose free variables are $y_0, \ldots, y_{l-1}$. The satisfaction relation is defined by means of winning strategies for an infinite game with players $\forall$ and $\exists$. This is done in detail in Keisler [9] where two extensions $\mathscr{L}(\omega)$ and $\mathscr{L}(\infty)$ of $\mathscr{L}_{\infty G}$ are defined and studied. The logics $\mathscr{L}_{\infty G}$ form together a system of logics which is absolute with respect to KP + Σ_1-separation. (See Barwise [1].) The concepts of a well-ordering and a well-founded relation can be defined in $\mathscr{L}_{\omega_1 G} = \mathscr{L}_{\infty G} \cap H(\omega_1)$ since

$$\neg \exists x_0 \exists x_1 \ldots \bigwedge_{n<\omega} x_{n+1} E x_n$$

is a sentence of $\mathscr{L}_{\omega_1 G}$ when $E \in \mathscr{L}$ is a binary predicate symbol. It is shown in Takeuti [15] that one can express in $\mathscr{L}_{\omega_1 G}$ things which cannot be expressed even in $\mathscr{L}_{\infty\infty}$. We do not need here all the expressive power of game quantifiers. What is enough for our purposes is the sublogic $\mathscr{L}_{\infty\omega}(\mathrm{WF})$ of $\mathscr{L}_{\infty G}$ which is obtained from $\mathscr{L}_{\infty\omega}$ by adding for each binary predicate symbol E a new atomic sentence, which says that E is well founded. For uniqueness, let these new atomic sentences be of the form (n, E) for suitable n.

Let * be a system of logics. Let $\mathscr{L}$ be an alphabet. We say that a class **C** of $\mathscr{L}$-structures is $\Delta_0(\mathscr{L}^*, \models^*)$ if it is a *model class* of $(\mathscr{L}^*, \models^*)$, i.e. if there is a sentence φ s.t.

$$\mathbf{C} = \{\mathfrak{M} | \mathfrak{M} \models^* \varphi\}.$$

We say that **C** is $\Sigma_1^{\mathrm{RPC}}(\mathscr{L}^*, \models^*)$ if there is an alphabet $\mathscr{K} \supseteq \mathscr{L}$, a unary predicate symbol $U \in \mathscr{K} - \mathscr{L}$ and a $\Delta_0(\mathscr{K}^*, \models^*)$-class **D** s.t. for all $\mathscr{L}$-structures $\mathfrak{M}$

(+) $\mathfrak{M} \in \mathbf{C}$ iff there is some $\mathfrak{N} \in \mathbf{D}$ s.t. $U^{\mathfrak{N}}$ is *$\mathscr{L}$-closed* and $(\mathfrak{N} \restriction \mathscr{L})^U = \mathfrak{M}$.

By saying that $U^{\mathfrak{N}}$ is $\mathscr{L}$-closed we mean that $U^{\mathfrak{N}}$ contains the interpretations of the constant symbols of $\mathscr{L}$ and is closed under the interpretations of the function symbols of $\mathscr{L}$. $\Sigma_1^{\mathrm{RPC}}(\mathscr{L}^*, \models^*)$-classes are called the Σ_1^1-

classes of $(\mathscr{L}^*, \models^*)$ in Barwise [2]. The $\Sigma_1^{\mathrm{RPC}}(\mathscr{L}^*, \models^*)$-classes are essentially the same as the PC-classes of $(\mathscr{L}^*, \models^*)$ of Feferman [8] because Feferman allows adding new sorts of objects in the definition of his PC-classes. The $\Sigma_1^{\mathrm{RPC}}(\mathscr{L}^*, \models^*)$-classes are often called the RPC-classes of $(\mathscr{L}^*, \models^*)$ in ordinary one-sorted model theory. Let $n \geq 1$. We say that **C** is $\Pi_n^{\mathrm{RPC}}(\mathscr{L}^*, \models^*)$ if

$$-\mathbf{C} = \{\mathfrak{M} | \mathfrak{M} \text{ is an } \mathscr{L}\text{-structure and } \mathfrak{M} \notin \mathbf{C}\}$$

is $\Sigma_n^{\mathrm{RPC}}(\mathscr{L}^*, \models^*)$. We say that **C** is $\Sigma_{n+1}^{\mathrm{RPC}}(\mathscr{L}^*, \models^*)$ if there is $\mathscr{K} \supseteq \mathscr{L}$, a unary predicate symbol $U \in \mathscr{K} - \mathscr{L}$ and a $\Pi_n^{\mathrm{RPC}}(\mathscr{K}^*, \models^*)$-class **D** s.t. (+) above holds. We say finally that **C** is $\Delta_n^{\mathrm{RPC}}(\mathscr{L}^*, \models^*)$ if it is both $\Sigma_n^{\mathrm{RPC}}(\mathscr{L}^*, \models^*)$ and $\Pi_n^{\mathrm{RPC}}(\mathscr{L}^*, \models^*)$. The $\Sigma_n^{\mathrm{PC}}(\mathscr{L}^*, \models^*)$-, $\Pi_n^{\mathrm{PC}}(\mathscr{L}^*, \models^*)$- and $\Delta_n^{\mathrm{PC}}(\mathscr{L}^*, \models^*)$-classes are defined in a similar way so that extended expansions of $\mathfrak{M}$ are replaced by expansions of $\mathfrak{M}$. The $\Sigma_1^{\mathrm{PC}}(\mathscr{L}^*, \models^*)$-classes are usually called the PC-classes of $(\mathscr{L}^*, \models^*)$. These hierarchies are called the *RPC-* and the *PC-hierarchy* of second order model theoretic definability of $(\mathscr{L}^*, \models^*)$. We are here mainly interested in the RPC-hierarchy. The following lemmas list some trivial properties of the RPC- and PC-hierarchies. When an analogous claim holds for both hierarchies, we state the claim only for the RPC-hierarchy.

LEMMA 1.1. If **C** is $\Delta_0(\mathscr{L}^*, \models^*)$ or $\Sigma_n^{\mathrm{(R)PC}}(\mathscr{L}^*, \models^*)$, $\Pi_n^{\mathrm{(R)PC}}(\mathscr{L}^*, \models^*)$ or $\Delta_n^{\mathrm{(R)PC}}(\mathscr{L}^*, \models^*)$ for some $n \geq 1$ then **C** is *closed under isomorphism*, i.e. if $\mathfrak{M} \in \mathbf{C}$ and $\mathfrak{M} \simeq \mathfrak{N}$ then $\mathfrak{N} \in \mathbf{C}$. □

We say that a logic $(\mathscr{L}^*, \models^*)$ is *closed under negation* if for all $\varphi \in \mathscr{L}^*$ there is some $\psi \in \mathscr{L}^*$ s.t. for all $\mathscr{L}$-structures $\mathfrak{M}$

$$\mathfrak{M} \models {}^*\psi \text{ iff not } \mathfrak{M} \models {}^*\varphi$$

and we say that $(\mathscr{L}^*, \models^*)$ is *closed under conjunction* if for all $\varphi, \psi \in \mathscr{L}^*$ there is some $\sigma \in \mathscr{L}^*$ s.t. for all $\mathscr{L}$-structures $\mathfrak{M}$

$$\mathfrak{M} \models {}^*\sigma \text{ iff } \mathfrak{M} \models {}^*\varphi \text{ and } \mathfrak{M} \models {}^*\psi.$$

LEMMA 1.2. (i) Let $n \geq 1$. A class **C** of $\mathscr{L}$-structures is $\Pi_{n+1}^{\mathrm{RPC}}(\mathscr{L}^*, \models^*)$ if and only if there is $\mathscr{K} \supseteq \mathscr{L}$, a unary predicate symbol $U \in \mathscr{K} - \mathscr{L}$ and a $\Sigma_n^{\mathrm{RPC}}(\mathscr{L}^*, \models^*)$-class **D** s.t. for all $\mathscr{L}$-structures $\mathfrak{M}$

(++) $\mathfrak{M} \in \mathbf{C}$ iff for all $\mathfrak{N}$ s.t. $U^{\mathfrak{N}}$ is $\mathscr{L}$-closed and $(\mathfrak{N} \upharpoonright \mathscr{L})^U = \mathfrak{M}$ it holds that $\mathfrak{N} \in \mathbf{D}$.

(ii) Suppose that the logic $(\mathscr{K}^*, \models^*)$ is closed under negation for each alphabet $\mathscr{K}$. Now $\mathbf{C}$ is $\Pi_1^{\mathrm{RPC}}(\mathscr{L}^*, \models^*)$ if and only if there is $\mathscr{K} \supseteq \mathscr{L}$, a unary predicate symbol $U \in \mathscr{K} - \mathscr{L}$ and a $\Delta_0(\mathscr{K}^*, \models^*)$-class $\mathbf{D}$ s.t. $(++)$ holds. □

If all the logics $(\mathscr{K}^*, \models^*)$ are closed under negation then by Lemma 1.2 we can define the $\Sigma_n^{\mathrm{RPC}}(\mathscr{L}^*, \models^*)$- and $\Pi_n^{\mathrm{PRC}}(\mathscr{L}^*, \models^*)$-classes in a natural way in terms of alternating strings of our model theoretic quantifications. In this case we may say that a class is $\Sigma_n^{\mathrm{RPC}}(\mathscr{L}^*, \models^*)$- or $\Pi_n^{\mathrm{RPC}}(\mathscr{L}^*, \models^*)$-defined by some unary predicate symbols $U_0, \ldots, U_{n-1}$, alphabets $\mathscr{L}_0 \subseteq \ldots \subseteq \mathscr{L}_{n-1}$ s.t. $\mathscr{L} \subseteq \mathscr{L}_0$ and $U_i \in \mathscr{L}_i$ for $i < n$ and a sentence $\varphi \in (\mathscr{L}_{n-1})^*$. For example $\mathbf{C}$ is the $\Pi_2^{\mathrm{RPC}}(\mathscr{L}^*, \models^*)$-class defined by U_0, U_1, $\mathscr{L}_0$, $\mathscr{L}_1$ and $\varphi \in (\mathscr{L}_1)^*$ if and only if for all $\mathscr{L}$-structures $\mathfrak{M}$

$\mathfrak{M} \in \mathbf{C}$ iff for all $\mathscr{L}_0$-structures $\mathfrak{N}_0$ s.t.
$(\mathfrak{N}_0 \restriction \mathscr{L})^{U_0} = \mathfrak{M}$ and $U_0^{\mathfrak{N}_0}$ is $\mathscr{L}$-closed
there is an $\mathscr{L}_1$-structure $\mathfrak{N}_1$ s.t.
$(\mathfrak{N}_1 \restriction \mathscr{L}_0)^{U_1} = \mathfrak{M}_0$, $U_1^{\mathfrak{N}_1}$ is $\mathscr{L}_0$-closed
and $\mathfrak{N}_1 \models {}^*\varphi$.

The definitions of $\Sigma_n^{\mathrm{CP}}(\mathscr{L}^*, \models^*)$- and $\Pi_n^{\mathrm{PC}}(\mathscr{L}^*, \models^*)$-classes can always be expressed by a simple string of second order quantifiers in the way as was done in the preface.

LEMMA 1.3. Suppose that the logic $(\mathscr{K}^*, \models^*)$ is closed under negation and conjunction for each alphabet $\mathscr{K}$ and suppose that for each unary predicate symbol $U \in \mathscr{K}$ there is a sentence of $\mathscr{K}^*$ which has the same models as $\forall x U(x)$. Now each $\Delta_0(\mathscr{L}^*, \models^*)$-class is a $\Delta_n^{\mathrm{RPC}}(\mathscr{L}^*, \models^*)$-class for each $n \geq 1$. Moreover each $\Sigma_n^{\mathrm{PRC}}(\mathscr{L}^*, \models^*)$- or $\Pi_n^{\mathrm{RPC}}(\mathscr{L}^*, \models^*)$-class is a $\Delta_m^{\mathrm{RPC}}(\mathscr{L}^*, \models^*)$-class for all $m > n$. In this case every $\Sigma_n^{\mathrm{PC}}(\mathscr{L}^*, \models^*)$-, $\Pi_n^{\mathrm{PC}}(\mathscr{L}^*, \models^*)$ or $\Delta_n^{\mathrm{PC}}(\mathscr{L}^*, \models^*)$-class is respectively a $\Sigma_n^{\mathrm{RPC}}(\mathscr{L}^*, \models^*)$, $\Pi_n^{\mathrm{RPC}}(\mathscr{L}^*, \models^*)$ or $\Delta_n^{\mathrm{RPC}}(\mathscr{L}^*, \models^*)$-class.

The latter part of the first claim of Lemma 1.3 should be proved first for $m = n + 1$ by induction on n. The first claim holds for the PC-hierarchy without the requirements of 1.3.

LEMMA 1.4. Suppose that * is an absolute system of logics. Let $n \geq 1$. Now each $\Sigma_n^{\mathrm{RPC}}(\mathscr{L}^*, \models^*)$-, $\Pi_n^{\mathrm{RPC}}(\mathscr{L}^*, \models^*)$- or $\Delta_n^{\mathrm{RPC}}(\mathscr{L}^*, \models^*)$- class is re-, spectively Σ_n, Π_n or Δ_n. On the other hand every $\Sigma_n^{\mathrm{PC}}(\mathscr{L}^*, \models^*)$-or $\Pi_n^{\mathrm{PC}}(\mathscr{L}^*$ $\models^*)$-class is Δ_2.

Proof: The power set axiom implies that the PC-quantifiers can be restricted. □

This lemma and the results of section 3 show a deep difference between PC- and RPC-quantifications. However a word of warning is of order here. The model classes of absolute logics are Δ_1 and the classes of the PC-hierarchies of absolute logics are Δ_2. So one might be inclined to conjecture that the model classes of nth order logics are Δ_n. But if $\mathscr{L}^{\mathrm{III}}$ is the third order logic where one can use quantifiers over relations of relations, then exactly the same argument, which shows that the model classes of $\mathscr{L}^{\mathrm{II}}$ are Δ_2, shows that also the model classes of $\mathscr{L}^{\mathrm{III}}$ are Δ_2. The same holds for higher order logics. One consequence of these remarks is that the result of Lemma 1.4 does not compel us to reject RPC-quantification as a second order quantification.

Let $\pi : \mathscr{L} \to \mathscr{K}$ be a bijection between the alphabets $\mathscr{L}$ and $\mathscr{K}$. We say that π is a *name changer*, if it maps constant symbols onto constant symbols, n-ary predicate symbols onto n-ary predicate symbols and n-ary function symbols onto n-ary function symbols. If $\pi : \mathscr{L} \to \mathscr{K}$ is a name changer and $\mathfrak{M}$ is an $\mathscr{L}$-structure then the $\mathscr{K}$-structure $(M, f_{\mathfrak{M}} o \pi^{-1})$ is denoted by $\mathfrak{M}_\pi$.

We say that a system of logics * *allows name changers* if the following is satisfied for all alphabets $\mathscr{L}$, $\mathscr{K}$

(iii) if $\pi : \mathscr{L} \to \mathscr{K}$ is a name changer then for all $\varphi \in \mathscr{L}^*$ there is some $\varphi_\pi \in \mathscr{K}^*$ s.t. for all $\mathscr{L}$-structures $\mathscr{M}$
$\mathfrak{M} \models^* \varphi$ and $\mathfrak{M}\pi \models {}^*\varphi_\pi$
are equivalent.

Let κ be an uncountable cardinal. We say that * *allows name changers in $H(\kappa)$* if (iii) holds for $\pi, \varphi \in H(\kappa)$ and if in this case φ_π may be required to be found in $H(\kappa)$. The following theorem is an abstract form of a well known fact.

THEOREM 1.5. Let κ be an infinite cardinal and let * be a system of logics which allows name changers in $H(\kappa^+)$. Suppose that for all $\mathscr{L}$ every sentence of $\mathscr{L}_{\kappa^+\omega}$ has an equivalent in $\mathscr{L}^*_{\kappa^+} = \mathscr{L}^* \cap H(\kappa^+)$ and $\mathscr{L}^*_{\kappa^+}$ is closed under conjunctions, and has the following downward Löwenheim-Skolem property:

> If a sentence $\varphi \in \mathscr{L}^*_{\kappa^+}$ has a model $\mathfrak{M}$ and if $X \subseteq M$ is of cardinality $\leq \kappa$ then there is some $\mathfrak{N} \subseteq \mathfrak{M}$ s.t. $X \subseteq N$, $|N| = \kappa$ and $\mathfrak{N} \models \varphi$.

Let $\mathscr{L} \in H(\kappa^+)$. Now for every $\Sigma_1^{\mathrm{RPC}}(\mathscr{L}^*, \models^*)$-class $\mathbf{C}$, which is definable by some $\varphi \in \mathscr{K}^*_{\kappa^+}$ for some $\mathscr{K} \in H(\kappa^+)$ s.t. $\mathscr{L} \subseteq \mathscr{K}$, there is a $\Sigma_1^{\mathrm{PC}}(\mathscr{L}^*, \models^*)$-class $\mathbf{D}$ defined by some $\mathscr{K}' \in H(\kappa^+)$ and $\psi \in (\mathscr{K})^*_{\kappa^+}$ which coincides with ω on structures of cardinality $\geqslant \kappa$. Moreover, if $\kappa = \omega$ and if $\mathscr{L}^*$ has an equivalent of each sentence of $\mathscr{L}_{(2^\omega)^+\omega}$ which is a disjunction of sentences of $\mathscr{L}_{\omega_1\omega}$, then $\mathbf{C}$ is $\Sigma_1^{\mathrm{PC}}(\mathscr{L}^*, \models^*)$.

Proof: Let $\mathscr{K} \in H(\kappa^+)$, $U \in \mathscr{K} - \mathscr{L}$ and $\varphi \in \mathscr{K}^*_{\kappa^+}$ be such that $\mathfrak{M} \in \mathbf{C}$ if and only if there is some $\mathscr{K}$-structure $\mathfrak{N}$ s.t. $(\mathfrak{N} \restriction \mathscr{L})^U = \mathfrak{M}$ and $\mathfrak{N} \models \varphi$. Let $\pi : \mathscr{K} \to \mathscr{K}'$ be a name changer s.t. $\pi \restriction (\mathscr{K} - \mathscr{L})$ is the identical map and that $\mathscr{K} \cap \pi(\mathscr{L}) = \phi$. Let φ_π be obtained from φ by $_\pi$. Let $\mathscr{K}'' = \mathscr{L} \cup \mathscr{K}' \cup \{F\}$ where F is a new unary function symbol. Let ψ correspond to a conjunction of φ_π and

> "F is an 'isomorphism' between the $\mathscr{L}$-restriction and the $\pi(\mathscr{L})$, U-restriction of the $\mathscr{K}''$-structure considered".

For a more detailed analogy of the latter sentence see the proof of theorem 3.2 in [1]. We show that the following two conditions are equivalent for $\mathscr{L}$-structures $\mathfrak{M}$ of cardinality $\geqslant \kappa$:

(i) There is some $\mathscr{K}$-structure $\mathfrak{N}_0$ s.t. $(\mathfrak{N}_0 \restriction \mathscr{L})^U = \mathfrak{M}$ and $\mathfrak{N}_0 \models {}^*\varphi$.

(ii) There is some $\mathscr{K}''$-structure $\mathfrak{N}_1$ s.t. $\mathfrak{N}_1 \restriction \mathscr{L} = \mathfrak{M}$ and $\mathfrak{N}_1 \models {}^*\psi$.

(i) ⇒ *(ii)*: Let $\mathfrak{N}_0$ satisfying (i) be given. Now there is some $\mathfrak{N}'$ of cardinality κ s.t. $M \subseteq N'$ and $\mathfrak{N}' \models \varphi$. So there is a bijection $f : N' \to M$. Define $\mathfrak{N}_1$ s.t. $f : \mathfrak{N}'_\pi \simeq \mathfrak{N}_1 \restriction \mathscr{K}$ and let $F^{\mathfrak{N}_1}$ be $f \restriction M$. It is obvious that $\mathfrak{N}_1 \models {}^*\psi$.

(ii) ⇒ *(i)*: Let $\mathfrak{N}_1$ satisfy (ii). So $f = F^{\mathfrak{N}'}$ is a bijection between M and $(\pi(U))^{\mathfrak{N}'}$. Let g be a bijection which extends f^{-1} to N_1. Let $N_0 = rng(g)$. It is easy to use g to define $\mathfrak{N}_0$ s.t. $\mathfrak{N}_0 \models^* \varphi$.

The last claim follows from the fact that finite structures can be characterized up to isomorphism in $\mathscr{L}_{\omega_1\omega}$. □

Remark: The theorem holds for all absolute systems of logics definable by a parameter $a \in H(\kappa^+)$ by an obvious strengthening of the downward Löwenheim-Skolem theorem of Barwise [1].

COROLLARY 1.6. The claim of theorem 1.5 holds also for $\Pi_1^{\mathrm{RPC}}(\mathscr{L}^*, \models *)$- and $\Delta_1^{\mathrm{RPC}}(\mathscr{L}^*, \models *)$-classes and respectively $\Pi_1^{\mathrm{PC}}(\mathscr{L}^*, \vdash *)$- and $\Delta_1^{\mathrm{PC}}(\mathscr{L}^*, \vdash *)$-classes.

We do not here study the general properties of these hierarchies any further. What we are interested in is the connection between these hierarchies and the Lévy hierarchy. As a corollary of our main result we obtain a hierarchy theorem for the RPC-hierarchy of $\mathscr{L}_{\infty\omega}(\mathrm{WF})$ or $\mathscr{L}_{\infty G}$. This result can also be modified for $\mathscr{L}_{\infty\omega}$.

2. A CHARACTERIZATION FOR THE Σ_1-CLASSES

From now on we shall consider the model theoretic hierarchies of $\mathscr{L}_{\infty\omega}(\mathrm{WF})$ and $\mathscr{L}_{\infty\omega}$. The reader should notice that the results concerning $\mathscr{L}_{\infty\omega}(\mathrm{WF})$ hold for every absolute logic like $\mathscr{L}_{\infty G}$ extending $\mathscr{L}_{\infty\omega}$ where one can say that a relation is well-founded.

Let A be a transitive set. We write $\mathscr{L}_A(\mathrm{WF})$ for the fragment $\mathscr{L}_{\infty\omega}(\mathrm{WF}) \cap A$. A $\Sigma_n^{(\mathrm{R})\mathrm{PC}}(\mathscr{L}_{\infty\omega}(\mathrm{WF}))$-class is $\Sigma_n^{(\mathrm{R})\mathrm{PC}}(\mathscr{L}_A(\mathrm{WF}))$ if it is defined by some φ, $U_0, \ldots, U_{n-1}$ and $\mathscr{L}_0, \ldots, \mathscr{L}_{n-1}$ which all are elements of A. The $\Pi_n^{(\mathrm{R})\mathrm{PC}}(\mathscr{L}_A(\mathrm{WF}))$- and $\Delta_n^{(\mathrm{R})\mathrm{PC}}(\mathscr{L}_A(\mathrm{WF}))$-classes are defined in a similar way.

The class of *WF-positive* sentences of $\mathscr{L}_{\infty\omega}(\mathrm{WF})$, denoted by $\mathscr{L}_{\infty\omega}(\mathrm{WF}^+)$, is the class of those sentences which can be obtained from sentences of $\mathscr{L}_{\infty\omega}$ and sentences of the form 'E is well-founded' using the connectives $\bigvee$ and $\bigwedge$. The class of *WF-negative* sentences of $\mathscr{L}_{\infty\omega}(\mathrm{WF})$, denoted by $\mathscr{L}_{\infty\omega}(\mathrm{WF}^-)$, is the class of those sentences that can be obtained from sentences of $\mathscr{L}_{\infty\omega}$ and sentences of the form 'E is not well-founded' using the connectives $\bigvee$ and $\bigwedge$. Clearly the negation of a WF-positive sentence is WF-negative; and conversely the negation of WF-negative sentence is WF-positive. The following lemma shows how the results of this and the next chapter give information about fragments of $\mathscr{L}_{\infty\omega}$ and especially $\mathscr{L}^{\mathrm{II}}$.

LEMMA 2.1. Let A be an admissible set s.t. $\mathscr{L} \in A$.

(i) Every model class of $\mathscr{L}_A(\mathrm{WF}^+)$ is $\Pi_1^{\mathrm{PC}}(\mathscr{L}_A)$ and every model model class of $\mathscr{L}_A(\mathrm{WF}^-)$ is $\Sigma_1^{\mathrm{PC}}(\mathscr{L}_A)$.

(ii) Let n be even. Every $\Sigma_n^{(\mathrm{R})\mathrm{PC}}(\mathscr{L}_A(\mathrm{WF}^-))$-class is $\Sigma_{n+1}^{(\mathrm{R})\mathrm{PC}}(\mathscr{L}_A)$, and every $\Pi_n^{(\mathrm{R})\mathrm{PC}}(\mathscr{L}_A(\mathrm{WF}^+))$-class is $\Pi_{n+1}^{(\mathrm{R})\mathrm{PC}}(\mathscr{L}_A)$.

(iii) Let n be odd. Every $\Sigma_n^{(R)PC}(\mathscr{L}_A(WF^+))$-class is $\Sigma_{n+1}^{(R)PC}(\mathscr{L}_A)$, and every $\Pi_n^{(R)PC}(\mathscr{L}_A(WF^-))$-class is $\Pi_{n+1}^{(R)PC}(\mathscr{L}_A)$.

Proof: We show first that if $\varphi \in \mathscr{L}_A(WF^+)$ then there is some $\mathscr{K}$ obtained from $\mathscr{L}$ by adding new unary predicate symbols and some $\psi \in \mathscr{K}_A$ s.t. for all $\mathscr{L}$-structures $\mathfrak{M}$.

$\mathfrak{M} \models \varphi$ iff for all $\mathscr{K}$-structures $\mathfrak{N}$ s.t. $\mathfrak{N}\restriction\mathscr{L} = \mathfrak{M}$ it holds that $\mathfrak{N} \models \psi$.

This is proved by induction on φ. If φ is of the form 'E is well-founded' then let ψ be

$$\exists x P(x) \rightarrow \exists x(P(x) \wedge \forall y(P(y) \rightarrow \neg\, yEx))$$

for any $P \in A - \mathscr{L}$. The rest of the induction is obvious. By taking negations we see that for every $\varphi \in \mathscr{L}_A(WF^-)$ there is some $\mathscr{K}$ obtained from $\mathscr{L}$ by adding new unary predicate symbols and a sentence $\psi \in \mathscr{K}_A$ s.t. for all $\mathscr{L}$-structures

$\mathfrak{M} \models \varphi$ iff there is some $\mathscr{K}$-structure $\mathfrak{N}$ s.t. $\mathfrak{N}\restriction \mathscr{L} = \mathfrak{M}$ and $\mathfrak{N} \models \psi$.

The claims (ii) and (iii) follow from this by replacing the sentence φ defining the $\Sigma_n^{(R)PC}(\mathscr{L}_A(WF))$- or the $\Pi_n^{(R)PC}(\mathscr{L}_A(WF))$-classes considered by the above $\Pi_1^{PC}(\mathscr{K}_A)$- or $\Sigma_1^{PC}(\mathscr{K}_A)$-class, depending on whether φ is WF-positive or WF-negative. □

THEOREM 2.2. Let **C** be a class of $\mathscr{L}$-structures which is closed under isomorphism. Now **C** is Σ_1 if and only if it is $\Sigma_1^{RPC}(\mathscr{L}_{\infty\omega}(WF))$. Moreover let A be an admissible set s.t. $\mathscr{L} \in A$. Now **C** is Σ_1 in parameters from A if and only if it is $\Sigma_1^{RPC}(\mathscr{L}_A(WF))$.

Proof: In both cases the implication from right to left follows from Lemma 1.4.

Suppose that **C** is Σ_1. Now there is a Δ_0-formula $\mathscr{Q}(x, y, z, u)$ and a set a s.t. for all $\mathscr{L}$-structures $\mathfrak{M}$

$$\mathfrak{M} \in \mathbf{C} \text{ iff } \exists u \mathscr{Q}(\mathfrak{M}, \mathscr{L}, u, a)$$

because we may code finitely many parameters by the sequence which lists these parameters. Let $X = TC(\{\mathscr{L}, a\})$ and

$$\mathscr{K} = \mathscr{L} \cup \{E, U, c, \mathscr{M}\} \cup \{\check{b} | b \in X\}.$$

where E is a binary predicate symbol, U a unary predicate symbol; and c, $\mathscr{M}$ and the $\check{b}$'s are constant symbols. We suppose, of course, that all these added symbols are distinct and that none of them is in $\mathscr{L}$. Let φ be the conjunction of the following statements

(1) (Extensionality, pair and union)$_E$

(2) 'E is well-founded'

(3) $\bigvee_{b \in X} \forall x(xE\check{b} \leftrightarrow \bigwedge_{d \in b} x = \check{d})$

(4) 'U is closed under the interpretations of the function symbols of $\mathscr{L}$'

(5) '$\mathscr{M}$ is a pair $(x, y)_E$ which is an $\mathscr{L}$-structure$_E$ and the identity map is an 'isomorphism' between the $\mathscr{L}$, U-restriction of the structure considered and $\mathscr{M}$'

(6) $\mathscr{Q}(\mathscr{M}, \check{\mathscr{L}}, c, \check{a})_E$

The subscript E indicates that set theoretic statements are expressed in terms of E instead of ε. The reader should see the proof of theorem 3.2 in Barwise [1] to get a better idea of how to write (5) in full detail. We shall now show that for all $\mathscr{L}$-structures $\mathfrak{M}$

$$\mathfrak{M} \in \mathbf{C} \text{ iff there is some } \mathfrak{N} \text{ s.t. } (\mathfrak{N} \restriction \mathscr{L})^U = \mathfrak{M} \text{ and } \mathfrak{N} \models \varphi.$$

'$\Rightarrow$': There must be some b s.t. $\mathscr{Q}(\mathfrak{M}, \mathscr{L}, b, a)$ because $\mathfrak{M} \in \mathbf{C}$. Now let B be an admissible set s.t. $\omega, \mathfrak{M}, \mathscr{L}, b\ a \in B$. Define the $\mathscr{K}$-structure $\mathfrak{N}$ by setting

$$\begin{aligned} N &= B, \\ E^{\mathfrak{N}} &= \in_B, \\ U^{\mathfrak{N}} &= M, \\ c^{\mathfrak{N}} &= b, \\ \mathscr{M}^{\mathfrak{N}} &= \mathfrak{M}, \\ \check{d}^{\mathfrak{N}} &= d \text{ for } d \in X. \end{aligned}$$

The function and predicate symbols of $\mathscr{L}$ are interpreted arbitrarily outside M. It is now rather easy to see that $(\mathfrak{N} \restriction \mathscr{L})^U = \mathfrak{M}$ and that $\mathfrak{N} \models \varphi$. To show that $\mathfrak{N}$ satisfies part (6) of φ we notice that

$$(B, \in_B) \models \mathscr{Q}[\mathfrak{M}, \mathscr{L}, b, a]$$

because B is transitive, $\mathscr{Q}$ is Δ_0 and $\mathfrak{M}, \mathscr{L}, b, a \in B$.

'$\Leftarrow$': Let $\mathfrak{N}$ be a $\mathscr{K}$-structure s.t. $(\mathfrak{N}\restriction\mathscr{L})^U = \mathfrak{M}$ and $\mathfrak{N} \models \varphi$. Let $p:(N, E^{\mathfrak{N}}) \simeq (B, \in_B)$ be the collapsing isomorphism given by Mostowski's isomorphism theorem. So B is a transitive set. Now one can easily see that $p(\check{b}^{\mathfrak{N}}) = b$ for each $b \in X$. It can also be shown that $\mathfrak{M}_0 = p(\mathscr{M}^{\mathfrak{N}})$ is an $\mathscr{L}$-structure isomorphic with $\mathfrak{M}$. Now

$$(B, \in_B) \models \mathscr{Q}[\mathfrak{M}_0, \mathscr{L}, p(c^{\mathfrak{N}}), a]$$

and hence $\exists u \mathscr{Q}(\mathfrak{M}_0, \mathscr{L}, u, a)$ because B is transitive and $\mathscr{Q}$ is Δ_0. So $\mathfrak{M}_0 \in \mathbf{C}$ and hence $\mathfrak{M} \in \mathbf{C}$ because $\mathbf{C}$ is closed under isomorphism.

For the latter part of the theorem it is enough to remark that if the parameters $a_0, \ldots, a_{n-1}$ are taken from an admissible set A then also $\langle a_0, \ldots, a_{n-1}\rangle \in A$ and if $\mathscr{L} \in A$ then also $\varphi \in A$ where φ is as before. □

COROLLARY 2.3. Let $\mathbf{C}$ be a class of $\mathscr{L}$-structures which is closed under isomorphism. Now $\mathbf{C}$ is Π_1 if and only if it is $\Pi_1^{\mathrm{RPC}}(\mathscr{L}_{\infty\omega}(\mathrm{WF}))$. Moreover let A be an admissible set s.t. $\mathscr{L} \in A$. Now $\mathbf{C}$ is Π_1 in parameters from A if and only if it is $\Pi_1^{\mathrm{RPC}}(\mathscr{L}_A(\mathrm{WF}))$. □

Corollary 2.3 is an immediate consequence of the previous theorem, but it is also easy to give a direct proof for this corollary: It can be shown that for all $\mathscr{L}$-structures $\mathfrak{M}$

$\mathfrak{M} \in \mathbf{C}$ iff for all $\mathfrak{N}$ s.t. $(\mathfrak{N}\restriction\mathscr{L})^U = \mathfrak{M}$ it holds that
$\mathfrak{N} \vdash \varphi \to \mathscr{Q}(\mathscr{M}, \check{\mathscr{L}}, c, \check{a})_E$

where φ is the conjunction of the statements (1), . . ., (5) of the proof of theorem 2.1.

Let κ be an infinite cardinal s.t. $\mathscr{L} \in H(\kappa^+)$. Theorem 2.2. and corollary 2.3. imply together with theorem 1.5 that if $\mathbf{C}$ is a class of $\mathscr{L}$-structures closed under isomorphism which is Σ_1 or Π_1 in parameters from $H(\kappa^+)$ then there is respectively a $\Sigma_1^{\mathrm{PC}}(\mathscr{L}_{\kappa^+\omega}(\mathrm{WF}))$- or a $\Pi_1^{\mathrm{PC}}(\mathscr{L}_{\kappa^+\omega}(\mathrm{WF}))$-class $\mathbf{D}$ which coincides with $\mathbf{C}$ on $\mathscr{L}$-structures of cardinality $\geq \kappa$. Moreover, if $\kappa = \omega$ then $\mathbf{C}$ is $\Sigma_1^{\mathrm{PC}}(\mathscr{L}_{2^\omega\omega}(\mathrm{WF}))$ or $\Pi_1^{\mathrm{PC}}(\mathscr{L}_{2^\omega\omega}(\mathrm{WF}))$. In the Δ_1-case. we can have a better result. But first we give an analogue of corollary 2.3

COROLLARY 2.4. Let $\mathbf{C}$ be a class of $\mathscr{L}$-structures which is closed under isomorphism. Now $\mathbf{C}$ is Δ_1 if and only if it is $\Delta_1^{\mathrm{RPC}}(\mathscr{L}_{\infty\omega}(\mathrm{WF}))$. Moreover let A be an admissible set s.t. $\mathscr{L} \in A$. Now $\mathbf{C}$ is Δ_1 in parameters from A if and only if it is $\Delta_1^{\mathrm{RPC}}(\mathscr{L}_A(\mathrm{WF}))$. □

This immediate consequence of theorem 2.1 and corollary 2.2. is some kind of an analogue of corollary 3.5. in Barwise [1]. It is somewhat interesting to notice that we do not have to require $A \subseteq H(\omega_1)$ here.

So far our results about the PC-hierarchy have been mainly about large structures. In the Δ_1-case it is possible to use Scott-sentences to characterize the small structures. But the Scott-sentence $\sigma_{\mathfrak{M}}$ of an $\mathscr{L}$-structure $\mathfrak{M}$ characterizes $\mathfrak{M}$ only up to $\equiv_{\infty\omega}$; that is

$$\mathfrak{N} \models \sigma_{\mathfrak{M}} \text{ iff } \mathfrak{M} \equiv_{\infty\omega} \mathfrak{N}$$

(See proposition 7 of Chang [4]). The classes **C** which our results characterize in terms of $\mathscr{L}_{\infty\omega}(\mathrm{WF})$ are supposed to be closed under isomorphism, not under $\equiv_{\infty\omega}$. So we must be able to reduce our considerations to the countable case where $\simeq$ and $\equiv_{\infty\omega}$ coincide. The simplest possible extra supposition is that the parameter needed in the definition of **C** belongs to $H(\omega_1)$. A more general supposition is that there is a Σ_1 collection definable in parameters from $H(\omega_1) = f$ parameters a s.t. the class $\mathbf{C}_a$ defined using a is closed under isomorphism. These remarks are expressed perhaps in the most elegant way in the following result. The proof will make the above discussion clearer. This discussion gives some new motivation to the (generalized) notion of an absolute logic.

COROLLARY 2.5. If $\mathscr{L} \in H(\omega_1)$ and $(\mathscr{L}^*, \models^*)$ is absolute in a parameter $H(\omega_1)$, every model class of $(\mathscr{L}^*, \models^*)$ is $\Delta_1^{\mathrm{PC}}(\mathscr{L}_{\infty\omega}(\mathrm{WF}))$. If * is a system of logics absolute in a parameter from $H(\omega_1)$ and if $\mathscr{L}$ is arbitary, then every model class of $(\mathscr{L}^*, \models^*)$ is $\Delta_1^{\mathrm{PC}}(\mathscr{L}_{\infty\omega}(\mathrm{WF}))$. Moreover, every $\Sigma_1^{\mathrm{PC}}(\mathscr{L}^*, \models^*)$-and $\Pi_1^{\mathrm{PC}}(\mathscr{L}^*, \models^*)$-class is respectively in latter case $\Sigma_1^{\mathrm{PC}}(\mathscr{L}_{\infty\omega}(\mathrm{WF}))$ or $\Pi_1^{\mathrm{PC}}(\mathscr{L}_{\infty\omega}(\mathrm{WF}))$.

Proof: Suppose that $\mathscr{L} \in H(\omega_1)$ and let **C** be a model class of $\varphi \in \alpha^*$. Let $\kappa = \mathrm{card}(\mathrm{TC}(\varphi))$. Theorem 1.5 and the previous results of this chapter imply that there is some $\mathscr{K} \supseteq \mathscr{L}$ and sentences φ and ψ of $\mathscr{K}_{\omega 1\omega}(\mathrm{WF})$ s.t. the following three requirements are equivalent for all $\mathscr{L}$-structures $\mathfrak{M}$ of cardinality $\geq \kappa$:

(i) $\mathfrak{M} \in \mathbf{C}$,

(ii) there is some $\mathfrak{N}$ s.t. $\mathfrak{N}\restriction\mathscr{L} = \mathfrak{M}$ and $\mathfrak{N} \models \varphi$,

(iii) for all $\mathfrak{N}$ s.t. $\mathfrak{N}\restriction\mathscr{L} = \mathfrak{M}$ holds that $\mathfrak{N} \models \varphi$.

Moreover, we may suppose that all models of φ and ψ are of cardinality $\geq \kappa$. Every $\mathscr{L}$-structure $\mathfrak{M} \in \mathbf{C}$ of cardinality $\leq \kappa$ is isomorphic with an

$\mathscr{L}$-structure $\mathfrak{M}'$ which belongs to the set X of those $\mathscr{L}$-structures in $\mathbf{C}$ whose domain is some cardinal $< \kappa$. We show that for all $\mathscr{L}$-structures $\mathfrak{M}$ the following three conditions are equivalent:

(i′) $\mathfrak{M} \in \mathbf{C}$,

(ii′) there is some $\mathfrak{N}$ s.t. $\mathfrak{N}\restriction\mathscr{L} = \mathfrak{M}$ and $\mathfrak{N} \models \varphi \vee \bigvee_{\mathfrak{M}' \in X} \sigma_{\mathfrak{M}'}$,

(iii′) for all $\mathfrak{N}$ s.t. $\mathfrak{N}\restriction\mathscr{L} = \mathfrak{M}$ it holds that $\mathfrak{N} \models \psi \vee \bigvee_{\mathfrak{M}' \in X} \sigma_{\mathfrak{M}'}$.

If $\mathfrak{M}$ is of cardinality $<\kappa$ and $\mathfrak{M} \in \mathbf{C}$ then (ii′) and (iii′) are satisfied because $\mathfrak{M} \models \sigma_{\mathfrak{M}'}$ for some $\mathfrak{M}' \in X$. Because the models of φ and ψ are of cardinality $\geq \kappa$, it is enough to show that if $\mathfrak{M} \models \sigma_{\mathfrak{M}'}$ for some $\mathfrak{M}' \in X$, then $\mathfrak{M} \in \mathbf{C}$. But if $\mathfrak{M} \models \sigma_{\mathfrak{M}'}$, then $\mathfrak{M} \equiv_{\infty\omega} \mathfrak{M}'$. Proposition 2.5 of Barwise [1] obviously generalizes to our notion of an absolute logic, when the parameter a is an element of $H(\omega_1)$. So $\mathfrak{M}$ and $\mathfrak{M}'$ satisfy the same sentences of $\mathscr{L}^*$. Hence $\mathfrak{M} \in \mathbf{C}$.

The claim about an absolute system of logics is proved in a similar way by generalizing proposition 2.5 of [1] to arbitary alphabets when * is a system of logics absolute in a parameter from $H(\omega_1)$. This is done simply by quantifying the Σ_1-sentence analysed in the proof of 2.5 of [1] existentially over $\mathscr{L}$.

The last claim follows now immediately by translating the definition of a given $\Sigma_1^{\mathrm{PC}}(\mathscr{L}^*, \models^*)$- or $\Pi_1^{\mathrm{PC}}(\mathscr{L}^*, \models^*)$-class. □

The preceding results have several model theoretic applications. If $\mathscr{L}$ and the other parameters are elements of HF, then the above results are results about $\mathscr{L}^{\Pi}$ and its RPC analogy. (To identify $\mathscr{L}^{\Pi}$ with the PC-hierarchy of $\mathscr{L}_{\omega\omega}$ we must suppose that the underlying model of ZFC is standard.) So for instance every Σ_1-class of $\mathscr{L}$-structures closed under isomorphism is $\Sigma_1^{\mathrm{RPC}}(\Pi_1^1)$ and coincides for infinite structures with a $\Sigma_1^{\mathrm{PC}}(\Pi_1^1)$-class; that is with a model class of a Σ_2^1-sentence. Here Π_1^1 is the logic of Π_1^1-sentences of $\mathscr{L}^{\Pi}$.

On the other hand every $\Sigma_n^{\mathrm{(R)PC}}$- or $\Pi_n^{\mathrm{(R)PC}}$-class of an absolute logic is respectively a $\Sigma_n^{\mathrm{(R)PC}}(\mathscr{L}_{\infty\omega}(\mathrm{WF}))$- or a $\Pi_n^{\mathrm{(R)PC}}(\mathscr{L}_{\infty\omega}(\mathrm{WF}))$-class. Moreover it is possible in the RPC-case to control the size of the fragment of $\mathscr{L}_{\infty\omega}(\mathrm{WF})$ needed. An interesting corollary is that $\mathscr{L}_{\infty G}$ and $\mathscr{L}_{\infty\omega}(\mathrm{WF})$ have the same $\Sigma_n^{\mathrm{(R)PC}}$- and $\Pi_n^{\mathrm{(R)PC}}$-classes. In the RPC-case the equivalence holds for the corresponding admissible fragments too. Theorem 1 of Swett [13] is a result of a similar character, but our results are in many ways much stronger. Our first results in this field were analogies of the present corollary 2.4 and hence analogous to Swett's results, but our work was independent.

Cutland [6] characterizes the Σ_1-compact fragments of $\mathscr{L}_{\infty\omega}(\mathrm{WF})$. This result together with our above results yields a characterization of the Σ_1-compact fragments of those absolute logics like $\mathscr{L}_{\infty G}$ whose admissible fragments contain those of $\mathscr{L}_{\infty\omega}(\mathrm{WF})$, and a sufficient condition for the Σ_1-compact fragments of other absolute logics. The proof is just an application of the fact that the translation in the proof of theorem 2.2 is effective.

The proof of the following slightly artificial result will show what kind of difficulties we would meet if we wished to generalize the above arguments directly to higher levels of the Lévy hierarchy. We say that a Σ_2-class $\mathbf{C}$ of $\mathscr{L}$-structures is *strongly closed under isomorphism* if there is a Δ_0-formula $\mathscr{Q}$ and a set a s.t. for all $\mathscr{L}$-structures $\mathfrak{M}$

$$\mathfrak{M} \in \mathbf{C} \text{ iff } \exists v_1 \forall v_0 \mathscr{Q}(\mathfrak{M}, \mathscr{L}, v_0, v_1, a)$$

and the classes $\mathbf{C}_b$ are closed under isomorphism where

$$\mathbf{C}_b = \{\mathfrak{M} | \forall v_0 \mathscr{Q}(\mathfrak{M}, \mathscr{L}, v_0, b, a)\}$$

Obviously $\mathbf{C}$ is closed under isomorphism if it is strongly closed under isomorphism. To get a better idea of the difference between these two concepts we notice that $\mathbf{C}$ is closed under isomorphism if and only if for all $\mathscr{L}$-structures $\mathfrak{M}$ and $\mathfrak{N}$

$$\mathfrak{M} \simeq \mathfrak{N} \text{ implies } \exists v_1 \forall v_0 \mathscr{Q}(\mathfrak{M}, \mathscr{L}, v_0, v_1, a) \rightarrow \exists v_1 \forall v_0 \mathscr{Q}(\mathfrak{N}, \mathscr{L}, v_0, v_1, a)$$

and that $\mathbf{C}$ is strongly closed under isomorphism if and only if for all $\mathscr{L}$-structures $\mathfrak{M}$ and $\mathfrak{N}$

$$\mathfrak{M} \simeq \mathfrak{N} \text{ implies } \forall v_1 (\forall v_0 \mathscr{Q}(\mathfrak{M}, \mathscr{L}, v_0, v_1, a) \rightarrow \forall v_0 \mathscr{Q}(\mathfrak{N}, \mathscr{L}, v_0, v_1, a)).$$

THEOREM 2.6. Let $\mathbf{C}$ be a Σ_2-class of $\mathscr{L}$-structures which is strongly closed under isomorphism. Now $\mathbf{C}$ is $\Sigma_2^{\mathrm{RPC}}(\mathscr{L}_{\infty\omega}(\mathrm{WF}))$. Moreover let A be an admissible set s.t. $\mathscr{L} \in A$. Now $\mathbf{C}$ is $\Sigma_2^{\mathrm{RPC}}(\mathscr{L}_A(\mathrm{WF}))$ if the parameters used in the definition of $\mathbf{C}$ are elements of A.

Proof: We consider only the first claim because the strengthening is obtained as before. Let $\mathscr{Q}$ and a be as in the above definition. Again let $X = \mathrm{TC}(\{\mathscr{L}, a\})$ and let

$$\mathscr{K}_0 = \mathscr{L} \cup \{E, U, c_1\}$$

and

$$\mathscr{K}_1 = \mathscr{K}_0 \cup \{U_1, c_0, \mathscr{M}\} \cup \{\check{d} | d \in X\}.$$

Here E is a binary predicate symbol; U and U_1 are unary predicate symbols; and c_0, c_1, $\mathscr{M}$ and the $\check{b}$'s are constant symbols. We suppose of course that all the added symbols are new and distinct. Let ψ be the sentence

$$\begin{aligned}&\text{'}E{\restriction}U_0 \text{ is well-founded and extensional'} \wedge\\ &\text{'}E{\restriction}U_0 \subseteq_{\text{end}} E\text{'} \wedge \forall x(U(x) \to U_0(x)) \wedge\\ &(\varphi \to \mathscr{Q}(\mathscr{M}, \check{\mathscr{L}}, c_0, c_1, \check{a})_E)\end{aligned}$$

where φ is the conjunction of parts (1), . . ., (5) of the proof of theorem 2.1. Let **D** be the $\Sigma_2(\mathscr{L}_{\infty G})$-class defined by U, U_0, $\mathscr{K}_0$, $\mathscr{K}_1$ and ψ. We shall show that **C** = **D**.

We show first that $\mathbf{D} \subseteq \mathbf{C}$. So let $\mathfrak{M} \in \mathbf{D}$. Now there is some $\mathscr{K}_0$-structure $\mathfrak{N}_0$ s.t. $(\mathfrak{N}_0{\restriction}\mathscr{L})^U = \mathfrak{M}$ and for all $\mathfrak{N}_1$ s.t. $(\mathfrak{N}_1{\restriction}\mathscr{K}_0)^{U_0} = \mathfrak{N}_0$ it holds that $\mathfrak{N}_1 \models \psi$. Let $p : (N_0, E^{\mathfrak{N}_0}) \simeq (B, \in_B)$ be the collapsing isomorphism. Let $b = p(c_1^{\mathfrak{N}_0})$. We shall see that $\mathfrak{M} \in \mathbf{C}_b$. Let d be an arbit ary set. We can easily find a $\mathscr{K}_1$-structure $\mathfrak{N}_1$ s.t. $(\mathfrak{N}_1{\restriction}\mathscr{K}_0)^{U_0} = \mathfrak{N}_0$,- $\mathfrak{N}_1 \models \varphi$; and if $q: (N_1, E^{\mathfrak{N}_1}) \simeq (C, \in_C)$ is the collapsing isomorphism, then $q(\mathscr{M}^{\mathfrak{N}_1}) = \mathfrak{M}$ and $q(c_0^{\mathfrak{N}_1}) = d$. Now $q(\check{\mathscr{L}}^{\mathfrak{N}_1}) = \mathscr{L}$ and moreover $q(c_1^{\mathfrak{N}_1}) = p(c_1^{\mathfrak{N}_0}) = b$ because $(N_0, E^{\mathfrak{N}_0}) \subseteq_{\text{end}} (N_1, E^{\mathfrak{N}_1})$. Since $\mathfrak{N}_1 \models \psi$ by the choice of $\mathfrak{N}_0$ and so $\mathfrak{N}_1 \models \mathscr{Q}(\mathscr{M}, \check{\mathscr{L}}, c_0, c_1, \check{a})_E$. So

$$(C, \in_C) \models \mathscr{Q}[\mathfrak{M}, \mathscr{L}, d, b, a]$$

and hence $\mathscr{Q}(\mathfrak{M}, \mathscr{L}, d, b, a)$ because C is transitive and $\mathscr{Q}$ is Δ_0. So $\mathfrak{M} \in \mathbf{C}_b$ because d was arbitrary.

Next we shall show that $\mathbf{C} \subseteq \mathbf{D}$. So let $\mathfrak{M} \in \mathbf{C}$. Now there is some b s.t. $\mathfrak{M} \in \mathbf{C}_b$. It is again easy to define a $\mathscr{K}_0$-structure $\mathfrak{N}_0$ s.t. $(\mathfrak{N}_0{\restriction}\mathscr{L})^U = \mathfrak{M}$, $E^{\mathfrak{N}_0}$ is well-founded and extensional and if $p : (N_0, E^{\mathfrak{N}_0}) \simeq (B, \in_B)$ is the collapsing isomorphism then $p(c_1^{\mathfrak{N}_0}) = b$. We have to show that $\mathfrak{N}_1 \models \psi$ holds for all $\mathfrak{N}_1$ s.t. $(\mathfrak{N}_1{\restriction}\mathscr{K}^0)^{U_0} = \mathfrak{N}_0$. It is enough to show that these $\mathfrak{N}_1$ satisfy $\mathfrak{N}_1 \models \varphi \to \mathscr{Q}(\mathscr{M}, \check{\mathscr{L}}, c_0, c_1, \check{a})_E$. So let $(\mathfrak{N}_1{\restriction}\mathscr{K}_0)^{U_0} = \mathfrak{N}_0$ and $\mathfrak{N}_1 \models \varphi$. Now $E^{\mathfrak{N}_1}$ is well-founded and extensional. So let $q: (N_1, E^{\mathfrak{N}_1}) \simeq (C, \in_C)$ be the collapsing isomorphism. Now $Q: (c_1^{\mathfrak{N}_1}) = b$ because $(N_0, E^{\mathfrak{N}_0}) \subseteq_{\text{end}} (N_1, E^{\mathfrak{N}_1})$. Moreover $q(\check{\mathscr{L}}^{\mathfrak{N}_1}) = \mathscr{L}$, $q(\check{a}^{\mathfrak{N}_1}) = a$

and $q(\mathcal{M}^{\mathfrak{N}_1}) = \mathfrak{M}_0$ is an $\mathscr{L}$-structure isomorphic with $\mathfrak{M}$. Now $\forall v_0 \mathscr{Q}(\mathfrak{M}_0, \mathscr{L}, v_0, b, a)$ because $\mathfrak{M}_0 \simeq \mathfrak{M}$ and $\mathbf{C}_b$ is closed under isomorphism. So

$$(C, \in_C) \models \forall v_0 \mathscr{Q}[\mathfrak{M}_0, \mathscr{L}, v_0, b, a]$$

because C is transitive and $\mathscr{Q}$ is Δ_0. Hence
$\mathfrak{N}_1 \models \mathscr{Q}(\mathcal{M}, \check{\mathscr{L}}, c_0, c_1, \check{a})_E$. □

It turns out to be useful to restrict our considerations to classes of set theoretic structures to avoid the above difficulties with isomorphic structures. Then we shall be able to strengthen 2.6 in section 3. We can also generalize our considerations to arbitrary classes. Let $\mathbf{C}$ be any class. We define

$$\mathbf{C}^s = \{(\mathrm{TC}(\{a\}), \in_{\mathrm{TC}(\{a\})}) | a \in \mathbf{C}\}$$

and call $\mathbf{C}^s$ the *structure class* of $\mathbf{C}$. We suppose of course that there is some fixed binary predicate symbol P s.t. $(\mathrm{TC}(\{a\}), \in_{\mathrm{TC}(\{a\})})$ denotes the $\{\mathrm{P}\}$-structure whose domain is $\mathrm{TC}(\{a\})$ and whose interpretation of P is $\in_{\mathrm{TC}(\{a\})}$. The transition from $\mathbf{C}$ to $\mathbf{C}^S$ will be important in the study of the relations between the Lévy hierarchy and the hierarchy of model theoretic definability. The same transition will be needed also when we prove the hierarchy theorem for the RPC-hierarchy defined in terms of $\mathscr{L}_{\infty\omega}(\mathrm{WF})$.

LEMMA 2.7. Let $n \geq 1$. The class $\mathbf{C}$ is Σ_n, Π_n or Δ_n if and only if $\mathbf{C}^s$ is respectively Σ_n, Π_n, or Δ_n; the same parameters can be used in both definitions.

Proof: The implication from right to left is very easy. One must only consider the Σ- and Π-cases separately. The implication from left to right becomes easy when one recalls that the relation $b = \mathrm{TC}(\{a\})$ can be written in a Σ_1-form. □

Let $\mathscr{L}$ be an alphabet and $\mathbf{C}$ be a class of $\mathscr{L}$-structures. We use $\mathbf{C}^{\sim}$ to denote the *closure of* $\mathbf{C}$ *under isomorphism*, i.e.

$$\mathbf{C}^{\sim} = \{\mathfrak{M} | \exists \mathfrak{N}(\mathfrak{M} \simeq \mathfrak{N} \wedge \mathfrak{N} \varepsilon \mathbf{C})\}.$$

LEMMA 2.8. Let $n \geq 1$. Let $P \in \mathscr{L}$ be a binary predicate symbol and let $\mathbf{C}$ be a class of $\mathscr{L}$-structures s.t. for each $\mathfrak{M} \in \mathbf{C}$ its domain M is a transitive set and $P^{\mathfrak{M}}$ is $\in_M$. $\mathbf{C}$ is Σ_n, Π_n or Δ_n if and only if $\mathbf{C}^{\sim}$ is respectively Σ_n, Π_n or Δ_n; the same parameters can be used in both definitions.

Moreover let **D** be any class. **D** is Σ_n, Π_n or Δ_n if and only if $\mathbf{D}^{s\sim}$ is respectively Σ_n, Π_n, or Δ_n; the same parameters can be used in both definitions.

Proof: If **C** is Σ_n, $\mathbf{C}^\sim$ is trivially Σ_n. If **C** is Π_n then it holds by Mostowski's isomorphism theorem that for all $\{P\}$-structures $\mathfrak{M}$

$$\mathfrak{M} \in \mathbf{C}^\sim \text{ iff } \forall f, A((\text{`}f \text{ is a function whose domain is } M \text{ and whose range is } A\text{'} \wedge \text{`}A \text{ is transitive'} \wedge \forall x \varepsilon M(f(x) = \{f(y) | y P^{\mathfrak{M}} x\})) \rightarrow (f(\mathfrak{M}) \in \mathbf{C})).$$

where $f(\mathfrak{M})$ is the image of $\mathfrak{M}$ under f. This implies that $\mathbf{C}^\sim$ is Π_n. For the converse implications it is enough to notice that, by Lemma 2.7.

$$\mathfrak{M} \in \mathbf{C} \text{ iff } \mathfrak{M} \in \mathbf{C}^\sim \wedge \text{`}M \text{ is a transitive set and } P^{\mathfrak{M}} \text{ is } \in_M\text{'}.$$

The other claim follows from the first and Lemma 2.7. □

COROLLARY 2.9. Let **C** be any class. **C** is Σ_1, Π_1 or Δ_1 if and only if $\mathbf{C}^{s\sim}$ is respectively $\Sigma_1^{\mathrm{RPC}}(\mathscr{L}_{\infty\omega}(\mathrm{WF}))$, $\Pi_1^{\mathrm{RPC}}(\mathscr{L}_{\infty\omega}(\mathrm{WF}))$ or $\Delta_1^{\mathrm{RPC}}(\mathscr{L}_{\infty\omega}(\mathrm{WF}))$ where $\mathscr{L} = \{P\}$. Moreover let A be an admissible set s.t. $\mathscr{L} \in A$. Now **C** is Σ_1, Π_1 or Δ_1 if and only if $\mathbf{C}^{s\sim}$ is respectively $\Sigma_1^{\mathrm{RPC}}(\mathscr{L}_A(\mathrm{WF}))$, $\Pi_1^{\mathrm{RPC}}(\mathscr{L}_A(\mathrm{WF}))$ or $\Delta_1^{\mathrm{RPC}}(\mathscr{L}_A(\mathrm{WF}))$.

Proof: The claim follows immediately from theorem 2.2 together with its corollaries and Lemma 2.8. □

3. A CHARACTERIZATION FOR THE LEVY HIERARCHY

In this section we shall present our main result. We shall actually prove the following theorem:

THEOREM 3.1. Let $P \in \mathscr{L}$ be a binary predicate symbol. Let **C** be a class of $\mathscr{L}$-structures s.t. for all $\mathfrak{M} \in \mathbf{C}$ M is a transitive set and $P^{\mathfrak{M}} = \in_M$. Let $n \geq 1$. Now **C** is Σ_n, Π_n or Δ_n if and only if $\mathbf{C}^\sim$ is respectively $\Sigma_n^{\mathrm{RPC}}(\mathscr{L}_{\infty\omega}(\mathrm{WF}))$, $\Pi_n^{\mathrm{RPC}}(\mathscr{L}_{\infty\omega}(\mathrm{WF}))$ or $\Delta_n^{\mathrm{RPC}}(\mathscr{L}_{\infty\omega}(\mathrm{WF}))$. Moreover let A be an admissible set s.t. $\mathscr{L} \in A$. Now **C** is Σ_n, Π_n or Δ_n in parameters from A if and only if $\mathbf{C}^\sim$ is respectively $\Sigma_n^{\mathrm{RPC}}(\mathscr{L}_A(\mathrm{WF}))$, $\Pi_n^{\mathrm{RPC}}(\mathscr{L}_A(\mathrm{WF}))$ or $\Delta_n^{\mathrm{RPC}}(\mathscr{L}_A(\mathrm{WF}))$.

Classes of admissible structures and ordinals together with their natural orderings are important examples of classes mentioned in the above theorem. Theorem 3.1 yields immediately the following generalization of corollary 2.7:

THEOREM 3.2. Let $n \geq 1$. Let $\mathbf{C}$ be any class. Now $\mathbf{C}$ is Σ_n, Π_n or Δ_n if and only if $\mathbf{C}^{s\sim}$ is respectively $\Sigma_n^{\mathrm{RPC}}(\mathscr{L}_{\infty\omega}(\mathrm{WF}))$, $\Pi_n^{\mathrm{RPC}}(\mathscr{L}_{\infty\omega}(\mathrm{WF}))$ or $\Delta_n(\mathscr{L}_{\infty\omega}(\mathrm{WF}))$. Moreover let A be an admissible set s.t. $\mathscr{L} \in A$. Now $\mathbf{C}$ is Σ_n, Π_n or Δ_n in parameters from A if and only if $\mathbf{C}^{s\sim}$ is respectively $\Sigma_n(\mathscr{L}_A(\mathrm{WF}))$, $\Pi_n(\mathscr{L}_A(\mathrm{WF}))$ or $\Delta_n(\mathscr{L}_A(\mathrm{WF}))$. Here $\mathscr{L} = \{P\}$.

Theorem 3.2 follows obviously from theorem 3.1 and Lemma 2.7.

Proof of theorem 3.1: We consider for simplicity only the first claim. The strengthening is obtained as before. It is also enough to prove the implication from left to right because the other implication follows from Lemma 1.4.

Let $\mathscr{Q}(x, y, v_0, \ldots, v_{k-1})$ be a Δ_0-formula. We may suppose that for all $\mathfrak{M}$ and all $a_0, \ldots, a_{k-1}$ $\mathscr{Q}(\mathfrak{M}, \mathscr{L}, a_0, \ldots, a_{k-1})$ implies that $\mathfrak{M}$ is an $\mathscr{L}$-structure of the form $(M, \in_M, \ldots)$ where M is a transitive set. We shall consider the Σ_n- and Π_n-classes of structures defined by

$$\exists v_{n-1} \forall v_{n-2} \ldots \overset{\exists}{\underset{\forall}{}} v_0\, \mathscr{Q}(\mathfrak{M}, \mathscr{L}, v_0 \ldots, v_{n-1}, a_0, \ldots, a_{m-1})$$

and

$$\forall v_{n-1} \exists v_{n-2} \ldots \overset{\forall}{\underset{\mathrm{E}}{}} v_0\, \mathscr{Q}(\mathfrak{M}, \mathscr{L}, v_0, \ldots, v_{n-1}, a_0, \ldots, a_{m-1})$$

where n $+ m = k$. To be able to characterize these classes model theoretically we have to find first the sentences needed in the definitions of the corresponding $\Sigma_n^{\mathrm{RPC}}(\mathscr{L}_{\infty\omega}(\mathrm{WF}))$- and $\Pi_n^{\mathrm{RPC}}(\mathscr{L}_{\infty\omega}(\mathrm{WF}))$-classes. Then we shall be able to prove the claim by induction on n.

Let $U, U_1, U_2, \ldots$ and $c_0, c_1, \ldots$ be sequences of new unary predicate and constant symbols. Let

$$\mathscr{K} = \mathscr{L} \cup \{E, U, U_1, \ldots, \mathscr{M}, c_0, \ldots\}.$$

Here E is a new binary predicate symbol, and $\mathscr{M}$ is a new constant symbol. Let $\check{a}$ be a name for each set a so that $\check{a} \notin \mathscr{K}$. So we may set for instance $\check{a} = (0, (\mathscr{K}, a))$. Let φ be the conjunction of the following statements:

(1) (Extensionality, pair and union)$_E$
(2) 'E is well-founded'
(3) $\forall xy(U(y) \rightarrow (P(x, y) \wedge U(x) \leftrightarrow xEy))$
(4) 'All the values of the function symbols of $\mathscr{L}$ are in U'
(5) '$\mathscr{M}$ is a pair $(x, y)_E$ which is an $\check{\mathscr{L}}$-structure$_E$ and the identity map is an 'isomorphism' between the $\mathscr{L}$, U-restriction of the structure considered and $\mathscr{M}$'.

Let $a_0, \ldots, a_{n-1}$ be sets. Let $\psi_{a_0 \ldots a_{n-1}}$ be the sentence

$$\bigwedge_{b \in \mathrm{TC}(\{a_0, \ldots, a_{n-1}\})} \forall x(xE\hat{b} \leftrightarrow \bigvee_{d \in b} x = \hat{d}).$$

We shall now define the formulas $\sigma_n(v_0, \ldots, v_{k-1})$ and $\Pi_n(v_0, \ldots, v_{k-1})$ for $1 \leq n \leq k$ in the following way:

$\sigma_1(v_0, \ldots, v_{k-1})$ is
$\varphi \wedge \mathscr{Q}(\mathscr{M}, \check{\mathscr{L}}, v_0, \ldots, v_{k-1})_E$,
$\Pi_1(v_0, \ldots, v_{k-1})$ is
$\varphi \rightarrow \mathscr{Q}(\mathscr{M}, \check{\mathscr{L}}, v_0, \ldots, v_{k-1})_E$,
$\sigma_{n+1}(v_0, \ldots, v_{k-1})$ is
('$U \subseteq_{\text{end}} U_n$ $\wedge$ '$U_n \subseteq_{\text{end}} U_{n-1}$' $\wedge$ '$E \restriction U_n$ is well-founded and extensional' $\wedge$ 'all the values of the function symbols of $\mathscr{L}$ are in U' $\wedge$ $\forall xy(U(y) \rightarrow P(x, y) \wedge U(x) \leftrightarrow xEy)) \wedge \forall x(U_n(x) \leftrightarrow U(x) \vee$ 'x is in the transitive closure$_E$ of $\{v_n, \ldots, v_{k-1}\}_E$')) $\wedge$ $\Pi_n(v_0, \ldots, v_{k-1})$,
$\Pi_{n+1}(v_0, \ldots, v_{k-1})$ is
('$U \subseteq_{\text{end}} U_n$' $\wedge$ '$U_n \subseteq_{\text{end}} U_{n-1}$' $\wedge$ '$E \restriction U_n$ is well-founded and extensional' $\wedge$ 'all the values of the function symbols of $\mathscr{L}$ are in U' $\wedge$ $\forall xy(U(y) \rightarrow P(x, y) \wedge U(x) \leftrightarrow xEy)) \wedge \forall x(U_n(x) \leftrightarrow U(x) \vee$ 'x is in the transitive closure$_E$ of $\{v_n, \ldots, v_{k-1}\}_E$')) $\rightarrow$
$\sigma_n(v_0, \ldots, v_{k-1})$.

It is completely straightforward to write these statements in $\mathscr{L}_{\infty\omega}(\mathrm{WF})$. If $n = 1$ then the part '$U_n \subseteq_{\text{end}} U_{n-1}$' of σ_{n+1} and Π_{n+1} should be replaced by the statement 'the domain of the whole structure is an end extension of U_1'.

We shall now prove that

$$\psi_{\mathscr{L} a_0, \ldots, a_{m-1}} \wedge \sigma_n(c_0, \ldots, c_{n-1}, \check{a}_0, \ldots, \check{a}_{m-1})$$

together with $U, U_{n-1}, \ldots, U_1$ and the alphabets

$\mathscr{L} \cup \{E, U, c_{n-1}\} \cup \{\check{b} | b \in \mathrm{TC}(\{\mathscr{L}, a_0, \ldots, a_{m-1}\})\}$,
$\mathscr{L} \cup \{E, U, U_{n-1}, c_{n-1}, c_{n-2}\} \cup \{\check{b} | b \in \mathrm{TC}(\{\mathscr{L}, a_0, \ldots, a_{m-1}\})\}$,

⋮

$\mathscr{L} \cup \{E, U, U_{n-1}, \ldots, U_2, c_{n-1}, \ldots, c_1\} \cup$
$\{b | b \in \mathrm{TC}(\{\mathscr{L}, a_0, \ldots, a_{m-1}\})\}$,
$\mathscr{L} \cup \{E, U, U_{n-1}, \ldots, U_1, \mathscr{M}, c_{n-1}, \ldots, c_0\} \cup$
$\{\check{b} | b \in \mathrm{TC}(\{\mathscr{L}, a_0, \ldots, a_{m-1}\})\}$

gives a $\Sigma_n^{\mathbf{RPC}}(\mathscr{L}_{\infty\omega}(\mathrm{WF}))$-definition for the class $\mathbf{C}^{\sim}$ where $\mathbf{C}$ is the Σ_n-class defined by

$$\exists v_{n-1}\forall v_{n-2}\ldots \overset{\exists}{\forall} v_0\, \mathscr{Q}(\mathfrak{M}, \mathscr{L}, v_0, \ldots, v_{n-1}, a_0, \ldots, a_{m-1}).$$

And we shall prove that

$$(\psi_{\mathscr{L}a_0\cdots a_{m-1}} \to \Pi_n(c_0, \ldots, c_{n-1}, \check{a}_0, \ldots, \check{a}_{m-1})) \wedge$$
'$P\restriction U$ is well-founded and extensional'

together with $U, U_{n-1}, \ldots, U_1$ and the alphabets mentioned above gives a $\Pi_n^{\mathbf{RPC}}(\mathscr{L}_{\infty\omega}(\mathrm{WF}))$-definition for the class $\mathbf{C}^{\sim}$ where $\mathbf{C}$ is the Π_n-class defined by

$$\forall v_{n-1}\exists v_{n-2}\ldots \overset{\forall}{\exists} v_0\, \mathscr{Q}(\mathfrak{M}, \mathscr{L}, v_0, \ldots, v_{n-1}, a_0, \ldots, a_{m-1}).$$

Here $n + m = k$, of course. From now on we shall list only the sentences and the unary predicate symbols used in the definitions of $\Sigma_n^{\mathbf{RPC}}(\mathscr{L}_{\infty\omega}(\mathrm{WF}))$- and $\Pi_n^{\mathbf{RPC}}(\mathscr{L}_{\infty\omega}(\mathrm{WF}))$-classes. The alphabets will be obvious from the context.

We proceed by induction. The Σ_1-case is essentially the proof of theorem 2.2. We prove now the Π_1-case: Suppose that $\mathfrak{M} \in \mathbf{C}^{\sim}$. We have to show that for all $\mathfrak{N}$ s.t. $(\mathfrak{N}\restriction\mathscr{L})^U = \mathfrak{M}$, $\mathfrak{N} \models \psi_{\mathscr{L}a_0\ldots a_{m-1}} \to \Pi_1(c_0, \check{a}_0, \ldots, \check{a}_{m-1})$. Clearly we may suppose that $\mathfrak{M} \in \mathbf{C}$. Now let $\mathfrak{N}$ be s.t. $(\mathfrak{N}\restriction\mathscr{L})^U = \mathfrak{M}$ and $\mathfrak{N} \models \psi_{\mathscr{L}a_0\ldots a_{m-1}} \wedge \varphi$. Let $p : (N, E^{\mathfrak{N}}) \simeq (B, \in^*)$ be the collapsing isomorphism. Now $\mathfrak{M}_0 = p(\mathscr{M}^{\mathfrak{N}})$ is again an $\mathscr{L}$-structure isomorphic with $\mathfrak{M}$. But parts (3) and (5) imply moreover that M_0 is a transitive set and that $P^{\mathfrak{M}_0} = \in_{M_0}$. So $(M, P^{\mathfrak{M}}) = (M_0, P^{\mathfrak{M}_0})$. Now it is easy to see that $\mathfrak{M} = \mathfrak{M}_0$ because $p(b) = b$ for $b \in M$. Also $p(\check{\mathscr{L}}^{\mathfrak{N}}) = \mathscr{L}$ and $p(\check{a}_i^{\mathfrak{N}}) = a_i$ for $i < m$. On the other hand $\mathscr{Q}(\mathfrak{M}, \mathscr{L}, a, a_0, \ldots, a_{m-1})$ holds for all a. So

$$(B, \in_B) \models \mathscr{Q}[p(\mathscr{M}^{\mathfrak{N}}), p(\check{\mathscr{L}}^{\mathfrak{N}}), p(c_0^{\mathfrak{N}}), p(\check{a}_0^{\mathfrak{N}}), \ldots, p(\check{a}_{m-1}^{\mathfrak{N}})$$

because B is transitive and $\mathscr{Q}$ is Δ_0. This implies that $\mathfrak{M}$ belongs to the $\Pi_1^{\mathbf{RPC}}(\mathscr{L}_{\infty\omega}(\mathrm{WF}))$-class $\mathbf{D}$ defined by U and

$$(\psi_{\mathscr{L}a_0\ldots a_{m-1}} \to \Pi_1(c_o, a_0, \ldots, a_{m-1})) \wedge$$
'$P\restriction U$ is well-founded and extensional'.

Now suppose that $\mathfrak{M}$ belongs to $\mathbf{D}$. So there is some $\mathfrak{M}_0 \simeq \mathfrak{M}$ s.t. M_0 is transitive and $P^{\mathfrak{M}_0} = \in_{M_0}$. Let a be arbitrary. We shall show that $\mathscr{Q}(\mathfrak{M}_0, \mathscr{L}, a, a_0, \ldots, a_{m-1})$, which implies the claim. It is easy to find an $\mathfrak{N}$ s.t.

$(\mathfrak{N}\restriction\mathscr{L})^U = \mathfrak{M}_0$, and $\mathfrak{N}_{\mathscr{L}}\ \psi_{\mathscr{L}a_0\cdots a_{m-1}} \wedge \varphi$, and $E^{\mathfrak{N}}$ is well-founded and extensional, and if $p : (N, E^{\mathfrak{N}}) \simeq (B, \in_B)$ is the collapsing isomorphism then $p(c_0^{\mathfrak{N}}) = a$. Now $\mathfrak{N} \models \mathscr{Q}(\mathscr{M}, \check{\mathscr{L}}, c_0, \check{a}_0, \ldots, \check{a}_{m-1})_E$ because $\mathfrak{M} \simeq \mathfrak{M}_0$ and $\mathfrak{M}$ belongs to **D**. $\mathfrak{M}_1 = p(\mathscr{M}^{\mathfrak{N}})$ is an $\mathscr{L}$-structure and $\mathfrak{M}_0 = \mathfrak{M}_1$ as before. Also $p(c_0^{\mathfrak{N}}) = a$, $p(\mathscr{L}^{\mathfrak{N}}) = \mathscr{L}$ and $p(a_i^{\mathfrak{N}}) = a_i$ for $i < m$. So

$$(B, \in_B) \models \mathscr{Q}[\mathfrak{M}_0, \mathscr{L}, a, a_0, \ldots, a_{m-1}]$$

and hence $\mathscr{Q}(\mathfrak{M}_0, \mathscr{L}, a, a_0, \ldots, a_{m-1})$ because B is transitive and $\mathscr{Q}$ is Δ_0. This completes the proof of the Π_1-case.

Next we shall consider the transition from the Π_n-case to the Σ_{n+1}-case. We suppose that $n > 1$. The case where $n = 1$ is slightly simpler. The transition from the Σ_n-case to the Π_{n+1}-case would be rather similar. So let

$$\mathbf{C} = \{\mathfrak{M} | \exists v_n \forall v_{n-1} \ldots {}^{\forall}_{\exists} v_0\, \mathscr{Q}(\mathfrak{M}, \mathscr{L}, v_0, \ldots, v_n, a_0, \ldots, a_{m-1})\}$$

and

$$\mathbf{C}_a = \{\mathfrak{M} | \forall v_{n-1} \ldots {}^{\forall}_{\exists} v_0\, \mathscr{Q}(\mathfrak{M}, \mathscr{L}, v_0, \ldots, v_{n-1}, a, a_0, \ldots, a_{m-1})\}.$$

Here $m + n + 1 = k$, of course. Now the inductive hypothesis easily implies that for all $\mathscr{L}$-structures $\mathfrak{M}$

$\mathfrak{M} \in \mathbf{C}^{\sim}$ iff there is some a s.t. $\mathfrak{M} \in \mathbf{C}_a^{\sim}$
iff there is some a s.t. $\mathfrak{M}$ belongs to
(+) the $\Pi_n^{\mathrm{RPC}}(\mathscr{L}_{\infty\omega}(\mathrm{WF}))$-class defined by U, $U_{n-1}, \ldots, U_1$ and
$(\psi\, \mathscr{L}_{aa_0} \ldots {}_{a_{m-1}} \rightarrow \Pi_n(c_0, \ldots, c_{n-1}, \check{a}, \check{a}_0, \ldots, \check{a}_{m-1})) \wedge$
'$P\restriction U$ is well-founded and extensional'.

Suppose that (+) holds for a. We may suppose that M is a transitive set and that $p^{\mathfrak{M}} = \in_M$. Let $\mathfrak{N}$ be s.t. $(\mathfrak{N}\restriction\mathscr{L})^U = \mathfrak{M}$ and

$$\begin{aligned} N &= M \cup TC(\{\mathscr{L}, a, a_0, \ldots, a_{m-1}\}), \\ E^{\mathfrak{N}} &= \in^N \\ c_n^{\mathfrak{N}} &= a \\ \check{b}^{\mathfrak{N}} &= b \text{ for } b \in \mathrm{TC}(\{\mathscr{L}, a_0, \ldots, a_{m-1}\}). \end{aligned}$$

Let $\mathfrak{N}'$ differ from $\mathfrak{N}$ only in that in $\mathfrak{N}'$ $\check{a}$ is the name of a instead of c_n and $\check{b}$ is interpreted to be b for $b \in \mathrm{TC}(a)$. Now all extended expansions of $\mathfrak{N}'$ are that of $\mathfrak{M}$. So $\mathfrak{N}'$ belongs to the $\Pi_n^{\mathrm{RPC}}(\mathscr{L}_{\infty\omega}(\mathrm{WF}))$-class defined by U_n, $U_{n-1}, \ldots, U_1$ and

$$\psi_{\mathscr{L}aa_0\ldots a_{m-1}} \to \Pi_n'(c_0, \ldots, c_{n-1}, \check{a}, \check{a}_0, \ldots, \check{a}_{m-1})$$

where

$$\mathscr{L}' = \mathscr{L} \cup \{E, U\} \cup \{\check{b} | b \in \mathrm{TC}(\{\mathscr{L}, a, a_0, \ldots, a_{m-1}\})\}$$

and Π_n' is obtained from Π_n by replacing U by U_n except in φ. But $\mathfrak{N}' \models \psi_{\mathscr{L}aa_0\ldots a_{m-1}}$. So $\mathfrak{N}'$ belongs to the $\Pi_n^{\mathrm{RPC}}(\mathscr{L}'_{\infty\omega}(\mathrm{WF}))$-class defined by $U_n, \ldots, U_1$ and

$$\psi_{\mathscr{L}aa_0\cdots a_{m-1}} \wedge \Pi_n'(c_0, \ldots, c_{n-1}, \check{a}, \check{a}_0, \ldots, \check{a}_{m-1}).$$

But now $\mathfrak{N}'$ belongs to the $\Pi_n^{\mathrm{RPC}}(\mathscr{L}'_{\infty\omega}(\mathrm{WF}))$-class defined by $U_n, \ldots, U_1$ and

$$\psi_{\mathscr{L}aa_0\cdots a_{m-1}} \wedge \sigma_{n+1}(c_0, \ldots, c_{n-1}, \check{a}, \check{a}_0, \ldots, \check{a}_{m-1})$$

where again σ_{n+1} is obtained from σ_{n+1} by replacing Π_n by Π_n. So it is easy to see that $\mathfrak{M}$ belongs to the $\Sigma_{n+1}(\mathscr{L}_{\infty\omega})$-class defined by $U, U_n, \ldots, U_1$ and

$$(++) \quad \psi_{\mathscr{L}a_0\cdots a_{m-1}} \wedge \sigma_{n-1}(c_0, \ldots, c_n, \check{a}_0, \ldots, \check{a}_{m-1}).$$

Finally we have to show that if $\mathfrak{M}$ belongs to the $\Sigma_{n+1}^{\mathrm{RPC}}$-class defined by $U, U_n, \ldots, U_1$ and $(++)$ then $(+)$ holds. There is an $\mathfrak{N}$ s.t. $(\mathfrak{N} \restriction \mathscr{L})^U = \mathfrak{M}$ and $\mathfrak{N}$ belongs to the $\Pi_n^{\mathrm{RPC}}(\mathscr{L}''_{\infty\omega}(\mathrm{WF}))$-class defined by $U_n, \ldots, U_1$ and $(++)$ where

$$\mathscr{L}'' = \mathscr{L} \cup \{E, U, c_n\} \cup \{\check{b} | b \in \mathrm{TC}(\{\mathscr{L}, \check{a}_0, \ldots, \check{a}_{m-1}\})\}.$$

Now $E^{\mathfrak{N}}$ is well-founded and extensional. So let $p : (N, E^{\mathfrak{N}}) \simeq (B, \in_B)$ be the collapsing isomorphism. Let $p(c_n^{\mathfrak{N}}) = a$. We shall show that $(+)$ holds for this a. By the definition of Π_n it is enough to show that if $\mathfrak{N}_0$ is s.t. $(\mathfrak{N}_0 \restriction \mathscr{L})^U = \mathfrak{M}$, $\mathfrak{N}_0 \models \psi_{\mathscr{L}aa_0\cdots a_{m-1}}$, $(M, P^{\mathfrak{M}}) \subseteq_{\mathrm{end}} (N_0, E^{\mathfrak{N}_0})$, $E^{\mathfrak{N}_0}$ coincides with $P^{\mathfrak{M}}$ on M and is well-founded and extensional, and if each $b \in N_0$ is either in M or in the transitive closure of

$$\{\mathscr{L}^{\mathfrak{N}_0}, c_{n-1}^{\mathfrak{N}_0}, a^{\mathfrak{N}_0}, a^{\mathfrak{N}_0}, \ldots, a_{m-1}^{\mathfrak{N}_0}\}$$

relative to $E^{\mathfrak{N}_0}$; then $\mathfrak{N}_0$ belongs to the $\Sigma_{n-1}^{\mathrm{RPC}}(\mathscr{L}'''_{\infty\omega}(\mathrm{WF}))$-cass defined by $U_{n-1}, \ldots, U_1$ and $\Pi_n(c_0, \ldots, c_{n-1}, \check{a}, \check{a}_0, \ldots, \check{a}_{m-1})$ where

$$\mathscr{L}''' = \mathscr{L} \cup \{E, U, c_{n-1}\} \cup \{\check{b} | b \in \mathrm{TC}(\{\mathscr{L}, a, a_0, \ldots, a_{m-1}\})\}.$$

Let $b \in M$. If $b = \check{d}^{\mathfrak{N}}$ for some $d \in \mathrm{TC}(\{\mathscr{L}, a, a_0, \ldots, a_{m-1}\})$ then

$$(\mathrm{M}, \mathrm{P}^{\mathfrak{M}}) \models \forall x(P(x, y) \leftrightarrow \bigvee_{c \in d} x = \check{c})[b]$$

and so $b = \check{d}^{\mathfrak{N}_0}$. Similarly if $b = \check{d}^{\mathfrak{N}_0}$ then $b = \check{d}^{\mathfrak{N}}$. So it is easy to transform $\mathfrak{N}_0$ to an extended expansion of $\mathfrak{N}$. Hence $\mathfrak{N}_0$ belongs to the required $\Sigma_{n-1}^{\mathrm{RPC}}(\mathscr{L}''_{\infty\omega}(\mathrm{WF}))$-class because σ_{n+1} contains Π_n as a conjunctive member. This completes the proof of the theorem. □

Now we can easily obtain our main result:

THEOREM 3.3. Let $\mathscr{L}$ be an alphabet and $\mathbf{C}$ a class of $\mathscr{L}$-structures closed under isomorphism. $\mathbf{C}$ is Σ_n, Π_n or Δ_σ if and only if it is respectively $\Sigma_n^{\mathrm{RPC}}(\mathscr{L}_{\infty\omega}(\mathrm{WF}))$, $\Pi_n^{\mathrm{RPC}}(\mathscr{L}_{\infty\omega}(\mathrm{WF}))$ or $\Delta_n^{\mathrm{RPC}}(\mathscr{L}_{\infty\omega}(\mathrm{WF}))$. Moreover if A is an admissible set s.t. $\mathscr{L} \in A$, then $\mathbf{C}$ is Σ_n, Π_n or Δ_n in parameters from A and if only if it is respectively $\Sigma_n^{\mathrm{RPC}}(\mathscr{L}_A(\mathrm{WF}))$, $\Pi_n^{\mathrm{RPC}}(\mathscr{L}_A(\mathrm{WF}))$ or $\Delta_n^{\mathrm{RPC}}(\mathscr{L}_A(\mathrm{WF}))$.

Proof: First notice that the claim can be reduced to the case where the alphabet $\mathscr{L}$ contains some binary predicate symbol P s.t. in all models $\mathfrak{M}$ of $\mathbf{C}$, $P^{\mathfrak{M}}$ is well-founded and extensional. But in this case $\mathbf{C} = \mathbf{C}_0^{\sim}$ where

$$\mathbf{C}_0 = \{\mathfrak{M} \in \mathbf{C} \mid M \text{ is transitive and } P^{\mathfrak{M}} \text{ is } \in_M\}.$$

So the claim follows immediately from lemma 2.8 and theorem 3.1. □

Theorem 3.3 has some algebraic implications of which we give just one example:

COROLLARY 3.4. Every Σ_2 class $\mathbf{C}$ of $\mathscr{L}$-structures which is closed under isomorphism is strongly closed under isomorphism.

This result is interesting because every Σ_2 definition of a class of $\mathscr{L}$-structures closed under isomorphism does not satisfy the requirements made in the definition of a Σ_2 class strongly closed under isomorphism. Theorem 3.2 yields as a corollary of the hierarchy theorem of Lévy [12] a hierarchy theorem for the present hierarchy defined in terms of $\mathscr{L}_{\infty\omega}(\mathrm{WF})$. It may, however, be of some interest to prove this fact without referring to the hierarchy theorem of the Lévy hierarchy. So we shall prove the following theorem:

THEOREM 3.3. Let $n \geq 1$. Let $\mathscr{L} = \{P, c\}$ where P is a binary predicate symbol and c is a constant symbol. There is a $\Sigma_n^{\mathrm{RPC}}(\mathscr{L}_{\infty\omega}(\mathrm{WF}))$-class which is not $\Pi_n^{\mathrm{RPC}}(\mathscr{L}_{\infty\omega}(\mathrm{WF}))$.

Proof: We can code the definitions of $\Sigma_n^{\mathrm{RPC}}(\mathscr{L}_{\infty\omega}(\mathrm{WF}))$- and $\Pi_n^{\mathrm{RPC}}(\mathscr{L}_{\infty\omega}(\mathrm{WF}))$-classes as $\mathscr{L}$-structures. As a matter of fact we can easily write down a sentence φ of $\mathscr{L}_{\infty\omega}(\mathrm{WF})$ s.t. for all $\mathscr{L}$-structures $\mathfrak{M}$ it holds that $\mathfrak{M} \models \varphi$ if and only if $P^{\mathfrak{M}}$ is well-founded and extensional and if $p : (M, P^{\mathfrak{M}}) \simeq (A, \in_A)$ is the collapsing isomorphism then $p(c^{\mathfrak{M}})$ is of the form

$$\langle U_0, \ldots, U_{n-1}, \mathscr{L}_0, \ldots, \mathscr{L}_{n-1}, \psi\rangle$$

where $U_0, \ldots, U_{n-1}$ are unary predicate symbols, $\mathscr{L}_0, \ldots, \mathscr{L}_{n-1}$ are alphabets s.t. $\mathscr{L} \subseteq \mathscr{L}_0 \subseteq \ldots \subseteq \mathscr{L}_{n-1}$ and $U_i \in \mathscr{L}_i$ for $i < n$ and finally $\psi \in (\mathscr{L}_{n-1})_{\infty\omega}(\mathrm{WF})$. Now the models of φ obviously code the definitions of certain $\Sigma_n^{\mathrm{RPC}}(\mathscr{L}_{\infty\omega}(\mathrm{WF}))$-classes. We shall define

$\mathbf{C} = \{\mathfrak{M} \mid \mathfrak{M} \models \varphi$ and $\mathfrak{M}$ belongs to the $\Sigma_n^{\mathrm{RPC}}(\mathscr{L}_{\infty\omega}(\mathrm{WF}))$-class defined by $c^{\mathfrak{M}}\}$.

Now it is straightforward to see that $\mathbf{C}$ is Σ_n. So theorem 3.1 implies that $\mathbf{C}$ is $\Sigma_n^{\mathrm{RPC}}(\mathscr{L}_{\infty\omega}(\mathrm{WF}))$. It is also easy to see that $\mathbf{C}$ is the class we are looking for: If $\mathbf{C}$ is $\Pi_n^{\mathrm{RPC}}(\mathscr{L}_{\infty\omega}(\mathrm{WF})$ then $-\mathbf{C}$ is $\Sigma_n^{\mathrm{RPC}}(\mathscr{L}_{\infty\omega}(\mathrm{WF}))$. But now there must be some model $\mathfrak{N}$ of φ which codes the $\Sigma_n^{\mathrm{RPC}}(\mathscr{L}_{\infty\omega}(\mathrm{WF}))$-definition of $-\mathbf{C}$. So

$\mathfrak{N} \in -\mathbf{C}$ iff $\mathfrak{N}$ belongs to the $\Sigma_n^{\mathrm{RPC}}(\mathscr{L}_{\infty\omega}(\mathrm{WF}))$-class defined by $\mathfrak{N}$

iff $\mathfrak{N} \in \mathbf{C}$

because $\mathfrak{N} \models \varphi$. This contradiction proves the claim. □

The above hypothesis about $\mathscr{L}$ was made just for simplicity. We recall that the results of this chapter hold about $\mathscr{L}_{\infty G}$, too. They can also be applied to $\mathscr{L}_{\infty\omega}$ by means of lemma 2.1. By choosing $A = HF$ it is possible to apply the results to $\mathscr{L}^{\mathrm{II}}$. But these applications require the full strength of the results of this chapter, because in the proof of theorem 3.1 quantifiers were replaced by parameters which may have large transitive closures.

As a final remark we notice that the theory of this paper can obviously be generalized to speak about set theoretic definability relative to some class **C**; that is about $\Sigma_n(\mathbf{C})$, $\Pi_n(\mathbf{C})$ and $\Delta_n(\mathbf{C})$, where $x \in \mathbf{C}$ is taken to be $\Delta_0(\mathbf{C})$. This covers, of course, relative definability in some set theoretical relation or operation. Again some results about $\mathscr{L}^{\mathrm{II}}$ will be obtained as special cases. These generalizations will be presented in detail in the author's doctoral dissertation.

Department of Mathematics,
University of Helsinki

BIBLIOGRAPHY

[1] Barwise, J., 'Absolute logics and $L_{\infty\omega}$', *Ann. of Math. Logic* **4** (1973), pp. 309–340.
[2] Barwise, J., 'Axioms for abstract model theory', *Ann. of Math. Logic* **7** (1974), pp. 221–266.
[3] Barwise, J., *Admissible Sets and Structures*, Springer-Verlag, Berlin, 1975.
[4] Chang, C. C., 'Some remarks on the model theory of infinitary languages', in Barwise (ed.): *The Syntax and Semantics of Infinitary Lcnguages*, Springer Lecture Notes in Math. 72 (1968), pp. 36–63.
[5] Chang, C. C. and Keisler, H. J., *Model Theory*, North-Holland Publ. Co., Amsterdam, 1973.
[6] Cutland, N., 'Σ_1-compactness in infinitary languages with infinite quantification', to appear in the proceedings of the ASL European Summer Meeting in Oxford 1976.
[7] Drake, F., *Set Theory*, North Holland Publ. Co., Amsterdam, 1974.
[8] Feferman, S., 'Two notes on abstract model theory, I', *Fund. Math.* **82** (1974), pp. 153–155.
[9] Keisler, H. J., 'Formulas with linearily ordered quantifiers', in the same volume as Chang [4], pp. 96–130.
[10] Keisler, H. J., *Model Theory for Infinitary Logic*, North-Holland Publ. Co., Amsterdam, 1971.
[11] Krynicki, M., 'On the expressive power of the language using the Henkin quantifier' this volume, pp. 259–265.
[12] Lévy, A., 'A hierarchy of formulas in set theory', *Memoirs of AMS* **57** (1965).
[13] Makowsky, J., Shelah, S. and Stavi, J., 'Δ-logics and generalized quantifiers', *Ann. of Math. Logic* **10** (1976), pp. 155–192.
[14] Swett, A., 'Absolute logics: Δ-closures and the number of countable models', *Notices of AMS* **21** 6 (1974), A553.
[15] Takeuti, G., 'A determinate logic', in the same volume as Chang [4], pp. 237–264.
[16] Vaught, R., 'Descriptive set theory in $\mathscr{L}_{\omega_1\omega}$', in *Cambridge Summer School in Math. Logic*, Springer Lecture Notes in Math. 337 (1973), pp. 574–598.

JUSSI KETONEN

OPEN PROBLEMS IN THE THEORY OF ULTRAFILTERS

INTRODUCTION

The purpose of this paper is to present a list of open questions in the theory of ultrafilters. Most of them seem almost impenetrable by the usual methods of set-theory. Needless to say, the list of such questions is infinite, and the topics chosen for this paper reflect the personal tastes and prejudices of the author. Our notation and terminology follows that of the most recent set-theoretic literature; for example $|x|$ denotes the cardinality of the set x, small Greek letters $\alpha, \beta, \gamma, \ldots$ denote ordinals, cardinals are initial ordinals, the set ${}^{y}x$ or x^{y} denotes the set of all functions $y \to x$ etc. For more, we refer the reader to Mathias [20]. By an ultrafilter over a set x we mean here a maximal filter in the field of subsets of x; that is:

1. *Definition:* D is an ultrafilter over a set x if D is a collection of subsets of x so that

(a) D is a filter:

$$z \in D, y \supseteq z \to y \in D$$
$$x_1 \ldots x_n \in D \to x_1 \cap \ldots \cap x_n \in D$$
$$z \in D \to z \neq 0$$

(b) D is maximal:

$$z \subseteq x \to z \in D \text{ or } x - z \in D$$

D is non-principal if in addition

(c) $\qquad y \in x \to \{y\} \notin D.$

It is sometimes useful to think of an ultrafilter D as a finitely additive measure ρ attaining only values 0 and 1: Define $\rho(z) = 1$ if $z \in D$ and 0 otherwise.

The notion of an ultrafilter gained some attention from the Polish mathematicians of the 30's, and it eventually became an object of study in general topology:

J. Hintikka, I. Niiniluoto, and E. Saarinen (eds.), Essays on Mathematical and Philosophical Logic, 227–247.

2. *Definition:* The Stone-Čech compactification βx of a set x is the set of all ultrafilters over x with the topology generated by the sets

$$\{D | y \in D\} \quad (y \subseteq x).$$

The first major work on the structural theory of ultrafilters was done by W. Rudin [28] in 1956. This line of research was continued by M. E. Rudin [26], [27] and later by D. Booth [3] and A. Blass [1].

Another major breakthrough was the work of Los on ultraproducts which showed that ultrafilters have a fundamental importance in foundations of mathematics. If D is an ultrafilter over a set I and for every $i \in I$ we have a set M_i, we can form the ultraproduct $\Pi_D M_i$ in the following style:

Let

$$M = \Pi_{i \in I} M_i.$$

For $f, g \in M$ define

$$f \equiv_D g \leftrightarrow \{i | f(i) = g(i)\} \in D.$$

Let $[f]_D$ denote the equivalence class of f under $\equiv_D$. Define the ultraproduct of the M_i's to be the collection of all these equivalence classes. Given relations $R_i \subseteq M_i^n (n < \omega)$ we can then define a corresponding relation R on $\Pi_D M_i$ by setting

$$R([f_1] \ldots [f_n]) \leftrightarrow \{i | R_i(f_1(i) \ldots f_n(i))\} \in D.$$

It is then a theorem of Los that statements expressible in first order logic behave well with respect to this operation. For more on ultraproducts, see Keisler-Chang [8]. The most important work on this line has been done by Keisler.

The application of modern set-theoretic methods has radically altered the picture. It was shown that there is an intimate connection between axioms postulating the existence of very large cardinal numbers and certain ultrafilters:

3. *Definition:* A filter D is λ-complete for a cardinal λ if for every collection $Y \subseteq D$ of cardinality less than λ

$$\cap Y = \cap\{y | y \in Y\} \in D.$$

It is then obvious that every ultrafilter D is ω-complete. However, at ω_1 we are already confronted with an entirely different phenomenon:

If a non-principal ultrafilter D is ω_1-complete then there is a inaccessible cardinal λ so that D is λ-complete and there is a non-principal λ-complete ultrafilter over λ (Keisler-Tarski [9]). Most of our paper is devoted to ultrafilters over fairly large cardinals. We have divided our paper into three parts: Ultrafilters over ω, countably complete ultrafilters and finally countably incomplete ultrafilters over uncountable cardinals.

1. ULTRAFILTERS OVER ω

The theory of ultrafilters over ω arose principally from the interest in Stone-Čech compactifications. Given a regular cardinal λ it is a classical result (W. Rudin [28]) that $\beta\lambda$ has 2^{2^λ} elements. That an individual ultrafilter has an enormously complex structure, is evident from the following result:

1.1. Theorem (*Sierpinski* [29]): If D is a nonprincipal ultrafilter over ω, then if we consider D as a subset of the Cantor space 2^ω, D is a non-measurable set.

To evaluate the complexity of D, it turns out to be useful to consider partial orderings on ultrafilters.

1.2. Definition: (a) (M. E. Rudin-Keisler) Given two ultrafilters D, U over a set x_0 we say: U is projectible onto D if there is a function $f\colon x_0 \to x_0$ so that for all $x \subseteq x_0$

$$x \in D \leftrightarrow f^{-1}(x) \in U.$$

In symbols:

$$D \leqslant_{\mathrm{RK}} U \text{ and } D = f_*(U).$$

If both $D \leqslant_{\mathrm{RK}} U$ and $U \leqslant_{\mathrm{RK}} D$ we say that D, U are isomorphic, in symbols

$$D \cong U.$$

(b) If D is an ultrafilter over x and f, g functions with domain x, define $f \leqslant_{\mathrm{RK}} g$ if and only if there is a function ϕ so that $f = \phi \circ g$ on a set belonging to D. Similarly, set

$$f \cong g \leftrightarrow f \leqslant_{\mathrm{RK}} g \text{ and } g \leqslant_{\mathrm{RK}} f.$$

(c) Given ultrafilters D over I, U_i over $x(i \in I)$, define the D-limit of U_i's to be the ultrafilter U defined by

$$y \in U \leftrightarrow \{i | y \in U_i\} \in D.$$

(d) Given D, U_i as in (c), say that the family $\{U_i\}$ is D-separated if there is a family $\{x_i | i \in J\}$ of pairwise disjoint sets so that $J \in D$ and for $i \in J$

$$x_i \in U_i.$$

(e) (M. E. Rudin-Frolik). Given D, U_i as in (c): If the U_i are D-separated, then define U to be the D-sum of U_i's:

$$U = \Sigma_D U_i$$

and set

$$D \leqslant_{\mathrm{RF}} U.$$

1.3. Theorem (*M. E. Rudin* [3]); Given two ultrafilters D, U:

(a) $D \cong U$ if and only if there is a one-to-one function f so that $D = f_*(U)$.

(b) If $D \leqslant_{\mathrm{RF}} U$, then $D \leqslant_{\mathrm{RK}} U$.

(c) The ultrafilters form a tree under $\leqslant_{\mathrm{RF}}$.

The following concepts are due to W. Rudin who used them to show that the space $\beta\omega - \omega$ (i.e. the space $\beta\omega$ with all the non-principal ultrafilters deleted) is not homogeneous:

1.4. Definition: Let λ be a fixed cardinal, D an ultrafilter over λ.

(a) A function f: $\lambda \to \lambda$ is almost one-to-one if for every $\alpha < \lambda$ $|f^{-1}(\{\alpha\})| < \lambda$. A function f is almost one-to-one (mod D) if there is an $x \in D$ so that $f|x$ is almost one-to-one.

(b) An ultrafilter D is a P-point if every function f: $\lambda \to \lambda$ which is not bounded by a constant $< \lambda$ is almost one-to-one (mod D).

(c) An ultrafilter D is selective if every function f: $\lambda \to \lambda$ which is not bounded by a constant $< \lambda$ is one-to-one (mod D).

(d) Given two ultrafilters D, E over λ, D, E are equivalent if there is a homeomorphism $\beta\lambda - \lambda \to \beta\lambda - \lambda$ mapping D into E.

To avoid trivial cases, we shall assume from now on that the phrase 'ultrafilter D over a cardinal λ' refers to a uniform ultrafilter over λ; i.e. an ultrafilter so that

$$x \in D \to |x| = \lambda.$$

It is then easy to see that for a fixed cardinal λ, P-points form an initial segment of ultrafilters over λ in the Rudin-Keisler order and that selective ultrafilters are exactly those which are minimal in the Rudin-Keisler order.

Most of the theory of countably indexed ultrafilters evolve around these concepts; there are no deep results and counter-examples abound. It is fair to say that under the continuum hypothesis, if there is no obvious proof of an assertion, there is a counterexample to it constructed by fairly simple-minded induction (with a few notable exceptions due to M. E. Rudin and K. Kunen). Most of these results have simple extensions to non-CH situations assuming Martin's axiom. One important exception is known to this rule:

W. Rudin [28] showed under CH that any two P-points over ω are equivalent. M. E. Rudin [26] generalized this to show that, assuming CH, given any two countable families $\{D_i\}_{i<\omega}$, $\{E_i\}_{i<\omega}$ of distinct P-points there is a homeomorphism $\phi: \beta\omega - \omega \to \beta\omega - \omega$ so that for any $i < \omega$

$$\phi(D_i) = E_i.$$

1.5. Problem: Being a P-point is in a sense a first order property in $\beta\omega$. The above result of W. Rudin shows that (under CH) all P-points are mutually equivalent; equivalence being on the surface a second order property in $\beta\omega$. Is there a first order characterization of equivalence classes of ultrafilters? Kunen conjectures that the answer is no.

The next category of non-CH questions is the question of existence of P-points. It was shown in W. Rudin [28] that assuming CH, there are lots of P-points. The following result is a direct extension of this:

1.6. Theorem (*Ketonen* [14]): Let (H) stand for the following hypothesis:

(H) If $\{f_\alpha | \alpha < \lambda\}$ is a family of functions $\omega \to \omega$ of power less than 2^ω then there is a function g so that for no $k < \omega$, $\alpha < \lambda$: For every $n \geq k$

$$g(n) \leq f_\alpha(n).$$

Then (H) holds if and only if every filter generated by less than 2^ω elements can be extended to a P-point.

A moment's reflection then shows that the most natural models to look for non-existence of P-points are iterated Sacks-forcing models and the random real model. The following is then known:

1.7. Theorem (*Kunen*): Let M be a countable standard model of $ZFC + CH$.

(a) If M' is the result of adjoining ω_1^M mutually generic Cohen reals and M'' is M' with ω_2^M random reals adjoined, then there are no selective ultrafilters in M''. Moreover, there is a P-point in M''.

(b) If D is a selective ultrafilter over ω in M, and r is a real random over M, then D cannot be extended to a P-point in $M[r]$. In particular, there are no selective ultrafilters over ω in the models obtained by adjoining ω_2^M random reals.

(c) There is a P-point D over ω in M so that if r is a real random over M, then D can be extended to a P-point in $M[r]$.

1.8. Theorem (*Shelah*): There is a model of ZFC without any P-points. Shelah's model is obtained by iterated Grigorieff forcing.

1.9. Problem: Does (H) imply that there are no ultrafilters over ω generated by less than 2^ω elements? (We say that an ultrafilter D is *generated* by a set $S \subseteq D$ if for every $x \in D$ there is a $y \in S$ so that $y \subseteq x$.)

For more on this question, see Ketonen [14]. A further question on the structure of an ultrafilter over ω as a subset of 2^ω:

1.10. Problem: Say that an ultrafilter D over ω is perfect if for every perfect closed set $P \subseteq 2^\omega$ the following holds:

If for every perfect closed $Q \supseteq P$, $Q \cap D \neq 0$, then there is a set $x \in D$ so that the set $\{y | y \subseteq x\} \cap P$ is uncountable.

It is then easy to see that every selective ultrafilter is perfect.

(a) If U is perfect and $D \leqslant_{RK} U$, is D perfect?

(b) If D is a P-point, is D perfect?

(c) If D has $\leqslant \omega$ isomorphism classes of functions, is D perfect?

(d) If D, U_i are perfect, is $\Sigma_D U_i$ perfect?

2. COUNTABLY COMPLETE ULTRAFILTERS

In the theory of countably complete ultrafilters our great aid is well-foundedness: If D is a nonprincipal λ-complete ultrafilter over a cardinal κ for some $\lambda > \delta$, we may form in the manner indicated in the introduction the ultrapower (V^D, E) of the whole set-theoretic universe (V, E). It is then easy to see that there are no infinite descending E-chains and that therefore (V^D, E) is isomorphic to a transitive structure (M^D, ε). We shall often confuse the distinction between these two structures. We also have an elementary embedding $i_D: V \to M^D$:

2.1. Theorem: (a) M^D is closed under λ-sequences: If $x_\alpha \in M^D(\alpha < \lambda)$, then $\langle x_\alpha | \alpha < \lambda \rangle \in M^D$.

(b) If $D \leqslant_{\mathrm{RK}} E$, there is an elementary embedding $M^D \to M^E$.

(c) More generally, if S is any collection of functions $\kappa \to V$, define

$$\mathrm{Gen}(S) = \{[f]_D | \exists f_1 \ldots f_n \in S : f \leqslant_{\mathrm{RK}} \langle f_1 \ldots f_n \rangle\}.$$

Then $\mathrm{Gen}(S)$ is isomorphic to a transitive class-model of set-theory M_S^M and the inclusion map induces an elementary embedding $M_S^M \to M^D$.

We have then the following two large cardinal axioms:

2.2. Definition: (1) (Solovay-Reinhardt [31]) κ is λ-supercompact if there is an elementary embedding $i: V \to M$ where $M \subseteq V$ is a transitive class so that

(a) $i(\alpha) = \alpha$ for $\alpha < \kappa$

(b) $i(\kappa) > \kappa$

(c) M is closed under λ-sequences.

κ is measurable if κ is κ-supercompact.

(2) (Keisler-Tarski [9]) κ is λ-strongly compact if every κ-complete filter on λ can be extended to a κ-complete ultrafilter.

2.3. Theorem: (a) (Solovay-Reinhardt [31]) If κ is 2^λ-supercompact, then κ is λ-strongly compact. Also, if κ is 2^κ-supercompact, then κ is the κ'th measurable cardinal.

(b) (M. Magidor [19]) If it is consistent with the axioms of set theory to have a strongly compact cardinal, then there is a model in which the first measurable cardinal is strongly compact.

(c) (Ketonen [11]) If for every regular cardinal $\mu \leqslant 2^\lambda$ there is a uniform κ-complete ultrafilter over μ, then κ is λ-strongly compact.

2.4. Problem: If λ is the first inaccessible over κ and there is a κ-complete uniform ultrafilter over λ, is κ μ-strongly compact for every $\mu < \lambda$?

In the general theory of countably complete ultrafilters one can investigate all the questions referred to in Section 3, hoping that the additional property of countable completeness will give more usable information. Unfortunately this has not yet turned out to be the case.

The rest of this section is devoted to uniform κ-complete ultrafilters over cardinals $\kappa > \omega$. To simplify our notation, we call such objects κ-ultrafilters. Here the situation is analogous to ultrafilters over ω. We have a few advantages here: Given a κ-ultrafilter D we can associate to it an ordinal $\delta(D)$ which is invariant under Rudin-Keisler isomorphism in the following way: Let $f: \kappa \to \kappa$ be the least (mod D) function which is one-to-one (mod D). Let $\delta(D) = [f]_D$.

2.5. Theorem (Solovay): The Keisler order on countably complete ultrafilters is well-founded. As a matter of fact, if D, E are κ-ultrafilters, then

$$D <_{\mathrm{RK}} E \to \delta(D) < \delta(E).$$

If $\delta(D) = [f]_D$, call $f_*(D) = D^*$ the normalization of D. Thus we have a canonical way of picking a member from each Keisler-isomorphism class.

2.6. Definition: (a) An ultrafilter D over a cardinal λ is a Q-point over λ if D is isomorphic to an ultrafilter extending the closed unbounded filter on λ. Equivalently, D is a Q-point if and only if every almost one-to-one function is one-to-one (mod D).

(b) D is normal if D is the normalization of a selective κ-ultrafilter: Equivalently, if f is any pressing down function $\kappa \to \kappa$ (i.e. for any $\alpha > \omega$ $f(\alpha) < \alpha$) then f is constant (mod D).

Note that a κ-ultrafilter is selective iff it is both a P-point and a Q-point.

2.7. Theorem (Solovay): If κ is 2^κ-super-compact, then there are 2^{2^κ} normal κ-ultrafilters on κ.

2.8. Problem: If κ is 2^κ-supercompact, how many normal κ-ultrafilters are there extending $\{\alpha | \alpha \text{ not measurable}\}$? (A result of Solovay states that there is at least one.)

2.9. Problem: If κ is strongly compact, how many normal ultrafilters are there on κ? It is not even known whether there is more than one.

2.10. Theorem: (Kunen-Paris-Mitchell [15], [16], [21]) If there is a measurable cardinal κ in V, there is a model in which κ carries exactly λ normal κ-measures where λ can be any one of 4, ω, κ, 2^κ, 2^{2^κ}.

In the theory of ultrafilters over ω the continuum hypothesis gave us a 'uniform' method of constructing ultrafilters. Here we have nothing comparable to it and consequently the first few levels of the Rudin-Keisler hierarchy are very little understood.

2.11. Theorem (Ketonen [12]): (a) If κ is measurable and κ is the κ'th measurable, then there is a κ-ultrafilter which is a non-selective P-point.

(b) If κ is κ-strongly compact, then for every κ-ultrafilter D there is a κ-ultrafilter E so that E is a Q-point and $D \leqslant_{\mathrm{RK}} E$.

2.12. Problem: If there is a κ-ultrafilter which is a non-selective P-point, is κ the κ'th measurable cardinal? If κ is κ-strongly compact, are there such ultrafilters?

2.13. Problem: If D is a κ-ultrafilter so that D is a P-point with exactly two Rudin-Keisler isomorphism classes of non-constant functions, is D a sum?

If D is a Q-point, let $\Gamma(D) = \{[f]_D | f_*(D)$ extends the closed unbounded filter$\}$.

2.14. Theorem (Kanamori): If D is a κ-ultrafilter, $\Gamma(D)$ is a closed subset of $i_D(\kappa)$.

Let $\lambda(D)$ = ordertype of $\Gamma(D)$. By taking sums of normal ultrafilters, one can construct κ-ultrafilters D so that $\lambda(D) = 1, 2, 3, \ldots$ and so that the D's are *hereditary* Q-points: Every ultrafilter below in the Rudin-Keisler order is also a Q-point. (For details, see Ketonen [12]). On the other hand, Theorem 2.11(b) gives us another method of getting Q-points with $\lambda(D) < \omega$.

2.15. Problem: Are there any hereditary Q-points with $\lambda(D) \geqslant \omega$?

2.16. Problem: If D is a Q-point with $\lambda(D) < \omega$ is D a hereditary Q-point?

2.17. Problem: If D is a Q-point with $\lambda(D) = 2$ is D isomorphic to a normal sum of normal ultrafilters?

2.18. Problem: If D extends the closed unbounded filter and $\lambda(D) = 2$, does

$$\{\alpha | \alpha \text{ regular}\} \in D?$$

2.19. Problem: Does there exist a κ-ultrafilter with exactly ω isomorphism classes of functions?

2.20. Problem: Does there exist a κ-ultrafilter D extending the closed unbounded filter with a sequence of functions $f^i\colon \kappa \to \kappa (i < \omega)$ so that

(1) $f^i_*(D)$ is a normal ultrafilter

(2) $\sup_{i<\omega} f^i(\alpha) = \alpha \pmod{D}$.

Observe here the following result:

2.21. Theorem (Solovay): If κ is κ-strongly compact, then an ultrafilter of the above type exists if and only if the following principle holds:

$\diamondsuit_\omega(\kappa)$: There are sets $s_\alpha \subseteq \alpha (\alpha < \kappa)$ so that for every $X \subseteq \kappa$ the set

$$\{\alpha | cf(\alpha) = \omega \text{ and } X \cap \alpha = S_\alpha\}$$

is Mahlo.

2.22. Theorem (Rowbottom [25]): Suppose that D is a selective κ-ultrafilter, $n < \omega$, $\lambda < \kappa$ and $F\colon [\kappa]^n \to \lambda$ where $[\kappa]^n$ denotes the set of all unordered n-tuples from κ. Then there is a set $x \in D$ so that F is constant on $[x]^n$.

This very important result gives rise to a few natural objects of study. For an arbitrary ultrafilter D, $n < \omega$, let $P_n(D)$ be the filter on the set

$$\Delta_n = \{(\alpha_1 \ldots \alpha_n) | \alpha_1 < \alpha_2 < \ldots < \alpha_n < \kappa\}$$

generated by sets of the form

$$(X \times \ldots \times X) \cap \Delta_n \, (X \in D).$$

In general, given κ-ultrafilters $D_1 \ldots D_n$, let $D_1 \otimes D_2 \otimes \ldots \otimes D_n$ be the filter generated by

$$(X_1 \times X_2 \times \ldots \times X_n) \cap \Delta_n \, (X_i \in D_i).$$

It is then easy to see that $D_1 \otimes \ldots \otimes D_n$ always has at least one atom A and that the ultrafilter induced by A is the product ultrafilter $D_1 \times D_2 \times \ldots \times D_n$ defined inductively by:

$$z \in D_1 \times D_2 \times \ldots \times D_n \leftrightarrow$$
$$\{(\alpha_1 \ldots \alpha_{n-1}) | \{\gamma | (\alpha_1 \ldots \alpha_{n-1}\gamma) \in z\} \in D_n\} \in D_1 \times \ldots \times D_{n-1}.$$

2.23. *Definition:* Say a κ-ultrafilter D is Rowbottom if for every $n < \omega$ $P_n(D)$ is completely atomic; i.e., every set of positive $P_n(D)$-measure contains an atom.

2.24. *Problem*: Is every κ-ultrafilter Rowbottom?

2.25. *Problem:* If $D_1 \ldots D_n$ are Rowbottom, is $D_1 \otimes D_2 \otimes \ldots \otimes D_n$ completely atomic?

For any κ-ultrafilter D, define

$$\rho_n(D) = \begin{cases} \text{number of atoms in } P_n(D) \\ 0 \text{ if completely atomic} \\ \infty \text{ otherwise.} \end{cases}$$

It is then easy to see that

$$\rho_i(D) \leqslant \rho_2(D) \leqslant \rho_3(D) \leqslant \ldots$$

and

$$D \leqslant_{\text{RK}} U \rightarrow \rho(D) \leqslant \rho(U).$$

By Rowbottom's theorem, if D is selective, then for any $n < \omega$

$$\rho_n(D) = 1.$$

Similarly, for $n < \omega$

$$\rho_n(D \times D) = 2^n - 1.$$

Say that D is super-Rowbottom if

$$\rho(D) = \sup_{n<\omega} \rho_n(D) < \kappa.$$

2.26. *Problem:* If $\rho_2(D) < \kappa$, is D super-Rowbottom? Does

$$\rho_2(D) < \kappa \rightarrow \rho_3(D) < \kappa?$$

It is easy to see that D is super-Rowbottom with $\rho(D) = \lambda$ if and only if for every $F: [\kappa]^n \to \mu$, $\mu < \kappa$ there is a $x \in D$ so that $F([x]^n)$ has cardinality λ if and only if the associated notion of Prikry-forcing (see Prikry [22]) does not collapse λ.

Another simple observation: If $t(D)$ denotes the number of isomorphism classes of non-constant functions then

$$t(D) \leqslant 2^{\rho_2(D)}.$$

2.27. Problem: Find a closer tie between $t(D)$ and $\rho_2(D)$. If $t(D) < \kappa$, is D super-Rowbottom?

2.28. Problem: If $\rho(D) \leqslant \kappa$, is D super-Rowbottom?

2.29. Problem: Is the super-Rowbottom sum of super-Rowbottom ultrafilters super-Rowbottom?

2.30. Problem: If D is a Q-point with $\lambda(D) = 2$, is D super-Rowbottom?

Next, a few questions about the Rudin-Frolik order. David Booth was the first to notice that it forms a well-founded tree.

2.31. Problem: Are there any Rudin-Frolik chains of κ-ultrafilters of length $\omega + 1$?

The following observation is useful:

2.32. Theorem: There are no $\omega + 1$-Rudin-Frolik chains if and only if there is a map θ from all κ-ultrafilters into ordinals so that:

$$\text{If } D = \Sigma_U E, \text{ then } \{\alpha \mid \theta(E_\alpha) < \theta(D)\} \in U$$

if and only if there is another map θ^1 from all κ-ultrafilters into ordinals so that:

$$\text{If } D = \Sigma_U E_\alpha, \text{ then } \{\alpha \mid \theta^1(E_\alpha \times E_\alpha) < \theta^1(D \times D)\} \in U.$$

It would also be of interest to get a characterization of Rudin-Frolik minimal ultrafilters.

2.33. Theorem (Solovay): If κ is 2^κ-super compact, then there is Rudin-Frolik minimal κ-ultrafilter which is not a P-point.

The following problem has a lot of relevance to the well-foundedness of certain partial orderings on κ-ultrafilters:

2.34. Problem: Are there any κ-ultrafilters on κ which are limits of themselves: i.e. does there exist κ-ultrafilters U, $U_\alpha(\alpha < \kappa)$ which are pairwise distinct so that

$$x \in U \leftrightarrow \{\alpha | x \in U_\alpha\} \in U.$$

For more on this question, see Glazer [5]. A weaker version:

2.35. Problem: Are there κ-ultrafilters U, $D_\alpha(\alpha < \kappa)$ so that the D_α are not U-separated? What if the D_α are all mutually isomorphic?

It is easy to see that any family of P-points can be separated with respect to any ultrafilter and that any family of κ-ultrafilters can be separated with respect to any P-point. Solovay has shown that any family of mutually isomorphic Q-points can be separated.

2.36. Theorem (Solovay): If D is a selective κ-ultrafilter over κ and $x_\alpha \in D(\alpha < \kappa^+)$ then there is a set $S \subseteq \kappa^+$ of power κ so that

$$\cap\{x_\alpha | \alpha \in S\} \in D.$$

2.37. Problem: Extend this to other κ-ultrafilters. Does it hold for selective ultrafilters over ω?

2.38. Problem: Can there exist a κ-ultrafilter generated by less than 2^κ elements?

2.39. Problem: Can W. Rudin's result referred to in 1.5 be extended to κ-ultrafilters; i.e. given two κ-ultrafilters D, E over κ so that both of them are P-points, does there exist a κ-complete homomorphism

$$\phi: P(\kappa)/_J \to P(\kappa)/_J$$

so that

$$x \in D \leftrightarrow \phi(x) \in E$$

where J is the ideal of sets of cardinality less than κ.

2.40. Problem: If D is a κ-ultrafilter work $x \subseteq \kappa$, what is the structure of $L[D, x]$ (all sets constructible relative to $\langle D, x \rangle$ in the sense of Levy)? Can one, for example, have exactly two normal ultrafilters in this universe by choosing x suitably generic?

Here Kunen [15] is relevant.

3. COUNTABLY INCOMPLETE ULTRAFILTERS

To measure the 'width' of an ultrafilter, the following notion, due to Keisler, is used:

3.1. Definition: An ultrafilter D is (λ, μ)-regular (where $\lambda \subseteq \mu$ are cardinals) if there is a family S of elements of D of power μ so that

$$T \subseteq S \text{ and } |T| \geq \lambda \rightarrow \cap T = 0.$$

3.2. Theorem (Keisler): (a) For any cardinal λ there is a (ω, λ)-regular ultrafilter over λ.

(b) If D is a (λ, μ)-regular ultrafilter then $\mathrm{Cof}(\Pi_D(\lambda, <)) \geqslant \mu^+$.

(c) If D is (λ, μ)-regular and $E \geqslant_{\mathrm{RK}} D$ then so is E.

For details, see Keisler-Chang [8].

3.3. Problem (Keisler-Gillman): Is it consistent with the axioms of set-theory to have a uniform, non-(κ, κ^+)-regular ultrafilter over a cardinal κ^+?

3.4. Theorem (Prikry [23]): In the constructible universe every uniform ultrafilter over a cardinal κ^+ is (κ, κ^+)-regular.

3.5. Theorem (Jorgenson [6]): An ultrafilter D over κ^+, κ regular, is not (κ, κ^+)-regular if and only if

$$\mathrm{Cof}(\Pi_D(\kappa, <)) = \kappa^+.$$

3.6. Theorem (Ketonen [2]): If D is a non-(κ, κ^+)-regular ultrafilter over κ^+, then D is a P-point; moreover, if f: $\kappa^+ \rightarrow \kappa^+$ is unbounded (mod D), then there is a set $x \in D$ so that for every $\alpha < \kappa^+$:

$$|f^{-1}(\{\alpha\}) \cap x| < \kappa.$$

3.7. Problem: Show that D is a P-point from a weaker form of non-regularity. This would give an extension of 3.10.

3.8. Definition: An ultrafilter D over a cardinal λ is weakly normal if every pressing down function has range of cardinality $<\lambda$ on a set of D-measure 1.

This is the correct generalization of normality to this situation. If D is a countably complete uniform ultrafilter over λ then there is a weakly normal ultrafilter below D in the Rudin-Keisler order (Ketonen [11]). In case λ is regular, this simply says that every pressing down function is bounded by a constant $<\lambda \pmod{D}$.

3.9. Theorem (Kanamori): If D is a non-(ω, λ)-regular ultrafilter over λ, λ is regular, then there is a Q-point below D in the Rudin-Keisler order.

Combining theorems 3.6 and 3.9 we get

3.10. Theorem (Kanamori): If D is a non-(κ, κ^+)-regular ultrafilter over κ^+, then there is a weakly normal ultrafilter below D in the Rudin-Keisler order.

For weakly normal ultrafilters we have the following characterization of nonregularity:

3.11. Theorem (Ketonen [11]): If D is a weakly normal ultrafilter over a regular cardinal λ, then D is (μ, λ)-regular if and only if

$$\{\alpha \mid cf(\alpha) < \mu\} \in D.$$

3.12. Problem: Get a similar characterization for singular cardinals.

By combining theorems 3.10 and 3.11 we get:

3.13. Theorem (Kanamori): If κ is a singular cardinal, then every uniform ultrafilter over κ^+ is (κ, κ^+)-regular.

Non-(κ, κ^+)-regular ultrafilters act in many ways like large cardinal axioms:

3.14. Theorem (Ketonen [2]): If there is a non-(κ, κ^+)-regular ultrafilter over κ^+, then

$$2^\kappa = \kappa^+ \rightarrow 2^{\kappa^+} = \kappa^{++}.$$

3.15. Theorem (Magidor): Con (ZFC + there exists a huge cardinal number) implies Con (ZFC + there exists a (ω_1, ω_3): non-regular ultrafilter over ω_3).

Most of the above results depend on the following technique of Benda, generalizing the results of Prikry and Jorgenson.

3.16. Theorem (Benda [2]*)*: Suppose f_α: $\kappa^+ \to \kappa(\alpha < \kappa^+)$ is a family of eventually different functions. Then an ultrafilter D over κ^+ is non-(κ, κ^+)-regular if and only if the f_α are cofinal in the ultrapower of κ. In particular, if the Kurepa Hypothesis on κ^+ holds, then every uniform ultrafilter over κ^+ is (κ, κ^+)-regular.

3.17. Theorem (Benda [2]*)*: If there is a non-(κ, κ^+)-regular ultrafilter over κ^+, then κ^{++} is inaccessible in L.

The following result represents a considerable improvement of 3.17.

3.18. Theorem (Ketonen): If there is a non-(κ, κ^+)-regular ultrafilter over κ^+, then $0^\#$ exists.

Using the technique of the proof 3.18., Jensen and Koppleberg have shown the following.

3.19 Theorem (Jensen-Koppelberg): If there is a weakly normal ultrafilter over a regular cardinal λ such that $2^\lambda = \lambda$, then $0^\#$ exists.

3.20 Theorem (Jensen-Koppelberg): If there is a weakly normal ultrafilter over a regular cardinal λ which is not (μ, λ)-regular for any $\mu < \lambda$, then $0^\#$ exists.

3.21. Problem: If D is a weakly normal non-(κ, κ^+)-regular ultrafilter, what is the structure of $L[D]$?

Other, slightly weaker notions of regularity have been studied with very little success.

3.22. Definition: A uniform ultrafilter D over κ^+ is

(1) $[\kappa, \kappa^+]^2$-regular if there are sets $X_s (s \in [\kappa^+]^2)$ so that

$$A \subseteq \kappa^+ \text{ and } |A| \geqslant \kappa \to \cap\{X_s | s \in [A]^2\} = 0$$

(2) weakly (κ, κ^+)-regular if there are sets $X_\alpha \in D(\alpha < \kappa^+)$ so that

$$A \subseteq \kappa^+ \text{ and } |A| \geqslant \kappa \to |\cap\{X_\alpha | \alpha \in A\}| < \kappa^+$$

(3) (α, κ^+)-regular for an ordinal α if there are sets $X_\gamma \in D(\gamma < \kappa^+)$ so that

$$A \subseteq \kappa^+ \text{ and ordertype of } A \geqslant \alpha \to \cap\{X_\gamma | \gamma \in A\} = 0.$$

The only known result of significance here is the following:

3.23. Theorem (Prikry): If $2^\kappa = \kappa^+$, then every uniform ultrafilter over κ^+ is weakly (κ, κ^+)-regular.

3.24. Problem: If D is non-(ω, ω_1)-regular over ω_1 what is the cardinality of $\Pi_D \omega$? Does there exist a function $f: \omega_1 \to \omega$ so that the functions

$$\{\phi \circ f | \phi: \omega \to \omega\}$$

are cofinal in $\Pi_D \omega$?

3.25. Problem: If D is a non-(ω, ω_1)-regular ultrafilter over ω_1, how many isomorphism classes of ultrafilters over ω are there below D in the Rudin-Keisler order?

Observe that using the usual Lévy-Solovay collapsing algebra for an inaccessible cardinal κ, we always have a way of 'transferring' a non-(κ, κ^+)-regular ultrafilter over κ^+ to a non-(ω_1, ω_2)-regular ultrafilter over ω_2 (Ketonen [11]).

3.26. Problem: Does Con(ZFC + $\exists$ non-(κ, κ^+)-regular ultrafilter) imply Con(ZFC + $\exists$ non-(ω, ω_1)-regular ultrafilter)?

3.27. Problem: If D is a non-(ω, ω_1)-regular ultrafilter over ω_1, does there exist a non-trivial Boolean extension of the universe in which every extension of D is a non-(ω, ω_1)-regular ultrafilter?

If we are to look for only (ω, λ)-nonregular ultrafilters over weakly inaccessible cardinals, we have plenty to choose from:

3.28. Definition: A filter F over λ is μ-saturated for a cardinal μ if there is no family of sets $X_\alpha (\alpha < \mu)$ of positive F-measure so that

$$\alpha < \beta \rightarrow -(X_\alpha \cap X_\beta) \in F.$$

3.29. Theorem (Silver-Kunen): If F is a λ-complete λ-saturated filter over λ, then every ultrafilter extending F is not (ρ, λ)-regular for any $\rho < \lambda$.

Using methods of Ketonen [10], Kunen has shown that in this case we have a lot of freedom:

3.30. Theorem (Kunen [10]): If F is a ω_1-saturated λ-complete filter over $\lambda \leqslant 2^\omega$, then there is an ultrafilter D extending F so that if $\mathfrak{A}$ is any countable model in a countable language, the ultrapower

$$\Pi_D \mathfrak{A}$$

is saturated.

3.31. Problem: What are the logical relationships between the following statements:

(a) There is a non-(ω, ω_1)-regular ultrafilter over ω_1.

(b) The closed unbounded filter on ω_1 is ω_2-saturated.

(c) Chang's conjecture: If F: $[\omega_2]^2 \rightarrow \omega_1$ then there is a set $S \subseteq \omega_2$ of power ω_1 so that $F''[S]^2$ is countable.

For more on this, see Ketonen [13] and Kunen [17].

To replace the completeness properties of countably complete ultrafilters, we have the following notion:

3.32. Definition (Keisler): If λ is any cardinal, an ultrafilter D is

(a) descending λ-complete if for any decreasing sequence $\langle x_\alpha | \alpha < \lambda \rangle$ of elements of D, (i.e. $\alpha < \beta \rightarrow x_\beta \subseteq x_\alpha$), we have

$$\cap \{x_\alpha | \alpha < \lambda\} \in D$$

(b) λ-decomposable if there is a function f into λ so that for every $y \in D$

$$|f''(y)| = \lambda.$$

It is easy to see that for regular λ, λ-indecomposability is equivalent to descending λ-completeness and that for singular λ, λ-decomposability implies $cf(\lambda)$-descending incompleteness.

The following theorem was proved by C. C. Chang [4] and extended to its present form by K. Prikry [24].

3.33. Theorem (*Chang-Prikry, GCH*)*:* Let λ be a cardinal, D an ultrafilter

(a) If λ is regular and D is λ^+-descendingly incomplete, then D is λ-descendingly incomplete.

(b) If $\mu = cf(\lambda) < \lambda$, and D is λ^+-descendingly incomplete, then either

(1) D is λ-decomposable

or

(2) there is a $\alpha < \lambda$ so that D is β-descendingly incomplete for all regular β such that $\alpha < \beta < \lambda$.

3.34. Problem (*Chang* [4])*:* (1) If an ultrafilter D is $\omega_{\omega+1}$-descendingly incomplete is D

(a) ω-descendingly incomplete

(b) ω_n-descendingly incomplete for all n?

(2) If D is a uniform ultrafilter over ω_ω is D

(a) ω_1-descendingly incomplete?

(b) ω_n-descendingly incomplete for all n?

3.35. Theorem (*Jensen-Silver-Prikry* [24])*:* Assume the axiom constructibility. Suppose that $\kappa < \omega$ is regular but not weakly compact. Then any D is λ-descendingly incomplete for all regular $\lambda \leqslant \kappa$.

Thus we have a positive answer to problem 3.34 (1) assuming the axiom of constructibility. On the other hand, an answer to question (2) is not known even in this case. This is equivalent to:

3.36. Problem: Are all uniform ultrafilters over ω_ω (ω, ω_ω) regular?

Prikry [23] has shown that all uniform ultrafilters over $\omega_n (n < \omega)$ are (ω, ω_n)-regular.

3.37. Theorem (*Silver* [30])*:* Suppose there is a uniform ultrafilter D over an inaccessible cardinal κ so that D is λ-indecomposable for all regular $\lambda \leqslant \kappa$. Then $0^{\#}$ exists.

3.38. Problem: Is κ measurable under these circumstances? Is κ measurable in $L[D]$?

Department of Mathematics, University of Hawaii
and Department of Mathematics, Stanford University.

BIBLOGRAPHY

[1] A. Blass, Doctoral dissertation, Harvard University, 1970.

[2] M. Benda and J. Ketonen, 'On regularity of ultrafilters', *Israel J. of Mathematics* **17** (1974), 231–240.

[3] D. Booth, 'Ultrafilters on a countable set', *Annals of Mathematical Logic* **2** (1970).

[4] C. C. Chang, 'Descendingly incomplete ultrafilters', *Trans. Am. Math. Soc.* **126** (1967), 108–118.

[5] S. Glazer, Doctoral dissertation, University of California, Berkeley, 1974.

[6] M. Jorgenson, 'Regular ultrafilters and long ultrapowers' (1971). To appear.

[7] H. J. Keisler, 'A survey of ultraproducts', in *Logic, Methodology and Philosophy of Science, Proceedings of the 1964 International Congress*, ed. by Y. Bar-Hillel, North-Holland, Amsterdam, 1965, pp. 112–126.

[8] H. J. Keisler and C. C. Chang, *Model Theory*, North-Holland, Amsterdam, 1973.

[9] H. J. Keisler and A. Tarski, 'From accessible to inaccessible cardinals', *Fund. Math.* **53** (1969), 225–308.

[10] J. Ketonen, 'On nonregular ultrafilters', *J. of Symbolic Logic* **37** (1972), 71–74.

[11] J. Ketonen, 'Strong compactness and other cardinal sins', *Annals of Mathematical Logic* **5** (1972), 47–76.

[12] J. Ketonen, 'Ultrafilters over measurable cardinals', *Fundamenta Mathematicae* **77** (1970), 257–269.

[13] J. Ketonen, 'Some combinatorial principles', *Trans. Am. Math. Soc* **188** (1974), 387–394.

[14] J. Ketonen, 'On the existence of *P*-points', to appear in *Fund. Math.*

[15] K. Kunen, 'Some applications of iterated ultrapowers in set theory', *Annals of Mathematical Logic* **1** (1970), 179–227.

[16] K. Kunen, 'Ultrafilters and independent sets', *Trans. Am. Math. Soc.* **172** (1972), 299–306.

[17] K. Kunen, 'Saturated ideals'. To appear.

[18] K. Kunen and J. Paris, 'Boolean extensions and measurable cardinals', *Annals of Mathematical Logic* **2** (1971), 359–377.

[19] M. Magidor, 'The first strongly compact can be the first measurable'. To appear.

[20] A. Mathias, 'How large is the first strongly compact cardinal? Or: A study in identity crisis', *Annals of Mathematical Logic* **10** (1976), 33–58.

[21] W. Mitchell, 'Sets constructible from sequences of ultrafilters', *J. of Symbolic Logic* **39** (1974), 57–66.

[22] K. Prikry, 'Changing measurable to accessible cardinals', *Rozprawy Matematyczne* **68** (1970), 1–52.

[23] K. Prikry, 'On a problem of Gillman and Keisler', *Annals of Mathematical Logic* **2** (1970), 179–187.

[24] K. Prikry, 'On descendingly complete ultrafilters'. *Proceedings of the Cambridge Summer School in Mathematical Logic*, Springer Lecture Notes **337** (1973), 459–488.

[25] F. Rowbottom, 'Some strong axioms of infinity incompatible with the axiom of constructibility', *Annals of Mathematical Logic* **3** (1971).

[26] M. E. Rudin, 'Types of ultrafilters', *Topology seminar, Wisconsin*, ed. by R. H. Bing and Ralph J. Bean, Princeton University Press, Princeton, 1966.

[27] M. E. Rudin, 'Partial orders on types in βN', *Trans. Am. Math. Soc.* **155** (1971), 353–362.
[28] W. Rudin, 'Homogeneity problems in the theory of Čech compactifications', *Duke Math. J.* **23** (1956).
[29] W. Sierpinski, *Hypothèse du continu*, second edition, Chelsea, New York, 1956.
[30] J. Silver, 'Indecomposable ultrafilters and $0^{\#}$', in *Proceedings of the Tarski Symposium*, ed. by D. Scott, AMS Minographe, Berkeley, 1972.
[31] R. Solovay, W. Reinhardt and A. Kanamori, 'Strong axioms of infinity and elementary embeddings', *Annals of Mathematical Logic* **13** (1978), 73–116.

PART IV

GENERALIZED QUANTIFIERS

J. A. MAKOWSKY*

THE REALS CANNOT BE CHARACTERIZED TOPOLOGICALLY WITH STRICTLY LOCAL PROPERTIES AND COUNTABILITY AXIOMS

ABSTRACT. A logic is presented which enables us to speak about topological properties which depend only on a basis for the topology as well as about countability conditions. We prove a compactness theorem and a Löwenheim Skolem theorem for this logic and conclude that within this logic the reals cannot be characterized.

INTRODUCTION

The aim of this paper is to propose a logical and finally syntactical analysis of topological concepts. Topology is usually formulated in second order logic. But no reasonable model theory is available for second order logic. And only recently fragments of second order logic have been studied more carefully, in particular by Shelah (unpublished). Useful references are Makowsky-Shelah-Stavi [4] and Makowsky-Shelah [3]. One of these fragments shall be proposed here to clarify two concepts in topology:

'strictly local properties' and
'basic countability axioms'.

Our presentation will focus around one theme: The reals cannot be characterized with strictly local properties and basic countability axioms. This is not surprising. But what we achieve is a logical explanation why this is so. Our modest hope is that this example of a successful logical analysis of topological concepts will attract the attention for further research in the field of logic and topology. For a survey of recent work in topological model theory we refer to [2].

I am indebted to J. Stavi for many inspiring and clarifying discussions.

* Research done during my stay at the Hebrew University in Jerusalem as a visiting research fellow.

J. Hintikka, I. Niiniluoto, and E. Saarinen (eds.), Essays on Mathematical and Philosophical Logic, 251–257.

1. STRICTLY LOCAL PROPERTIES IN TOPOLOGY AND THE LOGIC L^t

Hausdorff's separation axiom T_2 is usually stated as

(T_2) 'Two distinct points can be separated by disjoint open sets.'

This is equivalent to

(T_2') Two distinct points can be separated by disjoint sets of a given basis for the topology'.

In a way this shows that T_2 is a property which depends only on small open sets, i.e. it is a local property in a very strong sense (which holds here at every point). Let us call such properties *basic* or *strictly local.* (There might be some confusion: What about locally compact or locally connected? They will not be strictly local and they are better described as *localized* properties.)

To make our definition of strictly local precise we need some formal definitions.

A structure $\mathbf{A} = \langle A, \tau \rangle$ is called *topological* if A is a relational structure for predicate calculus and τ is a topology on the universe of A. It is called *basic* if τ is a basis for a topology. The logic L^2_{Top} is the two-sorted logic appropriare for topological or basic structures ($x_1, x_2, \ldots$, point variables, $U_1, U_2, \ldots$, variables for neighborhoods, atomic formulas are $x_1 = x_2$ and $x_1 \in U_1$ and the usual atomic formulas for A). A formula ϕ of L^2_{Top} is said to be B-invariant if for every topological $\mathbf{A} = \langle A, \tau \rangle \models \phi$ iff $\langle A, \beta \rangle \models \phi$ where β is a basis for τ.

Let L^t be the set of B-invariant sentences in L^2_{Top}.

Let $\phi(U_1)$ be a formula in L^2_{Top} ,U_1 a free variable (for neighborhoods).

The occurrence of U_1 in ϕ is called *B-invariant* if for every basic structure $\mathbf{A} = \langle A, \beta \rangle$ whenever $\langle A, U \rangle \models \phi(U)$ and $V \in \beta$ and $V \subset U$ then $\langle A, V \rangle \models \phi(V)$. The idea of B-invariant sentences is independently due to McKee, Garavaglia and Ziegler ([5], [1], [6]).

THEOREM 1.1. (Garavaglia, McKee and Ziegler)

(i) If T is an arbitrary set of sentences of L^t such that every finite subset S of T has a topological model then T has a topological model (compactness of L^t).

(ii) If T is a set of sentences of L^t of cardinality $\leq \kappa$ which has a topological model then T has a topological model of cardinality $\leq \kappa$ which has weight $\leq \kappa$.

Ziegler has also given a syntactical characterization of the sentences of L^t (cf. [6]). Here it is sufficient to know that L^t is closed under conjunction, disjunction, negation, existential quantification over point-variables and existential quantification of set variables provided their occurrence is B-invariant.

Our formal definition of a strictly local property in topology therefore is: A property is strictly local if it is expressible by a sentence in L^t.

EXAMPLES

1. The separation axioms T_0, T_1, T_2 and T_3 but not $T_{31/2}$ or T_4.
2. The statement 'R is open (closed)' where $R \subset A^n$ is a definable relation.
3. The statement 'x is in the interior of R (closure of R)' with R as in Example 2.
4. The statement 'x is an accumulation point (isolated point) in R' with R as above.
5. Generally the sentence $\exists U_1(x \in U_1 \wedge \phi(U_1))$ is in L^t provided U_1 does not occur positively in ϕ (i.e. is not within the scope of an even positive number of negation symbols).
6. Not expressible in L^t are compactness, local compactness, connectedness, local connectedness, separability, dimension zero etc.

2. BASIC COUNTABILITY AXIOMS

Basic countability axioms assert the existence of countable sets which behave like big sets. Big here may be a dense subset of the space or a basis for the topology or a cofinal set in an ordering etc.

Typical countability axioms are

1. Separability: The existence of a countable dense subset.
2. The first countability axiom: The existence of a countable basis at every point.
3. The second countability axiom: The existence of a countable basis for the topology (= weight ω).

I propose to call these statements *basic countability axioms* because their truth does not depend on the topology in question but only on an arbitrary basis for the topology.

To formalize this concept we shall proceed in two steps: We first define *the cofinally invariant logic* introduced in [4] and then combine it with L^t.

3. COFINALLY INVARIANT SENTENCES

Let A be a set. We denote by $\mathrm{CP}(A)$ the set of countable subsets of A. The analog of a basis in topology is a cofinal subfamily in $\mathrm{CP}(A)$. $S \subset \mathrm{CP}(A)$ is *cofinal* if for every $X \in \mathrm{CP}(A)$ there is a $Y \in S$ with $X \subset Y$. Let us consider in this section structures $\langle A, S \rangle$ where A is a first order structure and S is cofinal in $\mathrm{CP}(A)$. We call such structures cofinal. If $S = \mathrm{CP}(A)$ we call the structure *standard.*

Let L^2_{count} be the corresponding two-sorted logic with a membership relation $\in$, $x_1, x_2 \ldots$ as point variables, $X_1, X_2, \ldots$ for set variables ranging over S etc.

A formula ϕ of L^2_{count} is called *C-invariant* if for every cofinal structure $\mathbf{A} = \langle A, S \rangle$, $\mathbf{A} \models \phi$ iff $\langle A, \mathrm{CP}(A) \rangle \models \phi$. The occurrence of X_1 in $\phi(X_1)$ is called C-invariant if for every cofinal structure $\mathbf{A}$ and $X, Y \in S$, $X \subset Y$ if $\langle \mathbf{A}, X \rangle \models \phi(X)$ then $\langle \mathbf{A}, Y \rangle \models \phi(Y)$.

Let L^c be the logic consisting of the C-invariant sentences in L^2_{count}. L^c is closed under the boolean operations, existential quantification of point variables and existential quantification of set variables provided their occurrence is C-invariant.

THEOREM 3.1. (Makowsky-Shelah-Stavi)

(i) If T is a countable set of sentences of L^c such that every finite subset S of T has a standard model then T has a standard model. (Countable compactness)

(ii) If T is a set of sentences in L^c of cardinality ω_1 which has a standard model then T has a standard model of cardinality $\leqslant \omega_1$.

The significance of L^c for logic is discussed in [4].

4. THE LOGIC L^{tc}

We are now ready to combine the two notions of invariance. Let $\mathbf{A}$ be a basic structure. We can look at it as a two-sorted first order structure. We now want to talk about countable sets of points and countable families of open sets. Therefore our appropriate language is a 4-sorted language for structures of the form $\langle A, \beta, \mathrm{CP}(A), \mathrm{CP}(\beta) \rangle$ with point variables $x_1, x_2, \ldots$, variables for basic open sets $U_1, U_2, \ldots$, for countable subsets of A, $X_1, X_2, \ldots$, and for countable families of basic open sets, $\alpha_1, \alpha_2, \ldots$, and three membership relations which will all be denoted by $\in$.

Extra predicates are only over relations on A. Let us denote this logic with L^3_{tc}. Non-standard structures here are basic and cofinal, standard structures for L^3_{tc} are topological and standard in the sense of L^2_{count}.

A sentence ϕ in L^3_{tc} is BC-invariant if for every structure $\mathbf{A} = \langle A, \beta, S_1, S_2 \rangle$ which is basic and cofinal we have $\mathbf{A} \models \phi$ iff $\langle A, \tau, \text{CP}(A), \text{CP}(\tau) \rangle \models \phi$ where τ is the topology generated by β. Now let L^{tc} be the logic of BC-invariant sentences.

THEOREM 4.1. (i) L^{tc} is countably compact;

(ii) Let T be a set of sentences of L^{tc} of cardinality ω_1 which has a topological standard model. Then T has a topological standard model of cardinality $\leqslant \omega_1$ and weight $\leqslant \omega_1$.

Proof: First we note that BC-invariant sentences are C-invariant sentences if we disregard the topology. Therefore we can apply theorem 2.1.

(i) Let T be a countable set of sentences of L^{tc} such that every finite subset S of T has a topological standard model. By the above remark and theorem 2.1 (i) T has a model $\langle A, \beta, \text{CP}(A), \text{CP}(\beta) \rangle$ where β is a basis for a topology and the model is standard in the sense of L^2_{count}. So by the BC-invariance of T, T has a topological standard model.

(ii) Let T be a set of sentences as in the hypothesis. Again using the C-invariance of T and theorem 2.1 (ii) T has a model $\langle A, \beta, \text{CP}(A), \text{CP}(\beta) \rangle$ where A has cardinality ω_1 and β is a basis for a topology of cardinality ω_1. Using the BC-invariance of T now we get the desired result.

QED

Note that the definition of B-invariant (C-invariant) occurrence of a variable $U_i(X_i)$ also applies to L^{tc}. Also every L^t-sentence is an L^{tc}-sentence. L^{tc} is closed under all the logical operations listed for L^t or L^c. The problem with L^{tc} is that we do not have a straightforward way to define BC-invariant occurrences for variables α_i. Nevertheless we have

PROPOSITION 4.2. The following sentences are BC-invariant

(i) $\exists \alpha_1$('α_1 is a basis' $\wedge \forall U_1(U_1 \in \alpha_1 \Rightarrow \phi(U_1))$) where the occurrence of U_1 in ϕ is B-invariant.

(ii) $\exists \alpha_1$('α_1 is a basis at x' $\wedge \forall U_1(U_1 \in \alpha_1 \Rightarrow \phi(U_1))$) where the occurrence of U_1 in ϕ is B-invariant.

(iii) 'There is a countable dense subset'

$$\exists X_1(\forall x_1(\forall U_1(x_1 \in U_1 \Rightarrow \exists x_2(x_2 \in X_1 \wedge x_2 \in U_1))))$$

Proof: (i) Let Φ_1 be the sentence in (i). We have to show that

(1) $\langle A, \beta, S_1, S_2 \rangle |= \Phi_1$

iff

(2) $\langle A, \tau, \mathrm{CP}(A), \mathrm{CP}(\tau) \rangle |= \Phi_1$

where S_1, S_2 are cofinal in CP(A), CP(β) resp. and τ is the topology generated by β. (1) $\Rightarrow$ (2) is trivial. Now assume (2) and let (1) be a requested basis. By a standard argument from topology there exists a basis β_0 for τ $\beta_0 \subset \beta$ and for each $U \in \beta_0$ there exists a $V \in \alpha$ with $U \subset V$. Since α is countable so is β_0. Since the occurrence of U_1 in ϕ is B-invariant, every $U \in \beta_0$ satisfies ϕ and we conclude (1). The proof of (ii) is similar. (iii) is left as an exercise. QED

Proposition 4.2 justifies our claim that basic countability axioms are expressible in L^{tc}. What we propose is a formal definition: anything expressible by a single sentence in L^{tc} is a strictly local property or a basic countability axiom. A good example to illustrate this is 'metrizability in a bounded manner' which is equivalent to 'T_3 and separable'.

5. CONCLUSION AND OUTLOOK

Using the standard compactness argument of logic we get

THEOREM 5.1. No countable set of sentences of L^{tc} characterizes the reals up to homeomorphism (even with extra predicates).

Proof: If the continuum hypothesis is false this follows directly from theorem 4.1 (ii). Otherwise let c be a new constant symbol for a point and let S be the set of sentences asserting that c is bigger than any natural number. Let T be the set of sentences (over a countable language) in L^{tc} true in the reals. Then every finite subset of $T \cup S$ has a topological standard model. Therefore $T \cup S$ has a topological standard model, which must be non-archimedian and hence not connected. QED

According to the proposed definition of strictly local or basic countability axiom this tells us that the reals cannot be characterized with such properties. As said in the introduction this is not surprising. The question is if we can strengthen the logic L^{tc} to an even more expressive logic where similar compactness theorems do hold? Another question is if we can use the logic L^{tc} to get a hierarchy of countability axioms and local properties according to the complexity of the sentence describing it? To do so one should first have syntactic characterizations of the notions such as *BC*-invariant formula, *B*-invariant (*C*-invariant) occurrence of a variable etc. As one sees this paper is just (hopefully) a beginning.

Mathematisches Institute II,
Freie Universität Berlin

REFERENCES

[1] S. Garavaglia, *Model theory of topological structures*, Thesis, Yale, 1976.
[2] J. A. Makowsky, 'Topological model theory: a survey', in P. Maugani (ed.), *Model Theory and Applications*, Rome, 1975.
[3] J. A. Makowsky and S. Shelah, 'The theorems of Beth and Craig in abstract model theory II: compact logics', *Arch. f. Math. Logic.* **20** (1979).
[4] J. A. Makowsky, S. Shelah and J. Stavi, 'Δ-logics and generalized quantifiers', *Ann. Math. Log.* **10** (1976), 155–192.
[5] T. McKee, Thesis, Madison, 1974.
[6] M. Ziegler, 'A language for topological structures which satisfies a Lindström theorem', *BAMS* **82** (1976), 568–570.

M. KRYNICKI

ON THE EXPRESSIVE POWER OF THE LANGUAGE USING THE HENKIN QUANTIFIER

Investigations of languages with additional quantifiers have a rather long history. This history starts in 1957 with Mostowski's paper [9]. Next, Fuhrken [1, 2] Keisler [6] and Slomson [10] investigated properties of the language with such quantifiers as 'there exist infinitely many', 'there exist uncountably many', 'there exist as many as all elements' etc. In [3] Henkin proposed to investigate the quantifier which arose by writing the existential and universal quantifier parallel. Some people (see e.g. [4]) used this idea to construct a language and investigate its properties, mainly the syntactical properties. Only in [11] we find remarks about a connection between the quantifier defined by Henkin and other quantifiers. In this paper I want to throw some light on the semantical properties of the language with the quantifier defined in [3] and especially on its connection with the second order logic. These results were obtained in collaboration with Professor A. W. Lachlan and will be published in detail. I use the standard model theoretical notation as in [11].

As I have said, by the Henkin quantifier I mean the quantifier defined in [3] in the following way:

$$(Q_H x_1 x_2 y_1 y_2)\varphi \Leftrightarrow \begin{pmatrix} \forall x_1 \exists y_1 \\ \forall x_2 \exists y_2 \end{pmatrix} \varphi$$

Thus the formula $(Q_H x_1 x_2 y_1 y_2)\varphi$ can be read as meaning that for every x_1 there is a y_1 and for every x_2 there is a y_2 – depending only on x_2 – such that φ holds. So, this formula is equivalent to the second order formula:

$$(\exists fg)(\forall x_1 x_2)\varphi(x_1, x_2, f(x_1), g(x_2))$$

There is a connection between this quantifier and other quantifiers.
By Q_L I denote the quantifier defined as follows ([12])

$$(Q_L x)(\varphi, \psi) \Leftrightarrow \overline{\overline{\{x: \varphi(x)\}}} \leqslant \overline{\overline{\{x: \psi(x)\}}}$$

Q_L is defined by Q_H. Namely

J. Hintikka, I. Niiniluoto, and E. Saarinen (eds.), Essays on Mathematical and Philosophical Logic, 259–265.

$$(Q_L x)(\varphi, \psi) \Leftrightarrow (Q_H x_1 x_2 y_1 y_2)[(x_1 = x_2 \leftrightarrow y_1 = y_2) \wedge (\varphi(x_1) \rightarrow \psi(y_1))]$$

The right side of this equivalence says that there exists a one-one function f such that if $\varphi(x)$ then $\psi(f(x))$. So, it is really the same as the left side.

By Q_0 I denote the quantifier 'there exist infinitely many':

$$(Q_0 x)\varphi(x) \Leftrightarrow \overline{\overline{\{x: \varphi(x)\}}} \geq \aleph_0$$

By Q_c the Chang quantifier:

$$(Q_c x)\varphi(x) \Leftrightarrow \overline{\overline{\{x: \varphi(x)\}}} = \overline{\overline{\{x: x = x\}}}$$

By Q_I the Härtig quantifier

$$(Q_I x)(\varphi, \psi) \Leftrightarrow \overline{\overline{\{x: \varphi(x)\}}} = \overline{\overline{\{x: \psi(x)\}}}$$

Q_0 and Q_c are definable by Q_I:

$$(Q_c x)\varphi(x) \Leftrightarrow (Q_I x)(\varphi(x), x = x)$$
$$(Q_0 x)\varphi(x) \Leftrightarrow (\exists x)[\varphi(x) \wedge (Q_I y)(\varphi(y), \varphi(y) \wedge x \neq y)]$$

It is obvious that Q_I is definable by Q_L. So we have

FACT. Q_0, Q_c, Q_I are definable by Q_H.

Thus Q_H is a rather strong quantifier.

To sum up we have

$$L(Q_c) < L(Q_I) < L(Q_H).$$

From the above it follows that $L(Q_H)$ has no axiomatization property, compactness property or Skolem-Löwenheim property.

The first question which I would like to answer is what can be said about the $L^1(Q_H)$ monadic fragment of such a language.

In 1968 Slomson showed that $L^1(Q_c)$ is decidable and that its Hanf and Skolem number is $\aleph_0$.

Issell in 1969 ([5]) showed that the Hanf and Skolem number of $L^1(Q_I)$ is $\aleph_\omega$. Now I can show that $L^1(Q_H)$ is decidable and its Hanf and Skolem number is $\aleph_\omega$. Moreover, the expressive power of $L^1(Q_H)$ is the same as of $L^1(Q_I)$. More precisely; I can show the following:

THEOREM 1. Let $\mathfrak{A} = \langle A, U_0, \ldots, U_{k-1} \rangle$, $\mathfrak{Z} = \langle B, U_0', \ldots, U_{k-1}' \rangle$ be models of the same unary signature. Let for $\alpha \in 2^k$ $V_\alpha = \bigcap_{i<k} U_i^{\alpha(i)}$ and

$V_\alpha' = \bigcap_{i<k} U_i'^{\alpha(i)}$ where $U_i^0 = U_i$ $U_i'^0 = U_i'$, $U_i^1 = A - U_i$, $U_i'^1 = B - U_i'$.

Then
$\mathfrak{A}$ and $\mathfrak{Z}$ are elementary equivalent with respect to the language $L(Q_H)$ if and only if the following conditions hold:

(a) $\overline{\overline{V}}_\alpha < \overline{\overline{V}}_\beta \Leftrightarrow \overline{\overline{V}}{}_\alpha' < \overline{\overline{V}}{}_\beta'$ for all $\alpha, \beta \in 2^k$

(b) $\overline{\overline{V}}_\alpha < \aleph_0 \Leftrightarrow \overline{\overline{V}}{}_\alpha' < \aleph_0$ for all $\alpha \in 2^k$

(c) $\overline{\overline{V}}_\alpha = \overline{\overline{V}}{}_\alpha'$ for all $\alpha \in 2^k$ s.t. $\overline{\overline{V}}_\alpha < \aleph_0$

The first method of the proof of this theorem is by applying the game-theoretical characterization of elementary equivalence with respect to the language with the Henkin quantifier (see [7]).

The second method is by elimination of the Henkin quantifier. To show the decidability of the $L^1(Q_H)$ we eliminate the Henkin quantifier replacing it by the basis sentence of the form as in (a), (b), (c). This elimination is rather complicated and long. To find it in details see [8].

From theorem 1 we get the following corollaries:

COROLLARY 1. The Hanf and Skolem number of the language with the Henkin quantifier and with k unary predicates is $\aleph_{2k}$. Thus this number for $L^1(Q_H)$ is $\aleph_\omega$.

COROLLARY 2. Q_α is not definable by Q_H for each $\alpha > 0$.

COROLLARY 3 (Lindström). $L(Q_H)$ has no interpolation property.

Proof: To prove the last corollary let us take the following sentences

φ: $(Q_0 x)U(x) \wedge \neg(Q_c x)U(x)$
ψ: R is a linear discrete ordering with a first element $\rightarrow (\exists x)(Q_0 y)R(y, x)$

φ says that U is infinite but of power less than the power of the universe. ψ says that each linear discrete ordering with a first element has an infinite proper segment. It is clear that we have $\vDash \varphi \rightarrow \psi$.

But there is no χ such that $\vDash \varphi \rightarrow \chi$ and $\vDash \chi \rightarrow \psi$ and χ contains no predicate. Because such χ must say that the universe is uncountable.

The fact that the theory of unary relations is decidable in $L(Q_H)$ gives rise to the following question: What other theories are decidable in $L(Q_H)$? To investigate this problem let us consider the theory of one successor.

First, we observe that the standard model of the theory of one successor is characterizable in $L(Q_H)$. Let σ_0 be a first order system of axioms. Let us take the following sentence σ_1:

$$\neg(\exists t)(Q_H x_1 x_2 y_1 y_2)[(x_1 = x_2 \leftrightarrow y_1 = y_2) \wedge (x_1 = 0 \rightarrow y_1 = 0) \wedge (x_1 = Sx_2 \rightarrow y_1 = Sy_2) \wedge (x_1 = t \rightarrow y_1 \neq x_1)].$$

σ_1 says that there is, except for identity, no one-one function f such that $f(0) = 0$ which commutes with the successor function. So $\sigma_0 \wedge \sigma_1$ characterizes the system $\langle \omega, S \rangle$. To show the undecidability of this theory we first show the following fact.

FACT. Assume that the function $F\colon \omega^2 \rightarrow \omega$ is implicitly definable in $\langle \omega, S \rangle$ by a $L(Q_H)$ sentence of the form $(\forall xyz)\varphi(x, y, F(x, z), F(y, z), z)$. Then F is explicitly definable in $\langle \omega, S \rangle$.

Proof: Let's take $\phi(a, b, c)$: $(Q_H x_1 x_2 y_1 y_2)[(x_1 = x_2 \rightarrow y_1 = y_2) \wedge \varphi(x_1, x_2, y_1, y_2, b) \wedge (a = x_1 \rightarrow c = y_1)]$.
Thus ϕ explicitly defines the function F.

From this fact we get

COROLLARY 4. Addition and multiplication is definable in $\langle \omega, S \rangle$ by a formula of the language $L(Q_H)$.

For the formula which implicitly defines addition and multiplication we may indeed take their usual inductive definition.

COROLLARY 5. The theory of one successor is undecidable in $L(Q_H)$. The theory of well ordering is undecidable in $L(Q_H)$.

On the other hand we may show that the theory of addition and of one successor is decidable in the language with the Härtig quantifier. But the theory of one equivalence relation is undecidable.

Let us finally consider the connections of $L(Q_H)$ with the second order logic.

By the second order logic L^2 I mean the language in which there are two sorted variables: for individuals and for sets of individuals. Except the membership relation all other relations hold only between individuals.

Because of the definition of the Henkin quantifier the following fact is rather obvious.

FACT. Each elementary class of $L(Q_H)$ is pseudoelementary class of L^2. We denote it by $L(Q_H) \leqslant_{PC} L^2$.

I shall now show that in one sense the converse inequality also holds.

There is a natural translation of the formulas of the higher order language to first order formulas but in some extended language. The idea of it is as follows: We add one unary predicate, say U, and one binary predicate $\in$. We add to the usual axioms some special axioms of set theoretic type. For instance we will need the following three axioms:

$$(\forall xy)(x \in y \rightarrow U(y) \wedge \neg U(x))$$
$$(\forall xy)[U(x) \wedge U(y) \wedge (\forall z)(z \in x \leftrightarrow z \in y) \rightarrow x = y]$$
$$(\forall x)[U(x) \rightarrow (\exists y)(U(y) \wedge (\forall z)(\neg U(z) \rightarrow (z \in y \leftrightarrow z \notin x)))]$$

The translation $\varphi \rightarrow \varphi^*$ of the formulas of L^2 into the formulas of $L(Q_H)$ is defined by induction as follows

$\varphi^* = \varphi$ for atomic formulas

.

.

.

$$((\exists x)\varphi)^* = (\exists x)(\neg U(x) \wedge \varphi^*)$$
$$((\exists X)\varphi)^* = (\exists X)(U(X) \wedge \varphi^*)$$

Now let us take the following sentence σ_p:

$$(\exists t_0 t_1)(Q_H x_1 x_2 y_1 y_2)[t_0 \neq t_1 \wedge (y_1 = t_0 \vee y_1 = t_1) \wedge$$
$$(U(x_1) \rightarrow y_1 = t_0) \wedge \neg U(y_2) \wedge (x_1 = y_2 \wedge U(x_2) \rightarrow$$
$$(y_1 = t_0 \leftrightarrow x_1 \in x_2))]$$

LEMMA. $\langle A, U, E \rangle \models \sigma_p \Leftrightarrow$ there exists a $B \subseteq A - U$ such that there is no $a \in U$ such that for all $b \in A - U$ we have $b \in B \leftrightarrow \langle b, a \rangle \in E$.

Proof: ⇒ By the definition of satisfaction we have that there are $a_0, a_1 \in A$ and $f, g \in A^A$ such that $\mathrm{Rng}(f) \subseteq \{a_0, a_1\}$, $\mathrm{Rng}(g) \subseteq A - U$ and for all $b \in A$ and $a \in U$ if $b = g(a)$ then

$$f(b) = a_0 \Leftrightarrow \langle b, a \rangle \in E$$

Let us take $B = f^{-1}(\{a_1\})$. B is the needed set.

⇐ Assume that the right side holds. We have found two elements and two functions. Let us take $a_0, a_1 \in A$ such that $a_0 \neq a_1$. Let f be the function defined as follows

$$f(x) = \begin{cases} a_0 \text{ if } x \notin B \\ a_1 \text{ if } x \in B \end{cases}$$

Now, observe that for a given $a \in U$ there is a $b \in B$ such that $\langle b, a \rangle \notin E$ or there is a $b' \in (A - U) - B$ such that $\langle b', a \rangle \in E$. If the first case holds we put $g(a) = b$. If not, then the second case holds and we put $g(a) = b$. For $a \in A - U$ we let $g(a)$ be arbitrary. It is easy to check that f and g are functions which are needed.

From the lemma we have:

COROLLARY 6

If $\langle A, U, E, R \ldots \rangle \models \neg\sigma_p \wedge \varphi^*$ then $\langle A - U, R, \ldots \rangle \models \varphi$.
If $\langle B, R, \ldots \rangle \models \varphi$ then $\langle B \cup 2^B, 2^B, \in, R, \ldots \rangle \models \neg\sigma_p \wedge \varphi^*$.

So, it means that the second order logic is in some sense interpretable in $L(Q_H)$.

From the above also we get the following:

THEOREM 2. The Hanf number of $L(Q_H)$ is equal to the Hanf number of L^2. The Skolem number of $L(Q_H)$ is equal to the Skolem number of L^2.

Corollary 6 says that the notion of the power set is definable in $L(Q_H)$. Finally we can infer that the language with the Henkin quantifier is closer to the second order logic than we expected.

Department of Mathematics,
University of Warsaw

BIBLIOGRAPHY

[1] Fuhrken, G., 'Skolem - type normal forms for first order languages with a generalized quantifier', *Fund. Math.* **54** (1964), 291–302.

[2] Fuhrken, G., 'Languages with the added quantifier "there exist at least $\aleph_\alpha$', in *The Theory of Models*, ed. by J. Addison, L. Henkin and A. Tarski, North-Holland, Amsterdam, 1965, 121–131.

[3] Henkin, L., 'Some remarks on infinitely long formulas', in *Infinitistic Methods*, Pergamon Press, Oxford, 1961, 167–183.

[4] Hintikka, J., 'Quantifiers vs. Quantification Theory', *Dialectica* **27** (1973), 329–358.

[5] Issell, W., 'Semantische Untersuchungen über Quantoren I', *ZMLGM* **15** (1969), 353–358.

[6] Keisler, H. J., 'Logic with the quantifier "there exist uncountably many"', *Annals Mathematical Logic* **1** (1970), 1–91.

[7] Krynicki, M. 'Henkin quantifiers and decidability', in *Proceedings of the Symposia on Mathematical Logic, Oulu 1974 and Helsinki 1975*, ed. by S. Miethinen and J. Väänänen, Reports from the Department of Philosophy, University of Helsinki, 1977.

[8] Krynicki, M. and Lachlan, A. 'Semantical properties of language with the Henkin quantifier'. To appear.

[9] Mostowski, A., 'On a generalization of quantifiers', *Fund. Math.* **44** (1957), 12–36.

[10] Slomson, A., 'The monadic fragment of predicate calculus with the Chang quantifier and equality', in *Proceedings of the summer school in logic*, ed. by Löb (1968), 279–302.

[11] Väänänen, J., Lectures at the summer school of mathematical logic in Oulu (1974).

[12] Yasuhara, M., 'Incompleteness of *Lp* languages', *Fund. Math.* **66** (1970), 147–152.

JOUKO VÄÄNÄNEN

REMARKS ON FREE QUANTIFIER VARIABLES

We consider the logic L_q, introduced by Thomason [8], which is obtained from $L_{\omega\omega}$ by adding monadic quantifier variables. Thus if $A_1, \ldots, A_n$ are formulae of L_q, $x_1, \ldots, x_n$ individual variables and Q a *quantifier variable* of *type* $\langle n \rangle$, then

$$Qx_1 \ldots x_n A_1(x_1) \ldots A_n(x_n)$$

is a formula of L_q. An *interpretation* of L_q is obtained from an interpretation of $L_{\omega\omega}$ by assigning to the quantifier variables *monadic generalized quantifiers* in the sense of Lindström [6]. Thus a quantifier variable Q of type $\langle n \rangle$ is assigned in a given domain I a class $\mathcal{Q}$ of structures of the form $\langle I, X_1, \ldots, X_n \rangle$ such that $X_i \subseteq I (i = 1 \ldots n)$ and every isomorphic copy of a structure in $\mathcal{Q}$ is also in $\mathcal{Q}$. In this assignment

$$\langle I, R_1, \ldots, R_m \rangle \models Qx_1 \ldots x_n A_1(x_1, \mathbf{y}) \ldots A_n(x_n, \mathbf{y})[\mathbf{a}]$$

iff

$$\langle I, X_1, \ldots, X_n \rangle \in \mathcal{Q}$$

where

$$X_i = \{b \in I \mid \langle I, R_1, \ldots, R_m \rangle \models A_i(x_i, \mathbf{y})[b\mathbf{a}]\}.$$

Strictly speaking, Thomason's logic which we denote by L_q' has quantified expressions of the form $QxA(x)$ only, but this restriction is inessential for all our considerations, as we shall see later.

Thomason asked whether his logic is axiomatizable and what are the Hanf and Löwenheim numbers of L_q. Yasuhara [9] proved that L_q is not axiomatizable by characterizing the natural numbers. He also remarked that the successor powers are characterizable whence the Hanf and Löwenheim numbers cannot exist in the strong sense. We shall improve these remarks by demonstrating that L_q shares many of the problematic features of the second order logic.

J. Hintikka, I. Niiniluoto, and E. Saarinen (eds.), Essays on Mathematical and Philosophical Logic, 267–272.

We start by relating L_q to a logic with just one quantifier constant instead of quantifier variables. Härtig [5] introduced the quantifier

$$Q_I xyA(x)B(y) \Leftrightarrow_{df} \text{card}\{x|A(x)\} = \text{card}\{x|B(x)\}.$$

If L is a similarity type, we denote by LQ_I the logic obtained from $L_{\omega\omega}$ by adding the quantifier Q_I. The same convention is applied to L_q. Thus if $K \subseteq L$, then K_q consists of those L_q-sentences whose non-logical symbols are in K.

There is a simple translation of LQ_I to K_q for some $K \supseteq L$. To obtain this translation we let L be a finite similarity type and ϕ an LQ_I-sentence. Let Q, F, U, $<$, λ, S and I be new symbols and $K = L \cup \{Q, F, U, <, \lambda, S, I\}$. For any formula $\pi(x)$ let $\Phi(u, \pi(x))$ be a formula saying that for some t: $F(t, \cdot, \cdot)$ defines a bijection between $\pi(\cdot)$ and $U(u, \cdot)$. Now we construct ψ' for any subformula ψ of ϕ by replacing $Q_I xy\pi_1(x)\pi_2(y)$ repeatedly by the formula $(Eu)(\Phi(u, \pi_1(x)) \;\&\; \Phi(u, \pi_2(x)))$. Finally let θ be the conjunction of the following K_q-sentences:

'$<$ *is a linear ordering with the last element* λ',
'S *is a successor function and* $(Ux,\cdot) \subset U(y,\cdot)$ *if* $x<y$',
$(x)[U(\lambda, x) \;\&\; \{x \neq \lambda \Rightarrow \neg(QyU(x, y) \Leftrightarrow QyU(Sx, y)) \;\&$
$\neg(QyU(x, y) \Leftrightarrow QyU(\lambda, y))\}]$,
$(z_1 \ldots z_n)(Eu)(\Phi(u, \psi'(x, z_1, \ldots, z_n))$ *for all subformulae* ψ *of* ϕ.

Now one can prove that the following conditions are equivalent for any L-structure A:

(i) $A \models \phi$,
(ii) *There is a K-expansion of A which satisfies* $\theta \;\&\; \phi'$,
(iii) *Any K-expansion of A satisfies* $\theta \Rightarrow \phi'$.

Using the Δ-operation of e.g. Feferman [3] we can write this equivalence as $LQ_I \subseteq \Delta(L_q)$. We understand $\Delta(L_q)$ as a logic which does not have quantifier variables although they are allowed as extra symbols in Δ-definitions.

A similar translation of L_q to KQ_I is defined as follows. We consider the logic L'_q at first. Let L be a finite similarity type and ϕ an L'_q-sentence. Let $K = L \cup \{F, U, <, \lambda, I, T\}$. For any subformula ψ of ϕ a formula ψ' is constructed by replacing $Qx\pi(x)$ repeatedly by $(Etu)(T(t, u) \;\&\; \Phi(t, \pi(x)) \;\&\; \Phi(u, \neg\pi(x)))$. Let θ be the conjunction of the following KQ_I-sentences:

'$<$ *is a linear ordering on* I *with the last element* λ',
$(xy)(x < y \Rightarrow (U(\lambda, x) \,\&\, (z)(U(x, z) \Rightarrow U(y, z)) \,\&$
$\neg Q_I tuU(x, t)U(y, u)))$,
$(tu)(T(t, u) \Rightarrow \Phi(t, \neg U(u, x)))$,
$(z_1 \ldots z_n)(Etu)(\Phi(t, \psi'(x, z_1, \ldots, z_n) \,\&\, \Phi(u, \neg\psi'(x, z_1, \ldots, z_n)))$
for all subformulae $\psi(x, z_1, \ldots, z_n)$ *of* ϕ.

Now the following conditions are equivalent for any L-structure A:

(i) $(A, \mathscr{Q}) \models \phi$ for some generalized quantifier $\mathscr{Q}$,
(ii) There is a K-expansion of A which satisfies $\theta \,\&\, \phi'$.

So we have proved $\Delta(L_q') \subseteq \Delta(LQ_I)$. The construction easily generalizes to $\Delta(L_q) \subseteq \Delta(LQ_I)$. We have obtained the

THEOREM. $\Delta(L_q) = \Delta(L_q') = \Delta(LQ_I)$.

COROLLARY. *The decision problem of* L_q *is recursively isomorphic to that of* LQ_I.
L_q *and* LQ_I *have the same Hanf and Löwenheim numbers.*

Let us denote the decision problem of L_q by V_q and the recursive bijection of the decision problem of LQ_I onto V_q by f. It follows from Yasuhara's characterization of natural numbers in L_q that V_q is not hyperarithmetical. Lindström [6] remarked that wellordering is (relatively) characterizable in LQ_I, whence V_q is not Σ^1_2. A stronger result is obtained as follows. By Shoenfield's absoluteness Lemma $\Sigma^1_2 \cup \Pi^1_2$-formulae are absolute for transitive models of ZF containing all countable ordinals. So if $\phi(x)$ is a $\Sigma^1_2 \cup \Pi^1_2$-formula, then $\phi(n)$ holds if and only if $f(\ulcorner\theta \Rightarrow \phi(n)^{(M)}\urcorner) \in V_q$, where θ says that M is a well founded uncountable model of a suitable finite set theory. It follows that $\Sigma^1_2 \cup \Pi^1_2$-sets are reducible to V_q. The second part of the following theorem is proved similarily

THEOREM. V_q *is not* $\Sigma^1_2 \cup \Pi^1_2$. *If* $\mathscr{P}\omega \subseteq L$, *then* V_q *is not* Σ^1_n *for any* $n < \omega$.

If we assume $V = L$, and θ is a sentence of LQ_I saying that M is a well founded model of a suitable finite set theory, in which (*the power set of* $A(\cdot))^M$ is $B(\cdot)$ and the power of which is bigger than that of $A(\cdot)$, then θ characterizes structures of the form $\langle N, \in, A, B\rangle$ where $B(\cdot)$ is the power set of $A(\cdot)$. This allows us to translate the full second order logic into LQ_I. So we obtain the following theorem, suggested to us by Peter Aczel:

THEOREM. *If* $V = L$, *then* $\Delta(L_q) \supset$ *the full second order logic.*

This combined with a result of Montague [7] gives:

COROLLARY. *If* $V = L$, *then* V_q *is not* Σ_n^a *for any* $n < \omega$ *and any* a *which is definable in a finite level of the hierarchy of higher order logics.*

The proof of the theorem gives a provably recursive g such that the following is a theorem of ZF + $V = L$: 'ϕ *is valid in the second order logic if and only if* $g(\phi) \in V_q$'. There is no provably recursive g such that this is a theorem of ZFC + GCH. If $0^{\#}$ exists, then $V \neq L$, in fact then $V_q \notin L$, but it is still consistent that $\Delta(L_q) \supset$ *the second order logic*. Similar results hold for other assumptions, e.g. for the existence of a measurable cardinal. The details of these remarks will not be pursued here.

The above theorem can actually be proved in ZF_1, a set theory considered by Barwise [1] consisting of Zermelo's set theory with $\Sigma_1(\mathscr{P})$-*replacement* ($\mathscr{P}$ is the power set operation). This allows us to improve Fuhrken's [4] result about the Hanf and Löwenheim numbers of LQ_I. For '$V = L$' is a Π_2-axiom and Barwise proves that the existence of the Hanf number of the second order logic is not provable in ZF_1 + *any* $\Pi_3(\mathscr{P})$-*axiom*. Barwise also proves that no $\Sigma_1(\mathscr{P})$-formula can define in ZF the Löwenheim number of the second order logic. Such a $\Pi_1(\mathscr{P})$-formula exists trivially. So we have the

COROLLARY. *The existence of the Hanf number of* L_q(*or* LQ_I) *is not provable in* ZF_1, *or even in* $ZF_1 + V = L$. *There is no* $\Sigma_1(\mathscr{P})$-*formula which defines in* ZF *the Löwenheim number of* L_q(*or* LQ_I).

It is relatively clear that the *interpolation theorem* fails for L_q. We shall see that it fails also when restricted to quantifier variables. Let A, B, C, D be unary, $<$ and $\prec$ binary and F a ternary predicate symbol. Let ϕ_0 be a sentence which says (using F): $A \subseteq B$, $D \subseteq C$, $B = -C$, card A = card D, $\langle B, < \rangle$ and $\langle C, \prec \rangle$ are linearly ordered sets of power $\leqslant$ (card $A)^+$. Let $\phi(Q)$ be $(\phi_0 \,\&\, QxB(x) \,\&\, \neg QxC(x))$ and let $\psi(Q')$ be the sentence $(Q'xA(x) \Rightarrow (Q'xB(x) \vee Q'xC(x)))$. Then $\models \phi(Q) \Rightarrow \psi(Q')$ but there is no θ in which Q and Q' do not occur, which satisfies both $\models \phi(Q) \Rightarrow \theta$ and $\models \theta \Rightarrow \psi(Q')$.

Finally we consider some systems related to L_q. Let us denote by $L_q(\mathscr{K})$ the system which is obtained from L_q by restricting the range of the quantifier variables to $\mathscr{K}$, and by $L_q^-(\mathscr{K})$ the system obtained from $L_q(\mathscr{K})$ by dropping out the usual existential and universal quantifiers. Let $\mathscr{K}_0 = L_{\omega\omega}$-definable quantifiers. If $\mathscr{K} \cap \mathscr{K}_0$ is infinite and the Löwenheim number of $L_q(\mathscr{K})$ is ω (as e.g. in case $\mathscr{K}_0 = \mathscr{K}$) then $L_q(\mathscr{K})$ is not axiomatizable. $L_q(\mathscr{K}_0)$ has, however, trivially a Π_2^0-axiomatization. Let $\mathscr{K}_1 = \{Q_\alpha | \alpha \geqslant 0\}$. Bell & Slomson [2] proved that $L_q(\mathscr{K}_1)$ is axiomatizable but Yasuhara [10] showed that even $L_q^-(\mathscr{K}_1)$ is not decidable. Let $\mathscr{K}_2$ be the class of monotone quantifiers. $L_q^-(\mathscr{K}_2)$ is not axiomatizable, in fact, $\Delta(L_q) \subseteq \Delta(L_q^-(\mathscr{K}_2))$.

The author is indebted to his supervisor Peter Aczel for many helpful discussions during the preparation of this paper.

Department of Mathematics,
University of Manchester

Note added July 1977: The consistency of 'V_q is Δ_3^1 & the Löwenheim number of L_q is $< 2^\omega$' and other new results about L_q are announced in the author's abstract submitted to the Logic Colloquium '77.

REFERENCES

[1] J. Barwise, 'The Hanf number of second order logic', *J. Symb. Logic* **37** (1972), 588–594.

[2] J. L. Bell and A. B. Slomson, *Models and Ultraproducts: An Introduction*, North-Holland, Amsterdam, 1969, pp. ix + 322.

[3] S. Feferman, 'Applications of many sorted interpolation theorems', *Proceedings of Symposia in Pure Mathematics* XXV, AMS, Providence, Rhode Island, 1974, 205–224.

[4] G. Fuhrken, 'A remark on the Härtig-quantifier', *Z. Math. Logik u. Grundl. Math.* **18** (1972), 227–228.

[5] K. Härtig, 'Uber einen Quantifikator mit zwei Wirkungsbereichen', *Colloquy on the Foundations of Mathematics, Mathematical Machines and their Applications, Tihany (Hungary) 1964*, Budapest/Paris, 1965, 31–36.

[6] P. Lindström, 'First order logic with generalised quantifiers', *Theoria* **32** (1966), 186 195.

[7] R. Montague, 'Reductions of higher order logic', in J. W. Addison, L. Henkin and A. Tarski, eds., *The Theory of Models*, North-Holland, Amsterdam, 1967, 265–273.

[8] R. Thomason, 'A system of logic with free variables ranging over quantifiers (abstract)', *J. Symb. Logic* **31** (1966), 700.
[9] M. Yasuhara, 'Incompleteness of L_p languages', *Fund. Math.* **66** (1969), 147–152.
[10] M. Yasuhara, 'Syntactical and semantical properties of generalised quantifiers', *J. Symb. Logic* **31** (1966), 617–632.

PART V

RECURSION THEORY

DAG NORMANN

RECURSION IN ^3E AND A SPLITTING THEOREM

0. INTRODUCTION

In this paper we will make a survey of some results concerning recursion in a normal type-3-object as defined by Kleene [6]. We will use an alternative definition originally described in Normann [10] where we extend recursion theory to arbitrary sets.

The new theorem of the paper will be a splitting theorem for r.e. degrees of type-3-functions, and part B will contain a detailed proof of that theorem.

In part A we will look at some of the background material. Here we will skip details both in proofs and definitions. It is the main steps and ideas we want to give the reader some impression of.

By restricting ourselves to type-3-objects we omit certain difficulties in the theory, but also some of the finer points are lost.

Sufficient background for the proof in part B is found in Normann [10]. As an introduction to recursion in normal objects Moldestad [8] is recommended. For a general introduction to recursion theory, see Fenstad [2] or his forthcoming book [3]. Papers with results relevant to us are among others Harrington [5], MacQueen [7] and Sacks [11] and [12]. Fenstad and Hinman [4] contains several papers of general interest, and a list of further references.

A. A SURVEY

1. *E*-RECURSION

We will define a family of partially recursive functions, taking sets as arguments and sets as values. We use the schemes for rudimentary functions, given e.g. in Devlin [1], Normann [9] or [10]. To each scheme we give an index, a natural number e, telling how the rudimentary function is composed.

J. Hintikka, I. Niiniluoto, and E. Saarinen (eds.), Essays on Mathematical and Philosophical Logic, 275–285.

Examples from [10]:

(i) $f(x_1, \ldots, x_n) = x_i$ $\quad e = \langle 1, n, i \rangle$

(iv) $f(x_1, \ldots, x_n) \simeq \bigcup_{y \in x_1} h(y, x_2, \ldots, x_n)$ $\quad e = \langle 4, n, e' \rangle$ where e' is an index for h.

If f has an index e, we write $f = \{e\}$.

In addition to the schemes for rudimentary functions we give one scheme of diagonalization:

(vii) $f(e_1, x_1, \ldots, x_n, y_1, \ldots, y_m) \simeq \{e_1\}(x_1, \ldots, x_n)$ $e = \langle 7, n, m \rangle$

Here $\{e_1\}(x_1, \ldots, x_n)$ may not be defined (e.g. if e_1 is not an index), and schemes (i)–(vii) give all together a definition of an indexed class of partial functions, a class having the normal properties of a recursion theory.

The set of computation tuples $\langle e, \mathbf{x}, y \rangle$ such that $\{e\}(\mathbf{x}) \simeq y$ is inductively defined by the schemes (i)–(vii). From that induction we obtain a natural concept of the length $|e, \mathbf{x}|$ of the computation $\{e\}(\mathbf{x})$.

The following simple observation will give the theory the strength we need; let '$\{e\}(\mathbf{x})\downarrow$' mean '$\{e\}$ is defined on $\mathbf{x}$':

LEMMA 1. In E-recursion there is an index e such that for arbitrary $x, e_1, \mathbf{x}$

$$\{e\}(x, e_1, \mathbf{x}) \simeq \begin{cases} 0 \text{ if } \forall y \in x \{e_1\}(y, \mathbf{x}) \simeq 0 \\ 1 \text{ if } \forall y \in x \{e_1\}(y, \mathbf{x}) \downarrow \\ \quad \text{and } \exists y \in x \{e_1\}(y, \mathbf{x}) \neq 0 \end{cases}$$

Proof: We may assume that e_1 takes values $0 = \emptyset$ and $1 = \{\emptyset\}$ only. Let

$$\{e\}(x, e_1, \mathbf{x}) = \bigcup_{y \in x} \{e_1\}(y, \mathbf{x})$$

Remark: A consequence of Lemma 1 is that we may, in E-recursion, compare the lengths of two computations by a recursive function.

DEFINITION: Let X be a set, $\mathbf{x}$ an arbitrary sequence of sets.

(i) $Y \subseteq X$ is *E-recursive in* $\mathbf{x}$ *over* X if there is an index e

such that

$$\forall x \in X(\{e\}(x, \mathbf{x}) = 0 \text{ if } x \in Y \text{ and } \{e\}(x, \mathbf{x}) = 1 \text{ if } x \notin Y)$$

(ii) $Y \subseteq X$ is *semirecursive in* $\mathbf{x}$ *over* X if there is an index e

such that

$$\forall x \in X(\{e\}(x, \mathbf{x}) = 0 \Leftrightarrow x \in Y)$$

Another consequence of Lemma 1 is that we may recursively select an element from any nonempty semirecursive subset of ω (Gandy's selection).

E-recursion is relativized to $E(R)$-recursion by using the scheme

(vi) $\quad f(x_1, \ldots, x_n) \simeq x_i \cap R \qquad e = \langle 6, n, i \rangle$

We then write $\{e\}^R(x)$.

2. CONNECTION WITH RECURSION IN ^3E

Let $I = {}^\omega\omega, f\colon {}^\omega\omega \to \omega$. We define

$$^3E(f) = \begin{cases} 0 \text{ if } \exists x f(x) = 0 \\ 1 \text{ if } \forall x f(x) \neq 0 \end{cases}$$

Here f is assumed to be total.

It follows by Lemma 1 that E-recursion with I as an argument will be as powerful as recursion in ^3E. The two theories are actually equivalent, i.e.:

Given x, $\mathbf{x}$, the subsets of x recursive and semirecursive in $\mathbf{x}$, I in E-recursion, are exactly the subsets of x recursive and semirecursive in $\mathbf{x}$, ^3E, when $tp(\mathbf{x}) \leq 2$, $x \subseteq I \cup {}^I\omega$.

Thus, instead of dealing with recursion in normal type-3-objects, we may often work with E-recursion in I relative to arbitrary relations R.

Any type-3 functional is a relation. General methods for transforming results about E-recursion relative to relations into results about recursion in normal type-3 objects are developed in Normann [10].

3. THE SPECTRUM OF A RELATION

Let R be a relation. We define $Spec(R)$ = the spectrum of R over I as follows:

$$M_a(R) = \{x; \exists e(x = \{e\}^R(a, I))\}$$
$$\text{Spec}(R) = \langle M_a(R) \rangle_{a \in I}$$

The spectrum is an admissible-like structure, each $M_a(R)$ is non-transitive, countable and rudimentary closed. The spectrum is connected by the following Σ^*-collection principle:

Let $a \in I$, φ be a Δ_0-formula with parameters in M_a.

Assume

$$\forall b \in I \exists x \in M_{\langle a,b \rangle} \varphi(b, x)$$

Then

$$\exists y \in M_a \forall b \in I \exists x \in y \cap M_{\langle a,b \rangle} \varphi(b, x)$$

This is a consequence of Gandy's selection in E-recursion.

THEOREM 2 (Harrington [5]).
Let $a \in I$, let

$$c = \{\langle e_1, e_2 \rangle; \{e_1\}(a, I)\downarrow \ \&\ \{e_2\}(a, I)\downarrow \text{ and the first computation is shorter than the second}\}$$

(c is a complete subset of ω semirecursive in a, I)

Then M_a is a Σ_1-elementary substructure of $M_{\langle a,c \rangle}$. (Σ_1-reflection from $M_{\langle a,c \rangle}$ to M_a)

We will give an outline of the proof of this theorem:

Let φ be $\Delta_0(a)$. First we prove

$$\exists x \in M_{a,c} \varphi(x) \Rightarrow \exists y \in M_a \exists x \in y \varphi(x)$$

Let c_α be those tuples in c such that both computations are shorter than α.

Let WO be the codes for countable wellorderings (WO $\in \Pi^1_1$). Then

$$\forall w \in \mathrm{WO} \exists x \in M_{a,w} \begin{cases} \varphi(x) \text{ (if } \|w\| \geq \|c\|) \\ \text{or } x \text{ is an ordinal and} \\ \|w\| = \|c_x\| \text{ (if } \|w\| < \|c\|) \end{cases}$$

Let y be a collection of such x's, $y \in M_a$. Choosing w such that $\|w\| = \|c\|$ we see that $\exists x \in y \varphi(x)$. Now, by letting $\alpha = \mathrm{Sup}(\mathrm{On} \cap M_a)$ we see

$$M_{\langle a,c \rangle} \models \exists \alpha \exists x \in M_{a,c_\alpha} \varphi(x)$$

Let $y \in M_a$ be such that $\exists \alpha$, $x \in y (x \in M_{\langle a,c_\alpha \rangle}\ \&\ \varphi(x))$. In this case c_α is recursive in a, so $M_{a,c_\alpha} = M_a$. Thus $\exists x \in M_a \varphi(x)$. □

DEFINITION. Let $\mathrm{Spec}(R) = \langle M_a(R) \rangle_{a \in I}$. We define two natural domains for the spectrum:

$$M(R) = \bigcup_{a \in I} M_a(R),\ {}^1M(R)\{\langle a, x \rangle;\ x \in M_a(R)\}$$

It is semirecursive subsets over ${}^1M$ that are of interest when we translate results to results about Kleene-recursion.

4. E-RECURSION WHEN $V = L$

Now assume $V = L$. Let $\langle M_a\rangle_{a\in I} = \langle M_a(\emptyset)\rangle_{a\in I}$. Then $M = L_\beta$ for some non-admissible β.

There is a Δ^1_2-well-ordering $<$ of I of length $\aleph_1$. This is E-recursive in I.

For each a, $\{b; b \leqslant a\}$ is countable, and by $<$ we may pick a recursive enumeration of this set. This shows that $b < a \Rightarrow M_b \subseteq M_a$, and $M_a \cap I$ is an initial segment of I. Let a' (*a-jump*) be the minimal element of I not in M_a. By Theorem 2 it follows that

M_a is a Σ_1-elementary substructure of $M_{a'}$.

For each $a \in I$ we may well-order the set of convergent computations $\{e\}(a, I)$ by

$\{e_1\}(a, I) < \{e_2\}(a, I) \Leftrightarrow$ the first computation is shorter than the second, or they have the same length, but $e_1 < e_2$.

We then obtain a wellordering $<_a$ on M_a.
Define the following wellordering on ${}^1M$

$$\langle a, x\rangle <^2 \langle b, y\rangle \text{ if } a < b \text{ or } a = b \text{ and } x <_a y.$$

This wellordering is not recursive over ${}^1M$, so we will use recursive approximations:
Let $M^\alpha_a = \{\{e\}(a, I)$; the computation is shorter than $\alpha\}$

$$M^\alpha = \bigcup_{a\in I} M^\alpha_a,\ {}^1M^\alpha_a = \{\langle a, x\rangle : x \in M^\alpha\}$$

$<_\alpha$ is defined as $<^2$ relativized to ${}^1M^\alpha$.
Let $\| \|^2$ and $\| \|_\alpha$ be the associated norms. Given $\langle a, x\rangle \in {}^1M$ there will be an ordinal α in $M_a{}'$ such that for all $\beta \geq \alpha \|\langle a, x\rangle\|_\beta = \|\langle a, x\rangle\|^2$.

5. EXTENDED DEGREES OF RELATIONS

DEFINITION. Let M, ${}^1M$, $\langle M_a\rangle_{a\in I}$ be as above.
A subset X of ${}^1M$ is *acceptable* if
(i) $M = M(X)$
(ii) $\forall a \in I\ M_a$ is a Σ_1-elementary substructure of $M_a(X)$.

If R and Q are two relations, we say that R is *extended recursive* in Q if there is an $a \in I$ such that R is E-recursive in a relative to Q.

In Normann [10] priority-arguments are applied to prove:

THEOREM 3 ($V = L$).

(a) There are two semirecursive, acceptable subsets of ${}^1M$ such that neither is extended recursive in the other.

(b) There are two semi-recursive, acceptable, not extended recursive subsets of ${}^1M$, X and Y, such that for any $Z \subseteq {}^1M$, if Z is extended recursive in both X and Y, then Z is extended recursive.

Our task is now to prove that there is no minimal, acceptable semi-recursive and not extended recursive extended degree.

B. THE SPLITTING THEOREM

6. SOME NOTATION

If $R \subseteq {}^1M$ is semi-recursive over ${}^1M$, then

(a) There is an index e such that

$$x \in R \Leftrightarrow \{e\}(x, I)\downarrow$$

(b) There is a Δ_0-formula φ such that

$$\langle a, x\rangle \in R \Leftrightarrow \exists y \in M_a \varphi(a, x, y)$$

Let $x \in R^\alpha \Leftrightarrow (x \in {}^1M^\alpha$ and the computation $\{e\}(x, I)$ is shorter than $\alpha)$.

To any index e and $a \in I$, let $J_{e,a}$ be the partial set defined by

$$x \in J_{e,a} \text{ if } \{e\}(x, a, I) \simeq 0$$
$$x \notin J_{e,a} \text{ if } \{e\}(x, a, I) \simeq 1$$

whenever this makes sense.

We define the sets $J^\alpha_{e,a}$ as we defined R^α.

7. THE PROOF

LEMMA 4 (from Shoenfield [13]).

Let $X \subseteq {}^1M$ be acceptable. Let A, B be two disjoint sets semi-recursive in I over ${}^1M$ such that $X = A \cup B$. Then:

(i) X is semi-recursive in I over ^1M;

(ii) A and B are both recursive in I over ^1M relative to X;

(iii) A and B are acceptable.

Proof: (i) Let $x \in A \Leftrightarrow \{e_1\}(x, I) \simeq 0$; $x \in B \Leftrightarrow \{e_2\}(x, I) \simeq 0$;
Then $x \in X \Leftrightarrow \exists e \in \{e_1, e_2\}(\{e\}(x, I) \simeq 0)$.
By the Gandy selection operator, X is semi-recursive.

(ii) We see that $^1M \setminus A = (^1M \setminus X) \cup B$ will be semi-recursive in X. A is semi-recursive, so A is recursive over ^1M in X. The same argument holds for B.

(iii) is an immediate consequence of (ii).

THEOREM 5 ($V = L$).

Let $X \subseteq {}^1M$ be semi-recursive in I and not extended recursive. Assume that X is acceptable. Then there are subsets A and B of ^1M semi-recursive in I such that:

(i) $A \cup B = X$; (iii) X is not extended recursive in B;

(ii) $A \cap B = \emptyset$; (iv) X is not extended recursive in A.

Proof: We will construct the sets A^ξ and B^ξ by induction on ξ, and we will try to meet the following conditions:

1.n.a. $X \neq J_{n,a}(A)$ (A-conditions)

2.n.a. $X \neq J_{n,a}(B)$ (B-conditions)

In addition we will arrange the construction so that X is the disjoint union of A and B.

The conditions are ordered antilexicographically and are denoted by the associated ordinals $\nu < \aleph_1$.

The construction is by induction on the *position* $\xi = \langle \nu, \sigma \rangle$ where ν is a condition, σ, the *stage*, is an ordinal and the pairs are ordered antilexicographically. We identify a position with the associated ordinal.

The construction.

Let $A^o = \emptyset$, $B^o = \emptyset$, and at limit-positions λ,

$$A^\lambda = \bigcup_{\xi < \lambda} A^\xi, \; B^\lambda = \bigcup_{\xi < \lambda} B^\xi.$$

Through the construction we will create requirements for a condition ν consisting of elements to be kept out of A resp B to make the construction work. A requirement x for an A-condition is active at position ξ if $x \cap A^\xi = \emptyset$.

The requirements will be of three classes according to their power to meet the condition.

Now we will describe the successor step of the construction:

Position $\xi = \langle \nu, \sigma \rangle$.

1. Let y be element no ν in $<_\sigma$. If $y \notin X^\sigma$, or if $y \in A^\xi \cup B^\xi$, proceed to 2.

If $y \in X^\sigma \setminus (A^\xi \cup B^\xi)$ and y is not an element of any requirement active at position ξ, let $A^{\xi+1} = A^\xi \cup \{y\}$, $B^{\xi+1} = B^\xi$ and proceed to 2.

If y is in some active requirement, let ν_0 be the least condition such that y is in an active requirement for ν_0.

If ν_0 is an A-requirement, put y into $B^{\xi+1}$, if ν_0 is a B-requirement, put y into $A^{\xi+1}$. Then proceed to 2.

2. Assume $\nu = 1,n,a$ (2.n.a is analogous).

Ask: Is there an active requirement for ν of 1st class? If yes, proceed to the next position. If no:

Ask: $\exists y \in X^\sigma (y \notin J^\sigma_{n,a}(A^\xi))$? If yes, pick the least such y in $<_\sigma$ and let $x = {}^1M^\sigma \setminus A^\xi$ be a requirement for ν of 1st class. (If $x \cap A = \emptyset$, then $y \in X^\sigma$ and $y \notin J_{n,a}(A)$). If no:

Ask: Is there an active requirement z of 2nd class for ν such that some argument of z is not in X^σ? (We define arguments later.) If yes, proceed to the next position. If no, let $\sigma_1 \leq \sigma$ be the largest ordinal such that $J^\sigma_{n,a}(A^\xi)$ is total on ${}^1M^{\sigma_1}$. Let σ_2 be the least ordinal such that $J^{\sigma_1}_{n,a}(A^\xi)$ is total on ${}^1M^{\sigma_1}$, and:

Ask: Is there some $y \in {}^1M^{\sigma_1}$ such that $y \in J^\sigma_{n,a}(A^\xi)$ and $y \notin X^\sigma$? If yes, let all such y be arguments of a requirement x for ν of 2nd class consisting of ${}^1M^{\sigma_2} \setminus A^\xi$ (If $x \cap A = \emptyset$, then $J_{n,a}(A)$ is total on ${}^1M^{\sigma_1}$, and if in addition some argument of x is not in X, we obtain $X \neq J_{n,a}(A)$). If no, let the x described above be a requirement for ν of 3rd class.

Let

$$A = \cup A^\xi,\ B = \cup B^\xi.$$

This ends the construction.

CLAIM 1.

(i) $X = A \cup B$ and $A \cap B = \emptyset$

(ii) A and B are semi-recursive

Proof: (i) Let $y \in X$. Let σ_1 be such that $y \in X^{\sigma_1}$. Let $\sigma_2 \geq \sigma_1$ be such that $\sigma \geq \sigma_2 \Rightarrow \|y\|_\sigma = \|y\|_{\sigma_2}$. Let $\nu = \|y\|_{\sigma_2}$. Then $y \in A^{\langle \nu, \sigma_2 \rangle + 1} \cup B^{\langle \nu, \sigma_2 \rangle + 1}$. $A \cap B = \emptyset$ is clear from the construction, part 1.

(ii) Clearly A^ξ and B^ξ are recursive in ξ. If $y \in X$, $y = \langle a, x \rangle$, $\sigma_1 \in M_a$ and $\sigma_2 \in M_a'$ (see proof of (i)). Thus $\langle a, x \rangle \in A \Leftrightarrow \langle a, x \rangle \in X$ and $\exists \xi \in M_a'(y \in A^\xi) \Leftrightarrow \langle a, x \rangle \in X$ and $\exists \xi \in M_a(y \in A^\xi)$.

The last equivalence follows from Σ_1-reflection. By Gandy selection this defines A to be semi-recursive. Also B is semi-recursive by the same argument.

CLAIM 2. Let ν be a condition.

$$\{x;\ x \text{ is a requirement for } \nu\} \in M.$$

Proof: by induction on ν. Assume $\nu =$ l.n.a. First note that we cannot injure a ν-requirement with an element y unless for some $\nu_1 < \nu$, y is an element of some active ν_1-requirement. By the induction hypothesis and since M is closed under countable subsets,

$$Y = \cup\{x;\ x \text{ is a requirement for a condition } < \nu\} \in M.$$

Since X is acceptable, $X \cap Y \in M$, and thus

$$\exists \sigma \forall y \in X \cap Y(y \in A^{\langle o, \sigma \rangle} \cup B^{\langle o, \sigma \rangle})$$

Thus, after stage σ we will not injure any ν-requirement.

We must prove that we will stop producing new ones. If there at some stage $\geq \sigma$ is an active ν-requirement of 1st class, we will by construction stop producing new requirements. If we never stop producing requirements of 2nd and 3rd class, $J_{a,n}(A)$ will be total on ^1M, and any argument of any 2nd class requirement will be elements of X. To obtain a contradiction, assume that we never stop producing new ν-requirements.

SUBCLAIM. Let $\sigma_0 \geq \sigma$ and let $b \in$ I be such that $\nu, \sigma_0, \sigma \in M_b$. Then there is a $\sigma_1 \in M_b$ such that $\sigma_1 > \sigma_0$ and we create a new requirement at stage σ_1.

Proof: By Σ^*-collection there will be a $\sigma_2 > \sigma_0$ such that all arguments of any 2nd class requirement for ν active at stage σ_0 are in X^{σ_1}.

Now $J_{n,a}(A)$ is total and A is acceptable (by Lemma 4). Thus there is a stage σ_3 in $M_b(A)$ such that $J_{n,a}^{\sigma_3}(A)$ is total on $^1M^{\sigma_2}$. Let Z be all sets occurring in one of the computations $J_{n,a}^{\sigma_3}(A)(y)$, where $y \in {}^1M^{\sigma_2} \in M_b$. Z will be an element of $M_b(A)$. But then there is a $\sigma_4 \in M_b(A)$ such that

$Z \cap A \subseteq A^{\langle \nu, \sigma_4 \rangle}$, by Σ^*-collection over spec(A). We may assume $\sigma_4 > \sigma_3$. Then

$$J^{\sigma_4}_{n,a}(A^{\langle \nu, \sigma_4 \rangle}) \text{ is total on } {}^1M^{\sigma_2}.$$

Since M_b is a Σ_1-elementary substructure of $M_b(A)$, there is a $\sigma_1 > \sigma_2$, $\sigma_1 \in M_b$ such that

$$J^{1}_{n,a}(A^{\langle \nu, \sigma_1 \rangle}) \text{ is total on } {}^1M^{\sigma_2}.$$

If we have not produced any requirement since stage σ_0 we will then produce one at position $\langle \nu, \sigma_1 \rangle$.

ANOTHER SUBCLAIM. Let ν, $\sigma \in M_b$. For any $\langle c, x \rangle \in {}^1M$: $\langle c, x \rangle \notin X \Leftrightarrow \exists \sigma_0 \in M_{\langle c,b \rangle}$ (we create a requirement at position $\langle \nu, \sigma_0 \rangle$ and $\langle c, x \rangle \notin J_{n,a}(A)$ is protected by the requirement and $\sigma_0 \geq \sigma$).

Proof: By subclaim there will be a $\sigma_0 \in M_{\langle c,b \rangle}$ such that '$\langle c, x \rangle \in J_{n,a}(A)$' or '$\langle c, x \rangle \notin J_{n,a}(A)$' will be protected by a requirement created at stage σ_0.

If $\langle c, x \rangle \in J^{\sigma_0}_{n,a}(A^{\langle \nu, \sigma_0 \rangle})$, $\langle c, x \rangle$ will either be in X^{σ_0} or an argument for the requirement. In any case $\langle c, x \rangle \in X$, so $\Rightarrow$ holds.

Now, if we create a requirement at σ_0 and '$\langle c, x \rangle \notin J_{n,a}(A)$' is protected, this will never be injured. But if then $y \in X$, we will later create a 1st class requirement. This is against the assumption, and $\Leftarrow$ is verified.

From the other subclaim it follows that X is recursive in b, I over ${}^1M$. But X should not be extended recursive. This is a contradiction, and we may conclude that we will stop producing new requirements.

We may now end the proof:
Let $\nu = \langle 1, n, a \rangle$. Let σ be so that after Stage 5 we will not injure or create any requirement for ν.

If there is a requirement of 1st class active at stage σ, everything is O.K. If there is some argument of an active requirement of 2nd class that is not in x, everything is also O.K. If none of these holds, and if $\rho_{n,a}(A)$ is total, we see from the argument of the subclaim that we will produce more 3rd class requirements, thus $J_{n,a}(A)$ cannot be total and in particular not X.

8. CONCLUDING REMARKS

The arguments given above are valid for any level in the type-hierarchy. Thus we have proved the following theorem:

$V = L$: Let $I = tp(k)$. Let $A \subseteq tp(k + 1)$ be semi-recursive in ^{k+2}E and acceptable. Then either A is extended weakly recursive in ^{k+2}E or there are two sets $B \subseteq tp(k + 1)$, $C \subseteq tp(k + 1)$, semi-recursive in ^{k+2}E, both recursive in A, and neither extended recursive in the other.

Department of Mathematics,
University of Oslo

REFERENCES

[1] K. Devlin, *Aspects of Constructibility*, Springer Lecture Notes 354.

[2] J. E. Fenstad, 'Computation theories: An axiomatic approach to recursion on general structures', in Müller, Oberschelp and Potthoff (eds.), *Logic Conference, Kiel 1974*, 143–169, Springer Lecture Notes **499**.

[3] J. E. Fenstad, monograph on recursion theory, in preparation.

[4] J. E. Fenstad and P. G. Hinman (eds.), *Generalized Recursion Theory*, North-Holland, 1974.

[5] L. Harrington, *Contributions to Recursion Theory in Higher Types*, Ph.D. Thesis, MIT, 1973.

[6] S. C. Kleene, 'Recursive functionals and quantifiers of finite type I', *Trans. Amer. Math. Soc.* **91** (1959), 1–52, and II, **108** (1963), 106–142.

[7] D. B. MacQueen, *Post's Problem for Recursion in Higher Types*, Ph.D. Thesis, MIT, 1972.

[8] J. Moldestad, 'Computation Theory on Two Types', Preprint Series no. 2, Oslo, 1976.

[9] D. Normann, 'Imbedding of higher type theories', Preprint Series no. 16, Oslo, 1974.

[10] D. Normann, 'Degrees of functionals', Preprint Series no. 22, Oslo, 1975.

[11] G. E. Sacks, 'The 1-section of a type n object', in Fenstad and Hinman (eds.), *Generalized recursion theory*, 81–93, North-Holland, 1974.

[12] G. E. Sacks, 'The k-section of a type n object', to appear.

[13] J. R. Shoenfield, *Degrees of Unsolvability*, North-Holland, 1971.

I. A. LAVROV

RETRACTS OF POST'S NUMBERING AND EFFECTIVIZATION OF QUANTIFIERS

1. The class P of recursively enumerable sets is one of the main objects of investigations in recursion theory. It is well-known that the basic and important results about class P are easily proved when the class is provided with some computable numbering.

Let $\Pi_0, \Pi_1, \ldots$ be a standard computable Post's numbering of the class of all recursively enumerable sets P; this numbering is such that the relation $x \in \Pi_y$ is recursively enumerable. Denote $\mathscr{P} = \langle P, \Pi \rangle$.

Let $\mathscr{D}_0, \mathscr{D}_1, \ldots$ be a canonical numbering of the class of all finite sets. A subset $Q \subseteq P$ is a *subobject* of $\mathscr{P}$, if there exists a general recursive function f such that

$$Q = \{\Pi_{f(0)}, \Pi_{f(1)}, \ldots\}.$$

Different important kinds of subobjects of $\mathscr{P}$ have been studied. So a well-known Rice's theorem [1] gives description of the structure of the class Q such that $\Pi^{-1}Q$ is recursively enumerable. This subobject Q is called *e-subobject* of $\mathscr{P}$. Let f be a general recursive function, then $T = \{\mathscr{D}_{f(0)}, \mathscr{D}_{f(1)}, \ldots\}$ is a *strong array* of finite sets. Then Q is an *e*-subobject of $\mathscr{P}$ if and only if there exists a strong array of finite sets $T = \{\mathscr{D}_{f(x)}\}$ such that

$$Q = \{\Pi_x / \exists_y \mathscr{D}_{f(y)} \subseteq \Pi_x\}.$$

A subobject Q is *sn-subobject* (a special normal Lachlan's class [2]) if there exists general recursive function f such that

(1) $\quad Q = \{\Pi_{f(0)}, \Pi_{f(1)}, \ldots\}$;

(2) $\quad \Pi_x \in Q \to \Pi_{f(x)} = \Pi_x$ (elements of Q are fixed points);

(3) $\quad \Pi_{f(x)} \subseteq \Pi_x$.

Lachlan gave a complete description of *sn*-subobjects of class $\mathscr{P}$. His theorem asserts that Q is *sn*-subobject of $\mathscr{P}$ if and only if all elements of Q are the limits of increasing computable chains of some strong array T of finite sets and $\varnothing \in T$.

J. Hintikka, I. Niiniluoto, and E. Saarinen (eds.), Essays on Mathematical and Philosophical Logic, 287–292.

If we omit the condition $\varnothing \in T$, then we obtain a complete description of *wsn-subobjects* of $\mathscr{P}$ (weak special normal class [3]). In defining *wsn*-subobjects the function f from the definition of a *sn*-subobject has to be a partial recursive function.

The author investigated the problem of description factorizations and retracts in of $\mathscr{P}$ in [4].

Q is a *factorization* of $\mathscr{P}$, if there exists a general recursive function f such that

(1) $\quad Q = \{\Pi_{f(0)}, \Pi_{f(1)}, \ldots\};$

(2) $\quad \Pi_x = \Pi_y \rightarrow \Pi_{f(x)} = \Pi_{f(y)}.$

A factorization Q is a *retract* of $\mathscr{P}$, if all elements from Q are fixed points, in other words

(3) $\quad \Pi_x \in Q \rightarrow \Pi_{f(x)} = \Pi_x.$

This definition is a recursive analog of categorical or topological notion of a retract. Without any detail, $Q \subseteq P$ is a *retract* of $\mathscr{P}$, if there exists a mapping d: $\mathscr{P} \rightarrow Q$ such that $i \cdot d$ is an identical mapping $Q \rightarrow Q$ (this is a factorization), where i is an identical embedding of Q into $\mathscr{P}$ (it gives fixed points). There are a lot of important examples of retraction mappings: constructing subsets, constructing supersets and so on.

My description of factorization is as follows. Let $T_1 = \{\mathscr{D}_{f(x)}\}$ and $T_2 = \{\mathscr{D}_{g(x)}\}$ be two strong arrays of finite sets and $\phi \in T_1$. Class Q is a factorization if and only if there exist T_1 and T_2 such that

$$Q = \{\Pi_{p(x)}/\Pi_{p(x)} = \bigcup_{\mathscr{D}_{f(i)} \subseteq \Pi_x} \mathscr{D}_{g(i)}\}.$$

We can formulate necessary and sufficient conditions for T_1 and T_2 which generated some retract Q. These conditions include some dependences between T_1 and T_2 in terms of symbol $\subseteq$.

At present we have not similar descriptions for standard Lachlan's classes (*n-subobjects* [2]); it is a class Q such that there exists a general recursive function f such that

(1) $\quad Q = \{\Pi_{f(0)}, \Pi_{f(1)}, \ldots\};$

(2) $\quad \Pi_x \in Q \rightarrow \Pi_{f(x)} = \Pi_x.$

2. The aim of the second part of my paper is to point out a certain difficulty which arises under the construction of classification of recursively enumerable sets. The greatest Post's result is the introduction of notions

of creative sets, simple sets and so on [5]. According to Post a recursively enumerable set K is *creative*, if there exists a partial recursive function f such that

$$\forall x[\Pi_x \subseteq \bar{K} \rightarrow f(x) \in \overline{K \cup \Pi_x}].$$

We can easily replace in this definition a partial recursive function by a general recursive function.

It is obvious that for any creative set K there exists some infinite recursively enumerable set in $\bar{K}$ (it is a very special property of a creative set and it cannot characterize only the class of creative sets). Post introduced the notion of a simple set: nonrecursive recursively enumerable set S is simple if $\bar{S}$ does not include any infinite recursively enumerable set. Then Post and the other authors attempted to specify or to generalize the notion of a simple set. Here we have a wide spectrum of notions: hypersimple sets, hyperhypersimple sets, maximal sets and so on [6]. It is a pity these notions did not play such an important role as recursive or creative sets. We have only a lot of basic and special properties of these kinds of sets, but actually often they are not connected between each other and with other important objects of recursion theory. Most of natural conceptions of recursion theory (for example different kinds of reducibilities) are connected with these kinds of sets very weakly. Post's famous problem asks the question: is there a natural class of recursively enumerable sets which does not include any T-complete set? Friedberg [7] and Muchnik [8] gave a special solution of this problem: they constructed Turing incomparable recursively enumerable degrees. But the construction of a natural class of non-T-complete sets turned out to be a difficult task. Post constructed T-complete simple set. Yates [9] constructed a T-complete maximal set.

Recently S. S. Marchenkov [10] constructed a required class of sets. Without any details I shall explain his result.

Let $\mathfrak{A}$ be a class of recursively enumerable semi-recursive sets. A is a *semi recursive* set if A is an initial segment of some recursive linear order on N. This notion was introduced by Jockusch [11]. It is known that every T-degree includes some semi recursive set.

Let α be some positive equivalence on N. The set A is *α-hyperhypersimple* set if it is a recursively enumerable set, α-closed, set $\bar{A}$ contains an infinite number of equivalent α-classes and there is no general recursive function f such that

$$\forall x[\Pi_{f(x)} \text{ is } \alpha\text{-closed } \& \ \Pi_{f(x)} \cap \bar{A} \neq \varnothing] \ \&$$
$$\forall xy[x \neq y \to \Pi_{f(x)} \cap \Pi_{f(y)} \cap \bar{A} = \varnothing].$$

Let $\mathscr{L}$ be a class of α-hyperhypersimple sets, where α is some positive equivalence on N. It is a very important extension of the notion of hyperhypersimple set. The hyperhypersimple set possess some wonderful properties, and the best example is a remarkable Lachlan's theorem [12], which asserts that recursively enumerable set A is hyperhypersimple if and only if a lattice of recursively enumerable supersets of A is Boolean lattice. Among α-hyperhypersimple sets there are T-complete sets.

Class $\mathfrak{C} = \mathfrak{A} \cap \mathscr{L}$ is not empty and contains some nonrecursive set [13].

Marchenkov's theorem asserts that every set from $\mathfrak{C}$ cannot be T-complete.

I want to indicate one possible way of creating of notions recursively enumerable sets connected with the effectivization of quantifiers or with the existence of general recursive Skolem's functions. Uspenski [14] proved some interesting results in this direction. He offered effective variants of notions, namely nonfinite sets, nonrecursive sets and so on.

Our idea consists in consideration the effective negation for positive definitions.

Let $M_1, \ldots, M_n, \Pi_{x_1}, \ldots, \Pi_{x_m}$ be recursively enumerable sets. Let $\mathfrak{A}$ be some term, built from $\{M_1, \ldots, M_n, \Pi_{x_1}, \ldots, \Pi_{x_m}\}$ and operations $\cup$, $\cap$, —.

Positive definition is an expression of the form:

$$\exists x_1 \ldots x_m[\mathfrak{A}(M_1, \ldots, M_n, \Pi_{x_1}, \ldots, \Pi_{x_m}) = \varnothing].$$

Negative definition is an expression of the form:

$$\forall x_1 \ldots x_m[\mathfrak{A}(M_1, \ldots, M_n, \Pi_{x_1}, \ldots, \Pi_{x_m}) \neq \varnothing].$$

We shall call a property of recursively enumerably sets $M_1, \ldots, M_n$ *effectively negative defined*, if there exists a general recursive function f such that

$$\forall x_1 \ldots x_m f(x_1, \ldots, x_m) \in \mathfrak{A}(M_1, \ldots, M_n, \Pi_{x_1}, \ldots, \Pi_{x_m}).$$

EXAMPLE 1. Let $\mathfrak{A}(M_i\Pi_x) = (M \cap \Pi_x) \cup (\overline{M} \cap \overline{\Pi}_x)$. Then we have Myhill's notion of a complete creative set. It is known that notions of creative and complete creative sets coincide (Myhill's theorem).

EXAMPLE 2. Let $\mathfrak{A}(M_1, M_2, \Pi_{x_1}, \Pi_{x_2}) = \bigcup_{i=1,2} (M_i \cap \Pi_{x_i}) \cup (\overline{M}_i \cap \overline{\Pi}_{x_i})$.

Then we have notion of effective nonseparated pairs of recursively enumerable sets. These pairs play an important role in recursion theory.

We have seen that this approach gives us important, interesting and natural classes of recursively enumerable sets. It seems to me this approach can lead us to a good classification of sets.

We can change ways of construction of the term $\mathfrak{A}$, change a type of the question in a positive definition and so on.

I think that it is very interesting to prove some theorems of existence for effectively negative defined sets. There are quite a lot of other interesting questions in this direction.

Institute of Mathematics, Siberian Branch of the USSR Academy of Science, Novosibirsk

BIBLIOGRAPHY

[1] H. G. Rice, 'On completely recursively enumerable classes and their key arrays', *J. Symb. Logic* **21** (1956), 304–308.

[2] A. H. Lachlan, 'Standard classes of recursively enumerable sets', *Zeit. Math. Logik und Grund. Math.* **10** (1964), 23–42.

[3] А. И. Малъцев, 'К теории вычислимых семейств объектов', *Алгебра и логика* **3** (1964), 5–31.

[4] И. А. Лавров, 'Некоторые свойства ретрактов пумерации Поста', *Алгебра и логика* **13** (1974), 662–675.

[5] E. L. Post, 'Recursively enumerable sets of positive integers and their decision problems', *Bull. Amer. Math. Soc.* **50** (1944), 284–316.

[6] Х. Роджерс, *Теория рекурсивных функций и эффективная вычислимость*, Москва, 1972.

[7] R. M. Friedberg, 'Two recursively enumerable sets of incomparable degrees of unsolvability (solution of Post's problem 1944)', *Proc. Nat. Acad. Sci., USA* **43** (1957), 236–238.

[8] А. А. Мучник, 'Неразрешимостъ лроблемы сводимости теории алгоритмов', *Доклады АН СССР* **108** (1956), 194–197.

[9] C. E. M. Yates, 'Three theorems on the degree of recursively enumerable sets', *Duke Math. J.* **32** (1965), 461–468.

[10] С. С. Марченков, 'Об одном классе неполных множеств', *Матем. заметки* **20** (1976), 473–478.

[11] C. G. Jockusch, 'Semirecursive sets and positive reducibility', *Trans. Amer. Math. Soc.* **131** (1968), 420–436.

[12] A. H. Lachlan, 'On the lattice of recursively enumerable sets', *Trans. Amer. Math. Soc.* **130** (1968), 1–37.
[13] А. Н. Дегтев, 'О *tt*- и *m*-степенях', *Алгебра и логика* **12** (1973), 143–161.
[14] В. А. Успенский, *Лекции о вычислимых функциях*, Москва, 1960.

PART VI

LOGIC AND NATURAL LANGUAGE

JAAKKO HINTIKKA

QUANTIFIERS IN NATURAL LANGUAGES: SOME LOGICAL PROBLEMS, I

1. THE PROBLEM OF NATURAL-LANGUAGE QUANTIFIERS

Quantifiers are by common consent among the most important ingredients in the logical structure of both formalized and natural languages. As we all know, formalized quantifiers and their generalizations have been studied extensively by logicians. In contrast, relatively little close attention has been paid to the peculiarities of the logical behavior of quantifiers in natural languages, in spite of the fact that quantifiers have figured conspicuously in some of the most important recent methodological controversies in linguistics.[1] It is the purpose of this paper to help to change this situation by surveying some of the non-trivial logical phenomena that we find in examining natural-language quantifiers. In doing so, I am largely restricting my attention to the natural-language counterparts of logicians' existential quantifier and universal quantifier, neglecting such 'nonstandard' quantifier words as 'many' and 'most'. I shall cover in the survey a number earlier observations of mine which have already been partly published,[2] but I shall also try to concentrate on certain new insights which serve to extend and to consolidate the old results.

2. SCOPE PROBLEMS AND THE DIRECTION OF SEMANTICAL RULES

A heuristically useful place to begin to study the logic of natural languages are the differences between formal systems and natural languages with respect to the overt behavior of quantifiers. One of the most revealing differences lies in the behavior of what logicians would call the *scope* of quantifiers in natural languages. The way in which scope is handled in ordinary discourse is almost paradoxical. On the one hand, scope does not seem to be particularly important. It is not indicated by brackets, parentheses, dots, or by any comparable devices. Moreover, in some important sense absolute scope does not even seem to matter. The scope of a quantifier word can apparently extend arbitrarily far in a sentence and

J. Hintikka, I. Niiniluoto, and E. Saarinen (eds.), Essays on Mathematical and Philosophical Logic, 295–314

even in a text or discourse. For instance, one can introduce an individual by a quantifier phrase like 'an X' or 'some Y' and keep on referring to that 'randomly chosen' individual virtually indefinitely long by using repeatedly pronouns ('he', 'she', 'it', etc.) or anaphoric the-phrases 'the X' or 'the Y', respectively. In this respect, natural-language quantifier phrases look more like Hilbert's 'scopeless' ε-terms than formalized quantifiers. An example of a grammatical device which can be used to extend the scope of an existential-quantifier phrase arbitrarily far in a given sentence is found in Section 5 below in the form of the construction (*).

But although the absolute scopes of quantifier phrases thus seem to be irrelevant (or arbitrarily large), the *relative* scopes of quantifier phrases are vitally important for the proper understanding of many natural-language sentences. To find an example of this, one only has to consider the preferred readings of such sentences as the following:

(1) Every man loves some woman;

(2) Some woman is loved by every man.

The difference between the two lies precisely in a difference between the relative scopes of the two quantifier phrases 'every man' and 'some woman'.

This difference between absolute scope indications and relative scope indications has no obvious counterpart in formalized languages.

How, then, are the relative scopes of quantifiers (quantifier phrases) marked in ordinary discourse? A full answer to this question will of necessity be quite complicated. (Several of the sample sentences to be given later illustrate this complexity; see especially Sections 7–8 below.)

What is not equally complicated is to see what an appropriate framework might look like for studying the principles governing those relative scopes of natural-language quantifiers. Usually, the semantics of a formalized language is constructed in terms of recursive truth-clauses which show how the semantical interpretation of a complex expression is obtained step by step from the semantical interpretations of its component expressions.[3] However, the underdeterminacy of quantifier scopes tends to spoil the corresponding strategy in natural languages. The reason is that the underdeterminacy makes it impossible – or at least very hard – to decide what the appropriate component expressions of complex sentences are. Any component expression containing a quantifier has to take in all of the scope of this quantifier. But if the scope is in principle unrestricted (cf. above), we can scarcely find any component expression in the relevant sense of the word at all.

What a better semantical treatment of natural-language quantifiers might be like is not very hard to see, either, in the light of such observations. Instead of trying to build up the semantical interpretation of a sentence from the inside out, we can formulate our semantical rules so as to operate from the outside in. The reverse direction of the two procedures makes a difference because a step of analysis may involve changes in the remaining expression.

It is interesting to see what this possibility implies for the familiar principle that the meaning (semantical interpretation) of a complex expression must be a function of the meanings (semantical interpretations) of its components. What remains true in the kind of theory I am envisaging is that the meaning of a complex expression *S* is a function of the meanings of certain simpler expressions. They are no longer always components of *S*, however, in any syntactically relevant sense, even though they may perhaps be recovered from *S* by means a rule-governed procedure.

3. GAME-THEORETICAL SEMANTICS: QUANTIFIER RULES

My game-theoretical semantics for English quantifiers[4] and for a number of related expressions can be thought of as a system of rules of semantical analysis of the kind just indicated. It covers a fragment of English which for our present purposes needs to be delineated only approximately. The leading idea is to define the truth of each sentence *S* of this fragment by reference to an associated semantical game G(*S*). This game is a two-person, zero-sum game. The truth of *S* means that one of the players, called myself, has a winning strategy in G(*S*). The falsity of *S* means that the other player, called Nature, has a winning strategy in G(*S*). Certain sentences of our fragment – or of a mild extension of it – will have to be thought of as unanalyzable (atomic) ones, and their truth-values are assumed to be determined unproblematically. At each stage of the game an English sentence is being considered, beginning with *S*. I win a play of G(*S*) if it ends with a true atomic sentence, lose if it ends with a false one. In a sense, G(*S*) may thus be considered as an attempt to verify *S*, as far as my moves are concerned, against the schemes of a recalcitrant Nature. This helps to understand the intuitive motivation of the rules of my games. They will be presented by means of special cases which are readily generalizable. Here are some special cases of certain especially important rules.

(G.some) When the game has reached a sentence of the form

X – some Y who Z – W,

a person may be chosen by myself. Let the proper name of that person be '*b*'. Then the game is continued with respect to the sentence

X – *b* – W, *b* is a Y, and *b* Z.

An analogous rule can be set up for 'a(n)'.

(G.every) When the game has reached a sentence of the form

X – every Y who Z – W

a person may be chosen by Nature. Let the proper name of this person be '*d*'. Then the game is continued with respect to the sentence

X – *d* – W if *d* is a Y and if *d* Z.

Analogous rules can be set up for 'any' and 'each'.

Several comments are needed here. What has been formulated are special cases of certain much more general rules. The generalization does not present any essential problems, however, once we realize what the relevant directions of the different extensions are.

(i) Each wh-word of English comes with a range of individuals (in a logician's sense of the word) associated with it. In the case of 'who' they are persons, in the case of 'where' they are places, in the case of 'when' they are times, etc. For each wh-word, the choices envisaged in (G.some) and (G.every) are made from the associated domain.

(ii) It was assumed above that 'who' occupies the subject position in 'who Z'. If not, a slightly more complicated rule is needed. Essentially, the proper name '*b*' or '*d*' in the above rules goes into that place in Z where the initial wh-word was moved away from in generating the relative clause in question. (This transformation is often called wh-fronting.)

(iii) If the individual chosen has a proper name, this proper name can be used. If it does not have one, the players must give it a proper name which is used throughout the play of the game in question. Hence the final unanalyzable sentence of a game G(*S*) may belong to a mild extension of the original language in which *S* was expressed. Each such extension comprises as its new symbols only a finite number of proper names of individuals, however.

(iv) The rules (G.some) and (G.any) for 'some' and 'any', respectively, must be extended so as to cover also the plural uses of these quantifier words.

(v) Relations of pronominalization between words in X, Y, Z, and W may cause extra complications which will not be discussed in this paper.

(vi) The rule (G.a(n)) does not cover certain 'generic' (universal-quantifier) uses of the indefinite article.

4. GAME-THEORETICAL SEMANTICS: PROPOSITIONAL RULES

The rules for the particles 'and' and 'or' in their clause-binding uses are obvious.

(G.and) When the game has reached a sentence of the form

X and Y

where X and Y are clauses, Nature may choose X or Y, and the game is continued with respect to it.

(G.or) Likewise for (X or Y) except that the choice is made by myself.

Similar rules may be set up for 'if', 'but', etc.

In all such propositional rules depronominalization instructions must be built into the rule, that is, each pronoun in the chosen clause must be replaced by its antecedent (in the appropriate case) when this antecedent occurs in the other conjunct or disjunct. Typically, these rules must also be restricted to situations in which all such pronominal antecedents are proper names. There are interesting exceptions to this, studied in the paper 'Pronouns of Laziness in Game-theoretical Semantics' by Hintikka and Carlson.

Further rules can be set up to deal with the uses of 'and' and 'or' to form conjunctive and disjunctive noun phrases, respectively.

There are many ways of treating negation. One of them is the following.

(G.neg) When the game has reached a sentence of the form

$neg^{+}[X]$

the two players switch their roles, as defined by the game rules (including the rules for winning and losing), and the game is continued with respect to X.

Here $neg^{+}[-]$ denotes the operation of forming the (semantical) negation in English. It will not be studied in this paper, even though it will have to be presupposed in some of our game rules. For instance, when confronted by

Y if X

or

If X, then Y,

I have a choice according to a rule called (G.if) between Y and $neg^{+}[X]$.

The ways in which this rule will have to be developed further are studied in Carlson and Hintikka, 'Conditionals, Generic Quantifiers, and Other Applications of Subgames'.

An appropriate choice of the negation rule helps us to appreciate also the fact that all personification of Nature can in principle be dispensed with. All that is needed is a temporary switch by myself into the role of Devil's – or, rather, Nature's – advocate.[5]

5. GAME-THEORETICAL SEMANTICS: SUNDRY ADDITIONAL RULES

The appropriate rule or rules for the definite article 'the' can be fully understood only after the further development of (G.if) just mentioned. For the purposes of this paper, it suffices to register the interesting 'pronominal' uses of a phrase like 'the Y' to recover a reference to the individual which we are invited to consider by a quantifier phrase of the form 'some Y who Z' or 'a Y who Z'. The way to deal with such phrases is to require them to be replaced by a proper name as soon as their antecedent is replaced by this proper name.

One particular construction employing such pronominal the-phrases is the following.

(*) An X_1 who Y_1,
an X_2 who Y_2,
--- ---, and
an X_i who Y_i are such that Z

where in Z we may have phrases like 'the X_1', 'the X_2', . . ., 'the X_i'. (Instead of 'an' we may here have 'some'. Some of the quantifier phrases in (*) may also be replaced by proper names.)

The construction (*) can be dealt with semantically by substituting a proper name '*b*' for the phrase 'the X_j' as soon as '*b*' is substituted for the corresponding initial phrase 'an X_j who Y_j'. After the rule (G.a(n)) has been applied to all the initial quantifier phrases, we can simply drop the preface down to and including the words 'such that'.

By relying on suitable version of possible-worlds semantics, additional game rules can be formulated which apply to modal and epistemic concepts. (In this paper we shall use the term 'modal concept' in a narrow sense which excludes epistemic concepts, such as knowledge, belief, memory, perception, etc.).

6. GAME-THEORETICAL SEMANTICS: GENERAL REMARKS

It is easily seen how these game rules help to define a method for the semantical treatment of a class of English sentences. This method is of the kind sketched earlier as a *desideratum* for natural-language semantics. It begins with the given sentence G(*S*) in this class and proceeds to increasingly simpler sentences. By considering all the different things that might happen in applying the game rules (G.and) and (G.or) and by treating freshly introduced proper names as if they were variables bound to clause-initial quantifiers we can even obtain a partial translation of English quantifier sentences into a logical notation. (The translation is only a partial one because the atomic sentences cannot be translated in this way.)

Since there is another, slightly different translation from English to a logical notation obtainable from our game rules, we shall not comment on this first method of logical formalization in any detail here.

It is more important to note how game-theoretical semantics reflects the difference between absolute scope and relative scope mentioned earlier. The quantifier phrase which in rules like (G.some) is replaced by a proper name can pronominalize pronouns arbitrarily far away, and an application of (G.some) can thus have arbitrarily distant repercussions. In this sense, absolute scopes are indefinite. However, the relative scopes of quantifier phrases are indicated by the order in which game rules are applied to them.

Two main questions are left open by our game rules. They are

(i) the informational status of the different moves,

and

(ii) the order in which the different rules are to be applied (whenever more than one of them is applicable).

I shall discuss certain problems connected with (i) in this paper. In its companion paper, which shares Sections 1–6 with it, the other problem (ii) is examined briefly. This companion essay has appeared in *Linguistics and Philosophy* **1** (1977), pp. 153–172.

7. INFORMATIONAL INDEPENDENCE AND SKOLEM FUNCTIONS

The question (i) really means: Which of the earlier moves are assumed to be known to the players at a given stage of the game? The only plausible answer might at first sight seem to be: all of them. Yet it can be argued that in certain natural-language sentences this will not be the case but that we must countenance the possibility of informally independent moves.

It is important to see precisely what this failure of perfect information means. In general, our game rules will yield a partial translation of English into logical notation. First, we have to realize that the part of a player's strategy which governs the choice of the individual mentioned in the quantificational rules is expressible by a function that specifies the choice, given those earlier moves the player in question is aware of (i.e., has in his information set). These functions are known to logicians as *Skolem functions*. Then we can simply assert the existence of such Skolem functions as codify a winning strategy for myself in G(S) in order to find a translation of S. It is easily seen that Nature may be allowed to choose her moves arbitrarily.

For instance, (1) clearly translates as

(3) $(\exists f)(x)(x \text{ is a man} \supset$
$(f(x) \text{ is a woman} \wedge x \text{ loves } f(x)))$

while (2) translates as

(4) $(\exists y)(x)(y \text{ is a woman} \wedge$
$(x \text{ is a man} \supset y \text{ is loved by } x))$.

In (1)–(2), perfect information can clearly be assumed, and has been assumed in (3).

More interestingly, I have argued[6] that a sentence like

(5) Some novel by every novelist is mentioned in some survey by every critic

exhibits informational independence as far as the two pairs of quantifiers are concerned. If so, it translates as

(6) $(\exists f)(\exists g)(x)(y)[(x \text{ is a novelist} \wedge y \text{ is a critic}) \supset$
$(f(x) \text{ is a novel by } x \wedge g(y) \text{ is a survey by } y \wedge$
$f(x) \text{ is mentioned in } g(y))]$

Likewise,

(7) Every townsman admires a friend and every villager envies a cousin who have met each other

translates (if we take 'a friend' to be an ellipsis of 'a friend of his' and 'a cousin' an ellipsis of 'a cousin of his')

(8) $(\exists f)(\exists g)(x)(y)[(x \text{ is a townsman} \wedge y \text{ is a villager}) \supset$
$(f(x) \text{ is a friend of } x \wedge x \text{ admires } f(x) \wedge$
$g(y) \text{ is a cousin of } y \wedge y \text{ envies } g(y) \wedge$
$f(x) \text{ and } g(y) \text{ have met each other})]$

8. PUTATIVE OBJECTIONS CONFOUNDED

Since these translations are still claimed in the literature to be controversial,[7] a brief defense is perhaps in order. The main competing translations[8] of (5) and (7) are like (6) and (8), respectively, except for containing the two-argument function variables $f(x, y)$ and $g(x, y)$ instead of the one-variable function variables $f(x)$ and $g(y)$, respectively. We shall call these translations (9) and (10), respectively. They are obtained by giving the two universal quantifiers in (5) and (7) a maximal scope. They are to be distinguished from the 'mixed' readings in which one and only one of the two function variables f and g has two arguments. (Cf. (6) and (8) above.)

These mixed readings of sentences like (5) and (7) are ruled out by their obvious semantical asymmetry with respect to the two pairs of quantifiers. For it seems to me quite obvious that (5) and (7) are semantically symmetrical with respect to these two pairs of quantifiers. (In (5) each pair contains 'every' and 'some', in (7) 'every' and 'a'.) Additional arguments against these mixed readings are given in my paper, 'Quantifiers vs. Quantification Theory'. This leaves only the interpretation exemplified by (9) and (10) to be considered.

There is no need for me to deny that many speakers of English might be tempted to interpret (5) and (7) as (9) and (10), respectively, instead of (6) and (8). The difference between (6) and (9) or between (8) and (10) is a very subtle one, as is the precise force of (6) and (8). It is therefore entirely natural that people should be confused enough to prefer (9) and (10) as readings of (5) and (7), respectively, even though this is not what a truly competent speaker would do.

There is even a good general reason why (6) and (8) are normally avoided as interpretations of what a speaker of English is likely to have in mind in uttering (5) or (7), respectively. This principle is that of charity. It tells us to give the words of a speaker the interpretation which is most likely to make the intended meaning of his utterance true, even if what he says is apt to be literally speaking false. This principle is applicable here because (6) and (8) are so strong as to be unlikely to be true in most normal circumstances.

In order to see how this extra force comes about, we may note that the Skolem functions which embody parts of my winning strategies contain in (6) and (8) fewer arguments than on any of the competing interpretations. For this reason, (6) and (8) are stronger than all the comparable linear-quantifier readings of (5) and (7), respectively. (The less information I need to employ a winning strategy, the less likely the existence of such a strategy is, and the stronger is the claim that such a strategy exists.) In fact, (6) and (8) are so strong as to be somewhat unlikely to be true. Much of Fauconnier's argumentation amounts in fact to pointing out this greater logical force of the reading instantiated by (6) and (8) as compared with the interpretation illustrated by (9) and (10). From this it only follows that people are likely to mean (9) and (10) when they utter (5) of (7), respectively, if one assumes the principle of charity, and absolutely nothing follows from Fauconnier's observations concerning the basic semantical force of (5) or (7). Hence his observations do not tell against my theory in the least.

A somewhat parallel case may illustrate my point. Suppose someone tells me, 'This is a dangerous street. Someone is hit by a car every week in it.' For obvious reasons, in interpreting his meaning I am apt to assign in the second sentence the widest scope to 'every'. In so doing, I am not uncovering an exception to the left-to-right ordering principle of English quantifiers. I am merely applying the principle of charity. Fauconnier's arguments are not any more conclusive against my theory. All that follows from them is that competent speakers of English are unlikely to assert sentences of the type (5) or (7) with their literal meaning.

What is decisive here are not different speakers' more or less confused uneducated intuitions but the systematical questions which one can ask and answer here. They include the question as to which semantical representation preserves the entailments we all intuitively accept as well as the question which translation is based on a consistent, general set of rules of semantical interpretation. I shall deal with these two questions in the opposite order.

There does not seem to be any natural set of rules of semantical interpretation that yields (9) and (10) rather than (6) and (8).[9] The translations (9) and (10) are obtained by giving the two universal quantifiers in (5) and (7) maximal (linear) scope, a scope that includes both existential quantifiers. If this were the way to understand (5) and (7), then there does not seem to be any way to avoid similarly giving both universal quantifiers a maximal scope also in the following sentences.

(11) Some novel by every novelist is mentioned by every critic.
(12) Every townsman admires a friend who has met every villager.

Yet the preferred readings of (11) and (12) are clearly different from these maximal-scope interpretations. For instance, (11) obviously means

(13) $(x)[x \text{ is a novelist} \supset$
$(\exists y)(y \text{ is a novel by } x \wedge$
$(z)(z \text{ is a critic} \supset y \text{ is mentioned by } z))]$

which is equivalent with

(14) $(\exists f)(x)(z)[z$ is a novelist $\supset$
$(f(x)$ is a novel by $x \wedge$
$(z$ is a critic $\supset f(x)$ is mentioned by $z))]$

Clearly (13) is different from what we obtain by giving the two universal quantifiers a maximal scope in (11). Analogous remarks apply to (12). This suggests strongly that (9) and (10) are not viable readings of (5) and (7), respectively. This observation in turn provides support for the translations (6) and (8) of the same sentences.

This observation is, moreover, just what is to be expected on my theory. Using an obvious 'branching-quantifier' notation (to be discussed in the sequel) we can say that on my theory (7) has the form

(7)* $$\left.\begin{array}{l}(x)(\exists z)\\ \\ (y)(\exists u)\end{array}\right> M(x, y, z, u)$$

while by the same token (11) has the form

(11)* $$\left.\begin{array}{l}(x)(\exists z)\\ \\ (y)\end{array}\right> M(x, y, z)$$

However, while (7)* does not reduce in general to a linear-quantifier form, (11)* can easily be seen to do so, and in fact to be equivalent with

(11)** $(x)(\exists z)(y)M(x, y, z)$

(which is equivalent with (14)). Thus my theory, unlike the views of my critics, predicts precisely the right reading for both (7) and (11).

A small qualification is needed at this point. The dropping of a quantifier from (5) and (7) so as to obtain (11) and (12), respectively, may have consequences other than the step from the form (7)* to (11)*. One of them is that the branching quantifier interpretation becomes somewhat less likely as compared with the linear quantifier reading. For instance, the sentence

(15) Every writer is mentioned in some survey by every critic

has as its preferred reading (because of a left-to-right preference)

(16) $(x)(x$ is a writer $\wedge$
$(z)(z$ is a critic $\supset$
$(\exists u)(u$ is an essay by $z \wedge$
x is mentioned in $u)))$

The reason for this is that in (15) the quantifier pair 'some survey by every critic' is no longer taken to be informationally independent of 'every writer'. Why? Even though I have not discussed the conditions on which English quantifiers can be independent on each other, the behaviour of (15) is to be expected. The informational independence of quantifiers of each other goes hand in hand with their syntactical isolation from one another. Now in some clear sense the isolation is more marked in (5) and (7) than in (15). Hence a dependence and consequently a linear reading is only natural in (15). The logical form of (15) is therefore (cf. (7)* and (11)*).

(15)* $(y)(x)(\exists z)M(x, y, z)$

No counterevidence to my theory is hence forthcoming from examples like (15).

The most decisive refutation of (9) and (10) as semantical representations of (5) and (7), respectively, nevertheless turns on the obvious entailments between (5) and (7) on the one hand and certain other sentences on the other hand. If there is anything clear about (5), it is that (5) implies, together with the premise 'John is a writer', the sentence

(17) Some novel by John is mentioned in some survey by every critic.

On the reading (6) I am defending, this entailment is valid. However, on the interpretation (9) which Fauconnier and Stenius defend, (17) is *not* implied by (6). As you can easily see, (9) does not imply logically

(18) $(\exists z)(y)(\exists u)$[John is a novelist ⊃
(y is a critic ⊃
(z is a novel by John ∧
u is a survey by y ∧
z is mentioned in u))]

which obviously has the force of (17).

In other words, on the Fauconnier-Stenius view, (6) could be true while there is no single novel by John the writer which is mentioned in some survey by every critic. This consequence of their view is patently counter-intuitive, indeed so clearly as to belie the whole maximal-scope interpretation (9) which my critics have tried to substitute for the branching-quantifier interpretation (6) of (5). The maximal-scope interpretation simply does not preserve obviously valid entailments. The same applies immediately to other similar sentences, e.g., to the attempted interpretation of (7) as having the force (10). Even though the preservation of intuitively valid entailments is often overestimated as a criterion of correct semantical representation of natural-language sentences, here this criterion works extremely well.

In contrast, (6) obviously implies the desired consequence (17). Thus a simple test in terms of logical entailments is enough to show the superiority of my branching-quantifier reading of English sentences like (5) and (7).

9. PARTIALLY ORDERED QUANTIFIERS IN NATURAL LANGUAGES

One major reason why these observations are interesting is that while (9) and (10) are linear first-order sentences and while (13) admits of a translation into a linear first-order notation, sentences of the form (6) or (8) do not automatically do so. Rather, instead of a linear first-order translation they admit of a formulation in terms of branching (partially ordered) quantifiers, for instance as follows.

(19)
$$\begin{array}{l} (x)(\exists z) \\ \\ (y)(\exists u) \end{array} \!\!\!\begin{array}{l} \searrow \\ \nearrow \end{array} \begin{array}{l} [(x \text{ is a novelist} \wedge y \text{ is a critic}) \supset \\ (z \text{ is a novel by } x \wedge \\ u \text{ is a survey by } y \wedge \\ z \text{ is mentioned in } u)] \end{array}$$

The general form of this type of sentence is

(20)
$$\begin{array}{l} (x)(\exists z) \\ \\ (y)(\exists u) \end{array} \!\!\!\begin{array}{l} \searrow \\ \nearrow \end{array} \begin{array}{l} [(A(x) \wedge B(y)) \supset \\ (C(x, z) \wedge D(y, u) \wedge E(u, z))] \end{array}$$

Can all sentences of this form be translated into linear first-order notation? It is known[10] that not all sentences of the general form

(21)
$$\begin{array}{l} (x)(\exists z) \\ \\ (y)(\exists u) \end{array} \!\!\!\begin{array}{l} \searrow \\ \nearrow \end{array} M(x, y, z, u)$$

can be translated into a linear first-order form. Now we can so enlarge the domain of individuals presupposed in (21) as to allow also ordered pairs of original individuals into it.[11] By allowing our quantifiers to range over the enlarged domain and by re-interpreting its unanalyzable predicates (21) can be converted into the form (20) *salva* satisfiability. (The main step is of course to re-interpret each primitive predicate $P(x, y, z, u)$ in (17) involving x, y, and one or both of z and u as a predicate $P(\langle x, z\rangle, \langle y, u\rangle)$ of the two ordered pairs $\langle x, z\rangle$, $\langle y, u\rangle$.) Since there is in general no linear first-order translation for (21), there cannot always be one for (20), either.

Thus there seems to be plenty of English sentences whose semantical representation involves partially ordered (p.o.) quantifiers which do not reduce to linear ones. One can even argue that equivalents to all p.o. quantifier formulas are needed in English semantics. An argument to this effect is facilitated by the results of Walkoe according to which each p.o. quantifier sentence has an equivalent of the form

(22)

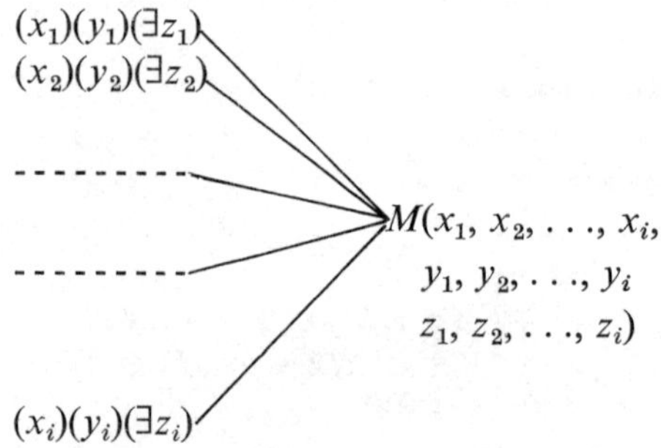

and an equivalent of the form

(23)

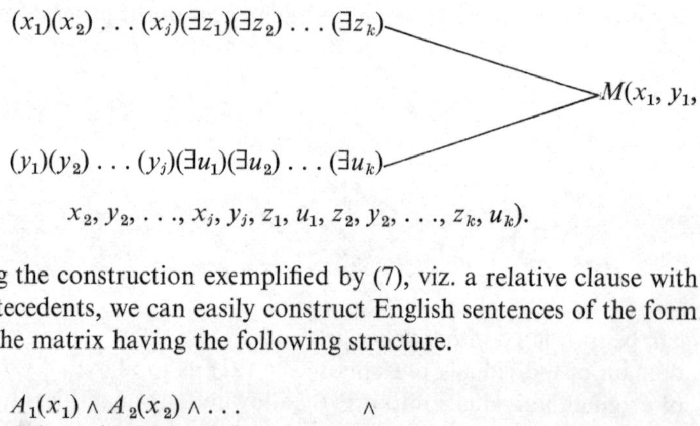

Now using the construction exemplified by (7), viz. a relative clause with several antecedents, we can easily construct English sentences of the form (22) with the matrix having the following structure.

(24)
$$\begin{array}{ll} A_1(x_1) \wedge A_2(x_2) \wedge \ldots & \wedge \\ B_1(x_1, y_1) \wedge B_2(x_2, y_2) \wedge \ldots &) \supset \\ (C_1(y_1, z_1) \wedge C_2(y_2, z_2) \wedge \ldots & \wedge \\ D_1(z_1, z_2) \wedge D_2(z_1, z_3) \wedge \ldots & \wedge \\ D_i(z_{i-1}, z_i)) & \end{array}$$

A case in point – illustrating the principles involved in constructing sentences of the form (24) – is the following:

(25) Every actor of each theatre envies a film star, every review of each critic mentions a novelist, and every book by each chess writer describes a grand master, of whom the star admires the grand master and hates the novelist while the novelist looks down on the grand master.

By an argument which is an extension of our earlier argument concerning (20) and (21) it can be shown that every sentence (22) can be re-interpreted so as to be of the form (24). First, the domain of individuals is enlarged by allowing ordered pairs and ordered triples of earlier individuals. The primitive predicates are reinterpreted as before. Finally, the sentence is reformulated (as in first-order reduction theory) so as to involve only dyadic relations.

This shows – or at least suggests very strongly – that the special form of the matrix of those p.o. quantifier formulas which are obtained as translations of suitable English quantifier sentences does not essentially restrict their generality. This in turn suggests that not only are there more complicated p.o. quantifier structures present in the semantics of English quantifier sentences. Literally *all* such structures are needed there.

A similar argument can be carried out in terms of the 'normal form' (23) instead of (22). Since in this case we are dealing with only two branches of quantifiers, the situation is simpler in that the reduction to dyadic predicates is dispensable in the argument.

One English construction that can be used to build sentences whose semantical representation has the form (23) is the repeated use of the particle 'of'. The following is a case in point.

(26) Some product of some subdivision of every company of every conglomerate is advertised in some page of some number of every magazine of every newspaper chain.

One reason why this is an interesting result is the observation – made in my earlier paper 'Quantifiers vs. Quantification Theory' – that the theory of p.o. quantifiers is as complicated as second-order logic (with standard interpretation) as far as its decision problem (degrees of unsolvability of the set of all valid sentences) is concerned. The two decision problems can be effectively (recursively) reduced to each other. An argument to this effect is given in 'Quantifiers vs. Quantification Theory'. Thus structures of enormous logical complexity must in effect be presupposed in the semantics of the English language.

10. A MORAL OF THE STORY

The application of game-theoretical semantics sketched above is calculated to illustrate a rather general thesis. Sight unseen, the logical and semantical behavior of natural-language quantifiers might perhaps be expected to be rather trivial from the vantage point a logician. In reality, there are problems of considerable complexity concerning natural-language quantifiers. The problems about branching quantifiers serve to exemplify such problems. It is to be hoped that more logicians will follow Richard Montague's example and become interested in the fascinating logical problems of natural-language semantics.[15]

This hope is enhanced by the striking consequences of our tentative findings for the methodology of theoretical linguistics. On the one hand, the logical strength of ordinary English is seen – at least tentatively – to come close to that of second-order logic. On the other hand, even the grammaticality of English quantifier sentences apparently is beyond the power of generative methods. Whatever the definitive status of these preliminary observations will turn out to be, there is no doubt about their tremendous potential interest and importance.

Academy of Finland and
Stanford University

NOTES

[1] See, for instance, Partee (1970), (1971) and Lakoff (1970), (1971).

[2] See Hintikka (1974), (1975a).

[3] See Davidson (1967), (1968), (1969), (1970), (1973), and cf. Hintikka (1975b), (1976b).

[4] See Hintikka (1974), (1975a), (1976a).

[5] Almost all that we need to do for the purpose of this depersonalization is to eliminate universal quantifiers and conjunctions in favor of existential quantifiers, disjunctions, and negations.

[6] See Hintikka (1974), and cf. Hintikka (1976c).

[7] See Fauconnier (1975) and Stenius (1976).

[8] See especially Fauconnier (1975).

[9] This is in fact a general methodological point of considerable importance both in semantics and in syntax. I have been amazed time and again by linguists who claim that they are dealing with competence and not performance and then go on to base their theories on people's uneducated and unanalysed reactions to complicated sentences.

[10] See Walkoe (1970).

[11] With the immediately following argument and with others like it in the sequel, please compare the procedures employed in the reduction theory of first-order logic, as e.g., in Suranyi (1959).

[12] Cf. Thomason (1974).

BIBLIOGRAPHY

Carlson, Lauri, and Jaakko Hintikka, 'Conditionals, Generic Quantifiers, and Other Applications of Subgames' in E. Saarinen (ed.), *Game-Theoretical Semantics,* D. Reidel, Dordrecht, 1978.

Davidson, Donald, 'Truth and Meaning', *Synthese* **17** (1967), pp. 304–323.

Davidson, Donald, 'On Saying That', *Synthese* **19** (1968), pp. 130–146.

Davidson, Donald, 'True to the Facts', *Journal of Philosophy* **66** (1969), pp. 748–764.

Davidson, Donald, 'Semantics for Natural Languages', in *Linguaggi nella società e nelle tecnica*, edizioni di Comunità, Milan, 1970, pp. 177–188.

Davidson, Donald, 'In Defense of Convention T', in H. Leblanc (ed.), *Truth, Syntax, and Modality*, North-Holland, Amsterdam, 1973, pp. 76–86.

Fauconnier, Gilles, 'Do Quantifiers Branch?', *Linguistic Inquiry* **6** (1975), pp. 555–567.

Hintikka, Jaakko, 'Quantifiers vs. Quantification Theory', *Linguistic Inquiry* **5** (1974), pp. 153–177.

Hintikka, Jaakko, 'On the Limitations of Generative Grammar', in the *Proceedings of the Scandinavian Seminar on Philosophy of Language*, Filosofiska Förening & Filosofiska Institutionen vid Uppsala Universitet, Uppsala (1975a), vol. 1, pp. 1–92.

Hintikka, Jaakko, 'A Counterexample to Tarski-type Truth-Definitions as Applied to Natural Languages', in Asa Kasher (ed.), *Language in Focus: Foundations, Methods and Systems Essays in Memory of Yehoshua Bar-Hillel*, D. Reidel, Dordrecht (1975b), pp. 107–112.

Hintikka, Jaakko, 'Impossible Possible Worlds Vindicated', *Journal of Philosophical Logic* **4** (1975c), pp. 475–484.

Hintikka, Jaakko, 'Quantifiers in Logic and Quantifiers in Natural Languages', in S. Körner (ed.), *Philosophy of Logic*, Oxford, Basil Blackwell (1976a), pp. 208–232.

Hintikka, Jaakko, 'The Prospects of Convention T', *Dialectica* **30** (1976b), pp. 61–66.

Hintikka, Jaakko, 'Partially Ordered Quantifiers vs. Partially Ordered Ideas', *Dialectica* **30** (1976c), pp. 89–99.

Hintikka, Jaakko *The Semantics of Questions and the Questions of Semantics* (*Acta Philosophica Fennica*, vol. 28, no. 4), North-Holland Publishing Company, Amsterdam, 1976d.

Hintikka, Jaakko and Lauri Carlson, 'Pronouns of Laziness in Game-Theoretical Semantics', *Theoretical Linguistics* **4** (1977), 1–29.

Hintikka, Jaakko, and Esa Saarinen, 'Semantical Games and the Bach-Peters Paradox', *Theoretical Linguistics* **2** (1975), pp. 1–20.

Lakoff, George, 'Repartee, or a Reply to "Negation Conjunction, and Quantifier"', *Foundations of Language* **6** (1970), pp. 398–422.

Lakoff, George, 'On Generative Semantics', in Danny D. Steinberg and Leon A. Jakobovits (eds.), *Semantics: An Interdisciplinary Reader*, Cambridge, Cambridge University Press, 1971, pp. 232–296.

Partee, Barbara Hall, 'Negation Conjunction, and Quantifiers: Syntax vs. Semantics', *Foundations of Language* **6** (1970), pp. 153–165.

Partee, Barbara Hall, 'On the Requirement that Transformations Preserve Meaning', in Charles J. Fillmore and D. Terence Langendoen (eds.), *Studies in Linguistic Semantics*, Holt, Rinehart and Winston, New York, 1971, pp. 1–21.

Rogers, Hartley, Jr., 'Certain Logical Reduction and Decision Problems', *Annals of Mathematics* **64** (1956), pp. 264–284.

Stenius, Erik, 'Comments on Jaakko Hintikka's Paper', *Dialectica* **30** (1976).

Suranyi, Janos, *Reduktionstheorie des Entscheidungsproblems im Prädikatenkalkul der ersten Stufe*, Verlag der ungarischen Akademie der Wissenschaften, Budapest, 1959.

Thomason, Richmond (ed.), *Formal Philosophy: Selected Essays of Richard Montague*, Yale University Press, 1974.

Walkoe, W. J., Jr., 'Finite Partially-Ordered Quantification', *Journal of Symbolic Logic* **35** (1970), pp. 535–555.

JENS ERIK FENSTAD

MODELS FOR NATURAL LANGUAGES

"The structure of every sentence is a lesson in logic." This often quoted statement of John Stuart Mill (1867) leads directly into a controversial area of theoretical grammar.

The relationship between logic and grammar has been a source of dispute all through the history of linguistics. Attitudes have ranged all the way from the marriage of grammar, logic, and ontology of the medieval *modistae* (Thomas of Erfurt) to the atheoretical descriptivism of American structural linguistics (Bloomfield, Harris).

Today the rôle of logic in linguistics is once more expanding. From being read out of linguistic theory as metaphysics semantics is once more a respected part of the grammatical enterprise.

Semantics has been a main ingredient in the development of formal logic from Frege *via* Tarski to the present time. And applications have been many and varied.

But applications of logic and formal semantics to the study of natural languages were slow in coming. Perhaps for good reasons. The rigid formalism of predicate logic seems to be a too blunt instrument for analyzing the fluid forms of a spoken tongue.

Formal logic is, however, much more than predicate logic. And model building is much more than mere formalization. One of the first who tried to apply the full range of tools supplied by modern logic to the analysis of natural languages was Hans Reichenbach who included one chapter on 'conversational languages' in his textbook from 1948, *Elements of Symbolic Logic*. Reichenbach used as his tool higher order logic extended by certain 'pragmatic' operators.

Dating from about the same time is a paper by Haskell B. Curry, 'Some logical aspects of grammatical structure. The paper was written in 1948 but was first published in 1961. In it Curry gives an analysis of the traditional parts of speech in terms of combinatory logic. It is closely related to, but independent of the categorial analysis of S. Łesniewski and K. Ajdukiewicz.

Curry and Reichenbach did not at that time seem to have any impact on

J. Hintikka, I. Niiniluoto, and E. Saarinen (eds.), Essays on Mathematical and Philosophical Logic, 315–340.

grammatical theory. Curry's paper was only privately circulated. And Reichenbach's analysis must to linguists in 1948 seem to be a step or two back to the *apriorism* of traditional doctrines.

But the linguists should, perhaps, have taken a closer look. What Reichenbach did was to provide a 'deep-structure' analysis of natural languages in terms of higher order logic. And his *pragmatic* operators was the best approximation at that time to a full intensional logic.

Present day intensional logic, thanks to the basic contributions of J. Hintikka, S. Kanger, and S. Kripke, is a much more powerful and versatile tool. But the beginnings of an analysis, very much in the current spirit, is clearly present in Reichenbach.

The more serious complaint against Reichenbach a linguist could bear, is that he did not give an analysis of the surface structure, and hence not of the connection between surface structure and the logical deep structure. But even here something did exist at the time.

Around 1950 Y. Bar-Hillel resurrected the categorial grammar of S. Łesniewski (1929) and K. Ajdukiewicz (1936). Few have read Łesniewski's paper, 'Grundzüge eines neuen Systems des Grundlagen der Mathematik'; his ideas were made more accessible through Ajdukiewicz's paper 'Die Syntaktische Konnexität'. Bar-Hillel had the idea of applying the categorial analysis in constructing a theory for machine translations.

But categorial grammar seemed only a brief episode to be submerged by the Chomskian revolution of *Syntactic Structures* (1957), which determined the shape of linguistic theory to come.

It was, however, a remarkable oversight of the times that one did not connect the categorial surface analysis of Ajdukiewicz with the deep-structure analysis of Reichenbach. Remarkable, since one of the explicit sources for Łesniewski's categories was indeed Russell's theory of types, the very logical formalism Reichenbach built on.

It is only today through the work of Richard Montague that the parts have been joined together and we see the outlines of a powerful method.

The present paper is divided in three parts. In Part I we give a brief introduction to Richard Montague's synthesis of categorial grammar and higher order (intensional) logic as presented in his paper 'The Proper Treatment of Quantification in Ordinary English' (1971). In Part II we indicate a non-transformational analysis of passive. It is somewhat different from and simpler than the standard Montague-style treatment.

In Part III we discuss certain extensions of the model of Part I.

Remark: Recently there has been published a book length study on

Reichenbach's grammatical work, W. E. McMahon, *Hans Reichenbach's Philosophy of Grammar*, 1976. The book has many interesting discussions, but its attempt to force Reichenbach into the Chomskian mould is clearly a distortion of Reichenbach ideas.

I. A MODEL FOR A FRAGMENT

We shall in this section present the *extensional* part of Montague's model in 'The Proper Treatment of Quantification in Ordinary English' (PTQ).

The model consists of three components:

A. An analysis of the *surface syntactic structure*.
B. A *logical deep structure*.
C. A *semantic interpretation*.

We shall give a brief description of each part separately before we discuss the translation from surface structure to deep structure.

A. *Extended categorial grammar* (*syntactic surface structure*).

a. We first define the class CAT of category symbols

1. $e, t \in \mathrm{CAT}$.
2. $A, B \in \mathrm{CAT} \Rightarrow A/B, A//B \in \mathrm{CAT}$.

Thus from two given elements e, t we generate an infinite class of symbols by means of the operations / and //.

b. To each $A \in \mathrm{CAT}$ is associated a class of words $\mathbf{B}_A$, the lexical items of category A. For all but finitely many A, $\mathbf{B}_A = \emptyset$.

$$\mathbf{B} = \bigcup_{A \in \mathrm{CAT}} \mathbf{B}_A \text{ is the lexicon.}$$

c. Derived syntactic rules: A finite list F_i of functions is given which generates a set $\mathbf{P}$ of (derived) phrases from the lexical basis $\mathbf{B}$. Typically a syntactic rule has the form

$$F_i(\alpha_{A/B}, \beta_B) = \gamma_A,$$

i.e. F_i is a binary operator which acts on a phrase α of category ${}^{A}/_{B}$ and a phrase β of category B and produces a phrase γ of category A.

We give some examples.

1. *Intransitive verbs:* $\mathrm{IV} = {}^{t}/_{e}$.

$\mathbf{B}_{\mathrm{IV}} = \{run, walk, \ldots\}$

2. *Terms:* $T = {}^{t}/_{\mathrm{IV}}$.

$$\mathbf{B}_T = \{John,\ Mary, \ldots, he_0,\ he_1, \ldots\}$$

There is a syntactic rule F_1 which acts on a term and an intransitive verb and produces a sentence, e.g.

$$F_1(John,\ run) = John\ runs.$$

We note that *John* is of category ${}^{t}/_{\mathrm{IV}}$, *run* is of category IV. The result of applying F_1 to *John* and *run* is a phrase of category t. Note that F_1(*John*, *run*) is not merely the concatenation of the arguments, the morpheme *s* has been introduced to mark third person singular in the verb.

We can write the analysis in tree form

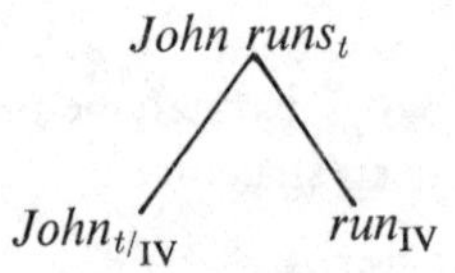

We finally note that this rule corresponds to the rule $S \to \mathrm{NP} + \mathrm{VP}$ of transformational grammar.

3. *Transitive verbs:* $\mathrm{TV} = {}^{\mathrm{IV}}/_{T}$

$$\mathbf{B}_{\mathrm{TV}} = \{find,\ love,\ be, \ldots\}$$

We give a simple example:

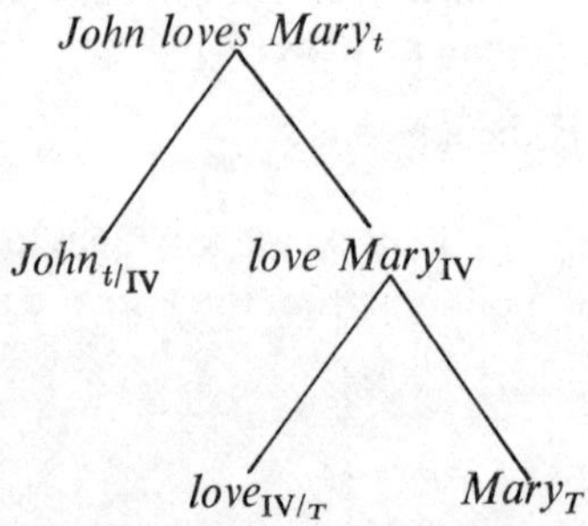

We have here a new syntactic rule F_2 which acts on a transitive verb and a term and produces an intransitive verb. This rule is purely categorical, i.e. a concatenation of the arguments. Instead of the tree we could write

$$F_1(John,\ F_2(love,\ Mary)) = John\ loves\ Mary$$

4. *Adverbs modifying intransitive verbs:* IAV $= {}^{IV}/_{IV}$.

$\mathbf{B}_{IAV} = \{slowly, \ldots\}$

A simple example:

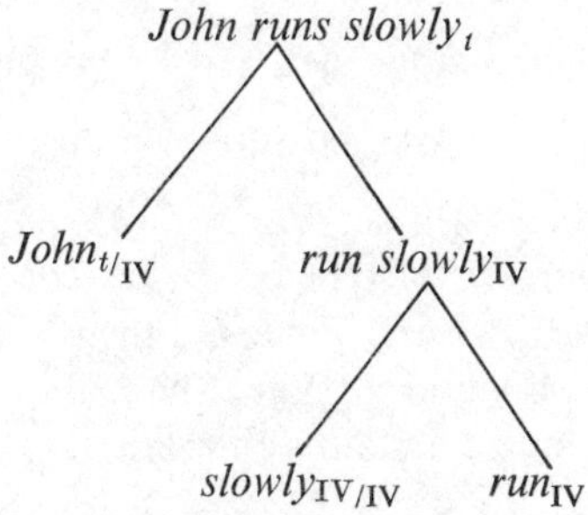

The new syntactic rule here $F_3(slowly_{IV/IV}, run_{IV}) = run\ slowly_{IV}$ inverts the order of the arguments.

5. *Common noun:* CN $= {}^{t}//_{e}$.

$\mathbf{B}_{CN} = \{man, woman, \ldots\}$

A simple example:

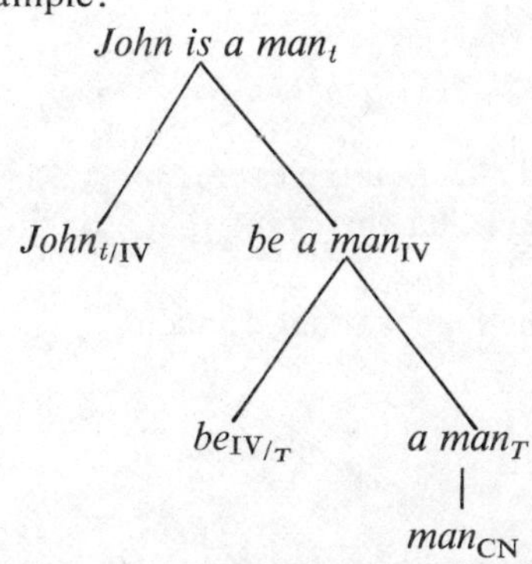

The new rule here acts on a single common noun and produces a term. The rule must specify which of *an, a* is to be chosen. Similar rules will produce: *the man, every man.*

This ends our few examples. One should note that the syntactic rules of the system may incorporate features which usually are located in the transformational part of TG (transformational grammar).

B. *Higher order logic as deep structure*

a. We first define the class TYP of type symbols

1. $e, t \in \mathrm{TYP}$
2. $a, b \in \mathrm{TYP} \Rightarrow \langle a, b \rangle \in \mathrm{TYP}$

b. The class of meaningful expressions $\mathrm{ME} = \bigcup_{a \in \mathrm{TYP}} \mathrm{ME}_a$ is defined as follows:

1. For each $a \in \mathrm{TYP}$ there is a set Var_a of *variables* of type a and a set Con_a of *constants* of type a. $\mathrm{Var}_a \cup \mathrm{Con}_a \subseteq \mathrm{ME}_a$.
2. *Application:* If $\alpha \in \mathrm{ME}_{\langle a,b \rangle}$ and $\beta \in \mathrm{ME}_a$, then $\alpha(\beta) \in \mathrm{ME}_b$.
3. *λ-abstraction:* If $\alpha \in \mathrm{ME}_b$ and $u \in \mathrm{Var}_a$, then $\lambda u . \alpha \in \mathrm{ME}_{\langle a,b \rangle}$.
4. Rules for identity, connectives, and quantifiers.

c. Application and λ-abstraction are connected through the important rule of λ-conversion:

$$(\lambda u . \alpha)(\beta) = \alpha[{}^u/_\beta],$$

where $\alpha[{}^u/_\beta]$ comes from α by replacing the free occurrences of u in α by β.

C. *A remark on the semantics for ME*. A model for extensional ME is a pair

$$\mathfrak{A} = \langle A, F \rangle$$

where $A \neq \emptyset$ and F is a map defined on the expressions of ME. (Since we are writing for logicians we feel that this somewhat 'rough' statement is precise enough.)

From the set A we generate the higher order domains

1. $D_e = A$
2. $D_t = \{0, 1\} = 2$
3. $D_{\langle a,b \rangle} = D_b^{D_a}$

For each $\alpha \in \mathrm{Con}_a$, $a \in \mathrm{TYP}$, we require that $F(\alpha) \in D_a$. Further, if $F(\alpha)$, $F(\beta)$ are defined and α and β are of appropriate types, then

$$F(\alpha(\beta)) = F(\alpha)(F(\beta)).$$

To interpret λ-abstraction we use a standard trick: Let $\mathfrak{A} = \langle A, F \rangle$ be a model for ME. Extend ME to a language $\mathrm{ME}(\mathfrak{A})$ by adding new constants $\boldsymbol{c}_h$, $h \in D_a$, $a \in \mathrm{TYP}$. Extend F to $\mathrm{ME}(\mathfrak{A})$ by setting

$$F(\mathbf{c}_h) = h.$$

We then define

$$F(\lambda u.\alpha)(h) = F(\alpha[{}^u/\mathbf{c}_h]).$$

From this the rest of the semantic is determined in the usual way.

D. *Translation from surface structure to deep structure*

The missing part of the model is the translation from **P** to ME.

a. First a map f: CAT $\rightarrow$ TYP

1. $f(e) = e.$
2. $f(t) = t.$
3. $f(A/B) = f(A//B) = \langle f(B), f(A)\rangle.$

b. Next a map g from a subset of the lexical base **B** to ME such that if $A \in$ CAT, $\alpha \in$ CAT, $\alpha \in \mathbf{B}_A$, and $g(\alpha)$ is defined, then $g(\alpha) \in \mathrm{Con}_{f(A)}$.

The idea here is that g associates with elements of the *open* word classes suitable constants in ME.

EXAMPLES: (i) $run \in \mathbf{B}_{t/e}, f({}^t/_e) = \langle e, t\rangle$
$\mathrm{run}' = g(run) \in \mathrm{Con}_{\langle e,t\rangle}$

(ii) $man \in \mathbf{B}_{t//e}, f(t//e) = \langle e, t\rangle$
$man' = g(man) \in \mathrm{Con}_{\langle e,t\rangle}.$

Given a model $\mathfrak{A} = \langle A, F\rangle$ we see that

$$run^{\mathfrak{A}} = F(run') \in 2^A$$
$$man^{\mathfrak{A}} = F(man') \in 2^A,$$

i.e. both correspond to properties, i.e. subsets of the domain of individuals.

c. For other items in **B** there are special translations, e.g.

$$John' = \lambda P.P(\mathrm{j}),$$

where $j \in \mathrm{Con}_e$. – We shall immediately explain why proper names have this type of translation.

d. Associated with every syntactic rule there is a special rule of translation. For $F = F_1$, F_2, or F_3 we impose

$$(F(\alpha, \beta))' = \alpha'(\beta'),$$

assuming that the translation of α and β are α' and β' respectively.

e. Finally, if $\alpha \in \mathbf{P}_{CN}$ translates to α', then

$$({}^{a}/_{an}\ \alpha)' = \lambda P.\exists x[\alpha'(x) \wedge P(x)].$$
$$(\text{every } \alpha)' = \lambda P.\forall x[\alpha'(x) \rightarrow P(x)].$$
$$(\text{the } \alpha)' = \lambda P.P(\iota x.\alpha').$$

With these specifications we can translate arbitrary elements of the set of phrases **P** to well-formed expressions in ME. We have only exhibited a few of the syntactic rules, and, correspondingly, only a small part of the translation rules of the Montague model in PTQ.

After having sketched the model we will apply it in three examples.

EXAMPLE 1. It is a syntactic fact that a proper name, *John*, and a quantifier phrase, *a man*, syntactically behave in exactly the same manner.

Semantically, however, they differ. We cannot at the same time assert *John runs* and *John does not run*. But we can say truthfully at the same moment *A man runs* and *A man does not run*.

Also notice the difference between the surface structure *Mary loves a man* and the logical structure $\exists x[man'(x) \wedge love(\mathbf{m}, x)]$.

How does the model handle this?

(i)

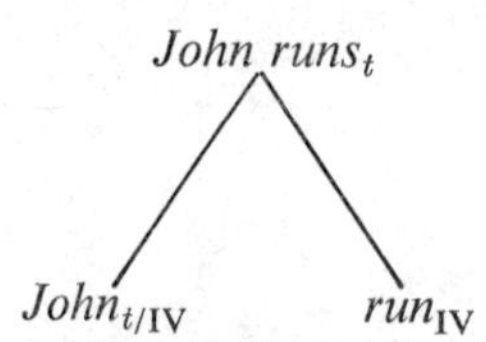

$$(John\ runs)' = (\text{John}')(\text{run}')$$
$$= (\lambda P.P(j))(run')$$
$$= run'(j)$$

(ii)

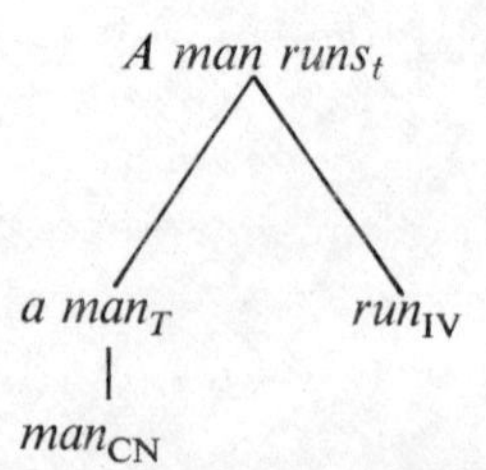

$$(A\ man\ runs)' = (a\ man)'(run')$$
$$= (\lambda P.\exists x[man'(x) \wedge P(x)])(run')$$
$$= \exists x[man'(x) \wedge run'(x)].$$

The example is, of course, a very simple one. But it comes out right. One simple test of adequacy of the model has been passed.

What Montague does here is to apply the more general notion of quantifiers well-known to logicians. Let $\mathfrak{A} = \langle A, F \rangle$ be a model for ME. Over the domain A the existential quantifier $\exists$ can be identified with the monotone family of sets

$$\exists_{\mathfrak{A}} = \{B \subseteq A \mid B \neq \emptyset\},$$

i.e. $\exists_{\mathfrak{A}}$ is a type-2 functional which tests for every property (i.e. subset of A) whether or not someone has that property. In the same way

$$\forall_{\mathfrak{B}} = \{A\}.$$

And for *John* we get over

$$John^{\mathfrak{A}} = \{B \subseteq A \mid j^{\mathfrak{A}} \in B\},$$

i.e. $John^{\mathfrak{A}}$ is the functional induced by an individual of the domain, i.e. 'point-evaluation'.

EXAMPLE 2. A common criticism of traditional grammar is its fixation on the basic subject/predicate form. In the present model the subject/predicate form is kept in the surface structure. But the translation from surface structure to logical deep structure can be arranged so that we in the deep structure have the full power of relational expressions.

The sentence *John loves Mary* has the following surface structure

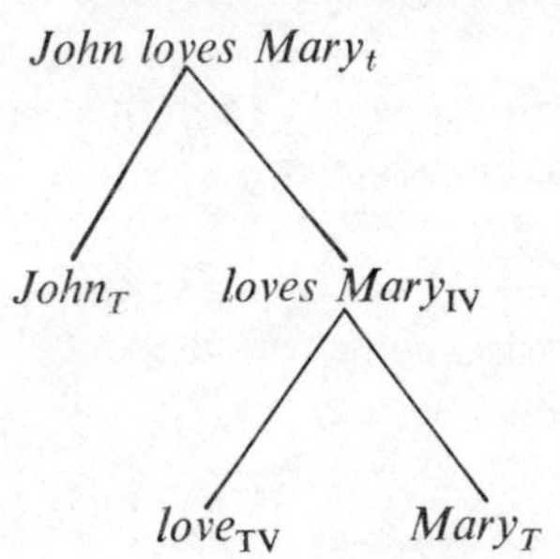

By our present rules we first get

$$(John\ loves\ Mary)' = (love'(\lambda Q\,.\,Q(m))(j).$$

There is a natural isomorphism $(2^A)^A \approx 2^{A \times A}$, hence,

$$love'(j, \lambda Q \,.\, Q(m)) \text{ for } (love'(\lambda Q \,.\, Q(m)))(j),$$

which again is standard working in higher types.

But we want *love'* to be a relation between individuals, not between an individual and a functional. This is solved by Montague through the introduction of certain *meaning postulates*. For certain relations, such as the transitive verb *love*, we assume that they come from relations between individuals, i.e. we assume that there is a constant $love'_*$ in ME such that for all x and y of type e and $\mathscr{P}$ of type $\langle\langle e, t\rangle, t\rangle$

(*) $\quad love'(x, \mathscr{P})$ iff $\mathscr{P}(\lambda y \,.\, love'_*(x, y))$.

Hence, if $\mathscr{P} = \lambda Q \,.\, Q(m)$, we get

$$\begin{aligned} love'(j, \lambda Q \,.\, Q(m)) &= (\lambda Q \,.\, Q(m))(\lambda y \,.\, love'_*(j, y)) \\ &= (\lambda y \,.\, love'_*(j, y))(m) \\ &= love'_*(j, m). \end{aligned}$$

Meaning postulates may sound a bit unfamiliar to the linguist. He would, perhaps, prefer a reformulation along the following lines. *love* as a transitive verb belongs to the lexical base. As a bivalent verb it has a natural interpretation as a relation of two argument, $love'_*(x, y)$. This relation should belong to (i.e. be a constant of) ME, the logical deep structure. So far according to the lexical rules.

And there will be no problem in proceeding to the syntactic analysis. Any relation between individuals can be lifted to relations between higher types. This we have done for *love* in (*). From this point of view the translation function g acting on *love* must give the lifted relation, which is now a defined element of ME.

Note. Not every transitive verb is to be treated in this way, one example being the *de dicto* reading of intentional verbs.

EXAMPLE 3. We shall, following Montague, give the translation of *be* as a transitive verb. The analysis should on the one hand show the similarity between

(i) *John runs.*
(ii) *John is a man.*

On the other hand it should also give a natural reading of identity statements such as

(iii) *John is the king.*

The translation of (i) is $run'(j)$. We want to translate (ii) as $man'(j)$ and (iii) as $j = \iota x king'(x)$. Recall that

$$(a\ man)' = \lambda P.\exists z[man'(z) \wedge P(z)]$$
$$(the\ king)' = \lambda P.P[\iota x.king'(x)]$$

We see that by choosing for P in both cases the property $E_j = \lambda y[j = y]$ we get the desired translation:

$$(a\ man)'(E_j) = \exists z[\text{man}'(z) \wedge j = z]$$
$$= man'(j)$$
$$(\text{the king})'(E_j) = (j = \iota x.king'(x))$$

The proper translation of be_{TV} is now obtained from the equation

$$be'(x, \mathscr{P}) = \mathscr{P}(E_x),$$

which gives

$$be'_{\text{TV}} = \lambda\mathscr{P}.\lambda x.\mathscr{P}(\lambda y[x = y]).$$

This completes our sketch of the extentional part of Montague's PTQ model. We shall now apply the model to an analysis of passive.

II. PASSIVE IN THE MODEL

From the logical point of view passive is the *converse* relation, i.e. given a binary relation (transitive verb) $R(x, y)$ the converse $\check{R}$ is defined by the equivalence

$$(x, y) \in \check{R} \text{ iff } (y, x) \in R.$$

In general given an n-ary relation $R(x_1, \ldots, x_n)$ there is $n!$ partially converse relations. Various types of passive formation give us some, but not necessarily all, of the converse relations.

So far from the logicians point of view. The grammarian is interested in how the different converse relations are realized on the syntactic level, and what the connection between syntactic form and semantic reading is.

We shall outline a non-transformational treatment of passive within the framework of the model of Part I.

A. Before giving our own treatment we shall briefly recall the standard Montague analysis, see R. Thomason and K. Lee, in particular the latter's 1974 thesis, *The Treatment of Some English Constructions in Montague Grammar*.

The analysis of the sentence

(1) *Mary is loved by John*

is parallel to the analysis of

(2) *Mary is a woman loved by John.*

It is straightforward to write down the structural tree for (2). We reproduce that part of the tree which is not covered by our previous examples

(3)

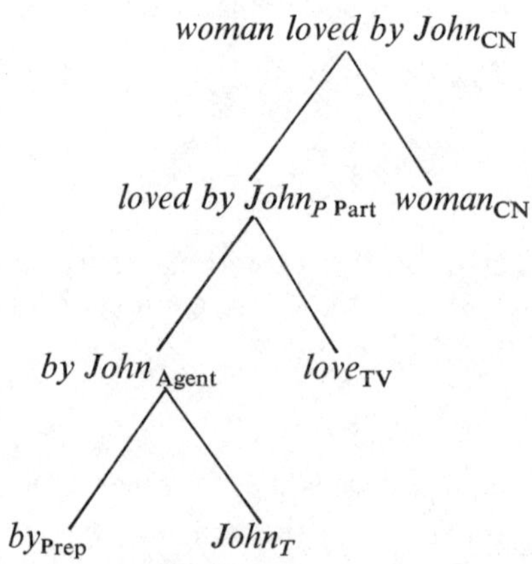

The tree gives the following category assignments

$$\begin{aligned} P\,\mathrm{Part} &= {}^{\mathrm{CN}}/_{\mathrm{CN}} \\ \mathrm{Agent} &= P\,\mathrm{Part}/_{\mathrm{TV}} \\ \mathrm{Prep} &= \mathrm{Agent}/_{T} \end{aligned}$$

The active form of (1) is

(4) *John loves Mary*

A reasonable active form of (2) is

(5) *John loves Mary and Mary is a woman*
(5′) *John loves Mary who is a woman*

A reasonable logical deep-structure for (5) (and also for (5′)) is

(6) $\mathscr{P}[\mathscr{W}(\lambda P.P(x))] \wedge Q(x)$,

i.e.

(7) *John*[*love*(*Mary*)] ∧ *woman*(*Mary*).

Given (3) and (6) we immediately read off the translation of *by*

(8) $by'_{\text{Prep}} = \lambda\mathscr{P}.\lambda\mathscr{W}.\lambda Q.\lambda x[[(\lambda P.P(x))] \wedge Q(x)]$.

Using the previous translation of be_{TV}, we get as translation of (2) the following particular instance of (6)

(9) $love'_{*}(j, m) \wedge woman'(m)$.

If we want an analysis of (4) parallel to (2) we have to work the effect of the part *a woman* into the copula *be*, i.e. we must read *be* as 'be some entity'. Being an entity is simply the function $\lambda z.[z = z]$.

The tree is now

(10)

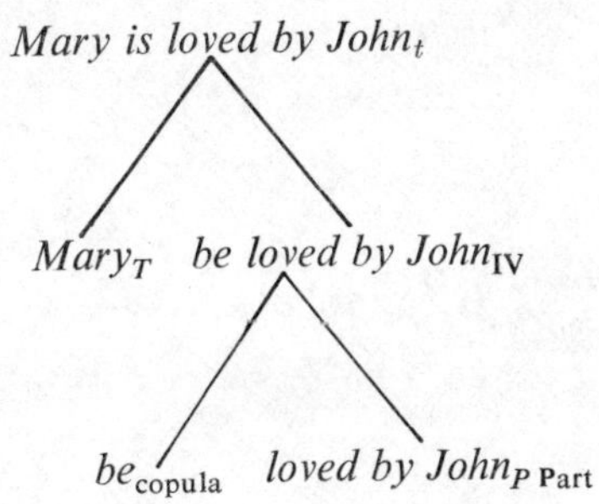

The category assignment is immediate: copula = $^{\text{IV}}/_{P\ \text{Part}}$. And the translation comes out as

(11) $be'_{\text{copula}} = \lambda\mathscr{Q}.\lambda x(\mathscr{Q}(\lambda z[z = z]))(x))$,

which gives the translation $love^*(j, m)$ of (1).

Note. The same analysis can be used to translate

(12) *Mary is pretty*,

which will then have the logical deep-structure "Mary is a pretty entity". The same analysis of be_{copula} can be used to interpret

(13) *Mary is loved.*

In this case we have to work the effect of an unspecified agent into the past participle *loved*. We are here led to

(14) $$loved'_{P\,\mathrm{Part}} = \lambda Q.\lambda x.\exists y[love'_*(y, x) \wedge Q(x)]$$

B. So far Lee. One may argue that the formulas (8) and (11) are rather complex. But they do the job.

However, taking our examples from Norwegian rather than English the analysis via (8), (11), and (14) becomes rather *ad hoc* and artificial.

In Norwegian there are several possibilities for forming passive, *bli*-passive where *bli* is an auxiliary verb different from *vaere* (*be*), and *s*-passive. Corresponding to (1) we have

(15) $$\mathit{Mary}\begin{Bmatrix}\mathit{blir\ elsket}\\ \mathit{elskes}\end{Bmatrix}\mathit{av\ John}$$

For (15) the 'natural' structural tree is

(16)

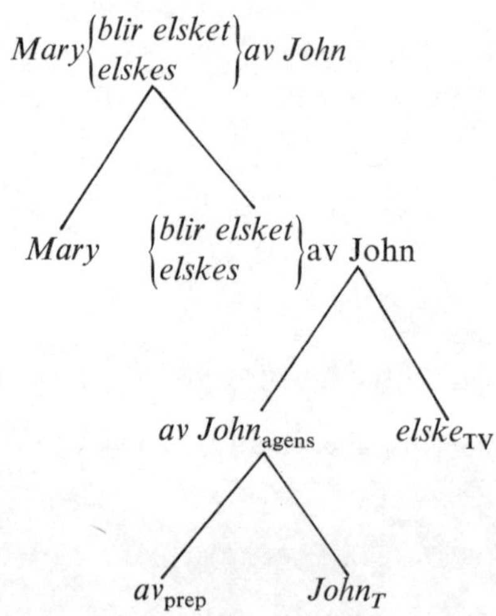

The category assignments come out as

$$\text{agens} = {}^{\mathrm{IV}}/_{\mathrm{TV}}$$
$$\text{prep} = {}^{\text{agens}}/_{T},$$

which should be compared to the categories of Agent, Prep, and *P* Part of Section **A**. In this case the translation of *av* (*by*) is

(17) $av'_{\text{prep}} = \lambda\mathscr{P}.\lambda\mathscr{W}.\lambda x\mathscr{P}[\mathscr{W}(\lambda P.P(x))]$

From this we easily get

(18) $$\left(\begin{Bmatrix} \textit{bli elsket} \\ \textit{elskes} \end{Bmatrix} \text{av John}\right)' = \lambda x.[elske'_*(j, x)],$$

which immediately gives the translation $elske'_*(j, m)$ of (15).

The main difference between this analysis and Lee's is that the auxiliary verb *bli* has not been given an independent translation. Corresponding to the tree (16) we have assumed a syntactic rule which acts on the *agens* term and the bivalent (transitive) verb as inputs and gives the desired participle as output. This seems to us even more natural when we do not have any auxiliary verb in the passive form, *elskes av John*.

Corresponding to (13) we have the following tree when *agens* is omitted.

(19)

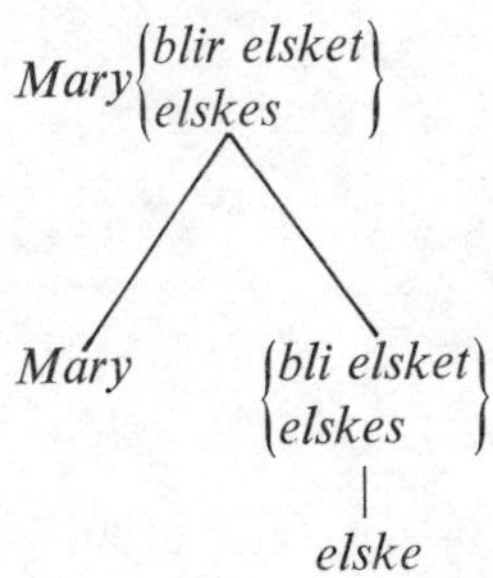

The translation of the converse relation is as direct as can be

(20) $$\begin{Bmatrix} \text{bli elsket} \\ \textit{elskes} \end{Bmatrix}' = \lambda x.\exists y[elske'_*(y, x)]$$

Note. In analogy to this analysis we can also give a similar one of *Mary er vakker* (Mary is pretty), i.e. not give an independent reading of the auxiliary verb *vaere* (*be*). This would give a more natural analysis of e.g. *Mary er laerer* (*Mary is teacher*), where we in Norwegian do not have any article, *en* (*a*).

Finally, note that this type of analysis is completely in line with Montague's treatment of the articles *a*/*an*, *every*, and *the*.

C. So far the standard case with a transitive verb. Consider now the following example

(21) *Det ble funnet et barn*

(there was found a child). How should the analysis of this proceed? Is it to be derived from

(22) *Et barn ble funnet*

by a suitable transformation? Or should it be treated analoguous to

(23) *Det ble rastet*

(approx. there was rested), which is a passive construction in connection with an intransitive verb *raste* (*to rest*).

A 'natural' tree for (23) is

(24)

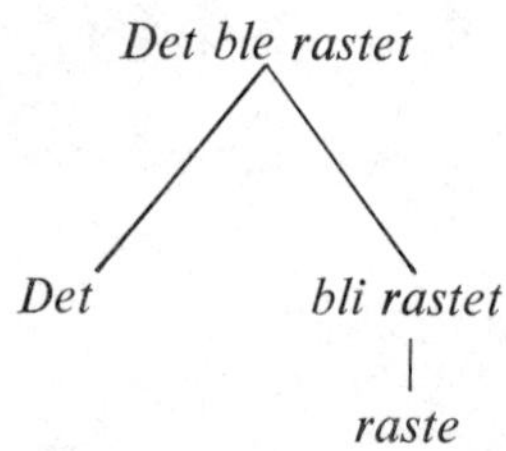

(25)

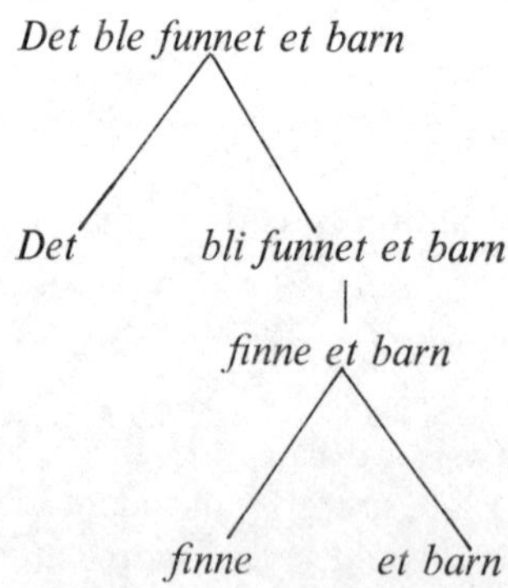

We note that in (24) and (25) we have intransitive verbs, *raste* and *finne et barn*, with translations *raste′* and

(26) $\lambda x.\exists z[barn'(z) \wedge finne'_{*}(x, z)]$.

Note that (26) follows from known rules. We now apply the same rule (20) which gave the transition from *elske* to *bli elsket* (*elskes*). The result is

(27) $(\textit{bli funnet et barn})' = \exists x[(\textit{finne et barn})'(x)]$
$= \exists x \exists z[\textit{barn}'(z) \wedge \textit{finne}'_{*}(x, z)]$

(28) $(\textit{bli rastet})' = \exists x[\textit{raste}'(x)]$

Omitting the *agens* in forming passive reduces the valency by one. Thus from the mono-valent verb phrases *raste* and *finne et barn* we derive the zero-valent forms of (27) and (28), which thus are of categorial type *t*.

The last step in the syntactic analysis is to add a formal subject (obligatory in Norwegian, but not in every language) to the zero-valent verb phrase. This we do by postulating a syntactic rule

(29) $F_{det}(\alpha) = det\alpha_1$,

where α_1 comes from α by changing *blir* to *ble/blir*. The associated rule of translation is simply

(30) $(F_{det}(\alpha))' = \alpha'$.

And we end up with the expected deep-structure analysis of (21) and (22).

Note. Our previous analysis of *av*, see (17), easily extends to cover such examples as

(31) *Det ble funnet et barn av John,*

(there was found a child by John). The only trick here is to lift the monovalent verb phrase *finne et barn* to a bivalent one

(32) $(\textit{finne et barn})'(x, \mathscr{P})$ iff $\mathscr{P}(\lambda z.\textit{finne et barn}'(x))$

to fit the format of (17)

D. We briefly indicate the case of a trivalent verb, i.e. a verb requiring both a subject, an object, and an indirect object. A typical example is

(33) *Petter ga Ola en bok*

(34) *Petter ga en bok til Ola*

(Petter gave a book to Ola.) There is a number of possible passive transforms. But common to each of them is first a reduction in the valency of the verb phrase.

Single reduction in valency.

(35) *En bok ble gitt Ola (av Petter)*

(36) *En bok ble gitt (av Petter) til Ola*

In these cases we assume first a transition from the verb *gi* to the verb phrase *gi (til) Ola*. But we could also have a transition from *gi* to *gi en bok*. This gives

(37) *Ola ble gitt en bok (av Petter)*

(38) *Ola ble gitt (av Petter) en bok*

Double reduction in valency.

(39) *Det ble gitt en bok til Ola*

(40) *Det ble gitt Ola en bok*

In these cases we assume a double reduction

$$\text{(41)}\quad gi \begin{cases} \nearrow\ gi\ en\ bok \rightarrow gi\ en\ bok\ til\ Ola \\ \searrow\ gi\ (til)\ Ola \rightarrow gi\ (til)\ Ola\ en\ bok \end{cases}$$

One may argue that the transitions described in (41) should belong to the lexical basis of the model and there have the following (derived) rules of translation:

(42) $(gi\ en\ bok)'_* = \lambda x . \lambda z \exists y[bok'(y) \wedge gi'(x, y, z)]$

(43) $(gi\ (til)\ Ola)'_* = \lambda x \lambda y[gi'(x, y, \boldsymbol{o})]$

(44) $\left.\begin{array}{l} (gi\ en\ bok\ til\ Ola)'_* \\ (gi\ (til)\ Ola\ en\ bok)'_* \end{array}\right\} = \lambda x \exists y[bok'(y) \wedge gi'(x, y, \boldsymbol{o})]$

Note. gi en bok is a bivalent verb phrase. (42) gives the 'reduced' (or, perhaps better: the lexical) translation. But as usual we can lift to the proper types

(45) $gi\ en\ bok'(x, \mathscr{P})$ iff $\mathscr{P}(\lambda z . gi\ en\ bok'_*(x, z))$

With these rules of translation the rest of the analysis follows in standard fashion.

(46)

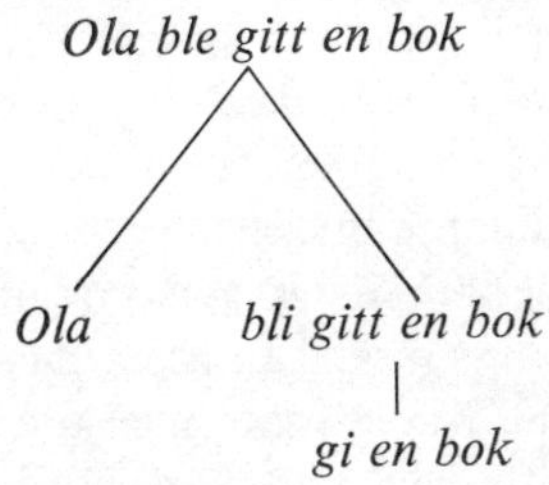

Using formula (20) we get the translation

$$\text{(47)} \qquad \textit{bli gitt en bok}' = \lambda z.\exists x[\textit{gi en bok}'_{*}(x, z)]$$
$$= \lambda z \exists x \exists y[\textit{bok}'(y) \wedge \textit{gi}'(x, y, z)]$$

This gives the end result

$$\text{(48)} \qquad (\textit{Ola ble gitt en bok})' = \exists x \exists y[\textit{bok}'(y) \wedge \textit{gi}'(x, y, \boldsymbol{o})]$$

The point to emphasize here is that we have made a distinction between the lexical analysis of a word and its (possibly) different syntactic roles. Thus the translation function g should really act on an ordered pair consisting of a lexical item and a category symbol; see also Example 2 of Part I.

We make one final remark on the use of the existential quantifier in the *agens*-less passive forms. It is perhaps reasonable to read *be loved* as *be loved by someone*. But when we say

Mary is liked

we may mean

Mary is generally liked
Mary is liked by many
Mary is liked by most,

it may not be sufficient that Mary is liked by just one. The meaning here depends on the context of the utterance.

But this can be taken care of using generalized quantifiers (see Example 1 of Part I). It is really an unspecified generalized quantifier that should be written into the translation of an *agens*-less form, such as (20), in passing from *surface structure* to *logical deep structure*. And it is the task of the F of an *interpretation* $\mathfrak{A} = \langle A, F \rangle$ to specify the meaning of the generalized quantifier: *one*, *several*, *many*, *most*, . . .

Remark. The generalized quantifier is simply a special kind of type-2 functional over the domain. This gives great flexibility in the type of analysis we are doing here. One more example: When we say that *the British are a friendly people*, we assert that the British in general, i.e. 'almost all' of them are friendly. But we are not prepared to assert of any given one that he is friendly. Over an infinite domain we can use a measure which gives every point measure zero but the whole space measure one. But if the space is finite, and the British, although many, are but finite, this is impossible.

The solution is simply to return to Bernoulli and give up the additivity of the measure – and we have a whole new range of generalized quantifiers at our disposal. (Non-additive measures have recently been studied by G. Shafter in *A mathematical theory of evidence*.)

III. THE MODEL EXTENDED

We have explained a three part formal model. The parts are:

(i) *Surface structure:* A language fragment generated from a lexical base by a set of syntactic rules.

(ii) *Deep structure:* A formal higher-order language being determined by a set of non-logical constants derived from the lexical base.

(iii) *Semantic interpretation:* A set-theoretic structure based on an arbitrary non-empty domain.

The model is formal, each of the three components is a formal construct. This is in itself not a serious or novel type of criticism. Any model in a natural science is a formal construct and yet can be used to describe, analyze, and predict phenomena in the 'outside world'.

The model uses in an essential way higher order logic, and this has been another source of criticism. But higher order structures are hardly any more objectionable in this type of model building than in other types of model construction for natural phenomena. Perhaps the reluctance against using higher order logic is connected with the incompleteness phenomena of these systems. But the question of completeness is very marginal to this type of analysis.

But there are other and more serious matters. The model lacks structure, the domain of the interpretation may be any non-empty set of 'individuals'. And the model is, at best, local.

We shall briefly discuss – or, rather, speculate about what kinds of extensions are necessary to overcome these difficulties.

A. We should not claim too much for the model. It is capable of describing *locally* or in an *isolated form* certain features or mechanisms of natural languages. This is brought out by the various examples of Parts I and II.

But language in use, the flow of speech in a conversation, is not a local but a *global* or *total* phenomenon.

Is it possible to extend the model of Part I in a simple step by step fashion to obtain a global analysis?

Probably not. It does not seem easier to explain the flow of speech than to explain globally the behaviour of the flow of 'dirty' (i.e. non-idealized) water in a river. Of course, locally present day theories of hydrodynamics can give an 'explanation'. But the synthesis of global behaviour from local analysis is still much of a mystery.

Keeping this picture in mind what is there to be said? Let us touch on some examples.

(i) When we say *the whale is a mammal*, we do not mean that there is in the world one and only one whale and this, unique, whale is a mammal.

But is not this what the model forces us to mean? *the whale is a mammal* is given a straightforward surface structure which translates into the deep structure as

$$mammal'(\iota x . whale'(x)).$$

This seems clear, and we should not like to give it up. And we need not. In the transition from deep structure to semantic interpretation, $\mathfrak{A} = \langle A, F \rangle$, we have a freedom of choice. The interpretation need not be a 'global interpretation'. Uttering the phrase *the whale is a mammal* we focus our attention on an interpretation $\mathfrak{A} = \langle A, F \rangle$ where there is just one 'whale' and this 'whale' has the property of being a mammal. If the phrase occurs as part of a longer conversation about whales, what we do at this point could be described as passing to a 'quotient' structure of the original interpretation obtained by the equivalence relation of lumping together all whales to one generic object. Continuing, we just pull back to the original structure what extra information we obtained by using the reduced structure.

This is one way of getting around the generic use of *the* while keeping the simple syntactic and deep structures of the phrase. There are other possibilities, e.g. the use of partial interpretations. In either way, we can keep the traditional surface structure, the usual deep structure with the description operator, and put all the burden on the choice of interpretation $\mathfrak{A} = \langle A, F\rangle$.

(ii) We mention another problem which has been treated extensively by J. Hintikka and L. Carlson in 'Conditionals, pronominalization by quantifiers and other applications of subgames'. Consider the sentence.

If John owns a book, he reads it.

In natural languages pronouns typically act as variables. The *he* refers back to *John*. And because of the uniqueness of reference of proper names, we can substitute *John* for *he* and not change the meaning. The pronoun *it* refers back to *a book*. This is a quantifier phrase with no unique reference. Hence we cannot with preservation of meaning substitute *a book* for *it*:

If John owns a book, he reads a book.

The book he reads may not be the book he owns. Thus the usual rule of substitution seems to break down. And if we try to keep the analysis within one interpretation, we will have trouble.

But lifting our eyes a bit from the formal details of the model we see that the common-sense solution to the puzzle is indeed the sound one. The premiss of a natural language conditional often acts to fix an interpretation to be used in the conclusion. *The* book John is reading is the one he owns.

Therefore the total interpretation of the phrase is a *functional* Φ which takes an interpretation F of the premiss and maps it to an interpretation $F' = \Phi(F)$ of the conclusion. In this case the functional Φ is induced by the so-called *specific reading* of a quantifier.

These are just two hints. Formal details can be supplied, but are not really appropriate in this discussion. Our examples are intended to show how the analysis can be extended. Not by forcing everything into one interpretation, but by *putting interpretations together in a more delicate manner*.

This leads from syntax and semantics to pragmatics. Recently, there has been a trend in intensional logic to force all of pragmatics into the index set I of a general model $\mathfrak{A} = \langle A, F, I\rangle$. This seems extremely artificial. In a speaker/listner situation we rather have to deal with *partial* interpretations, *conflicting* interpretations, the *building up* of an interpersonal interpretation as part of the conversations. And interpretations may be modified or determined by *functionals* which carry along the presuppositions introduced.

These are matters much discussed in recent pragmatics. Our remarks are intended to show that these discussions can be given a foundation in the formal (but local) models of syntax and semantics.

Remark. After having given our talk we read with great interest the paper by R. Bartsch, 'The role of categorial syntax in grammatical theory', in which she extends the formal model in a way compatible with the discussion here to account for certain pragmatic aspects.

B. An important criticism against the model of Part I is that an interpretation $\mathfrak{A} = \langle A, F \rangle$ has so little structure. The notion of interpretation derives in its technical form from logic. And the logicians are not interested in too much structure, because this will destroy the particular completeness theorem they may be interested in proving. But completeness, as remarked above, is only of marginal interest from the applied point of view.

As one example of how the interpretations lack structure take the transitive verb *love* of the lexical basis. In the deep structure we meet two versions, *love'* and $love'_*$, related in a certain canonical manner. In an interpretation we essentially have one binary relation *love* over the domain A. But nothing more. The model does not tell us what it 'means' for two objects of the domain to stand in the relation *love*, neither does it tell us how to 'decide' whether two objects of the domain belongs to the relation.

There seems to be one ready answer to this. Give the lexical basis more structure. And this is indeed of great importance, as we have hinted at several times in Parts I and II, i.e. forming the participle *bli elsket* from the active form *elske* is an example of a lexical analysis.

But lexical rules or meaning postulates give only part of the answer. It is equally important to enrich the structure of the domains. And to 'enrich' means to give more *geometric structure* to the interpretations.

Intensional logic today is full of dubious ontology and peculiar problems (such as cross-world identifications). And this may partly be due to the lack of structure which has been an inevitable consequence of the logician's almost excessive concern for completeness theorems.

The question of enriching the structure has not been totally absent from philosophical logic. As examples we may mention the essay 'Propositional Objects' by W. Quine and the discussion in Chapter 3 of Cresswell's book *Logic and Language*.

These discussions are, however, rather poor in geometric content. A contribution of a totally different order can be found in several papers by

R. Thom (see in particular his 'Topologie et linguistique' and 'Langage et Catastrophes: Eléments pour une Sémantique Topologique'.

Thom's starting point is the dependency grammar of L. Tesnier. Much of Tesnier's analysis can be captured on the level of our deep structure (*predicates* with their associated *actants* and *circonstants*). So there are obvious connections with the type of modelling expounded in this paper – a connection we plan to return to at some later occasion.

Remark. After having presented this paper I became acquainted with the recent book, *Language and Perception*, by G. A. Miller and P. N. Johnson-Laird where the authors lay the foundation for a field of inquiry called *psycholexicology*. Obviously, psycholexicology and the catastrophe theory of Thom are not unrelated activities, although it may not by easy at first to see the similarities through the 'surfaces' differences of the approaches.

If we believe that we use language not necessarily to describe the world as it is, but to describe the world as it is perceived, then the theories of Thom and of Miller and Johnson-Laird are bound to have an impact on the way we build models for natural languages.

The lesson to be drawn here is that the pure set theoretic models of the formal semanticians need to be augmented by a richer structure, a geometric structure.

Remarks. (1) A richer structure is, in particular, needed to give an adequate discussion of the semantics of *mass nouns*. For a recent discussion of mass nouns which points in this direction, see the paper 'The formal semantics of mass terms' by H. Bunt.

(2) Puzzles of cross-world identifications in intensional logic should also profit from some differentiable structure on the 'possible worlds'; an individual is not only a point, but a point with direction. See also Thom's notion of object in 'Langage et Catastrophes'; we plan to return to this topic at some later occasion.

We have indicated two types of extensions of the basic model, the need for a richer local structure and the need for a theory how local structures play together to give a global interpretation.

To avoid possible misunderstandings let me emphasize one point. The need for a richer local structure does not mean that the notion of set theoretic model is wrong. It is correct and adequate up to a certain point, as we believe was brought out by a number of examples in Parts I and II.

Perhaps the following example will illustrate the point better than any extended general discussion. In *model theory* we study structures of the form

$$\mathfrak{A} = \langle A, R_1, \ldots, R_n \rangle$$

where A is some non-empty set and $R_1, \ldots, R_n$ are relations over A. This is the setting for general model theoretic discussion, and the theory is rich in mathematical content.

Studying the theory of natural numbers in this setting we must construct *addition* as a relation between triples of natural numbers. And this is adequate up to a certain point. But if you want to know of three natural numbers a, b, and c if $a + b = c$, then, according to the theory, you have to ask if $\langle a, b, c \rangle$ belongs to the relation of addition. But how do you decide this? By checking if $a + b = c$. But this was the question we started with!

There is no paradox here. Relative to a general model we have to assume that the relations are there, unanalyzed, given in advance in the form of a (finite or infinite) list. But if we really want to *compute* the relation, we must go one step further. Typically, as in the case of addition, the relation will be introduced by a set of recursion equations

$$f(a, o) = g(a)$$
$$f(a, b + 1) = h(f(a, b), a, b),$$

where g and h are given or 'understood' functions. And, of course, this extended analysis is compatible with the set theoretic notion of relation and model.

Remark. Linguists are discussing the same phenomena. And the solutions they reach are often more similar than the external formalisms of the different schools would lead one to believe. As one example I was struck by the number of similarities between the present extended Montague-type model and the *Lexical-Interpretive Model of Transformational Grammar* recently proposed by J. Bresnan.

Department of Mathematics and *Department of Philosophy*,
University of Oslo

REFERENCES

Montague's papers on linguistics and related topics have been edited by R. Thomason, *Formal Philosophy*. A recent collection of studies is B. H. Partee, *Montague Grammar*. An introductory paper from the linguistic point of view is B. H. Partee, 'Montague Grammar and Transformational Grammar'. The present paper, which is a somewhat extended version of a survey lecture to the 4th Scandinavian Logic Symposium, is written from a logician/mathematician's point of view.

Ajdukiewicz, K. (1936), 'Die Syntaktische Konnexität', *Studia Philosophica* **1**.

Bartsch, R. (1976), 'The Role of Categorial Syntax in Grammatical Theory', in Kasher, A. (ed.), *Language in Focus*, D. Reidel, Dordrecht.

Bresnan, J. (1976), 'Toward a Realistic Model of Transformational Grammar', paper presented at the MIT-AT & T Convocation on Communications, Sessions on Language and Cognition.

Bunt, H. (1976), 'The Formal Semantics of Mass Terms', paper presented at the III Nordiska Lingvistmøtet.

Chomsky, N. (1957), *Syntactic Structures*, Mouton, The Hague.

Cresswell, M. J. (1973), *Logics and Languages*, Methuen, London.

Curry, H. B. (1961), 'Some Logical Aspects of Grammatical Structure', in Jakobson, R. (ed.), *The Structure of Language and its Mathematical Aspects*, American Math. Society, Providence, Rhode Island.

Hintikka, J. and Carlson, L. (1978), 'Conditionals, Pronominalization by Quantifiers, and Other Applications of Subgames' in E. Saarinen (ed.), *Game-Theoretical Semantics*, D. Reidel, Dordrecht.

Lee, K. (1974), *The Treatment of Some English Constructions in Montague Grammar*, Doctoral Dissertation, University of Texas, Austin.

McMahon, W. E. (1976), *Hans Reichenbach's Philosophy of Grammar*, Mouton, The Hague.

Miller, G. A. and Johnson-Laird, P. N. (1976), *Language and Perception*, Cambridge University Press.

Montague, R. (1973), 'The Proper Treatment of Quantification in Ordinary English', reprinted in Thomason (1974).

Partee, B. H. (1975), 'Montague Grammar and Transformational Grammar', *Linguistic Inquiry* **6**.

Partee, B. H. (ed.) (1976), *Montague Grammar*, Academic Press, New York.

Quine, W. V. (1969), 'Propositional Objects', in *Ontological Relativity and Other Essays*, Columbia University Press, New York.

Reichenbach, H. (1947), *Elements of Symbolic Logic*, Macmillan, New York.

Shafter, G. (1976), *A Mathematical Theory of Evidence*, Princeton University Press.

Thom, R. (1970), 'Topologie et Linguistique', in *Essays on Topology*, Springer-Verlag, Berlin.

Thom, R. (1973), 'Langage et Catastrophes: Eléments pour une Sémantique Topologique', in *Dynamical Systems*, Academic Press, New York.

Thomason, R. (ed.) (1974), *Formal Philosophy, Selected Papers of Richard Montague*, Yale University Press, New Haven, Connecticut.

ESA SAARINEN

BACKWARDS-LOOKING OPERATORS IN TENSE LOGIC AND IN NATURAL LANGUAGE*

INTRODUCTION

This paper contains two parts. In the first one, we shall argue that in the tense system of English there are particles which are best analysed as, what we call, backwards-looking operators. By means of these operators (whose formal counterparts are to a large extent new in the literature) we shall establish the following limiting thesis concerning an adequate semantics for English tenses: For all natural numbers *n*, the semantics should have a capacity to keep track of *n* points introduced earlier in an evaluation. In the second part of this paper, we shall present a formal language which contains operators of this new kind. We formulate explicit model theory for this formal language in Hintikka's game-theoretical semantics. This semantical approach is sufficiently rich to satisfy the condition laid down by the limiting thesis mentioned.

I

Limiting Theses

We shall start our discussion by presenting some *limiting theses* or *limiting proofs* concerning an adequate semantics for English tenses. A limiting thesis is a statement which gives a lower bound or minimum requirement for an adequate syntactical or semantical theory. There is a great deal of interest in this kind of result. If the point needs to be established let me quote Julius M. E. Moravcsik in his book *Understanding Language*.

> ... on the one hand, the construction of a model for the accomplishment of certain basic human skills is hardly of great interest; such a feat by itself gives no guarantee that we found the way in which humans naturally perform the actions in question. The same performance can be reached by quite different processes and structures. On the other hand, so-called *limiting-proofs*, i.e., arguments that show how a device of a certain

* I am indebted to Lauri Carlson, Jaakko Hintikka, and Krister Segerberg. Most of the material of the first part of this paper has appeared under the title 'How Complex is English Tense Structure?' in *Papers from the Third Scandinavian Conference of Linguistics*, Turku, Finland, pp. 337–348.

J. Hintikka, I. Niiniluoto, and E. Saarinen (eds.), Essays on Mathematical and Philosophical Logic, 341–367.

degree of complexity is incapable of accepting a language with a certain set of formal properties, carry great explanatory power; it is in virtue of such arguments that we can draw some inferences about the nature of human mind. (Moravcsik, 1975, p. 63.)

The limiting thesis or proof which we shall try to establish here deals with the so-called problem of 'memory'. A semantics for a tense logical language typically splits up a given formula into simpler and simpler ones. Hence we face a sequence of stages of evaluation, and at each stage we are lead to consider simpler and simpler formulas. Now a problem of memory arises when the truth value of a formula cannot be determined without knowledge of some aspects of an earlier stage of the evaluation.

The question we are here interested in can be put thus: What information should the memory of an adequate semantics for English tense system contain? Some minimum requirements are known from the literature, the first one being due to Hans Kamp (1971):

Limiting Thesis 1 (Kamp). The semantics should keep track of the original point.

(By 'original point' we here mean 'the point of utterance' or the point from which the evaluation started.) This observation of Kamp's can be appreciated by considering some typical uses of 'now'. Consider for instance the following sentences:

(1) I learned last week that there would now be an earthquake.
(2) Dick once thought that he would now be the president.

In these sentences, 'now' refers to the original point of time even though it occurs in them in the scope of other tenses. For instance in (1) 'now' refers to the moment of utterance and not to the time last week when I learned about the earthquake. The same holds *mutatis mutandis* of sentence (2) as well. (For further discussion, see Kamp, 1971; Prior, 1968b; and Gabbay, 1974.)

The history of the following limiting thesis is not clear to us but as far as we know it is due to Frank Vlach (1973):

Limiting Thesis 2 (Vlach). The semantics should keep track of the immediately preceding point.

Here the term 'immediately preceding point' refers to the point *in an evaluation* immediately preceding the point at which we are considering a formula. Vlach's sentence is this:

(3) John was once going to cite everyone then driving too fast.

The uses of 'then' exemplified by this sentence suggest other examples, such as:

(4) Every man who ever supported the Vietnam war now thinks that he was an idiot then.

In this sentence, 'then' refers to the time which immediately precedes (in the evaluation) the moment at which the sentence 'he was an idiot then' is considered. However, this point is not the original point. (An analogous point can be made concerning (3).) Hence limiting thesis 2 is established. (This limiting thesis is also discussed and argued for in Gabbay, 1974.)

Limiting Thesis 3 (*Gabbay and Moravcsik*). *It is not enough for our semantics to keep track of the original point and the immediately preceding point.*

The more convincing one of the two sentences Gabbay and Moravcsik present in favour of their conclusion is this:

(5) She will go to the school, that was her mother's alma mater, and it will become better than Harvard.

Here the last future operator is what we are interested in. It is essential that this operator refers to the future of 'her' attendance at the school; but this time is neither the original time nor the immediately preceding time, as counted from the time point at which the last futurity operator is evaluated. (In fact, this statement holds only if 'was' is analysed as a 'two-step' operator, first taking us to the original time and then to the past of this moment of time.)

Gabbay and Moravcsik raise doubts concerning (5) as evidence for their conclusion. Couldn't (5) be explained away on syntactic grounds, they ask? As they say, "we have here a co-ordinate structure and what we 'jump over' is a relative clause of the first conjunct. Thus explanation for the 'jumpings' might be that in a sequence of conjuncts the temporal reference points form a sequence that bypasses the sequence built into any relative clause." (Gabbay and Moravcsik as reported in Gabbay, 1974.)

The explanation Gabbay and Moravcsik offer for (5) might very well work in the case of that particular sentence. However, the conclusion can be established by other examples. Consider for instance

(6) Every man who ever supported the Vietnam war believes now that one day he will have to admit that he was an idiot then.

Here 'then' refers to the time when the supporting took place. This time is not the original time nor the one immediately preceding the one at which 'then' is evaluated. On the other hand, no explanation of the Gabbay-Moravcsik type is possible here.

Although (6) in our judgement already establishes limiting thesis 3, it is instructive to consider another type of example which establishes the same conclusion. This new type of example will illustrate a phenomenon that plays an important role in the argumentation below.

First, let us note that in

(7) A child was born who would become ruler of the world

'would' refers to the future of the moment of the birth. In

(8) Joe said that a child had been born who would become ruler of the world

'would' is ambiguous. It can refer to the future of the moment of the birth *or* to the future of the moment at which Joe's saying took place. The reason for this ambiguity is not hard to see. What is involved is the shifting of tenses that takes place in indirect speech. When one reports what someone said, expressed, though etc., on some previous occasion by using indirect speech, the tenses of the sentence being reported shift. (On this point, see e.g., Jespersen, 1933, p. 260.) This is why the main verb of (7) is shifted to its pluperfect form. However, the auxiliary 'would' of the relative clause remains unaffected (unmarked) because in English there is no special item to mark the shifted form of 'would'. This is what creates the ambiguity of (8). For 'would' in the relative clause of (8) can be either the unmarked shifted form of 'would' in (7) *or else* the shifted form of the auxiliary 'will'. In other words, the sentence being reported in (8) need not be (7) but could as well be

(7)′ A child was born who will become ruler of the world.

(The shifted form of the auxiliary 'will' is 'would'. See Jespersen, 1933, pp. 288–289.) Below, we shall make use of this observation that the auxiliary 'would' in indirect speech can be either the shifted form of the auxiliary 'would' or the auxiliary 'will' occurring in the direct speech statement being reported.

One application of the above observations is to note that in

(9) Bob mentioned that Joe had said that a child had been born who would become ruler of the world

'would' creates a three-way ambiguity. (There are possibly other sources of ambiguity as well. They are not relevant here.) In (9) we report what was mentioned by Bob on an earlier occasion. But what was mentioned by Bob on that occasion? There are two possibilities here. Either Bob mentioned sentence (8), or else he mentioned not (8) but

(8)′ Joe said that a child had been born who will become ruler of the world.

This shows that in contrast to (8), (9) allows the additional possibility that 'would' refers to the future of the time when Bob's mentioning took place. But the behaviour of 'would' under this reading cannot be grasped in a semantics that recalls only the original point and the immediately preceding one. Hence the sentence (9) shows that limiting thesis 3 expresses a statement true of the English tense system. Again, no explanation of the Gabbay-Moravcsik type is possible here.

The following generalization of limiting thesis 3 has been put forward by Gabbay and Moravcsik:

Limiting Thesis 4. The semantics should keep track of all points introduced in an evaluation as well as the operators that introduced them.

Nevertheless, Gabbay and Moravcsik do not present evidence for their claim. (See Gabbay, 1974.) We shall here try to argue in favour of their claim by presenting a 'proof' of the following somewhat weaker statement:

Limiting Thesis 5. The semantics should have a capacity to be able to keep track of any finite number of points that are introduced in an evaluation.

We shall establish our result by presenting an infinite chain of English sentences with more and more complex tense structure. The base of the construction is this sentence:

(10) A child was born.

Consider now

(11) Joe said that a child had been born.

(12) Bob mentioned that Joe had said that a child had been born.

(13) Jim denied that Bob had mentioned that Joe had said that a child had been born.

(14) Gil was sure that Jim had denied that Bob had mentioned that Joe had said that a child had been born.

We can continue the process suggested by these examples to produce a denumerable chain of sentences of increasing complexity. More precisely, we take an NP, a transitive (S-complement taking) verb in the imperfect tense, and the sentence thus far constructed with the main verb of the highest S changed to its pluperfect tense, forming the concatenation of these items and the lexical item 'that' in the obvious way. It is easy to select the NP's and V's in such a way that all the sentences are both semantically and syntactically acceptable English sentences. The construction thus defined is called construction A.

We shall now describe another construction of sentences which we shall call construction B. The starting point is again sentence (10). Consider now

(15) A child was born who would become ruler of the world.

(16) A child was born who would become ruler of the world and who would be widely recognized.

(17) A child was born who would become ruler of the world, who would be widely recognized and who would be considered to be one of the greatest of men.

(18) A child was born who would become ruler of the world, who would be widely recognized, who would be considered to be one of the greatest of men, and who would be killed.

A construction of sentences is suggested by these examples. The induction step is defined by stipulating that whenever *S* is the sentence thus far constructed, then the next sentence is obtained by concatenating *S* in the obvious way with

(19) and who would VP

(for some VP) and deleting the last occurrence of 'and' in *S*. Clearly we can select the VP's in such a way that all the sentences are both syntactically and semantically acceptable English sentences.

The two constructions can be combined to yield a construction C which generates for each *n*, *k* in Nat (the set of natural numbers without 0) an acceptable English sentence such that the sentence

(20) A child had been born

is in the scope of *n* past tense operators and the NP 'a child' has in its scope *k* relative clauses each of which contains one occurrence of 'would.'

What is important here is that each occurrence of 'would' in sentences thus constructed can refer to the future of the time of any of the past tense operators dominating it. (The 'time of an operator' is the time that operator introduces in an evaluation.) To see this, recall what was observed about the auxiliary 'would' in indirect speech, in particular that 'would' can mark either a shifted form of 'will' or 'would' itself. Then consider as an example

(21) Joe said that a child had been born who would become ruler of the world and who would be widely recognized.

In this sentence, either of the occurrences of 'would' can refer to the future of the moment when Joe's saying took place or to the future of the moment when the child's birth took place. In (21) we report, using indirect speech, what Joe said at an earlier occasion. Keeping in mind the two different roles 'would' can play in indirect speech, we see that the relevant sentences Joe might have said on that previous occasion are the following:

(22) A child was born who would become ruler of the world and who would be widely recognized.

(23) A child was born who would become ruler of the world and who will be widely recognized.

(24) A child was born who will become ruler of the world and who would be widely recognized.

(25) A child was born who will become ruler of the world and who will be widely recognized.

(We are not interested in other ambiguities there possibly are in (21).)

The above observations show that what marks a difference between the readings is the moments with respect to which the futurity operators operate. For instance, the readings which (22) and (25) give us differ from each other in that the latter is true only if the moments of becoming ruler of the world and becoming widely recognized are both in the future of the moment when Joe's saying took place. The former reading allows more circumstances because it is true if the moments of becoming ruler of the world and becoming widely recognized are in the future of the child's birth.

Clearly the above argument can be restated for any acceptable sentence obtained from the sentence

(26) Joe said that a child had been born

by construction B, i.e., no matter how many occurrences of 'would' we have in the scope of the two past tense operators in (26). As the number of such occurrences of 'would' increases, so also does the number of different readings increase.

The number of different readings increases also when we apply construction A to sentences considered, i.e., when the number of past tense operators having 'would' in their scope increases. The situation which arises is analogous to that found above in connection of sentence (9). Sentences given to us by construction A are indirect speech representations of what have been said, thought, expressed etc., on an earlier occasion. But when we paraphrase the latter in direct speech we get again an indirect report of what has been said, thought, expressed etc., on an earlier occasion. This argument can be reiterated until we come to the sentence that served as the base for construction A. At each stage of these 'jumpings' from indirect to direct speech, we must be careful with the tense shifts involved. In particular, we must recall that a 'would' of indirect speech can be shifted form of the auxiliary 'will' or simply of the auxiliary 'would' itself. This gives us the multiplicity of ambiguities in which we are interested.

An example might be helpful here. For this purpose, consider

(27) Bob mentioned that Joe had said that a child had been born who would become ruler of the world and who would be widely recognized.

In (27), we face quite a number of different readings. For (27) to be true at a certain moment, there must be an earlier moment at which Bob mentions a certain sentence. What is this sentence that we report in (27) Bob to have mentioned? Because of the tense shift, there (at least) the following alternatives for the sentence Bob could have mentioned:

(28) Joe said that a child had been born who would become ruler of the world and who would be widely recognized.

(29) Joe said that a child had been born who will become ruler of the world and who would be widely recognized.

(30) Joe said that a child had been born who would become ruler of the world and who will be widely recognized.

(31) Joe said that a child had been born who will become ruler of the world and who will be widely recognized.

Again, what marks a difference between the different readings is the moment with respect to which the futurities are counted. Notice that sentences (28)–(30) are themselves ambiguous; (28) is four ways ambiguous while (29) and (30) are two, as far as 'would' is concerned.

The above considerations show that in (27), both occurrences of 'would' can refer to the future of any of the past tense operators there. Hence either one of them can refer either to the future of the moment when either Bob's mentioning, Joe's saying or the child's birth took place. This gives us the nine different readings of (27) we are here interested in.

These observations prove our claim that each occurrence of 'would' in all sentences given by construction C can refer to the future of the time of any of the past tense operators dominating it.

To establish limiting thesis 5, we actually need only a somewhat weaker result. This weaker result can be expressed in the form of the following conclusion:

For each *n* in Nat, we can construct a semantically and syntactically acceptable English sentence *S* such that

(i) there are *n* occurrences of 'would' which are dominated by *n* past tense operators in *S*;
(ii) *S* has a reading such that for each occurrence of 'would' in *S* there is a past tense operator which introduces the time to the future of which that occurrence of 'would' refers;
(iii) no two occurrences of 'would' in *S* refer to the future of the time introduced by one and the same past tense operator.

This conclusion establishes the observation expressed by limiting thesis 5 that an adequate semantics for English tenses will have to have a memory that can keep track of unrestrictedly many points. Furthermore, this yields evidence for limiting thesis 4 also, even though it does not establish it.

'Would' and 'Then' as Backwards-Looking Operators

In the kind of uses studied above, 'then' is most naturally analysed as a 'backwards-looking' operator, an operator whose intuitive task is to step back to a point considered earlier. This is clearly seen e.g. from sentence (4):

(4) Every man who ever supported the Vietnam war now thinks that he was an idiot then.

What happens here is that 'then' refers to the time when the supporting took place. This time in turn is given by the first past tense operator in (4). It is hence natural to capture the meaning of (4) in the following lines:

(32) $(x)(x$ is a man $\supset H(x$ supports the Vietnam war $\supset NT_x D(x$ is an idiot$)))$.

Here D is an operator of a new kind. Its task is to refer to the point which the operator H introduced to us. This line of thought will be followed in the second part of this paper where we construct a formal language that contains backwards-looking operators. (The syntax of the new operators will nevertheless be more complicated than what is suggested by the above example.) This will give us a formal treatment of some typical uses of 'then'.

Reflection on the behaviour of 'would' (in uses of the kind considered above) suggests that something similar is also involved there. Take e.g. the different readings found in

(9) Bob mentioned that Joe had said that a child had been born who would become ruler of the world.

What marked a difference between the different readings was the moment to the future of which 'would' refers. In each case this moment is introduced by an operator in (9). It is hence natural to analyse 'would' to be an operator with two components. In this analysis, the first component would simply be a futurity operator of the usual Priorean kind that refers to the future of a given moment. The second component would be a backwards-looking operator that brings us back to a point introduced earlier. In the different readings of 'would' in (9), the first component would be the same in all of them but the second component would vary. This is the line of thought followed below. It will amount to a formal treatment of some typical uses of 'would'.

II

We shall now develop an explicit model theoretic account of the backwards-looking operators. This will enable us to formalize the uses of 'then' and 'would' studied above.

We shall spell out the semantics of backwards-looking operators using Hintikka's game-theoretical semantics. This semantical approach has turned out to be fruitful both in the study of some formal languages as well as in the study of the semantics of natural language. (Of the former, see Hintikka, 1968; 1974; 1975a; Hintikka and Rantala, 1976; Rantala, 1975; 1978. Of the latter, see Hintikka, 1974; 1975b; 1976; 1978; Hintikka and Carlson, 1977; 1978; Hintikka and Saarinen, 1975; Saarinen, 1978. See also Saarinen, 1977a.)

The work of Hans Kamp and especially that of Frank Vlach shows that usual Carnap-Tarski type semantics (as extended to intensional languages by Kripke and others) can be generalized so as to meet limiting theses 1 and 2. (See Kamp, 1971; Vlach, 1973.) This leads to the study of two-dimensional tense (and other intensional) logics. (Independently of Vlach, these logics were studied by Åqvist, 1973 and Segerberg, 1973. See also Gabbay, 1976a; 1976b.) Even two-dimensional tense logics are not sufficiently rich to meet limiting theses 4 and 5. However, the variable dimensional tense logics suggested in the appendix of Vlach's thesis and studied by Kuhn (1976) are sufficiently rich to satisfy the condition laid down by limiting thesis 5. This is hardly surprising for Kuhn shows that (under certain assumptions) the variable dimensional logics are simply first order predicate logics in another dress. In our judgement, these variable dimensional logics are to a large extent *ad hoc*. Backwards-looking operators, which is the phenomenon motivating these logics, can be handled in the game-theoretical framework without introducing any extra concepts over and above those we would have needed anyway. Indeed, usual game-theoretical semantics meets not only limiting thesis 5 but limiting thesis 4 as well. Consequently, in this approach the backwards-looking operators can be handled very naturally, as we hope to show below. This result can be considered as indirect evidence for this semantical approach. (On this topic, see also Saarinen, 1977b.)

In most cases, it is sufficient to introduce in informal terms the notions needed in game-theoretical semantics. In the case of backwards-looking operators this is not so, however. The presentation of the semantics of these operators will benefit considerably from an explicit account of the notions used. As there is not a formal treatment of game-theoretical semantics available in the literature, we shall have to present one. (Rantala, 1978, gives a formal game-theoretical semantics for some infinitary languages. His treatment does not suit our purposes, however.)

Below, we shall first define a first order predicate tense logical language $\mathscr{L}(\mathscr{O})$. To make our treatment sufficiently general, we shall allow $\mathscr{L}(\mathscr{O})$ to contain just any class $\mathscr{O}$ of tense operators of the usual kind, in a sense to be defined. For simplicity, all our tense operators will be one-place.

We shall then present game-theoretical semantics for $\mathscr{L}(\mathscr{O})$. After this, we are ready to extend our language $\mathscr{L}(\mathscr{O})$ so as to contain backwards-looking operators. Game-theoretical semantics for the new language $\mathscr{L}(\mathscr{O}, \mathscr{D})$ will then follow.

Syntax for a Tense-Logic

Definition 1. The *vocabulary* of the language $\mathscr{L}(\mathscr{O})$ consists of:

(i) A denumerable set $\mathscr{V}$ of individual variables $v_1, v_2, \ldots$

(ii) A (possibly empty) set $\mathscr{C}$ of individual constants $a_1, a_2, \ldots$

(iii) For each finite i in Nat, a (possibly empty) set $\mathscr{Q}(i)$ of i-place relation symbols $Q_1^i, Q_2^i, \ldots$ The union of all sets $\mathscr{Q}(i)$ is denoted by $\mathscr{Q}$.

(iv) The symbols $\sim$ (negation), $\vee$ (disjunction), $=$ (the identity symbol),) and ((parentheses).

(v) A (possibly empty) set $\mathscr{O}$ of one-place tense operators $O_1, O_2, \ldots$

Definition 2. An atomic formula of $\mathscr{L}(\mathscr{O})$ is defined by the following clauses:

(i) If Q^n is in $\mathscr{Q}$ and $t_1, \ldots, t_n$ are in the union of $\mathscr{C}$ and $\mathscr{V}$, then $Q^n(t_1, \ldots, t_n)$ is an atomic formula.

(ii) If t_1 and t_2 are in the union of $\mathscr{C}$ and $\mathscr{V}$, then $(t_1 = t_2)$ is an atomic formula.

(iii) Nothing else is an atomic formula.

Definition 3. A *formula* of $\mathscr{L}(\mathscr{O})$ is defined by the following clauses:

(i) Each atomic formula is a formula.

(ii) If A and B are formulas, then so are $\sim A$, $(A \vee B)$, and OA, for all $O \in \mathscr{O}$.

(iii) If A is a formula and v_n is in $\mathscr{V}$, then $(v_n)A$ is a formula.

(iv) Nothing else is a formula.

Game-Theoretical Semantics for a Tense Logic

In game-theoretical semantics we assume that we are dealing with an interpreted language and that a truth definition for all atomic formulas of the language has been given. For this objective, the concepts of a frame and model must be defined.

Definition 4. A *frame* F for $\mathscr{L}(\mathcal{O})$ is a triple (T, R, I), where T and I are non-empty sets and R is a subset of T^2. (Whenever U and V are sets, U^V is the set of all functions from V to U.) A *model* M for $\mathscr{L}(\mathcal{O})$ is a frame (T, R, I) together with an interpretation function V such that

for all c in $\mathscr{C}$, $V(c)$ is a function from T to I.
for all Q^n in $\mathscr{Q}$, $V(Q^n)$ is a function from T to the set of all subsets of I^n.

Definition 5. Let c be an element of $\mathscr{C}$ and v_n an element of $\mathscr{V}$. Then define for each model (T, R, I, V) for $\mathscr{L}(\mathcal{O})$ and for each x in I^ω,

$c[x] = V(c)$,
$v_n[x]$ is that function f from T to I such that for each $t \in T$, $f(t) =$ the nth element of the sequence x.

Definition 6. Let (T, R, I, V) be a model for $\mathscr{L}(\mathcal{O})$ and assume t is in T and x is in I^ω. Then *an atomic formula A of $\mathscr{L}(\mathcal{O})$ is true* at point t in model $M = (T, R, I, V)$ for $\mathscr{L}(\mathcal{O})$ with respect to sequence x in I^ω, in symbols $\overset{M}{\underset{t}{\models}} A[x]$ iff

(i) if A is $Q^n(t_1, \ldots, t_n)$ for some $Q^n \in \mathscr{Q}(n)$, $t_1, \ldots, t_n \in \mathscr{C} \cup \mathscr{V}$, then $(t_1[x](t), \ldots, t_n[x](t)) \in V(Q^n)(t)$,
(ii) if A is $(t_1 = t_2)$ for some $t_1, t_2 \in \mathscr{C} \cup \mathscr{V}$, then $t_1[x](t) = t_2[x](t)$.

Otherwise the atomic formula A is false at point t in model (T, R, I, V) with respect to sequence x, in symbols $\overset{M}{\underset{t}{\not\models}} A[x]$.

Definition 7. If $F = (T, R, I)$ is a frame for $\mathscr{L}(\mathcal{O})$, then we say that T is the *universe* of F and I is the *domain* of it.

The treatment of individuals adopted above (and further developed in the quantifier-rule below) is not the only possible one. In the judgement of the present author, it is not even the best or most natural one. Because the problem of individuals and quantifiers in intensional contexts is not the main issue of this paper, we have nevertheless chosen to use this kind of treatment which is rather pleasant to work with. Something similar is also often found in the literature. (See e.g. Kamp, 1971.)

We shall now extend the concept of truth of a formula of $\mathscr{L}(\mathcal{O})$ from the atomic formulas to all formulas of the language. A fundamental concept of our semantics is that of a semantical game tree. This concept is given by the following inductive definition:

Definition 8. Let $\mathscr{T} = (\mathscr{N}, \mathscr{R})$ be a binary structure. Then $\mathscr{T}$ is *the semantical game tree* of a formula A of $\mathscr{L}(\mathscr{O})$ with respect to model $M = (T, R, I, V)$ for $\mathscr{L}(\mathscr{O})$, point t in T, and x in I^{ω} iff $\mathscr{T}$ satisfies the following conditions:

(1) There is $r = (\mu_1, \ldots, \mu_4)$ in $\mathscr{N}$ such that $\mu_1 = A$, $\mu_2 = t$, $\mu_3 = x$, and $\mu_4 = (\phi, \phi)$. (Here ϕ is the empty set.)

(2) Assume $n_1 = (\mu_1, \ldots, \mu_4)$ is in $\mathscr{N}$ and $r\mathscr{R}^k n_1$ for some k in Nat. Then $n_1 \mathscr{R} n_2$ for any $n_2 = (\zeta_1, \ldots, \zeta_4)$ in $\mathscr{N}$ iff μ_1 is a non-atomic formula of $\mathscr{L}(\mathscr{O})$ and the following conditions are fulfilled:

(C.~) If $\mu_1 = {\sim}B$ for some formula B of $\mathscr{L}(\mathscr{O})$, then $\zeta_1 = B$, $\zeta_i = \mu_i$ for $i = 2, 3$, and $\zeta_4 = (\mu_4, n_1)$.

(C.v) If $\mu_1 = (B_1 \vee B_2)$ for some formulas B_1 and B_2 of $L(\mathscr{O})$, then $\zeta_1 = B_1$ or $\zeta_1 = B_2$, $\zeta_i = \mu_i$ for $i = 2, 3$, and $\zeta_4 = (\mu_4, n_1)$.

(C.O) If $\mu_1 = OB$ for some formula B of $\mathscr{L}(\mathscr{O})$ and some O in $\mathscr{O}$, then $\zeta_1 = B$, ζ_2 is any element of T, $\zeta_3 = \mu_3$, and $\zeta_4 = (\mu_4, n_1)$.

(C.U) If $\mu_1 = (v_n)B$ for some formula B of $\mathscr{L}(\mathscr{O})$ and some v_n in $\mathscr{V}$, then $\zeta_1 = B$, $\zeta_2 = \mu_2$, ζ_3 is any element of I^{ω} otherwise identical with μ_3 but which differs from μ_3 in that the nth element of ζ_3 is allowed to be any element of $\mathscr{T}$, and $\zeta_4 = (\mu_4, n_1)$.

(3) For each n in $\mathscr{N}$ other than r there is k in Nat such that $r\mathscr{R}^k n$, and for no n in $\mathscr{N}$, $n\mathscr{R}r$.

(4) There is no binary structure $\mathscr{T}' = (\mathscr{N}', \mathscr{R}')$ such that $\mathscr{T}'$ satisfies (1)–(3) and $\mathscr{N}$ is a proper subset of $\mathscr{N}'$.

When a binary structure $\mathscr{T}$ is the semantical game tree of a formula A with respect to model $M = (T, R, I, V)$, point t in T, and sequence x in I^{ω}, we write $\mathscr{T} = (\mathscr{N}, \mathscr{R})^{M,t}_{A,x}$. We say that $\mathscr{T}$ is a semantical game tree for $\mathscr{L}(\mathscr{O})$ when there are A, M, t and x such that $\mathscr{T} = (\mathscr{N}, \mathscr{R})^{M,t}_{A,x}$.

We observe without proof that for each formula A of $\mathscr{L}(\mathscr{O})$, each model $M = (T, R, I, V)$, each t in T, and each x in I^{ω}, $(\mathscr{N}, \mathscr{R})^{M,t}_{A,x}$ is uniquely determined.

The fourth element of the nodes n in $\mathcal{N}$ of our semantical game trees $(\mathcal{N}, \mathcal{R})$ are there just to guarantee that our game trees are always trees in the topological sense of the word. The meaning of a tree we have in mind here is given by the following definition:

Definition 9. A binary structure $(\mathcal{N}, \mathcal{R})$ is a *tree* (in our sense) iff (i) there is a distinguished element r in $\mathcal{N}$ (called the *root* of the tree) such that for no n in $\mathcal{N}$, $n\mathcal{R}r$; (ii) for all n in $\mathcal{N}$ other than r there is k in Nat such that $r\mathcal{R}^k n$; for all n in $\mathcal{N}$ other than r there is precisely one n' in $\mathcal{N}$ such that $n'\mathcal{R}n$.

It can be easily checked that whenever $\mathcal{T}$ is a semantical game tree for $\mathcal{L}(\mathcal{O})$, then $\mathcal{T}$ is also a tree in the sense of the above definition.

Definition 10. Let $\mathcal{T} = (\mathcal{N}, \mathcal{R})$ be a semantical game tree for $\mathcal{L}(\mathcal{O})$. Each element in $\mathcal{N}$ is a *node*. If n is in $\mathcal{N}$ and there is no n' such that n' is in $\mathcal{N}$ and $n\mathcal{R}n'$, then n is *an end node*. A sequence $(n_1, \ldots, n_k)$ with n_i in $\mathcal{N}$ for each i $(1 \leq i \leq k)$ is a *path* in $(\mathcal{N}, \mathcal{R})$ iff n_1 is the root of $(\mathcal{N}, \mathcal{R})$ and for each i $(1 \leq i \leq k)$ $n_i\mathcal{R}n_{i+1}$. If $(n_1, \ldots, n_k)$ is a path in $(\mathcal{N}, \mathcal{R})$ and n_k is an end node, then the path is a *maximal path*.

As is generally the case in mathematical game theory, the intuitive meaning of the concept of a semantical game tree is not hard to see. A maximal path of a game tree $\mathcal{T}$ represents a play of the corresponding game. Each node in the game tree represents a position in the game, the root representing the initial position. The branches issuing from a given node represent all the different positions that are possible at the next stage of the game. In the definition of a semantical game tree, condition (4) ensures that all the relevant alternatives are present.

It would now be natural to proceed to define which player's move is marked by a given node in a semantical game tree. However, this will not be done here. The reason for this departure from the game-theoretical practice is the general treatment of the usual tense operators (members of $\mathcal{O}$) adopted by us.

Assume that $(\mathcal{N}, \mathcal{R})_{A,t}^{M,x}$ is a semantical game tree for $\mathcal{L}(\mathcal{O})$ with $M = (T, R, I, V)$, t in T, and x in I^ω. Then note that whenever $n = (\mu_1, \ldots, \mu_4)$ is in $\mathcal{N}$, the following statements hold:

(1) If $\mu_1 = {\sim}B$ for some formula B of $\mathcal{L}(\mathcal{O})$, then there is precisely one node n' in $\mathcal{N}$ such that $n\mathcal{R}n'$.

(2) If $\mu_1 = (B_1 \vee B_2)$ for some formulas B_1 and B_2 of $\mathcal{L}(\mathcal{O})$ (not

identical with each other), then there are precisely two nodes n' in $\mathcal{N}$ such that $n\mathcal{R}n'$.

(3) If $\mu_1 = OB$ for some formula B of $\mathcal{L}(\mathcal{O})$ and some O in $\mathcal{O}$, then there are precisely card(T) many nodes n' in $\mathcal{N}$ such that $n\mathcal{R}n'$.

(4) If $\mu_1 = (v_n)B$ for some formula B of $\mathcal{L}(\mathcal{O})$ and some v_n in $\mathcal{V}$, then there are precisely card(I) many nodes n' in $\mathcal{N}$ such that $n\mathcal{R}n'$.

(Whenever U is a set, then card(U) is the cardinality of U.)

To complete the definition of a semantical game, it is necessary to define the concept of a semantical pay-off function. As is generally the case in game theory, a pay-off function assigns an outcome to a play of a game. In our case the outcomes are simply the two truth values:

Definition 11. Let $\mathcal{T} = (\mathcal{N}, \mathcal{R})_{A,t}^{M,x}$ be a semantical game tree for $\mathcal{L}(\mathcal{O})$. *The semantical pay-off function* $\Phi(\mathcal{T})$ is a function such that whenever $n = (\mu_1, \ldots, \mu_4)$ is an end node in $\mathcal{T}$, $\Phi(\mathcal{T})(n) = 1$ if the atomic formula μ_1 is true at point μ_2 in model M with respect to sequence μ_3. $\Phi(\mathcal{T})(n) = 0$ otherwise. We now get our fundamental concept:

Definition 12. Let $\mathcal{T}$ be a semantical game tree for $\mathcal{L}(\mathcal{O})$. Then the pair $(\mathcal{T}, \Phi(\mathcal{T}))$ is a *semantical game* (for $\mathcal{L}(\mathcal{O})$).

Our semantical games (in the sense of the above definition) are not strictly speaking games in the sense of mathematical game theory. (See e.g. Gale and Stewart, 1953; Owen, 1968.) The reason is the departure of the game theoretical practice mentioned above, in that our semantical game trees do not tell which player's move a given node marks. Given this extra information, our semantical games would be games in the standard sense of mathematical game theory. (More precisely, they would be two person zero-sum games with perfect information. See Rantala, 1978.) Our reasons for not following the usual game theoretical line of thought were already mentioned. It can be pointed out, moreover, that while we gain in generality by this departure, there is nothing much we lose by it, at least as far as the purposes of the present paper are concerned. For it is the concept of truth of an arbitrary formula of our language that we want to define by reference to our semantical games. And for this purpose the extra information mentioned is not needed, as is shown below.

Even though we want to keep the treatment of the usual tense operators (members of $\mathcal{O}$) as general as possible, the interpretation of these logical symbols cannot of course be left completely arbitrary. Hans Kamp has suggested a general characterization of a tense. We shall adopt Kamp's suggestion here and will define a tense as follows:

Definition 13. (*Kamp*) A (one-place) *tense* is a function which for each binary structure (T, R) assigns a function in $\{0, 1\}^T \to \{0, 1\}^T$.

(See Kamp, 1968; 1971.) Even if this definition were not to be quite adequate, it is general enough for our present purposes. Therefore, we shall make the following assumption:

Assumption. With each O in $\mathcal{O}$, there is a tense f_O.

We can now proceed to extending the concept of truth from atomic formulas of $\mathcal{L}(\mathcal{O})$ to all formulas of it. This can be done by the following value function which for each semantical game for $\mathcal{L}(\mathcal{O})$ assigns a truth value:

Definition 14. Let $(\mathcal{T}, \Phi(\mathcal{T}))$ be a semantical game for $\mathcal{L}(\mathcal{O})$, where $\mathcal{T} = (\mathcal{N}, \mathcal{R})^{M,x}_{A,t}$ is a semantical game tree for $\mathcal{L}(\mathcal{O})$ with root r. Then *the value function* $W(\mathcal{T}, \Phi(\mathcal{T}))$ for the semantical game $(\mathcal{T}, \Phi(\mathcal{T}))$ is $g(r)$, where g is that function from $\mathcal{N}$ to $\{0, 1\}$ which for each $n = (\sigma_1, \ldots, \sigma_4) \in \mathcal{N}$ satisfies the following conditions:

(W.atom) Assume σ_1 is an atomic formula of $\mathcal{L}(\mathcal{O})$. Then $g(n) = \Phi(\mathcal{T})(n)$.

($W.\sim$) Assume σ_1 is $\sim B$ for some formula B of $\mathcal{L}(\mathcal{O})$. Then $g(n) = 1$ iff $g(n') = 0$ for that node n' in $\mathcal{N}$ such that $n\mathcal{R}n'$.

($W.\vee$) Assume σ_1 is $(B_1 \vee B_2)$ for some formulas B_1 and B_2 of $\mathcal{L}(\mathcal{O})$. Then $g(n) = 1$ iff $g(n_1) = 1$ or $g(n_2) = 1$, where n_1 and n_2 are those two nodes n' in $\mathcal{N}$ such that $n\mathcal{R}n'$.

($W.O$) Assume σ_1 is OB for some formula B of $\mathcal{L}(\mathcal{O})$ and some O in $\mathcal{O}$. Then $g(n) = f_O((T, R))(\bar{B})(\sigma_2)$, where f_O is the tense associated with O and $\bar{B} = \{t'$ in T: there is a node $n' = (\mu_1, \ldots, \mu_4)$ in $\mathcal{N}$ such that $n\mathcal{R}n'$, $g(n') = 1$ and $\mu_2 = t'\}$. (Here we assume that $M = (T, R, I, V)$.)

(W.U) Assume σ_1 is $(v_n)B$ for some formula B of $\mathcal{L}(\mathcal{O})$ and some v_n in $\mathcal{V}$. Then $g(n) = 1$ iff for each n' in $\mathcal{N}$ such that $n\mathcal{R}n'$, $g(n') = 1$.

We now arrive at the usual semantical terminology by stipulating that a formula A of $\mathcal{L}(\mathcal{O})$ is *true at a point t* in the universe of a model $M = (T, R, I, V)$ for $\mathcal{L}(\mathcal{O})$ with respect to sequence x in I^{ω} iff $W(\mathcal{T}, \Phi(\mathcal{T})) = 1$, where $\mathcal{T} = (\mathcal{N}, \mathcal{R})^{M,t}_{A,x}$. In the same way we can define falsity as value 0 of the value function W. Other semantical concepts are defined as usually. A formula A of $\mathcal{L}(\mathcal{O})$ is *true* in model $M = (T, R, I, V)$ iff A is true at each point in the universe of M with respect to each sequence x in I^{ω}. A formula A is *valid in frame* $F = (T, R, I)$ iff A is true in the model (T, R, I, V) for each interpretation function V. A formula A is *valid* iff it is valid in each frame.

It is not difficult to show that the concept of tense logical validity defined above can be captured by using the more customary Carnap-Tarski type semantics (as developed for intensional languages by Kripke and others). Indeed, if we allow the vocabulary of our language $\mathcal{L}(\mathcal{O})$ to be the one presented by Kamp (1971, Section 4, p. 257) the set of valid formulas obtained is identical with Kamp's.

What is new here is indicated by the observation that game-theoretical semantics meets limiting thesis 4 while Carnap-Tarski type semantics as formulated e.g. by Kamp meets only limiting thesis 2. That game-theoretical semantics satisfies even the condition laid down by limiting thesis 4 suggests that in this framework it is not difficult to present semantics for backwards-looking operators. It is to this problem that we shall now turn.

Syntax for Backwards-Looking Operators

We shall now extend our tense logical language $\mathcal{L}(\mathcal{O})$ so that it contains backwards-looking operators. The extension of $\mathcal{L}(\mathcal{O})$ presented here is only one of several alternatives. Nevertheless, in our judgement it is intuitively the most natural one.

Definition 15. The *vocabulary* of our extended language $\mathcal{L}(\mathcal{O}, \mathcal{D})$ is that of $\mathcal{L}(\mathcal{O})$ together with the set $\mathcal{D}$ of backwards-looking operators defined by: For all O in $\mathcal{O}$, D_O is in $\mathcal{D}$; whenever d is in $\mathcal{D}$, then D_d is in $\mathcal{D}$.

Definition 16. The *formulas* of $\mathcal{L}(\mathcal{O}, \mathcal{D})$ are defined by clauses (i)–(iii) of Definition 3 together with the following extra clauses:

(iv) If A is a formula, then DA is a formula, for all D in $\mathcal{D}$.

(v) Nothing else is a formula.

In usual quantification theory, a sentence is defined to be a formula with no free variables. When defining sentences of $\mathscr{L}(\mathscr{O}, \mathscr{D})$ the situation is not quite as simple for now we have to take into account also 'free' occurrences of the new backwards-looking operators. What we mean by a free occurrence of a backwards-looking operator is not hard to see. For the very idea of these new operators is to 'look back' to a point introduced earlier in the same evaluation by an operator. Hence there typically must be such an operator in order for the backwards-looking operator to make clear sense. In case such an operator does exist, the occurrence of a backwards-looking operator considered is 'bound'.

In order for the above line of thought to be applicable, we must somehow correlate occurrences of backwards-looking operators in a formula with 'binding' operators in that formula. The kind of correlation we shall use below is intuitively the following. Given an occurrence of a backwards-looking operator in a formula, the correlated operator will be of the syntactic form given by the subscript of the backwards-looking operator. (The set $\mathscr{D}$ is defined in such a way that there always is such a subscript.) Not just any occurrence of an operator of the right syntactic form will do, however. Rather, the occurrence correlated with a *D*-operator, say D_d, in a given formula will be that occurrence of the operator *d* such that it has D_d in its scope and is not correlated with any other operator D_d with larger scope than the given one.

After thus syntactically correlating an occurrence of a *D*-operator with another operator having that *D* in its scope, we shall define a semantical requirement which demands that the former refers to the point where the latter was evaluated.

Semantics for Backwards-Looking Operators

In extending our game-theoretical semantics for $\mathscr{L}(\mathscr{O}, \mathscr{D})$ we shall first define the 'correlation' between *D*-operators and the operators 'binding' them in a precise way. For this purpose, the following concept will be needed:

Definition 17. Let *S* be any set of ordered pairs (k, O) with *k* in Nat and *O* in $\mathscr{O} \cup \mathscr{D}$ such that *S* is a function. Then *S determines the sequence sq* of operators in $\mathscr{O} \cup \mathscr{D}$ iff *sq* results from *S* by the following construction:

(i) Assume n_1 is the smallest member of Nat in the domain of S. Then let S_1 be S with the pair $(n_1, S(n_1))$ deleted and let sq_1 be $S(n_1)$.

(ii) Assume S_n and sq_n have been determined, for some n in Nat. Assume S_n is non-empty. Assume further that n_k is the smallest member of Nat in the domain of S_n. Then let S_{n+1} be S_n with the pair $(n_k, S_n(n_k))$ deleted and let sq_{n+1} be $sq_n S_n(n_k)$.

(iii) Assume S_n and sq_n have been determined, for some n in Nat and S_n is empty. Then sq_n is the resulting sequence.

Note that any non-empty sequence is determined by infinitely many different sets.

Definition 18. Let S be any set that satisfies the conditions of the previous definition. We now define *the grammatical antecedent* of a pair (i, D_d) in S.

(i) Assume k is the smallest i in Nat such that $(i, D_d) \in S$, for some $D_d \in \mathcal{D}$. Then $(j, d) \in S$ is the grammatical antecedent of (k, D_d) in S iff j is the largest member i of Nat such that $i < k$ and $(i, d) \in S$. If there is no i in Nat satisfying these conditions, then (k, D_d) doesn't have any grammatical antecedent in S.

(ii) Assume n is the smallest i in Nat such that $(i, D_d) \in S$, for some $D_d \in \mathcal{D}$, there is at least one i' in Nat such that $(i', D_d) \in S$ and $i' < i$, and the grammatical antecedent of (i, D_d) in S is not yet determined. Then $(j, d) \in S$ is the grammatical antecedent of (n, D_d) iff j is the largest i in Nat such that $i < n$, $(i, d) \in S$, and (i, d) is not the grammatical antecedent of any $(n', D_d) \in S$ such that $n' < n$. If there is no i in Nat satisfying these conditions, then (n, D_d) has no grammatical antecedent in S.

It can be shown that whenever S and S' are two sets that determine one and the same sequence of operators, then the following holds. For all $j \in$ Nat and all $D_d \in \mathcal{D}$: If (j, d) is the grammatical antecedent of (i, D_d) in S then there are i' and j' in Nat such that (j', d) is the grammatical antecedent of (i', D_d) in S'; and if (j, d) is the grammatical antecedent of (i, D_d) in S', then there are i' and j' in Nat such that (j', d) is the grammatical antecedent of (i', D_d) in S.

This shows that we could consider grammatical antecedents of occurrences of D-operators directly in sequences of operators, without going through the sets that represent sequences. The condition $(C.D_d)$ on semantical game trees nevertheless can be more easily formulated by using the above defined concept of a grammatical antecedent.

We now turn to defining the concept of a *semantical game tree* of a formula A of $\mathscr{L}(\mathcal{O}, \mathscr{D})$. The definition is obtained from Definition 8 by modifying it as follows. First, we will have to consider formulas of $\mathscr{L}(\mathcal{O}, \mathscr{D})$ everywhere where Definition 8 considered those of $\mathscr{L}(\mathcal{O})$. (Frames and models for the two languages are of course identical.) Second, we will have to add the following extra clause to the construction conditions (2) of the definition:

$(C.D_d)$ Assume $\mu_1 = D_d B$ for some formula B of $\mathscr{L}(\mathcal{O}, \mathscr{D})$ and for some $D_d \in \mathscr{D}$. We have two cases:

(1) Assume there is a node $n = (\delta_1, \ldots, \delta_4)$ in $\mathscr{N}$ such that
(i) $r\mathscr{R}^i n$ for some $i \in \mathrm{Nat}_0$ (the set of natural numbers with 0) such that $i < k$;
(ii) $n\mathscr{R}^l n_1$ for some $l \in \mathrm{Nat}$;
(iii) $\delta_1 = dB_1$ for some formula B_1 of $\mathscr{L}(\mathcal{O}, \mathscr{D})$ and $d \in \mathcal{O} \cup \mathscr{D}$; and
(iv) (i, d) is the grammatical antecedent of (k, D_d) in the set $S = \{(n, O): n \in \mathrm{Nat}_0\ O \in \mathcal{O} \cup \mathscr{D}$, there is $n' = (\chi_1, \ldots, \chi_4) \in \mathscr{N}$ such that $r\mathscr{R}^n n'$, $n'\mathscr{R}^m n_1$ for some m in Nat_0, and $\chi_1 = OB_2$ for some formula B_2 of $\mathscr{L}(\mathcal{O}, \mathscr{D})\}$.
Then $\zeta_1 = B$, $\zeta_2 = \delta_2$, $\zeta_3 = \mu_3$, and $\zeta_4 = (\mu_4, n_1)$.

(2) Assume there is no node n satisfying the above conditions (i)–(iv). Then $\zeta_1 = B$, ζ_2 is any element of T, $\zeta_3 = \mu_3$, and (μ_4, n_1).

In $(C.D_d)$, case (1) is the one where we deal with a 'bound' occurrence of a backwards-looking operator. The condition laid down there is fulfilled precisely when there is an operator of the right syntactic sort correlated with the D-operator we are studying, i.e. when the D-operator has a grammatical antecedent in the sequence of operators dominating it. The moment of time to which the D-operator then refers is then the time of the correlated operator. This is precisely what is stated in case (1) of $(C.D_d)$. It is not hard to see why we have chosen to handle D-operators as universal tense operators when there is no grammatical antecedent. For a formula of the form $D_d A$ (say) does not make any clear statement when we are not told as to which point the backwards-looking operator there refers. Hence it is natural to consider such a formula true only if it is true no matter to which point the problematic *D*-operator there refers. This stipulation is also motivated by the semantical treatment of formulas with free variables in usual quantification theory. There such formulas are considered true when they are true for all values of the free variables.

The concepts of a semantical game tree for $\mathscr{L}(\mathscr{O}, \mathscr{D})$, and the semantical pay-off function $\Phi(\mathscr{T})$ for a semantical game tree $\mathscr{T}$ for $\mathscr{L}(\mathscr{O}, \mathscr{D})$ are defined analogously to what we did in the case of $\mathscr{L}(\mathscr{O})$. The same remark holds for the concept of a semantical game for $\mathscr{L}(\mathscr{O}, \mathscr{D})$ as well. The concept of the value function (for $\mathscr{L}(\mathscr{O}, \mathscr{D})$) is obtained from Definition 14 by requiring attention to semantical games, semantical game trees and formulas of $\mathscr{L}(\mathscr{O}, \mathscr{D})$ instead of those of $\mathscr{L}(\mathscr{O})$. Moreover, the following extra clause must be added to the conditions that the function *g* has to satisfy:

($W.D_d$) Assume σ_1 is D_dB for some formula *B* of $\mathscr{L}(\mathscr{O}, \mathscr{D})$ and some $D_d \in \mathscr{D}$. Then $g(n) = 1$ iff $g(n') = 1$ for all $n' \in \mathscr{N}$ such that $n\mathscr{R}n'$.

All the other semantical concepts are defined as above. This is all we have to do to complete our semantics for the tense logical language $\mathscr{L}(\mathscr{O}, \mathscr{D})$ with backwards-looking operators.

Formalizing 'Would' and 'Then'

We can now go on to formalizing the uses of 'would' and 'then' exemplified in the sentences of the first part of this paper. (It goes without saying that both these particles have uses semantically essentially different from the uses which are of interest for our present purposes.) For this purpose, let *H* be the usual past tense operator 'always in the past', *P* its dual operator 'sometime in the past', *G* the usual future operator 'always in the future' and *F* its dual operator 'sometime in the future'. The semantics of these operators can be easily spelled out within the above framework. (It can be easily seen what the one-place tenses corresponding to these operators are.) Let further *N* be Kamp's 'now' operator. Its semantics can also be formulated easily in the game-theoretical framework.

We shall not bother to formalize all the sentences considered in the first part of this paper but will formalize some of them as examples. The reader can then write the remaining formalizations himself or herself.

The sentence (4) can be formalized as follows:

(33) $(v_1)(v_1$ is a man $\supset H(v_1$ supports the Vietnam war $\supset N\text{Thinks}_{v_1} D_N(v_1$ is an idiot))).

(We follow Hintikka's practice in symbolizing propositional attitudes, clauses such as '*a* knows that', '*b* believes that' etc. See e.g. Hintikka, 1962; 1969.)

Sentence (6) can be formalized as follows:

(34) $(v_1)(v_1$ is a man $\supset H(v_1$ supports the Vietnam war $\supset N\text{Believes}_{v_1} F\text{Has-to-admit}_{v_1} D_N(v_1$ is an idiot))).

In (33) as well as in (34) the backwards-looking operator D_N refer to the time where the formula 'starting' with *N* was considered. This time, in turn, is the one that the past tense operator *H* introduced, i.e. the time when the supporting took place. This is of course precisely how things ought to be here.

The three readings of (9) can be formalized thus:

(35) $P\text{Mentions}_{\text{Bob}} P\text{Says}_{\text{Joe}} P[(\text{E}v_1)(v_1$ is a child $\wedge$ v_1 is born $\wedge$ $F(v_1$ becomes ruler of the world))].

(36) $P\text{Mentions}_{\text{Bob}} P\text{Says}_{\text{Joe}} P[(\text{E}v_1)(v_1$ is a child $\wedge$ v_1 is born $\wedge$ $D_P F(v_1$ becomes ruler of the world))].

(37) $P\text{Mentions}_{\text{Bob}} P\text{Says}_{\text{Joe}} P[(\text{E}v_1)(v_1$ is a child $\wedge$ v_1 is born $\wedge$ $D_P D_P F(v_1$ becomes ruler of the world))].

Here we have assumed that clauses like '*a* mentions that' or '*b* says that' are to be analysed as sentential operators. We hence follow Hintikka's practice in symbolizing these clauses. If the reader finds this assumption dubious, he is welcome to try the same sentences with the clearly operator-type locutions '*a* knows that' and '*b* believes that' substituted for our clauses. The sentences one thus obtains are perfectly acceptable but not quite as natural as our original ones.

What marks a difference between the three readings (35)–(37) of (9) is the moment the future of which the futurity operator *F* refers to. In (35), *F* refers to the future of the moment of birth; in (36) it refers to the future of the moment when Joe's saying took place; and finally in (37) *F* refers to the future of the moment when Bob's mentioning took place. This is precisely what we found to be peculiar to the different readings of 'would' in (9).

In Part I of this paper, we found (21) to be four-ways ambiguous. (Cf. (22)–(25).) We can now formalize these readings as follows:

(38) $P\text{Says}_{\text{Joe}}P[(\text{E}v_1)(v_1$ is a child $\wedge$ v_1 is born $\wedge$ $F(v_1$ becomes ruler of the world) $\wedge$ $F(v_1$ is widely recognized))].

(39) $P\text{Says}_{\text{Joe}}P[(\text{E}v_1)(v_1$ is a child $\wedge$ v_1 is born $\vee$ $F(v_1$ becomes ruler of the world) $\wedge$ $D_PF(v_1$ is widely recognized))].

(40) $P\text{Says}_{\text{Joe}}P[(\text{E}v_1)(v_1$ is a child $\wedge$ v_1 is born $\wedge$ $D_PF(v_1$ becomes ruler of the world) $\wedge$ $F(v_1$ is widely recognized))].

(41) $P\text{Says}_{\text{Joe}}P[(\text{E}v_1)(v_1$ is a child $\wedge$ v_1 is born $\wedge$ $D_PF(v_1$ becomes ruler of the world) $\wedge$ $D_PF(v_1$ is widely recognized))].

Again, what is essential here are the moments the future of which the futurity operators refer to. The reader can easily ascertain that (38)–(41) differ from each other precisely in the way found characteristic of the different readings of 'would' in (21).

Other 'would'-sentences considered in Part I are only more complicated versions of (21). The reader can formalize them himself or herself without much difficulty.

GENERAL CONCLUSIONS

There is a more general moral which can be drawn from the above discussion. This moral can be expressed in the form of the following general conclusions:

Conclusion 1. Prior-type analysis of natural language tenses is essentially defective, at least in being insufficient.

Comment: The basic idea of Prior-type analysis of natural language tenses is to take them as operators. (See e.g. Prior, 1957; 1967; and 1968a.) This is not defective in itself, but it is defective together with the kind of semantical analysis of operators adopted by Prior. Prior's semantics for tenses do not satisfy even limiting thesis 1. And from limiting thesis 1 there is a long step to limiting thesis 5.

Conclusion 2. Even though Prior-type tense analysis is defective and too restricted, we need not analyse tenses as explicit quantifiers.

Comment: Paul Needham (1975; 1976) has also argued for Conclusion 1. His argument is based mainly on the claim that there are sentences of

English that cannot be expressed in Prior-type tense logic with tenses analysed as operators. From this observation Needham draws the conclusion that tenses should be analysed as explicit quantifiers. This conclusion is nevertheless unjustified. For it is the *semantics* that is defective in Prior, not the *syntax*. In other words, we can analyse tenses syntactically as operators if we make the semantics of operators powerful enough. Backwards-looking operators are syntactically operators while they are powerful enough to account for the phenomena on which limiting thesis 5 is based.

Conclusion 3. There is an essential similarity between the semantical behaviour of pronoun anaphora and tense anaphora.

Comment: Lauri Carlson has put forward a convincing treatment of pronoun anaphora. The framework Carlson is using is Hintikka's game-theoretical semantics, as developed for natural language directly. (See Carlson, 1975; 1978; and Hintikka and Carlson, 1978.) The game-theoretical import of Carlson's rule for anaphoric pronouns is precisely the same as that of my rule for backwards-looking operators. Backwards-looking operators were put forward above to capture in a formal language the behaviour of certain aspects of tense anaphora.

What has just been said has bearing on another question. The reader might ask why we wanted to use game-theoretical semantics for spelling out the semantics of our backwards-looking operators. On the basis of Conclusion 3, a partial answer to the question is forthcoming. Instead of game-theoretical semantics, we could have used a variable-dimensional intensional logic which meets limiting thesis 5. However, then the connection between tense anaphora and pronoun anaphora would have been missed. (The connection of variable dimensional logics to usual quantification theory is a nontrivial one.)

It could also be claimed that we have not 'really' solved the problem of the 'would'-sentences of Part I because we have not given any formalization rules. In other words, we have not explained as to how to arrive on formulas of $\mathscr{L}(\mathcal{O}, \mathscr{D})$ on the basis of the natural language sentences and knowledge of syntax only.

This criticism of our discussion is correct but points beyond the tasks of this paper. We quite agree that in the last analysis, if we want to study a natural language by trying to translate it into a formal language, we must of course make the connection between the two explicit. One particularly convincing way to do this is game-theoretical semantics, as developed for a natural language directly. We did not try to develop game-theoretical semantics in this way for the sentences of Part I mainly because this would have necessitated discussion of the game-theoretical treatment of English tenses. This in turn is a large problem-area which contains a number of open problems not of direct relevance to tense anaphora. It seems quite obvious, however, that once a game-theoretical treatment of English tenses is developed, our observations on tense anaphora can easily be incorporated into it.

Department of Philosophy,
University of Helsinki
(*Academy of Finland*)

BIBLIOGRAPHY

Åqvist, Lennart (1973), 'Modal Logic with Subjunctive Conditionals and Dispositional Predicates', *Journal of Philosophical Logic* **2**, 1–76.

Carlson, Lauri (1975), *Peliteoreettista semantiikkaa*, unpublished thesis submitted in partial fulfilment of the requirements for the fil.kand. degree in the University of Helsinki, Helsinki.

Carlson, Lauri (1978), 'Strategic Anaphora', forthcoming.

Gabbay, Dov M. (1974), 'Tense Logics and the Tenses of English', in J. M. E. Moravcsik (ed.) *Logic and Philosophy for Linguists*, Mouton, The Hague, 177–186.

Gabbay, Dov M. (1976a), 'Two Dimensional Propositional Tense Logics', in Asa Kasher (ed.) *Language in Focus: Foundations, Methods, and Systems*, D. Reidel Publishing Company, Dordrecht, 569–586.

Gabbay, Dov M. (1976b), *Investigations in Modal and Tense Logics with Applications to Problems in Philosophy and Linguistics*, D. Reidel Publishing Company, Dordrecht.

Gale, D. and F. M. Stewart (1953), 'Infinite Games with Perfect Information', *Annals of Mathematical Studies* **28**, 245–266.

Hintikka, Jaakko (1962), *Knowledge and Belief: An Introduction to the Logic of the Two Notions*, Cornell University Press, Ithaca.

Hintikka, Jaakko (1968), 'Language-Games for Quantifiers', in N. Rescher (ed.), *Studies in Logical Theory*, Basil Blackwell, Oxford. Reprinted with revisions as Chapter III of Jaakko Hintikka, *Logic, Language Games, and Information*, Clarendon Press, Oxford, 1973.

Hintikka, Jaakko (1969), *Models for Modalities*, D. Reidel Publishing Company, Dordrecht.

Hintikka, Jaakko (1974), 'Quantifiers vs. Quantification Theory', *Linguistic Inquiry* **5**, 153–177.

Hintikka, Jaakko (1975a), 'Impossible Possible Worlds Vindicated', *Journal of Philosophical Logic* **4**, 475–484.

Hintikka, Jaakko (1975b), 'On the Limitations of Generative Grammar', in *Proceedings of the Scandinavian Seminar on Philosophy of Language*, Philosophical Studies published by the Department of Philosophy, University of Uppsala, Uppsala, 1-92.

Hintikka, Jaakko (1976), 'Quantifiers in Logic and Quantifiers in Natural Language', in S. Körner (ed.), *Philosophy of Logic*, Basil Blackwell, Oxford, 208–232.

Hintikka, Jaakko (1978), 'Quantifiers in Natural Language. Some Logical Problems', this volume, pp. 295–314.

Hintikka, Jaakko and Lauri Carlson (1977), 'Pronouns of Laziness in Game-Theoretical Semantics', *Theoretical Linguistics* **3**, 1–29.

Hintikka, Jaakko and Lauri Carlson (1978), 'Conditionals, Generic Quantifiers, and Other Applications of Subgames', in E. Saarinen (ed.) *Game-Theoretical Semantics*, D. Reidel Publishing Company, Dordrecht.

Hintikka, Jaakko and Esa Saarinen (1975), 'Semantical Games and the Bach-Peters Paradox', *Theoretical Linguistics* **2**, 1–20.

Hintikka, Jaakko and Veikko Rantala (1976), 'New Approach to Infinitary Languages', *Annals of Mathematical Logic* **10**, 95–115.

Jespersen, Otto (1933), *Essentials of English Grammar*, George Allen & Unwin Ltd., London.

Kamp, Hans (1968), *Tense Logic and the Theory of Linear Order*, UCLA dissertation, University Microfilms, Ann Arbor, Michigan.

Kamp, Hans (1971), 'Formal Properties of "Now"', *Theoria* **37**, 227–273.

Kuhn, Steven (1976), *Many – Sorted Modal Logics*, Stanford dissertation, University Microfilms, Ann Arbor, Michigan.

Moravcsik, Julius M. E. (1975), *Understanding Language*, Mouton, The Hague.

Needham, Paul (1975), *Temporal Perspective*, Philosophical Studies published by the Department of Philosophy, University of Uppsala, Uppsala.

Needham, Paul (1976), 'The Speaker's Point of View', *Synthese* **32**, 309–327.

Owen, Guillermo (1968), *Game Theory*, W. Saunders Company, London.

Prior, A. N. (1957), *Time and Modality*, Clarendon Press, Oxford.

Prior, A. N. (1967), *Past, Present and Future*, Clarendon Press, Oxford.

Prior, A. N. (1968a), *Papers on Time and Tense*, Clarendon Press, Oxford.

Prior, A. N. (1968b), '"Now"', *Nous* **2**, 101–119.

Rantala, Veikko (1975), 'Urn Models: A New Kind of Non-Standard Modal for First-Order Logic', *Journal of Philosophical Logic* **2**, 455–474.

Rantala, Veikko (1978), 'Extension of Partial Isomorphism and Game-Theoretical Semantics', this volume, pp. 119–151.

Saarinen, Esa (1977a), 'Game-Theoretical Semantics', *The Monist* **60**, 406–418.

Saarinen, Esa (1977b), 'Backwards-Looking Operators in Intensional Logic and in Philosophical Analysis: A Summary of Dissertation', *Reports from the Department of Philosophy, University of Helsinki*, no. 7/1977.

Saarinen, Esa (1978), 'Intensional Identity Interpreted', *Linguistics and Philosophy* **2**, 151–224.

Segerberg, Krister (1973), 'Two-Dimensional Modal Logics', *Journal of Philosophical Logic* **2**, 77–96.

Vlach, Frank (1973), *'Now' and 'Then'. A Formal Study in the Logic of Tense Anaphora*, UCLA dissertation, University Microfilms, Ann Arbor, Michigan.

PART VII

PHILOSOPHICAL LOGIC

BENGT HANSSON

PARADOXES IN A SEMANTIC PERSPECTIVE

1. WHAT IS A SOLUTION?

Let us take Russell's set of all sets that do not contain themselves as a paradigm of paradox. We assume that there is such a set, and if so, it has to be either a member of itself or not. But, as we know, either way leads to a contradiction. However, *this* is not the paradox. So far, the argument is just like a *reductio ad absurdum* proof, and the conclusion to draw is that our assumption was wrong, i.e. there is no such set. The paradox enters when we nevertheless insist that there has to be one – our intuitions about the existence of sets are so strong that we are not prepared to give them up even if they lead to a contradiction. And we cannot deny that they do – the logic of that part is impeccable.

So the paradox is the clash between two convictions: the existence of a certain set and the correctness of the logical reasoning. If the clash leads us to revise our beliefs and to give up one of the convictions, there would be no paradox, only a contradiction based on false assumptions – a phenomenon much too common to merit particular attention. Only when both convictions remain firm does the situation become paradoxical. Therefore, the existence of the paradoxes is basically a psychological problem and a real solution is, accordingly, an argument that makes us understand why one (or both) of these convictions was rash, oversimplified or otherwise untenable in its full generality. This is the way in which I will try to solve the paradoxes: to present an analysis which persuades us to modify our convictions.

A consequence of this view is that a solution to the paradoxes is not something which is true or false, but more or less successful, just like an argument may be more or less successful in convincing an opponent, independently of its validity and its logical structure.

There is also another problem in connection with the paradoxes, viz. how to render them harmless in an axiomatic framework. But this is the question of how to block the paradoxes rather than how to understand them, i.e. it concerns the extension of the paradoxes, rather than their intension. Certainly, the two are connected – a good understanding helps

J. Hintikka, I. Niiniluoto, and E. Saarinen (eds.), Essays on Mathematical and Philosophical Logic, 371–385.

to find a way to block, but it is possible to block without much understanding.

In particular, some blocking devices block too much. A case in point is a general ban on self-reference or, more generally, on cyclic reference. To be sure, all paradoxes seem to have something to do with cyclic reference, so certainly it blocks enough. But it is not a solution, for it doesn't help us at all to understand why certain forms of cyclic reference are quite harmless, while others are malignant. The point of this paper is to sort out a certain type of cyclic reference as the harmful one.

But it may also be the case that blocking devices block too little. I will argue below that there are sets and sentences which do not lead to contradictions, but which are nevertheless as indeterminate and unacceptable as the genuinely paradoxical ones. I will therefore start by grouping what I consider to be inadmissible and harmful constructions into three groups, which I propose to call Ordinary Paradoxes, Prior-type Paradoxes and Truth-teller type Paradoxes (the latter term being borrowed from J. L. Mackie's *Truth, Probability and Paradox*, Oxford 1973).

2. THREE KINDS OF PARADOXES

The paradoxes usually discussed all fall within the class of *ordinary paradoxes*. Russell's paradox, mentioned above, is a case in point, and so is the *Liar*, the man who says 'This sentence is false'. Usually, these two paradoxes are classified differently, Russell's as a logical one and the Liar as a semantic one. I do not emphasise this common distinction and I will argue below that it is without significance for the solution of the paradoxes.

As an example of how the argument goes, let us take the paradoxical set of all founded sets and call it A. Either A is a member of itself or not. If it is, the infinite sequence consisting of A's only will form an infinite descending chain where each element is a member of the foregoing, thus making A into a non-founded set, contrary to the assumption. And if A is not a member of itself, there is an infinite sequence of sets $x_1, x_2, x_3, \ldots$, such that A has x_1 as a member, which has x_2 as a member, which has x_3 as a member, etc. But then there is an infinite descending chain starting with x_1 too, which means that x_1 is not founded, although it was a member of A. Again a contradiction.

It is characteristic of this type of paradox that it starts with a construction of some sort, that a question is asked about this construction, such that only two (occasionally three or more) answers are logically possible, and that each possibility in turn is found to lead to a contradiction.

This is no longer true for *Prior-type paradoxes*. The standard example is Epimenides, the Cretan, saying 'All Cretans are liars'. If to be a liar means to lie always, this sentence cannot be true, for it would then apply to Epimenides as well, thus creating a contradiction. But it can be false, for there may have been some other Cretan who sometimes told the truth, even if Epimenides himself were a liar. What is paradoxical is not that each logical possibility leads to a contradiction, but that the only non-contradictory possibility leads to an empirical fact – the existence of another Cretan – which in no way was included in the original sentence. Can we really prove something to be the case by such an argument?

Of course not! We have escaped the paradox by a trick. Prior-type paradoxes can be routinely derived from ordinary ones, so they do not constitute a new type of problem, but they serve as useful remainders that the impossibility of assigning truth-values is not a defining characteristic of paradoxes. The trick is this: to say 'All Cretans are liars' is to say the ordinarily paradoxical 'This is false' in conjunction with something more, say P. The compound sentence cannot be true, for then its ordinarily paradoxical first conjunct has to be true too, which is impossible. For the compound sentence to be false, however, only means that one of the conjuncts has to be false. Since the paradoxical one cannot be, it is P that has to be false. We have thus derived not-P by a concealed use of paradox. One can also use disjunctions and get a true compound sentence.

The general structure of a Prior-type paradox is the following. Consider the sentence 'If this sentence is true, then I will be in Tokyo tomorrow', understood as a material implication. Call this sentence S. S has the form of a material implication, say $P \rightarrow Q$. Here P means that S is true. Therefore P and S have identical truth-conditions and are either both true or both false. But they cannot be both false, for if P is false, S has a false antecedent and must then be true, being of the form $P \rightarrow Q$. Therefore S and P are both true. But when $P \rightarrow Q$ and P are both true, we are entitled to conclude Q, i.e. 'I will be in Tokyo tomorrow' (or whatever we choose to put in as the consequent of S).

The *truth-teller type paradoxes* are mirror images of the ordinary ones, like 'this sentence is true'. If we assume the truth-value of this sentence to be true, we find that this assumption is sustained by what seems to be the meaning of the sentence. But if we assume it to be false, its meaning becomes that it is false. We cannot decide its truth-value by looking at the two possibilities in turn. This alone, of course, does not call for explanation – normally we have to look at the subject matter of the sentence in addition. But the peculiar thing with 'this sentence is true' is that its subject matter is its own truth and that it therefore should be possible to determine its truth-value by looking at its truth-value.

The clash is here between the two convictions that a sentence must have a truth-value and that that truth-value depends on the subject matter of the sentence only. In this sense the truth-teller is as paradoxical as the other types.

Some may feel that the word paradox should be reserved for what I have here called ordinary paradoxes which lead to contradictions whatever truth-value we assign to them. The ground for such a claim may be that they are particularly dangerous because, if added to a formal system, they will make it inconsistent and useless. I do not wish to quibble about words and I would be perfectly happy to give up the term 'paradox' in my sense, and may even admit that the first type is the most harmful one. But that does not make the other types harmless. And the Prior type would make a formal system just as inconsistent as an ordinary paradox, for it makes it possible to derive an arbitrary sentence, and, applied twice to some sentence and its negation, it yields a formal contradiction. Any sentence that cannot be safely used in an argument is harmful, and it seems to be a good idea to use the label 'paradox' for this purpose. And if we remember that the present problem is to understand how the paradoxes arise, rather than to block them, it is clear that any acceptable solution will have to bear upon all three types.

3. SELF-REFERENCE AND IMPREDICATIVITY

It has often been argued that the root of the paradoxes is self-reference. Generally speaking, I share this view, but I find it important to qualify it, as I did in Section 1, to mean only that certain types of self-reference are harmful. In order to characterise the harmful types I will, in the next section, discuss the linguistic devices for referring. This will also prepare the ground for my proposed solution in the following section.

This does not mean that I take it for granted that some sort of self-reference is necessary for the paradoxes, directly or indirectly. That this is in fact the case will be argued later, but it is not presupposed. What I am doing here is to prepare the ground for that argument, and I will do that by discussing how expressions refer.

Strictly speaking, what I said about self-reference is oversimplified in two respects. First, *indirect* self-reference or cyclical reference is sufficient to produce the same sort of paradoxes as genuine self-reference. Secondly, in the case of set-theoretical paradoxes, it is a problem of self-*application* rather than self-reference.

It is obvious that referential cycles of arbitrary length can be made to function just like the sentence 'this sentence is false'. Consider e.g.

The next sentence is false.
The next sentence is true.
The next sentence is true.
. . . . . .
The n:th foregoing sentence is true.

Such cyclical reference could also be called indirect self-reference and it can be treated on par with genuine self-reference for the purpose of this paper.

There can be no question of genuine self-reference in the set-theoretical paradoxes. It is linguistic expressions that refer, and what they refer to are sets or classes, which is a distinct category from expressions. Rather, in this case we have a cycle where only some links are referential. We can picture the situation like this (A is a short for the expression $\{x;\ x \in x\}$):

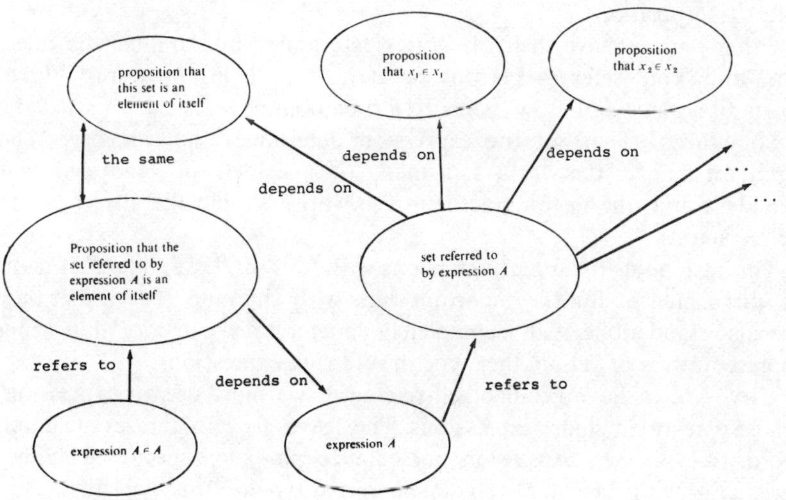

In view of these oversimplifications I propose to use the term 'impredicativity' instead of 'self-reference' and 'self-application'. This is only for notational conveniency and is not intended as a technicality to increase precision. By 'impredicativity' I understand the property of being given in such a form that the determinateness of the definition or description depends on the determinateness of something else, which depends on the determinateness of something else, which depends . . ., which depends on the determinateness of the first expression. This is enough precise for my purposes.

The concept of 'impredicativity' applies equally to so-called semantical and set-theoretical paradoxes, and if I am right in my approach, this will mean that there is no fundamental difference between the two types.

4. REFERENCE

Let us now take a look at referring expressions and phrases. They can be classified along two different dimensions: the category of entities they refer to and the method they utilise for referring.

Some phrases refer to certain entities by force of non-linguistic conventions. We have created canonical systems for e.g. times and positions, and therefore 'five to ten' refers to a time point and '5° north, 160° east' refers to a position.

Other phrases have their categories determined by the linguistic rules. 'me' and 'you' refer to persons or at least to living creatures, 'there' refers to a place and 'now' refers to a time point.

Sometimes a part of the expression determines the category. 'The strongest man in the world' is a man, 'the president of Venezuela' is a president and 'the fastest tractor in the world' should literally speaking be a tractor.

This also holds for some expressions with 'this' or 'that', like 'that man' or 'this sentence'. But the important thing with 'this' and 'that' is that they can also stand alone, with a completely undetermined category, thus being representatives of yet another type of referring expressions.

If we are to have genuine self-reference, we must use an expression, whose category includes expressions. This leaves us with phrases including words like 'phrase', 'expression' or 'sentence' plus some specifying characteristic, or with the completely uncommitted type of 'this' and 'that'.

Here we have another reason why the more general concept of impredicativity is more useful than self-reference. The full sentence 'this sentence is false' does *not* refer to itself. It is the phrase 'this sentence' – only a part of the sentence – which refers to the whole sentence. The phrase 'this phrase' is properly self-referring, but that doesn't lead to any paradoxes.

We are here lead to the distinction between term-like and sentence-like expressions. Only the former can refer in the every-day usage of 'refer'. But we can extend that use. If we think of that which is referred to as something that is singled out or pointed out by the expression or normally put into the focus of interest in one's mind by the use of an expression, it is the most natural extension to say that sentences refer to potential facts or propositions. I will adopt this usage in this paper, but, again, I do not wish to quibble about words – at the expense of some clumsyness some other, especially designed verb could have been used. Neither does the exact meaning and ontological status of propositions need any elaboration in this context, as long as we agree that there is some reality which is independent of languages and descriptions.

Let us now turn to the different ways in which an expression can refer. The most direct one is by way of a proper name or another conventional designation. Thus 'John' refers to the man, 'eleven o'clock' to a time point, and 'London' to the city.

Another way is to put something into relation to something else. 'Next day' refers to a day, and which day depends on which the present day is. 'Two hundred kilometres north of London' makes a reference to a place, conditional on the reference of 'London'. We can regard these as referential functions, taking one reference as an argument and producing another.

Other terms, particularly pronouns, are functions from utterance situations to references. What 'I' and 'you' refer to depends on who participates in the communication situation, and the corresponding holds for e.g. 'now', 'here', 'this' and 'that'. All these have as their linguistic function to provide such floating reference.

The last method of referring to be discussed here is by the-expressions. One typical structure is 'the *X* who *Y*', where *X* gives the category and *Y* a condition, which applies to one and only one object of category *Y*. Many other structures can easily be transformed into this one – 'the president of Venezuela' into 'the person who is president of Venezuela', etc. Very common is the simple structure 'the *X*', which partially belongs to the preceding class as well, because it presupposes that the utterance situation singles out one object of category *X* as the predominant one.

We have been conditioned to associate immediately from the-expressions to failure of successful reference by the failure of *Y* to apply to anything at all, but there is also the possibility that Y applies to two or many objects. In most of these cases the context of utterance helps to determinate the right reference, but if it fails to do so, we have another failure of reference, as serious as the other one.

But it is not only the-expressions that can fail to refer because of special empirical circumstances. This can happen in all the groups we have distinguished between. '0.30 a.m.' is a conventional designation for a time point, but it fails to refer successfully on the night when a nation switches to daylight saving time. 'A hundred kilometres north of X' can be no more successful in referring than 'X' (and possibly less, if X is close to the North Pole). 'You are a humbug, sir!' and 'another such victory' do not refer if uttered in a context with no addressee and with no mention of any victory respectively. Common to all these examples is that it is an empirical fact that determines whether the referring is successful.

5. INTERPRETATION AS EQUATION-SOLVING

Let us imagine ourselves in a situation where we have to interpret an expression. We presume that we know the references of proper names, the rules of the language and that we possess all other background knowledge that may be necessary. Is that sufficient for establishing an effective procedure for finding the correct reference?

The answer is no, and the main reason is the possibility of making dependent references. Let us assume that X is a referring expression, which is dependent on the reference of the expression Y which is part of X (like 'two hundred kilometres north of Y'). Y in turn may be dependent on some Z, and so on. This chain may end with a proper name or another expression, whose reference we know, and all we then have to do is to work our way back to X. But we may also reach an expression W, whose reference is dependent on some previous link in the chain, say X. This is a case of impredicative reference. But what does it mean for the determinateness of the references of X, Y, etc.?

It certainly means the failure of finding an effective, terminating procedure, yielding the correct reference for X. But that does not settle the existence question. We are here lead to a distinction of a type which is common in mathematics: between the existence of an algorithm to solve an equation and the existence of solutions. It is the latter problem that corresponds to the existence of correct references and truth-values. The problem can be stated as a system of equations: find the x such that

(a) *X* refers to *x*
(b) there is a *y*, such that *Y* refers to *y* and *x* is two hundred kilometres north of *y*.
(c) there is a *z*, such that *Z* refers to *z* and *y* stands in the relation to *z* that corresponds to the dependency of *Y* upon *Z*.

and so on.

There may be no solution, one solution or several solutions for *x*. Only in the case of one solution can we speak of *X* being successful in referring, but in that case it *does* refer, whether or not we have an effective procedure for finding this solution. We adopt this as a principle of reference: if there is one and only one thing that an expression *can* refer to, then it *does* refer to it.

Let us take the phrase 'the man who married his own mother' as an example. To find the reference of this phrase is to find the solution to the equation $x =$ the man who married *x*'s mother. We know that Oedipus is one solution to this equation. If there is no other solution within the universe of discourse, we can conclude that the phrase refers to Oedipus. It is clear that this type of reference is not 'grounded' in the sense of being ultimately based on something which refers directly. Yet it is perfectly uncontroversial and not at all uncommon: 'which of the men do you mean?', 'the one who scratches his head' and similar expressions are both useful and normal.

A similar kind of indirect reference is often found in arithmetical puzzles: 'I am thinking of the integral number which is less than some integral number by the amount of that latter number divided by the number I am thinking of. What number am I thinking of?' To know the answer to this is to know the solution to the equation

$$x = n - \frac{n}{x}.$$

This equation has a unique solution, viz. the number 2. The asker of the question therefore has succeeded in uniquely specifying the number he was thinking of and he has in fact made a successful reference to it. This fact must be clearly separated from the possible difficulties various individuals may have in verifying this.

The irrelevance of the existence of an effective procedure for identifying the reference also means that those who try to reduce the paradoxes to lack of 'groundedness' e.g. by requiring that all occurrences of impredicative references be replaced by proper names or predicatively referring phrases, are unnecessarily restrictive. Many normal and useful expressions would be ruled out by such measures. Incidentally, the easiest way to prove that the number 2 is the only solution to the equation above also provides a simple, but excellent illustration of a logical mistake which is often made in connection with substitution in referring phrases.

To solve the equation

$$x = n - \frac{n}{x}$$

one substitutes the value $n - \frac{n}{x}$ for the second occurence of x, thus getting

$$x = n - \frac{n}{n - \frac{n}{x}}.$$

Any x that satisfies this also satisfies

$$x = n - \frac{x}{x - 1}.$$

Since both x and n are integers, $\frac{x}{x-1}$ has to be an integer too. This is only possible for x equal to zero or two. But zero does not solve the original equation, so two is the only solution, as is easily verified.

The point here is that the logical structure of solving equations is that *if* x is a solution to the original equation, *then* it is also a solution to the derived one. But the converse need not hold – the derived equation may have solutions which do not satisfy the original one. This is a trivial and universally recognised observation as far as equations are concerned.

But exactly the same thing holds for linguistic expressions. Consider the equation:

(1) x = the man who married the mother of x

Now make a substitution for x:

(2) x = the man who married the mother of the man who married the mother of x

The logical structure of this is that any x that satisfies (1) also satisfies (2) (so Oedipus is a solution to (2)). But we are not entitled to assume that anything else carries over from (1) to (2). In particular, failure to satisfy (1) does not entail failure to satisfy (2). And it is certainly not the case that (2) has the same *meaning* as (1). E.g. someone who married his paternal grandmother would satisfy (2). Therefore, attempts to account for the paradoxes in terms of the termination or non-termination of successive substitutions must take this complicating fact into consideration.

It is typical for the paradoxes that they in some sense do not say anything about anything except themselves. But I do not think that this is a defining characteristic, as will be seen from the following example:

(1) This sentence has the same truth-value as the one immediately below.

(2) This sentence has the same truth-value as the one immediately below.

(3) At least one of this and the two immediately preceding sentences is false.

Are these sentences true or false? One can argue like this: suppose (1) is true. Then (1) and (2) have the same truth-value. Therefore (2) is true. Then (2) and (3) have the same truth-value. Therefore (3) is true. Then at least one of (1), (2) and (3) is false. But this contradicts our assumption, which hence was false. Therefore suppose instead that (1) is false. Then (1) and (2) have different truth-values. Therefore (2) is true. Then (2) and (3) have the same truth-value. Therefore (3) is true and at least one of (1), (2) and (3) is false, which is well in accordance with our assumption. So the only possibility is that (1) is false and (2) and (3) are true.

We can easily see this search for truth-values as the solving of a set of equations, or rather of arithmetic conditions: let x_1, x_2 and x_3 be the truth-values of (1), (2) and (3). Then the three sentences correspond to the following conditions:

$$x_1 = 1 \text{ iff } x_1 = x_2$$
$$x_2 = 1 \text{ iff } x_2 = x_3$$
$$x_3 = 1 \text{ iff } x_1 x_2 x_3 = 0$$

It is easy to verify that the only solution is $x_1 = 0$, $x_2 = x_3 = 1$.

One can argue that these sentences are uninteresting because they are completely useless and pointless, being in no way connected to anything outside themselves. This is of course true, but one should not be tempted to conclude that they are paradoxical just because of that. Some sentences of this general type are indeed harmful and paradoxical, but what I have just shown is that some are not, and to be merely useless falls far short of being harmful.

6. CONCLUSION

To interpret a sentence is thus to find the answer to the question: which proposition (possible fact, state of affairs) satisfies the condition that the sentence poses? It is illuminating to compare this question with a corresponding one for equations: which number x satisfies the condition posed by the equation?

In a great number of cases – in fact in all ordinary ones – this is an overly complicated way to describe a simple situation, because the condition is trivial. One hardly regards $x = 10$ as an equation, worthy of a solution, but the question 'which number x satisfies the condition of being equal to 10?' is meaningful, even if the asking of it is not. In the same way one does not see a problem in interpreting 'John ate an apple' because the answer to 'what proposition satisfies the condition of meaning that John ate an apple?' is too obvious.

But when the condition is *impredicative*, the situation is different. A typical case is when a number x is identified with some expression that depends on x, $x = f(x)$. As we know, the difficulty of solving such an equation may range from the trivial to the impossible, even if the unique existence of a solution is secured (which it need not be).

The parallel to the case of referring terms is easy to see, and examples like the one about Oedipus above are easy to understand and not uncommon. But the important thing for the understanding of the paradoxes is that they are examples of the same idea applied to sentences.

Now, it may happen that there is no solution to an equation $x = f(x)$. Sometimes it is easy to see, sometimes impossible to determine, but it is always easy to understand: f can be chosen in a way that makes solutions impossible. For propositions, some cases are also easy, like 'this sentence means the opposite of its own content', but it seems to be more difficult to accept and understand why this happens in the typical examples of *ordinary paradoxes*. What I have tried to do here is to point out that some linguistic referring devices are so similar in structure to algebraic ones (in fact, the latter are, historically speaking, abbreviations of the former) that one should not be more surprised at the possibility of constructing an inconsistent function f for propositions than for numbers. Probably, what has blocked our mind here is an idea to the effect that grammatical well-formedness somehow guarantees successful reference to one proposition or another. But this is just as wrong as to suppose that arithmetical well-formedness guarantees solvability for equations.

In the same way as the non-existence of solutions leads to ordinary paradoxes, several solutions lead to *truth-teller type paradoxes*. The mechanism behind this is not more difficult to understand than in the case of no solution. But one must be careful to make sure that one has listed all conditions, also those that emerge from the context (linguistic and extra-linguistic). 'The bald man' has several solutions, taken out of context, but certainly some contexts make this expression uniquely referring. Since such conditions may be more or less vague, it is often difficult to say whether we have successful reference. In order not to overuse the term 'paradox', I therefore suggest to identify truth-teller type paradoxes with sentences that have multiple referents in *all* conceivable contexts. Thus, we avoid having the concept of a paradox relativised to the context.

There is in this approach no need to use the concept of truth and falsify for deriving the paradoxes, unless, of course, the sentence under consideration makes use of these concepts. But the truth/falsify line of argument can be seen as a special case of a general technique to determine unsolvability: All possible solutions are divided into two (or any finite number) of mutually disjoint classes (e.g. true and false propositions, even and odd integers). If the assumption that a solution belongs to the first class leads to a contradiction and the assumption that it belongs to the second does likewise, we can conclude that no solution exists. In this way the typical dilemma of paradoxes is construed: if a sentence can express neither a true nor a false proposition, it is a paradox, i.e. it can express no proposition at all.

The same general technique is often useful in number theory: find the integral solutions to the equation

$$n(n^3 + 1) - n^2 = 3n^3 + 1.$$

If n is odd, $n(n^3 + 1) - n^2$ is odd too, and $3n^3 + 1$ is even. If n is even, $n(n^3 + 1) - n^2$ is even and $3n^3 + 1$ is odd. Thus, they can never be equal and the equation has no solution.

But the logical structure of this technique does *not* allow one to draw positive conclusions about solvability. If only the assumption that the solution belongs to the first class leads to a contradiction, we are *not* entitled to derive the existence of a solution in the second class, but only the hypothetical conclusion that *if* a solution exists, it has to be in the second class. A numerical example will make this clear: the equation

$$n(n^3 + 1) - n^2 = 2n^3$$

can have no odd solution, for then the two sides would have different parity. But for an even n, both sides will be even. From this follows that *if* a solution exists, it is even. But in fact no integral solution exists.

Therefore, a truth/falsify argument is *not* sufficient to classify something as truth-teller paradoxical, but an additional argument is needed – e.g. the discussion about subject matter in the definition of 'truth-teller type paradoxes' in Section 2 above.

This observation resolves the problem of *Prior-type paradoxes*. 'All Cretans are liars' cannot be true, but seems that it can be false. But this is no guarantee that there exists a false proposition expressed by it. In fact, since it can be regarded as a conjunction of the ordinary paradox 'I am lying' and 'All other Cretans are liars', there can be no such proposition.

A variation of the example with the three sentences referring to each other of last section will exemplify this:

(4) This sentence has the opposite truth-value as the one immediately below.

(5) This sentence has the same truth-value as the one immediately below.

(6) At least one of this and the two immediately preceding sentences is false.

These sentences correspond to the following truth-value conditions:

$$x_4 = 1 \text{ iff } x_4 \neq x_5$$
$$x_5 = 1 \text{ iff } x_5 = x_6$$
$$x_6 = 1 \text{ iff } x_4 x_5 x_6 = 0.$$

We have two solutions to this: either $x_4 = 1, x_5 = 0, x_6 = 1$ or $x_4 = x_5 = 0$, $x_6 = 1$. In either case $x_5 = 0$ and $x_6 = 1$. Are we therefore entitled to say that (5) and (6) are not paradoxical, while (4) is? No, we are not – only to say that *if* they are not paradoxical, they are false and true, respectively. But (5) depends on (6) and (6) depends on (4) which is truth-teller paradoxical and therefore (5) and (6) are truth-teller paradoxical as well.

One thing that may make a difference between propositions and numbers is that our ontology for propositions is vague. There may e.g. be some doubt whether there really exist propositions corresponding to sentences (1)–(3) of last section. This is a problem that has to be discussed in the setting of a more complete theory of language – here I have followed a generous line of admitting whatever does not lead to difficulties. Similar problems in arithmetic have been solved by introduction of new number domains – the equation $4x = x - 2$ has no integral solution, but it has a fractional one; $x^2 - 2x + 2 = 0$ has no real solutions, but the introduction of complex numbers makes it solvable. Maybe a similar distinction of levels will prove useful in semantics? But there are limits – $x \neq x$ is unsolvable, no matter how the domain is extended.

By admitting many forms of impredicativity the present theory also allows for the meaningfulness of performative utterances, e.g. 'I hereby promise to do *x*.' It refers back to itself, but is in no way abnormal or unpermitted.

The approach taken here also has many ramifications into the general theory of language. One of these is that interpretation is no mechanical activity and that there may be no effective procedure for it, even if one adopts a generating grammar of a constructive kind. Interpreting is not merely reading the grammar backwards.

Department of Philosophy, University of Lund
and
International Institute for Applied Systems Analysis,
Schloss Laxenburg, Austria

V. TSELISHCHEV

HINTIKKA'S POSSIBLE WORLDS AND RIGID DESIGNATORS

K. Barnes has observed that Hintikka's work on epistemic logic and S. Kripke's work on naming and necessity are incompatible in their conceptual schemes of philosophical interpretations of modal logic[1]. To oppose these names is not arbitrary: both scholars were the first to construct semantics for quantified modal logic and the difference between the two approaches to formal constructions is evident because they were inspired by different tasks. But the extent to which the motivation of formal constructions influenced the difference between philosophical interpretations is still a problem. In particular, S. Kripke's concept of 'a rigid designator' cannot be incorporated into Hintikka's framework of cross-identification.

The comparison and evaluation of these approaches is of indubitable interest only if one realizes their considerable divergence. The recent comparisons are spoiled precisely by the 'interference' of one conceptual approach on evaluation of the other[2].

Both G. Stine and K. Barnes try to show not only the difficulties resulting from some consequences of Hintikka's particular system of epistemic logic but also the inadequacy of applying semantics of possible worlds to the analysis of propositional attitudes in general. The arguments for both cases, however, coincide, because Hintikka's approach is considered a paradigm of that type of semantic analysis which makes use of the notion of a possible world.

The present paper is a critical analysis of G. Stine's and K. Barnes' arguments. It should be noted that Kripke's conception of rigid designators is not treated independently, but only as a conceptual approach incompatible with J. Hintikka's approach to the analysis of modalities.

No discussion of semantics for quantified modal logic can do without a form of essentialism. In fact the only scholar who refutes essentialism is J. Hintikka. G. Stine attempts to prove that (1) essentialism is incompatible with the semantics of possible worlds for propositional attitudes; and that (2) J. Hintikka's refutation of essentialism is actually illusory.

J. Hintikka, I. Niiniluoto, and E. Saarinen (eds.), Essays on Mathematical and Philosophical Logic, 387–399.

G. Stine believes that essentialism is a doctrine which states that for each entity x there is the property N and the property E such that x has N in some possible worlds and x has not-N in other worlds; however, x has E in every possible world where x exists and, moreover, not every y is such that y has E.

The problem G. Stine thus formulates consists in the fact that the semantics of possible worlds does not allow anybody to be ignorant of the essential properties.

Let:

(1) a know that b is F; where F is a contingent property such that the knowledge of this property excludes the knowledge of essential properties, and such that F is the only thing a knows.

According to the semantics for propositional attitudes the statement (1) entails the statement:

(2) There is a possible world compatible with the knowledge of a where b is F.

Let further the singular term b be a rigid designator. *Prima facie* this supposition does not contradict the premises taken as a basis for the statements (1) and (2). A rigid designator is a proper name designating one and the same object in all possible worlds; the assignment of a contingent property to an object named by a rigid designator can only result in falsity. Thus, we have:

(3) There is no possible world compatible with the knowledge of a where b is F; and therefore:

(4) It is incompatible with the knowledge of a, that b is F.

Thus, we have a contradiction here: the statements (1) and (4) are incompatible. It is strange because the statement (1) is intuitively quite plausible, and the statement (4) entails that a knows only essential properties of b, what is not plausible.

Before proceeding to the analysis of the contradiction it is necessary, I believe, to pay attention to the supposition that b is a rigid designator. Introduction of the notion of a rigid designator into the framework of the semantics of possible worlds for propositional attitudes provokes a number of difficulties[3].

In Hintikka's system the statements of the form

$$b = c \,.\, \sim K_a(b = c)$$

(where K is an operator of a propositional attitude 'knowledge') are defensible, i.e. logically possible. For instance, Tulli = Cicero, but John does not know that Tulli = Cicero. On the other hand, if a and b are rigid designators, then the statements of the form $a = b$ are necessary ones. It follows from the fact that a and b designate one and the same object in all possible worlds and the fact that necessity is precisely truth in all possible worlds.

We shall come back to the analysis of this contradiction later but now we shall dwell upon S. Kripke's position. Two theses are of interest here[4]. The thesis (T1) asserts existence of necessary *a posteriori* truths, the thesis (T2) states that proper names are rigid designators. It is quite possible to deduce the thesis (T1) from the thesis (T2) which means that all true statements of identity while being necessary can be necessary *a posteriori* statements. This fact proves very important in analysing the contradiction that is discussed here.

Let us analyse now the contradiction between statements (1) and (2). The statement (1) is intuitively plausible but it means in possible worlds semantics that one can be unaware of the essential properties of an object, and this is unsatisfactory. Thus an intuitively plausible statement in the semantics of possible worlds leads to the unsatisfactory aspect; so much the worse for the semantics. Therefore, it is necessary to solve the paradox within the semantics without taking into account intuitive considerations. This can be done by coordinating the truth values of (1) and (2) by in turn appealing to the other aspects of the notion of necessity and possibility interpretation.

One possible way of making such a coordination is the following one. Suppose that possible worlds semantics allow that somebody does not know the essential properties of an object. This implies the truth of (1) and (2). Is there any way to interpret necessity which would make possible the compatibility of the true statement (1) and (4)?

We could answer this question affirmatively going over from ontological modalities (aletic) to epistemic ones. Provided that b is a rigid designator the statement (1) is ontologically false because a statement in which a contingent property is assigned to a rigid designator is false, while in the case of epistemic modalities the statement (1) may readily prove true. Then, if the sentence (1) conveys a certain modal fact we have to admit that we have here an epistemic modality which is not an ontological modality.

G. Stine believes that this tactic of transition to epistemic modalities does not deserve attention because the epistemic modalities which are not at the same time ontological modalities make a mess of the notion of essence. Then the cross-identification of objects is accomplished not by means of the real properties of the object but with the help of those properties which occur in the field of vision of a certain subject. Those properties may not be among the real properties of the object.

One can agree with G. Stine that the notion of an entity thus becomes obscure. This does not, however, entail the conclusion she arrives at, viz. that possible worlds semantics cannot permit anybody not to know the essential properties. By only using the notion of an entity which is implicitly present in a rigid designator one can show that epistemic interpretation may turn out to be more important than logical interpretation in the discussion of modalities. J. Hintikka analyses an example with a rigid designator 'the first Earl of Avon'[5]. Let there be *a* number of facts hindering us from determining to whom this expression refers. According to S. Kripke's causal theory of reference the absence of a genuine object of reference in the interpretation is quite possible. At the same time if a certain subject does not know whom the expression 'the first Earl of Avon' refers to, his knowledge is compatible with possible worlds where this term does not refer to the genuine bearer of the title. J. Hintikka observes that in this case also the possible worlds are not certain ontological entities. They are rather such situations or states of affairs which would make it possible for a subject to find out a genuine bearer of the title.

Thus, we have here the interpretation of epistemic modalities which are not ontological even if we assume the notion of essence. It means that it would be better to refute the notion of essence than to refute the possible worlds semantics for propositional attitudes.

Let us consider now another way of fitting the statements (1) and (2). Suppose the statement (1) is false. The falsity of 'a knows that b is F' means that b is F from the point of view of a certain subject a (remember that F are contingent properties, E are essential properties of the object b, properties F exclude properties E and conversely). Within the framework of a certain interpretation of the notion of essence and essential properties the statement

b is E

is a logical truth and it follows that so is

b is not F.

A certain feature is characteristic of a number of systems of epistemic logic which makes it possible to understand why the statement 'a is not

F' is considered a logical truth and why the statement (1) can then be considered false. This characteristic feature is that a perceiving subject knows all logical truths and consequences of what he knows[6]. This fact is incompatible with intention and thus it becomes a subject of a special discussion about the so called 'perfect knower'.

The knowledge of the logical truth '*b* is *E*' for a perfect knower means the knowledge of the falsity of a sentence '*b* is *F*'. Thus the possible worlds semantics can be justified by an appeal to the perfect knower if he actually knows all truths relevant to predication of essential properties to an object.

G. Stine's strategy consists now in showing that a perfect knower does not, however, know all the truths of the kind. Its detailed argumentation amounts to the following: If '*b* is *E*' is a logical truth, then only rather trivial properties like self-identity can be essential. An admissible way to avoid this trivialization is the extension of the class of logical truths, which is again connected with rigid designators. If proper names are rigid designators then the statements of self-identity can be the necessary empirical truths of Kripke. The perfect knower does not know empirical truths. Therefore essentialism, if it makes sense at all, is connected with the assumption of necessary *a posteriori* truths. As for possible worlds semantics the appeal to the perfect knower does not make it compatible with essentialism.

This strategy of G. Stine's seems to be absolutely ungrounded. The reason for appealing to it lies in the use of the notion of a rigid designator for the refutation of possible worlds semantics. But there is no need to apply the notion of a perfect knower as a part of the conceptualization belonging to possible worlds semantics. Firstly, the epistemic logic within possible worlds semantics in no way commits us to the notion of a perfect knower. Secondly, the argumentation in favour of this view is based on the view that epistemic modalities are not logical ones. It means that statement (2) is not false on the ground of additional arguments in favour of the truth of (1). So we cannot consider (1) as false because of the simple alternative 'True-false' relative to (1).

J. Hintikka's arguments stating that possible worlds semantics do not commit us to the notion of a perfect knower[7] seem to be arguments of this kind. His main points are as follows:

(I) A sentence of the form '*a* knows that *p*' is true in the world *W*, if and only if *p* is true in all epistemic *a*-alternatives to *W*, i.e. in all epistemically possible worlds compatible with all that *a* knows in *W*.

The fact that the notion of a perfect knower is not a part of possible worlds semantics is expressed as follows:

(II) There are a, p and q such that a knows that p, p logically entails q, but a does not know that q.

Then, the notion used here of logical truth is:

(III) A statement is logically true, if and only if it is true in every logically possible world.

The description of the perfect knower as a consequence of possible worlds semantics is based on the apparent incompatibility of the statements (I)–(III). Let us assume that a, p, q are given. The fact that a does not know q means that there is a possible world W', alternative to the real world, where q is false. One can show in an analogous way that p is true in the world W'. Because W' is a logically possible world, the truth of $p \supset q$ means that q is true in every logically possible world where p is true. Thus, q must be true in W'. But this fact contradicts the accepted assumptions.

In this argument, J. Hintikka observes, the contradiction arises because of the implicit premise that every epistemically possible world is logically possible. And that is exactly why we suppose that W' is a logically possible world and that it is not simply an epistemic alternative to the real world. This confusion of logical and epistemic possible worlds is precisely what provoked the present situation with the perfect knower.

Thus, if we bring up some arguments for the existence of epistemic modalities which are not logical modalities then we do not have to appeal to the notion of the perfect knower for the evaluation of possible worlds semantics.

It is necessary to refer in this connection to the recent works of J. Hintikka[8] and V. Rantala[9] concerning an application of distributive normal forms. The epistemic worlds which are not logical are identical with urn models changing so little that they don't differ from invariant models up to a certain level of analysis. According to the definition of the perfect knower model and invariant model, it means that similar worlds are described by non-trivially inconsistent constituents of distributive normal forms[10]. The notion of the non-trivial inconsistency which J. Hintikka introduced in the theory of distributive normal forms, is analogous to the notion of the inextendable model in the theory of surface models[11]. The systematic comparison of surface models and constituents[12] can be proved to be rather important for understanding the mechanism of non-trivial inconsistency[13]. Therefore, the notion of a surface model is quite enough in itself for solving the problem of the perfect knower. The use of urn models reinforces the epistemic characteristics of the surface models. Since we are talking about changes in the world while describing it, as well as about corresponding notions in the apparatus of description, the epistemic element is clearly present. Though this element is rather evident in the notion of extendable constituent mentioned above, reinforcement shows well enough that possible worlds semantics does not commit us to a perfect knower.

G. Stine and K. Barnes are so confident that Kripke's conception of rigid designators is correct that they refuse to consider other theories of proper names. The distinctive feature of S. Kripke's theory is that the problem of identification of individuals through possible worlds is considered a pseudo-problem. J. Hintikka's conception where the notion of cross-identification is a central one is quite different as compared to Kripke's theory. In addition to Hintikka's arguments it is necessary to pay attention to other theories where the notion of cross-identification occupies a more modest position than in Hintikka's theory, and which are nevertheless alternatives to Kripke's theory.

Gibbard has recently presented an example of such a theory[14]. He has pointed out an important instance of identity between concrete objects which shows, in contrast to Kripke's theory, that some identities among proper names can be contingent. Without coming to a detailed analysis of Gibbard's theory, we shall describe here only those characteristics of the theory which are relevant for the present discussion.

Since a proper name of a concrete object designates a thing existing in the real world under some conditions, these conditions must be taken into account when stating the identity of concrete objects. In particular, the objects cannot be identical if they do not begin to exist, and cease to exist, simultaneously: for otherwise the conditions of their existence can be essentially different. The correlation between these conditions is such that the identity of objects can be contingent.

Gibbard proves that proper names of concrete things are not rigid designators in a number of cases. A thing referred to by a rigid designator cannot be different in another possible world from what it is in another. In order to determine whether the names of concrete things are rigid designators it is necessary to determine whether things under consideration are 'one and the same thing' in another possible world. But as Gibbard suggests: "What a thing would be, though, apart from the way it is designated, is a question without meaning"[15].

The way of designation depends on the kind of thing designated. The conditions of existence of concrete things belong to the kind taken as a whole. The identity is contingent if the thing belongs to two kinds at the same time. Then one proper name refers to a thing as a representative of one kind with definite conditions of existence corresponding to this kind, while another name refers to the thing as a representative of another kind, having other conditions of existence. The contingency of the identity is due to the fact that one condition of existence might be satisfied in one possible world and yet another condition might not be.

In consequence, the question whether a name is a rigid designator is not correct until the kind of thing is indicated. A certain name can be a rigid designator only relative to its own kind. In some situations proper names designate a thing in the real world and invoke into consideration the conditions of existence of the kind of thing.

The reference by a name in a possible world depends on the reference in the real world as well as on the conditions of existence determining the kind of thing.

Kripke refutes the problem of identification of individuals through the possible worlds because the possible worlds are descriptions of situations we stipulate ourselves. Situations can be stipulated in which we speak about definite objects and this stipulation is no less natural than stipulation of situations in which we speak about definite qualities[16]. Then the names of such objects will be the names of the same objects in all possible worlds.

But as Gibbard points out, even such a stipulation of situations does not guarantee that we are speaking about a definite entity, for even the most detailed description of things in the stipulated situation does not answer the question as to what kinds of things are stipulated. It is explained by meaninglessness of the question if the thing in our world is the same in another world without decision about the kind of the thing. All this means that "a situation, then, can be fully stipulated even though questions of identity across possible worlds remain unsettled"[17].

Thus, despite S. Kripke's arguments, the problem of the identification of an object through the possible worlds remains still unsolved. A. Gibbard's arguments do not show that the problem of cross-identification is solved in a definite way but only that the stipulation of rigid designators does not remove it.

The fact that Gibbard's theory is founded on the analysis of a rather special situation and, thus, the possible worlds represent a rather narrow class of possible situations (a possible world 'branches off' the real world at the moment when a thing which is referred to both in the real and in the possible worlds begins to exist) is of no importance.

Gibbard has constructed a rather interesting theory of proper names and modal properties of concrete things. The most interesting fact for us here is that an alternative way to view individuals in Gibbard's theory of proper names is to take them as Carnap's 'individual concepts', i.e. functions from possible worlds to the ordinary individuals[18]. But in this case the problem of cross-identification can be formulated in the general form which is characteristic of J. Hintikka's approach.

The use of the doctrine of essentialism when giving substance to quantified modal logic can be explained by pointing out that the essential properties of objects are designed to give the interpretation of cross-identification. The answer to the question whether the quantification in the presence of modalities is admissible depends on the philosophical evaluation of the essentialist doctrine. On the one hand, the essentialism is considered to be an acceptable price for being able to mix modalities with quantification and the identification of objects through the possible worlds in accomplished with the help of essential properties. On the other hand, W. Quine's argumentation against aletic modalities tries to show that the use of the notion of necessity and of possibility in a metaphysical sense provokes serious difficulties. R. Chisholm has put well the alternative character of the two points of view: either everything has its own essence providing for its uniqueness in every possible world, or else nothing in one possible world is ever identical with anything in another possible world[19].

There is, however, one more conception of the correlation between essentialism and quantified modal logic, according to which the identification of objects through possible worlds is not connected with its essence as a bundle of essential properties.

The similar refutation of essentialism characteristic of J. Hintikka's approach is connected with identification of individuals through the possible worlds by means of an individuating function. In this case the cross-identification is accomplished by means of those properties which permit the subject to make sure that he knows who or what the corresponding individual is while the knowledge of who and what the given individual is does not mean knowledge of all properties of individual essence; it means knowing a sufficient number of properties. In the present case a sufficient number of properties guarantees the uniqueness of reference to the individual in all epistemically possible worlds. In the formal language the requirement of uniqueness is of the form:

$$(\mathsf{E}x)\ [K_a(x = b).\ (x = b)]$$

I.e., there is an individual x such that from the point of view of the subject a, x is identical with individual b, and, moreover, in the real world it is identical with individual b. In the ordinary language it has a simple reading

a knows who (what) is *b*.

G. Stine believes[20] that knowledge by acquaintance, which J. Hintikka explicated, is just the kind of knowledge one has when one

knows who (what) is *b*.

without knowing any concrete properties[21]. It means that the condition of uniqueness is not connected with a bundle of properties, capturing the essence of individual and permitting to cross-identify it. One cannot agree with this statement of G. Stine's because in the case of knowledge by acquaintance the requirement of uniqueness is of the same form as in the previous case:

$$(\exists x)[K_a(x = b)\ .\ (x = b)]$$

There is however one important exception: the quantifier $(\exists x)$ is of a different nature from the quantifier $(\mathsf{E}x)$. The latter is an ordinary quantifier and its corresponding quantification is based on the so called descriptive individuation while the former is J. Hintikka's perception quantifier, and the corresponding quantification bases on the so called perceptual individuation. Both types of quantification presuppose the identification procedures compatible with the refutation of the essentialism.

G. Stine holds nevertheless that J. Hintikka's refutation of essentialism is only apparent. Suppose a certain individual b has properties F, G, and H, two of which are enough for identification of the individual through possible worlds. Then the property (further $\vee\wedge$-property)

$$Fb\ .\ Gb \vee Fb\ .\ Hb \vee Gb\ .\ Hb$$

is characteristic of the individual b in all possible worlds. Thus, again there is an essential property of the individual and, hence, we are committed to essentialism.

G. Stine asserts further that one can defend Hintikka's position in the following way. There are logically possible worlds which are not epistemically possible worlds. Let us consider, in particular, R. Chisholm's example[22]. In one possible world Methuselah lived 931 years, in the next possible world he lived 930 years, etc. By accumulating such little steps Methuselah may gradually become the Noah of a certain possible world, each step arousing no suspicions. The difficulty does not seem easy to overcome if one construes possible worlds metaphysically. The world line from Methuselah to Noah makes no sense: it is too 'long' because it goes through all worlds. It is necessary to 'shorten' this line so that Methuselah would not turn into Noah. The natural way for this 'truncation' is to limit the totality of the worlds through which the world-line runs by those worlds which are compatible with the knowledge of a certain subject. Indeed, though each step between possible worlds is quite plausible the true belief and knowledge do not permit us to assume that Methuselah is Noah.

Essential properties are meaningful only as regards logically possible worlds. The epistemic 'truncation' of the individual world-lines may be caused by $\wedge\vee$-properties, or, at least, it may be meaningful for them. Then, $\wedge\vee$-properties will not be essential ones and essentialism is not accepted.

In order to discredit this objection G. Stine analyses the following situation. Suppose subject *a* knows nothing of the nature of individual *b*, except that

b is *K*, i.e.

(7) As far as *a* knows, *b* is *K*, where having a property *K* implies absence of $\wedge\vee$-properties. Then we have:

(8) There exists a world compatible with the knowledge of *a*, where *b* is *K*.

As in the previous cases we assume that *b* is a rigid designator. Then the statement (8) is false. But because (7) is intuitively plausible, we here face a contradiction.

G. Stine's argumentation is to be considered perfectly groundless. As previously, she makes a presupposition which does not belong to the given context, viz. that proper names are rigid designators. And in the present case this presupposition is unacceptable for two reasons. If we admit that specific features of modal worlds are important as means for the refutation of essentialism, then it is evident that the same specific features are no less important for the refutation of the assumption about proper names as rigid designators. Indeed, since we take into account the properties from the ∧∨-bundle of properties as means of cross-identification, the character of this identification excludes the use of the notion of a rigid designator. It follows that the truth values of the statement (8) for the two ways of solving the problem of identification of individuals through possible worlds are different. If *b* is a rigid designator, the statement (8) is false. If, on the other hand, we mean a realistic procedure of identification of individuals through the possible worlds, then the statement (8) is not necessarily false. G. Stine acknowledges this fact and admits that *b* in (8) might designate in the possible world an individual different from that in the real world.

It is natural that G. Stine notes the shortcomings of J. Hintikka's approach in his theory of reference. For Hintikka's approach is unable, Stine believes, to compete with the theory of rigid designators. The point is that she does not agree with the main principles of J. Hintikka's theory of reference. K. Barnes has more precisely analysed the incompatibility of Hintikka's conception with the theory of rigid designators and, what is more important, he has explicitly preferred the latter. What this argument (the argument that in the conceptual scheme of semantics accepted by J. Hintikka, proper names are not rigid designators) shows, he says, is that either proper names are not rigid designators or the conceptual scheme accepted is not at all acceptable[23].

Thus, K. Barnes as well as G. Stine prefer from the very beginning the conception of rigid designators. One can agree with their arguments about its incompatibility with J. Hintikka's conception of reference. But if one's theory is based on the latter, their arguments seem to be groundless. Moreover, as we have tried to show in this article, the confusion of these conceptions does not permit one to make a satisfactory evaluation of possible worlds semantics for propositional attitudes.

Institute of History, Philology and Philosophy
Academy of Sciences, Novosibirsk

NOTES

[1] Barnes, K. T., 'Proper Names, Possible Worlds, and Propositional Attitudes', *Philosophia* **6** (1976), pp. 29–38.

[2] Stine, G., 'Essentialism, Possible Worlds and Propositional Attitudes', *Philosophical Review* **82** (1973), pp. 471–482; Barnes, K. T., *op. cit.*

[3] Barnes, K. T., *op. cit.*
[4] *ibid.*, p. 25.
[5] Hintikka, J., 'Quine on Quantifying In: A Dialogue', in *The Intentions of Intentionality and Other New Models for Modalities*, D. Reidel, Dordrecht, 1975, p. 123.
[6] Hintikka, J., *Knowledge and Belief*, Cornell University Press, Ithaca, N.Y., 1962; Hintikka, J., 'Individuals, Possible Worlds and Epistemic Logic', *Noûs* **1** (1967), pp. 33–62; Føllesdal, D., 'Knowledge, Identity, and Existence', *Theoria* **33** (1967), pp. 1–27; Sleigh, R., 'Restricted Range in Epistemic Logic', *Journal of Philosophy* **69** (1972), pp. 67–77.
[7] Hintikka, J., 'Impossible Possible Worlds Vindicated', *Journal of Philosophical Logic* **4** (1975), pp. 475–484.
[8] *ibid.*, pp. 475–484.
[9] Rantala, V., 'Urns Models: A New Kind of Non-Standard Model for First-Order Logic', *idem.*, pp. 455–474.
[10] Hintikka, J., *op. cit.*, p. 483.
[11] Hintikka, J., 'Surface Semantics: Definitions and its Motivation', in H. Leblanc (ed.), *Truth, Syntax, and Modality*, Amsterdam, 1973.
[12] Hintikka, J. and Niiniluoto, I., 'On the Surface Semantics of Quantificational Proof Procedures', *Ajatus* **35** (1973).
[13] Tselishchev, V. V., *Filosofskije problemy semantiki vozmozhnuch mirov*, Novosibirsk, 1977.
[14] Gibbard, A., 'Contingent Identity', *Journal of Philosophical Logic* **4** (1975), pp. 187–221.
[15] *ibid.*, p. 194.
[16] Kripke, S., 'Identity and Necessity', in M. Munitz (ed.), *Identity and Individuation*, New York University Press, N.Y., 1971, p. 148.
[17] Gibbard, A., *op. cit.*, p. 200.
[18] *ibid.*, p. 213.
[19] Chisholm, R., 'The Logic of Knowing', *Journal of Philosophy* **60** (1963), pp. 794–795.
[20] Stine, G., *op. cit.*
[21] Hintikka, J., 'Logic of Perception', in *Models for Modalities*, D. Reidel, Dordrecht, 1969.
[22] Chisholm, R., 'Identity through Possible Worlds: Some Questions', *Noûs* (1967), pp. 1–8.
[23] Barnes, K., *op. cit.*, p. 33.

I have recently learned about the premature death of Professor G. Stine. This paper, which is a comment on Professor Stine's ingenious work, is only a small tribute to her. Professor Stine's death is a great loss to the philosophy of logic.

A. A. STARCHENKO

ON THE CONTENT ANALYSIS OF TWO NORMATIVE NOTIONS

1. A number of adequately developed systems of deontic logic have been constructed during the past few decades. They are represented by G. H. von Wright, A. R. Anderson, A. N. Prior, G. Kalinovski, J. E. Fenstad, N. Rescher, H. N. Castañeda, A. Ivin and others. Initially, deontic studies were related to the tasks of logic itself, e.g. the problem of imperative inference. Subsequently an increasing attention has been paid to the practical application of deontic systems to moral, legal and other kinds of reasoning and interaction. It is manifested by the attempts to expand standard systems, to construct unparadoxical logics, etc.

Yet, it is well known that the results achieved in this field can hardly be considered as adequate. Further development of deontic logic is linked to overcoming two kinds of difficulties. The first is inherent in the complexity of the object of investigation alone. Both deontic and normative relations have been considered from different aspects. In the case of a real-ontological aspect, normative relations imply human activities, subjects of these activities, liability interrelations, etc. In the case of a personal will aspect, consideration must be given to different communicative contexts, the most important element of which is imperatives. The third aspect is the linguistic form of an imperative expression, presented in terms of prescriptions.

As a result, logicians deal with a multilayer cake, the stuffing or meaning of which they often try to express by simplified and, therefore, not always correct means.

The second difficulty is that many authors try to express the meaning of multidifferent prescriptions, i.e. command, order, norm, advice, instruction, etc. without an adequate analysis of a real content of these means of communication. Semantic generalizations in this field can be fruitful provided they are based on extensive comparative studies of contexts in natural languages instead of obscure and vague intuitions. It is appropriate to mention here that it is this path that many of our colleagues – Finnish logicians – follow in their semantic investigations.

I would like to make a number of remarks on basic notions of normative logic. Returning to basic notions by the method of thinking backwards

J. Hintikka, I. Niiniluoto, and E. Saarinen (eds.), Essays on Mathematical and Philosophical Logic, 401–404.

can sometimes be useful, as it eliminates some imaginary paradoxes and dilemmas. It seems that of particular interest among the fundamental notions of normative logic are such notions as 'norm', 'normative proposition' and their interrelations.

2. Comparative analysis of expressions of usual languages shows that the term 'norm' is used to denote a great variety of prescriptions which differ by their content, such as legal norm, technical norm, moral norm, norms of grammar, etc. What are the general properties characteristic of these expressions?

To begin with, all of them are closely connected with the process of communication, acting as a means of conveying certain information. Information expressed in norms consists of will acts containing universally valid imperatives acting as regulators of human conduct.

Therefore, essential indications of norms are as follows:

(1) a will act;
(2) a universally valid imperative;
(3) a regulative function;
(4) a prescriptive language form.

If (1)–(3) are obligatory (indispensable) indications of a norm, (4) is a possible indication only, as imperatives can also be expressed in terms of indicatives. Examples are verbal wordings of certain moral norms and legal rules expressed in the constitutions, criminal and civil codes of many states.

The imperative (2) nature of a normative will act makes it possible to distinguish a norm from such expressions of will as advice, request, recommendation and the like.

The valid regulative function of a norm implies its regular applicability (repetition) in a similar communicative context. Norms are thereby considerably different from such singular and, as a rule, personified imperatives as order, command and others.

It should also be noted that the presence of an imperative in a norm always implies either implicitly or explicitly the existence of a certain sanction. It is analytically connected with a notion of the imperative. Without a sanction the imperative stays as nonsense or means a simple wish, preference, etc. Following Anderson, by sanction we imply any type of effect, i.e. compulsion, censure, deterioration of affairs, that assures the fulfilment of prescription.

Therefore, the term 'norm' acts as a name to denote a special real entity, i.e. an imperative will act that prescribes a certain standard of conduct and provides for the application of a sanction in case a prescription is not fulfilled. The above-mentioned indications, with the exception of sanction, constitute what Prof. G. H. von Wright calls norm-kernel. It seems reasonable also to include sanction in the norm-kernel as it makes explicit the meaning of imperative.

In ordinary language, norms are expressed as imperative sentences, e.g. (1) 'No smoking', (2) 'Do not cross the street when the red light is on', (3) 'Render aid to victims of a shipwreck'.

The meaning of these prescriptions is simple enough. All of them include will prescriptions, each of which has its own name – 'norm (1)' (smoking is forbidden), 'norm (2)' (crossing the street against a red light is forbidden), 'norm (3)' (there is a duty to help the victims of a shipwreck). The term 'norm' implies each of the above-mentioned will acts.

3. The notion 'norm' differs much from the notion 'normative proposition'. The latter is a statement that includes the notion of norm as a component. Such expressions are often called 'proposition of a norm'. As distinguished from norm, such a proposition reveals the complete structure and the real meaning of norm.

A norm can perform its regulative function if the following are known: (a) the authority of the norm; (b) the addressee of the prescription; (c) the reasons for acceptance of the norm; (d) the action or deed; (e) the deontic characteristic of the prescription (obligation, forbidden, permission); (f) the sanction.

If symbols for individual variables are introduced:

authorities – $a_1, \ldots, a_n$
addressees – $b_1, \ldots, b_n$
reasons – $c_1, \ldots, c_n$
actions – $d_1, \ldots, d_n$
sanctions – $s_1, \ldots, s_n$

and also

deontic operators – $O, F, P,$

then, for instance, the complete structure of a forbidding normative proposition will be as follows: $\forall d \exists a \exists b \exists c \exists s\, F(a, b, c, d, s)$. In this formula the deontic operator plays the role of a multiplace predicate. The normative proposition will be meaningful provided all the variables are represented. Otherwise, if there is no information about the authority, or addressee, or action, or sanction, etc., then the norm cannot be considered to be completely meaningful and able to act as a regulator of conduct.

4. The above-mentioned differences make it possible to determine the logical status of the expressions 'norm' and 'normative proposition'.

If the term 'norm' denotes distinctive imperatives, there is no sense thereby in saying that a norm itself, as a will act, is true or false. It can be just or unjust, expedient or inexpedient, but neither true nor false.

The term 'normative proposition' denotes distinctive propositions which can thereby be either true or false. Therefore, the logic of norms must be constructed as a logical system of propositions and not as the logic of norms themselves: as in this case we shall deal with the system of norms, staying outside the framework of the calculus of propositions.

Department of Philosophy,
Moscow State University

YU. GLADKICH

SINGULAR TERMS, EXISTENCE AND TRUTH: SOME REMARKS ON A FIRST ORDER LOGIC OF EXISTENCE

I. It is true that there is a considerable variety of ways in which a semantics can be built for a first order logic of existence and denotationless singular terms. But it is also true that several such ways serve for several purposes in logical analysis, not all the same. The aim of this report is to show that if negations of singular existence sentences are to be explicated in a first order logic of existence, then one should be guided in semantics by reasonable principles, to wit (1) an atomic sentence is true only if every singular term in it denotes a member of the universe and (2) bound variables in terms or formulas range only over the given universe. In what follows we argue that some logics that have been fitted to a semantics without (1) or (2) encounter intuitive difficulties. On the other hand, under the guidance of (1) and (2) it is possible to offer an intuitively satisfactory general logical treatment of existence and denotationless singular terms.

II. The first order logic of existence is formulated in a formal language the basic symbols of which belong to one of the four categories: 1. logical constants: E, $=$, $\forall$, ι, $\neg$, $\rightarrow$ ('exists', 'is identical with', 'for every', 'the unique', 'not', 'if ... then'); 2. n-ary predicate symbols, for any n; 3. individual constants; 4. individual variables. The notions of a formula and of a term and the notions of a free (resp. bound) occurence of variable and of a free (resp. bound) variable in a formula or term are defined in the obvious manner for a first order languages. Formulas of the form '$\mathsf{E}(t)$' or '$(t = s)$' or '$P(t_1 \ldots t_n)$', where 'P' is an n-ary predicate symbol and s and t with or without subscripts are terms, are taken to be atomic. The predicate symbol 'E' is understood to be an explicate for the natural language 'exists' or 'there is an individual such as'.

In order to give a semantical interpretation for the formal language, one first gives a structure $M = \langle D, i \rangle$, where D is a set (the domain of individuals which one is assumed to speak about, or the universe) and i is an interpretation of n-ary predicate and individual constants. Then relative to the given structure one defines the values of formulas (they will be the truth-values) and of terms (they will be the individuals or members of D) corresponding to the values given to the free variables. In choosing

J. Hintikka, I. Niiniluoto, and E. Saarinen (eds.), Essays on Mathematical and Philosophical Logic, 405–411.

our precise model theoretical formulation we are guided above all by the following non-formal principles: (1) an atomic sentence is true only if every singular term in it denotes something in the given universe; in particular, E(t) is true if and only if t denotes something in the universe; (2) bound variables in terms or formulas range only over the given universe; (3) singular terms denote (if at all) only the members of the universe; (4) the universe may be empty. Under the guidance of (1) we are required to interpret all n-ary predicates as n-ary relations within the universe. In particular, the logical existence predicate is understood to be a unary relation of being in the universe; more exactly, i assigns to 'E' the function that determines the unique set D, or the universe of i, within which i assigns interpretations. The logical identity predicate is understood to be an identity relation within the universe. This characterization of predicates seems to be in full agreement with the fundamental semantical intuition about truth and with the pre-systematic conception of a simple 'real' predicate. For according to the latter, true predications at the most basic level express attributions of properties or relations to objects, and lacking objects, basic predications cannot be true, but their truth-value is invariably falsity. The principle (1) expresses the thesis (characteristic of the whole tradition from Plato, Aristotle and Avicenna to B. Russell and our days) that the truth of atomic sentences is contingent on the contained singular terms having denotation. The principle (2) is accepted by many authors, and it seems to be a reasonable one[1]. Under the guidance of (3) we are no longer required to give any values to singular terms in all their occurrences. For example, Pegasus does not exist, not here; not anywhere. And 'Pegasus' does not denote. Accordingly, we assign no value whatever to that terms which are intuitively regarded as denotationless. But whenever a value is assigned to a term (in a model), we regard that value as a member of D denoted by the term in the model. As distinct from Schock the function i (for individual constants) and the function α of assigning values to free variables are defined as partial functions: any individual constant or free variable is interpreted in the expected way, when it is within the domain of i or of α, and otherwise it is left without a value at all[2]. This choice of α and of i does not necessarily result in the assumption of truth-value gaps. Under every interpretation any sentence of our formal language has one of the truth values: 'true' and 'false'. The notions of a value for terms and formulas (for the given assignment α relative to the structure M) can be defined by recursive conditions. This definition is straightforward, and it is omitted here. As usual, a formula A

is understood to be true (in M) only if A is satisfied by every assignment in the M. We say that A is valid only if A is true in every M.

The class of valid formulas is, of course, axiomatizable. The logical axioms of our formulation are taken to be the (valid) formulas of following forms, for all component formulas A, B, C, for all n-ary predicate symbols R, variables x, y, and terms s, t: A1 $\vdash R(t_1 \ldots t_n) \rightarrow \mathsf{E}t_1 \wedge \ldots \wedge \mathsf{E}t_n$; A2 $\vdash \forall x \mathsf{E}x$; A3 $\vdash \mathsf{E}t \wedge \forall x A \rightarrow A[^x_t]$; A4 $\vdash \forall x(C \rightarrow A) \rightarrow . C \rightarrow \forall x A$, where x is not free in C; A5 $\vdash \mathsf{E}t \rightarrow t = t$; A6 $\vdash t = s \rightarrow . A[^x_t] \leftrightarrow A[^x_s]$; A7 $\vdash \mathsf{E}(\iota x A) \leftrightarrow \exists y \forall x(A \leftrightarrow x = y)$, where x and y are distinct variables and y is not free in A, A8 $\vdash \exists y \forall x(A \leftrightarrow x = y) \rightarrow A[^x_{\iota x A}]$, under the same condition that in A7, A9 $\vdash A$, if A is a propositional tautology. The rules of inference are modus ponens and universal generalization. The notions of provability and of provable (in this system) formula are defined in the usual manner. This formulation is shown to be sound and complete (relative to the given semantics), and descriptions are eliminable in it.

The present approach enables us to throw light on a number of problems concerning the possibility of meaningfully using 'existence as predicate', the possibility and the possible logical properties of denotationless singular terms, the possible logical relations between singular existential sentences and other sentences. It seems that some old puzzles can be brushed aside. First of all, in the presence of A3 the axiom A2 does not lead to the counterintuitive conclusion that sentences of the form $\mathsf{E}t$, for any t, are trivially true and, hence, their negations, $\neg\mathsf{E}t$, are trivially false. Thus, A2 accords with a widely held viewpoint that singular existential sentences are contingently true. The main existential principle for atomic formulas, A1, enables us to derive the Russell's equivalence $R(\iota x A) \leftrightarrow \exists y(\forall z(A \leftrightarrow z = y) \wedge Ry)$[3]. Thus, we agree with Russell about truth conditions. But we disagree with him about logical form. We take atomic sentences with ex hypothesi non-denoting singular terms as primitive in our formal language rather than as abbreviations for another language form, as Russell did. As a consequence, we give direct semantics for such sentences, while Russell did it only indirectly via semantical analysis of quantification theory with identity. Since A7 is something like Russell's existence definition for descriptions, our logic of 'E' resembles Russell's 'E!'. However, unlike 'E!', our 'E' is applicable to all terms and not just to descriptions. It is possible to give a complete logical treatment of existence and denotationless terms in the system without axioms for identity. Thus it is not necessary to explain existence in terms of the existential quantifier and identity. Moreover, our system gives a complete information about 'nonexistents', since atomic formulas are satisfied by none denotationless singular terms. In particular, 'nonexistents' are indiscernible, for we have $\vdash \neg \mathsf{E}t \wedge \neg \mathsf{E}s \rightarrow . A[^x_t] \leftrightarrow A[^x_s]$. It is easy to see that our characterization of identity satisfies the minimum obvious restrictions on any identity predicate; we have A6 and $\vdash \forall x(x = x)$. The axiom A5 is our principle of selfidentity for existents. In the presence of axioms for identity, we have $\vdash \mathsf{E}t \leftrightarrow t = t$ and also $\vdash \mathsf{E}t \leftrightarrow \exists x(t = x)$, where x is not free in it. We say that the universe is not empty just in the case where – to be a member of the universe is to be a value of bound variables and also just in the case where to be is to be self-identical.

III. We shall now discuss some proposed logics for existence and denotationless singular terms. We shall refer to D. Scott and J. T. Kearns as representatives (mutatis mutandis) for many authors. Our viewpoint is that their logics face intuitive difficulties, since they may lead from true premises to untrue conclusions. Thus, D. Scott uses the axiom $\neg \exists y(y = \iota x A) \rightarrow * = \iota x A$, where '*' is an empty constant for '$\iota x \neg(x = x)$', which enables us to identify all improper descriptions.[4] Since it is not the case that the square circle or the present king of France exists, we derive from this axiom: (a) the square circle is identical with the present king of France. But this sentence is untrue under its intended interpretation. Again, J. T. Kearns uses the axiom $\forall x A \rightarrow A[^x_t]$ in a language, wherein some singular terms fail to denote[5]. And it enables us to derive (b) there is an x such that x does not exist. But (b) is, of course, false under its intended interpretation. So that, given uncontroversial assumptions, each of these logics may imply sentences which are uncontroversially untrue.

It will be wise to take into account the reasons for accepting axioms which lead to untrue conclusions. D. Scott notes as a reason for identifying all nondenoting singular terms the resulting ability to have the 'highly useful' principle of extensionality for descriptions: $\forall x(A \leftrightarrow B) \rightarrow \iota x A = \iota x B$. This principle is useful because together with Leibniz's principle it provides substitutivity for singular terms whether they denote or not. As we know, it is such substitutivity that is needed in the adequacy proof of the truth theory for object language. But we have $\vdash \forall x(x = t \leftrightarrow x = s) \rightarrow A[^x_t] \leftrightarrow A[^x_s]$ and also $\vdash \forall x(A \leftrightarrow B) \rightarrow \forall x(x = \iota x A \leftrightarrow x = \iota x B)$ and, hence, all the substitutions which are needed in this proof are justified in our system. However, in contrast to Scott's logic, the logic of Section II does not lead to untrue conclusions. Again, J. T. Kearns notes as a reason for accepting the standard axiom-schemata $\forall x A \rightarrow A[^x_t]$ the resulting ability to eliminate the category of variables which take individuals as values. His system is valid in all domains, including the empty one; it is possible to bring out in it a logical difference between 'existents' and 'non-existents'. Our logic has the same effects, but it does not lead to untrue consequences, in contrast with Kearns' logic.

IV. It is the case that by translating each occurrence of '*' and each occurrence of '$t = s$' as '$\imath x \neg(x = x)$' and '$\forall x(x = t \leftrightarrow x = s)$' respectively, one can prove that a sentence A is provable in Scott's logic if and only if its translation is provable in our system. Thus, from the viewpoint of our system Scott's logic preserves truth only when '$t = s$' in it means that t and s do not differ in denotation, not that their denotations are the same. On the other hand, by translating each occurrence of '$\mathsf{E}t$' and each occurrence of '$R(t_1 \ldots t_n)$', where 'R' is any n-ary predicate, including identity, as '$\exists x(t = x)$' and $R(t_1 \ldots t_n) \wedge \exists x(t_1 = x) \wedge \ldots \wedge \exists x(t_n = x)$' respectively, one can prove that A is provable in our system if and only if its translation is provable in Scott's logic. This result shows that D. Scott offers a different interpretation of predication from our one. Again, by translating each occurrence of '$\forall xB$' as '$\forall xB \wedge B[^x_*]$' one can prove that a sentence A is provable in Kearns' logic if and only if its translation is provable in our system (plus A10 $\vdash \neg \mathsf{E}*$, where '*' is thought to be an abbreviation for '$\imath x \neg(x = x)$', since we consider the formulation without axioms for descriptions). This result shows that Kearns' logic preserves truth under given translation, but '$\forall xB$' in his system does not mean that 'for every individual, B'. On the other hand, by translating each occurrence of '$\mathsf{E}t$' and of '$\forall xB$' as '$t = t$' and '$\forall x(x = x \rightarrow B)$' respectively, one can prove that A is provable in our logic if and only if its translation is provable in Kearn's logic. Thus, J. T. Kearns offers a different interpretation of quantification from ours. These syntactical differences signalize the differences in respective semantics. So far as both Scott's logic and Kearns' logic are semantically sound, it must be a model for each one of them that makes all their sentences true. But in order to explain, in terms of the correspondence relation, the claimed truth of sentences of the form $t = t$, D. Scott need suggest that every singular term in it denotes in the model, though it may denote an object not actualized in the given universe. Only the actual objects exist, and bound variables should range over the given (possibly, empty) domain of actual objects. Again, in order to explain the claimed truth of $\exists x \neg \mathsf{E}x$, J. T. Kearns need suggest that bound variables of quantifications should range over not just the actual objects but over the non-actual (fictive) as well and that every singular term denotes, though it may denote an ad hoc chosen fictive object. But only the actual objects exist, and all n-ary predicates are constructed in Kearns' logic as relations within the given (possibly empty) domain of actual objects. We observe that semantics fitting these logics have the postulate (3) singular terms denote in all their occurrences, though they may denote a fictive object.

Being formally correct, both of these logics still leave open some old questions, in particular, the problem of negations of singular existential sentences. The point is that they do not allow for empty terms. But the classification between actual and fictive (non-actual) objects is irrelevant to this problem. For, as Russell noted in his criticism of Meinong, this classification does not offer a solution of the problem, but simply pushes it back one step, and this problem arises again as tangled as ever in sentences of the form 't does not exist', if one takes existence as a generic notion of which the actual and fictive are species. Thus, each of these logics simply shifts the problem from a realm of actual objects to a realm of fictions. We have observed that the postulation of (3) in semantics is forced by the way of characterizing predication, as in Scott's system, or by the way of characterizing quantification, as in Kearns' system. That is why neither of these constitutes an adequate strategy for the logical explication of negations of singular existential sentences. Accordingly, such plausible logical axioms as $t = t$ or $\forall xA \to A[^x_t]$ are not compatible with a straightforward means of avoiding the known difficulties. On the other hand, the semantics and the logic of Section II offer the satisfactory logical (syntactical and semantical) solution of the old puzzle. Hence if negations of singular existential sentences are to be explicated in the first order logic of existence, one should follow the principles (1) and (2) indicated at the beginning, and, if I am right, the thesis holds.

Department of Philosophy,
Rostov-on-the-Don State University

NOTES

[1] For an argument, see: K. J. J. Hintikka, 'Existential Presuppositions and their Elimination', in *Models for Modalities*, Dordrecht, Holland, 1969.

[2] Cf. R. Schock, *Logics without Existence Assumptions*, Stockholm, 1968. Schock gives a Frege-type model theory for his logic using the empty set as the denotation of intuitively denotationless singular terms. In non-empty universes, he construes all variables as having values.

[3] A. N. Whitehead and B. Russell, *Principia Mathematica*, Cambridge University Press, Cambridge, 1962. (See *56.)
[4] See: D. Scott, 'Existence and Description in Formal Logic', in *B. Russell. Philosopher of the Century* (ed. by R. Schoenman), London, 1967.
[5] See: J. T. Kearns, 'A Universally Valid System with no Existential Assumptions', *Logique et Analyse* **43** (1968).

PART VIII

TRUTHLIKENESS

DAVID MILLER

ON DISTANCE FROM THE TRUTH AS A TRUE DISTANCE

0. INTRODUCTION

False scientific theories, we imagine, can differ in their closeness to the truth. Popper's qualitative theory of verisimilitude ([1963], pp. 228–237) sought to explain this difference as a variation in the set of true consequences coupled with a countermodulation in the false. That is to say, B is more truthlike than A is if all A's true consequences are consequences of B, and all B's false consequences are consequences of A, and at least one of these inclusions is proper. Writing $Ct_T(A)$ and $Ct_F(A)$ respectively for the set of true consequences of A, its *truth content*, and the set of false consequences of A, its *falsity content*, we can express Popper's condition compactly thus:

(◇) B is more truthlike than A is if and only if
either $Ct_T(A) \subset Ct_T(B)$ *and* $Ct_F(B) \subseteq Ct_F(A)$
or $Ct_T(A) \subseteq Ct_T(B)$ *and* $Ct_F(B) \subset Ct_F(A)$.

Unfortunately no two false theories A and B are ever so related (Miller [1974a], Tichý [1974]). We are left wondering whether any other recommendation along these lines can ring more truly.

Professor Popper has himself suggested (in conversation) that truthlikeness might be assessed simply by truth content. One objection to this is that amongst false theories truth content varies directly with content, so that if we have a theory which is false we can without effort move on to one that is closer to the truth (Tichý [1974], p. 157, note 2). This is a familiar sort of 'tacking argument', and the response to it has to be that comparisons of truthlikeness are only of interest between theories that have been seriously proposed in some domain of inquiry. More unacceptable is the somewhat stronger result that, if the set T of true sentences is not axiomatizable, then amongst (recursively) axiomatizable theories, *whether true or false*, truth content varies directly with content. For this means that progress towards the truth is not only easy but quite unrelated to where the truth actually is; that is, progress towards the truth is at the

J. Hintikka, I. Niiniluoto, and E. Saarinen (eds.), Essays on Mathematical and Philosophical Logic, 415–435.

same time progress towards every other unaxiomatizable complete theory (Miller [1974a], pp. 167, 174f.). This objection, in particular, persuaded me not to pursue Professor Popper's proposal.

In this paper I shall present an alternative, intuitively most appealing, method of assessing the distance of a theory from the truth⁰. This method will turn out, given remarkably feeble assumptions, to be closely similar to Popper's truth-content suggestion, and to be equally vulnerable to the objections noted above. Nevertheless, I shall try to defend it against those objections, and to develop it in some detail as a theory of truthlikeness. I shall not attempt to conceal the fact, however, that there are further criticisms to which I cannot at present offer any adequate response.

1. AUTOMETRIC OPERATIONS IN LATTICES

If theories can be close to, or distant from the truth, then presumably *they can be close to or distant from one another*. I propose that we take quite literally this presumption, and that we seek to define a distance function $d(A, B)$ which satisfies the usual axioms for a metric operation; that is, $d(A, B)$ must be zero if and only if $A = B$, $d(A, B)$ must equal $d(B, A)$, and the triangle inequality must obtain. Now just as in the Euclidean plane (say) there is a copious supply of distance functions that satisfy these axioms, so too we can expect plenty of real-valued distance functions between theories. The uncontrollable arbitrariness that seems bound to arise in the choice of one distance function rather than another can, it appears, be magically diminished by requiring d to be not *real*-valued but *lattice*-valued, in particular by making it an *autometric* operation, taking values in the very lattice of theories on which it is defined. Let me hasten to explain what is at issue here.

The symbols 0, 1, +, ., −, ≤, apart from their arithmetical use, have a perfectly familiar use in lattice theory, and in particular in Boolean algebra. 0 is the zero or smallest element of the algebra, 1 the unit or widest element; + is the sum or join operation, . is the product or meet, − the complement, and ≤ the lattice (partial) ordering. It was observed by Ellis [1951a] that in a Boolean algebra the *symmetric difference* $a \triangle b$ of a and b, defined by $(a.-b)+(-a.b)$, satisfies the stated axioms for a metric provided that each of 0, +, and ≤ is interpreted in the Boolean manner. A proof of this fact is given in Theorem 1 below. If we define the distance between a and b as $a \triangle b$ we thereby obtain a metric operation whose values are themselves elements of the Boolean algebra. Ellis called such a Boolean algebra *autometrized.* Blumenthal [1970] speaks picturesquely of *Boolean geometry.* Shortly afterwards Elliott [1953] (see also Blumenthal [1970], pp. 114f.) established that *the symmetric difference is the only autometric operation in a Boolean algebra that has a number of further properties*; for example, it alone is a group or semi group operation. From our point of view there is little to be said for a distance function's having some of these extra properties, but this is not of much importance since, as we shall show, a single much weaker assumption, the assumption of *normality*, is sufficient almost to characterize the function *distance from the truth* within a Boolean algebra. These matters will occupy our attention in Section 2.

In Section 3 we shall extend the discussion to Brouwerian algebras. For although the lattice of sentences is Boolean, this no longer holds for the lattice of deductive theories. If we so label this lattice that the meet and join have the logical force of conjunction and disjunction respectively, then we obtain in general a Brouwerian algebra. In such an algebra there is an operation akin to *symmetric difference*; this operation is a metric operation; and it is possible to develop by these means *Brouwerian geometry* (Nordhaus and Lapidus [1954]). Again one weak assumption confines very severely the possibilities for the function *distance from the truth.*

In Section 4 we shall return to Boolean geometry, offering a defence of the proposed metric against the objections already adumbrated. That there are other objections that are not so easily answered is something that I hope to substantiate elsewhere.

2. NORMAL BOOLEAN GEOMETRY

Let 0, 1, $+$, $.$, $-$, $\leq$ be the zero, unit, join, meet, complement, and ordering relation of a Boolean algebra $\mathscr{B}$. A function $*$ from $\mathscr{B} \times \mathscr{B}$ to $\mathscr{B}$ will be called *autometric* if the following axioms are satisfied:

(1) $a * b = 0$ if and only if $a = b$,

(2) $a * b = b * a$,

(3) $a * c \leq (a * b) + (b * c)$.

THEOREM 1. The symmetric difference $a \triangle b$ is an autometric function.

Proof: It is trivial that $a \triangle a = 0$; and if $a \triangle b = 0$ then $a \,.\, -b = -a \,.\, b = 0$, so that $a = b$. This proves (1); and (2) is immediate. For (3) we note that since $a \,.\, -b + b = a + b$ we surely have $a \leq a \,.\, -b + b$.

Multiplying each side by $-c$ yields $a\ .\ -c \leq a\ .\ -b\ .\ -c + b\ .\ -c$, so that $a\ .\ -c \leq a\ .\ -b + b\ .\ -c$. Likewise $-a\ .\ c \leq -a\ .\ b + -b\ .\ c$, and the addition of these last two inequalities gives $a \triangle c \leq (a \triangle b) + (b \triangle c)$.

An alternative way of proving the triangle inequality (3) is this: let $\mathscr{B}$ be an algebra of sentences, so that $a \triangle b$ is identified with the sentence $a \not\leftrightarrow b$, the sentence which asserts that a and b have different truth values. What (3) says in this interpretation is that if a and c have different truth values, then either a and b have different truth values, or b and c do; and this is easily seen to be the case.

Following Ellis [1951b] and [1951c] we shall say that an autometric operation is *normal*, or that it has *basality*, if there is an element e satisfying $a_* e = a$ for all a. By (1) such a base point must be 0; and it is immediate that symmetric difference has basality. The following result concerning normal autometric functions is adapted from the proof (Elliott [1953], p. 325) that the symmetric difference is the only autometric function that is also a group operation.

THEOREM 2. If $_*$ is a normal autometric operation on $\mathscr{B}$ then $a \triangle b \leq a_* b \leq a + b$.

Proof: Applying the triangle inequality (3) to the triangle $\langle 0, a, b\rangle$ we obtain (using also (2))

$$a_* b \leq (a_* 0) + (0_* b) = a + b,$$

which settles the upper bound. For the lower bound we again apply (3) to triangle $\langle 0, a, b\rangle$ to give

$$a = a_* 0 \leq (a_* b) + (b_* 0) = (a_* b) + b.$$

Multiplication of the outside terms by $-b$ yields $a\ .\ -b \leq (a_* b)\ .\ -b$, so that $a\ .\ -b \leq a_* b$. In the same way, clearly, $-a\ .\ b \leq a_* b$, so that by addition $a \triangle b \leq a_* b$ as required.

Given any language that has been reasonably competently formalized we can identify any two sentences in it that are interderivable by the rules of elementary logic. By so doing we manufacture a Boolean algebra, the Lindenbaum/Tarski algebra of the language, where $\leq$ can be read as derivability, 0 and 1 as the contradiction and tautology respectively, and $+$, $.$, and $-$ as disjunction, conjunction, and negation. If the set of all true sentences of the language is finitely axiomatizable, then there is in the Lindenbaum/Tarski alegbra a narrowest true element t, which has a good claim to be called simply 'the truth'. Distance from the truth is then distance from t. As is well known, t is an *atom* in the algebra, so that if $a < t$ we must have $a = 0$. Moreover, for every a either $t \leq a$ (in which case a is *true*) or $t \leq -a$ (in which case a is *false*). These locutions can of course be used whenever t is an atom in a Boolean algebra $\mathscr{B}$, whether or not $\mathscr{B}$ is the Lindenbaum/Tarski algebra of some language. Their use gives point to the following theorem, which holds for every Boolean algebra $\mathscr{B}$ in which t is an atom.

THEOREM 3. Let $_*$ be a normal autometric operation on $\mathscr{B}$.

$$\text{If } a \text{ is false then } a \,_*\, t = a + t = a \triangle t;$$

$$\text{if } a \text{ is true then } a \,_*\, t = \begin{cases} a = a + t \\ a \,.\, -t = a \triangle t. \end{cases}$$

Proof: Suppose that a is false. Then $a \leq -t$, so $a \triangle t = a \,.\, -t + -a \,.\, t = a + t$. By Theorem 2, $a \,_*\, t = a + t$.

If a is true, $a \triangle t = a \,.\, -t$, so that, again by Theorem 2,

$$a \,.\, -t \leq a \,_*\, t \leq a + t = a.$$

We shall show that $a \,_*\, t$ must equal one or other of these bounds.

Whatever $a \,_*\, t$ is, it must be either true or false. In the former case, $t \leq a \,_*\, t$, so $a \,.\, -t + t \leq a \,_*\, t$, so $a + t \leq a \,_*\, t$; thus $a \,_*\, t = a + t$. In the latter case, $a \,_*\, t \leq -t$, so that $a \,_*\, t \leq a \,.\, -t$; thus $a \,_*\, t = a \,.\, -t$. So if $a \,_*\, t$ is true it equals the upper bound given; if false it equals the lower bound. This completes the proof of the theorem.

There are thus comparatively few normal autometric functions on a Boolean algebra $\mathscr{B}$.

The truth content of a theory, it will be recalled, is defined by Popper as the set of the theory's true consequences. The truth content of a theory is therefore itself a theory. Since $a + b \leq c$ if and only if $a \leq c$ and $b \leq c$, it is clear that, when the truth is an element t of the Lindenbaum/Tarski algebra $\mathscr{B}$, the *truth content* of a is represented by the element $a + t$. To emphasize the fact that a's truth content is here taken to be an element of $\mathscr{B}$ itself, we shall write it $ct_T(a)$. When a is true, $ct_T(a)$ is of course just a. By Theorem 3, accordingly, the distance of any particular a from the truth t is either $ct_T(a)$, or $a \triangle t$, or both. This is not to say, of course, that the function *distance from the truth* has to be identically equal to one or other of these two functions. Indeed, it cannot coincide everywhere with the truth content $ct_T(a)$, for in that case we would have $t * t = t$, instead of 0. On the other hand the postulate of normality is certainly not sufficient to ensure that $*$ is identical with $\triangle$. This can be established by means of the following simple example.

EXAMPLE: Let $\mathscr{B}$ be a Boolean algebra with only four elements. Since t is one of the atoms, the other is $-t$; and as $-t$ is false we have $t * -t = t + -t = 1$. Thus the following distances are determined by the requirement that $*$ be a normal metric:

$*$	0	t	$-t$	1
0	0	t	$-t$	1
t	t	0	1	
$-t$	$-t$	1	0	
1	1			0

If $*$ were identical with $\triangle$, the remaining squares would be completed by $t * 1 = 1 * t = -t$ and $-t * 1 = 1 * -t = t$. But it is in fact quite consistent to set the distance of 1 from the truth as its truth content; that is, $1 * t = 1$. This immediately makes $t * 1 = 1$, by (2). However, $-t * 1$ and $1 * -t$ can be consistently either t or 1. Thus there are three normal metrics on this algebra $\mathscr{B}$:

$*$	0	t	$-t$	1
0	0	t	$-t$	1
t	t	0	1	$-t$
$-t$	$-t$	1	0	t
1	1	$-t$	t	0

(a)

$*$	0	t	$-t$	1
0	0	t	$-t$	1
t	t	0	1	1
$-t$	$-t$	1	0	t
1	1	1	t	0

(b)

$*$	0	t	$-t$	1
0	0	t	$-t$	1
t	t	0	1	1
$-t$	$-t$	1	0	1
1	1	1	1	0

(c)

To establish that the two matrices (b) and (c) do indeed satisfy our postulates we need only check, of course, the triangle inequality (3). A degenerate triangle, one, that is, with two vertices identical, automatically satisfies (3), so the problem reduces to the four non-degenerate triangles $\langle 0\ t\ -t \rangle$, $\langle 0\ t\ 1 \rangle$, $\langle 0\ -t\ 1 \rangle$, $\langle t\ -t\ 1 \rangle$. It is easily seen that the lengths of the sides of these triangles are given by the following table, so that (3) is, as required, everywhere obeyed:

	(a)	(b)	(c)
$\langle 0, \; t, -t\rangle$:	$t, \; 1, -t$	$t, 1, -t$	$t, 1, -t$
$\langle 0, \; t, \; 1\rangle$:	$t, -t, \; 1$	$t, 1, \; 1$	$t, 1, \; 1$
$\langle 0, -t, \; 1\rangle$:	$-t, \; t, \; 1$	$-t, 1, \; t$	$-t, 1, \; 1$
$\langle t, -t, \; 1\rangle$:	$1, \; t, -t$	$1, t, \; 1$	$1, 1, \; 1$

I propose now to introduce a further condition on the normal autometric function $*$ which is sufficient to ensure that it is everywhere the same as the symmetric difference. We shall call an autometric operation *downward strictly monotone* if and only if

(4) $\quad c < b < a \rightarrow b \,{}_{*}\, a < c * a,$

and *upward strictly monotone* if and only if

(5) $\quad c < b < a \rightarrow c \,{}_{*}\, b < c \,{}_{*}\, a.$

THEOREM 4. If ${}_{*}$ is a normal downward strictly monotone autometric operation on $\mathscr{B}$ then for every a we have $a \,{}_{*}\, t = a \triangle t$.

Proof: If $a \,{}_{*}\, t \neq a \triangle t$, then by Theorem 3 it follows that a is true and $a \,{}_{*}\, t = a$. By normality, therefore, $a \,{}_{*}\, t = a \,{}_{*}\, 0$; in the presence of (2) this conflicts with (4). That $\triangle$ itself satisfies both (4) and (5) is routine.

It is worth noting that (5) is not sufficient for this purpose. Indeed, it is a trivial consequence of normality that all three metrics of the Example are upward strictly monotone.

Following Blumenthal ([1953], p. 317) we shall say that b lies *between* a and c in $\mathscr{B}$, or that $B(a, b, c)$, if and only if $(a \,{}_{*}\, b) + (b \,{}_{*}\, c) = a \,{}_{*}\, c$. (This is a 'weak' relation of betweenness, in that the identity of the various relata is not ruled out.) The following lemma and theorem incorporate and extend a number of results of Blumenthal, *op. cit.*, pp. 320, 333.

LEMMA. The following four conditions are equivalent in a Boolean algebra $\mathscr{B}$.

(i) $\quad (a \triangle b) + (b \triangle c) = a \triangle c;$
(ii) $\quad a \,.\, b + b \,.\, c = b = (a + b) \,.\, (b + c);$
(iii) $\quad a \,.\, c \leq b \leq a + c;$
(iv) $\quad b \triangle c \leq a \triangle c.$

Proof: The proof of the equivalence of (i), (ii), (iii) can be found in the work of Blumenthal just cited. To prove (iv) from (iii), we note that the former can be written $b \,.\, -c + -b \,.\, c \leq (a + c) \,.\, (-a + -c)$, which, after the elimination of trivialities, boils down to the conjunction of $b \,.\, -c \leq a + c$ and $-b \,.\, c \leq -a + -c$. That is, $b \,.\, -c \leq a + c$ and $a \,.\, c \leq b + -c$. The first of these is an immediate consequence of the second half of (iii), the second of the first.

That (iii) follows from (iv) can be proved in essentially the same way. It is, however, perhaps more illuminating to observe that as a and c are symmetrically placed in (iii), we can also derive from it that $b \triangle a \leq a \triangle c$. The addition of this last inequality to (iv) allows us to prove (i), as we know already from Theorem 1 that axioms (1) and (2) are satisfied by the symmetric difference.

THEOREM 5. If $_*$ is a normal downward strictly monotone autometric operation on $\mathscr{B}$

then $b \,_*\, t = a \,_*\, t$ if and only if $b = a$;
and $b \,_*\, t < a \,_*\, t$ if and only if
either $b < a$, and a and b have the same truth value,
or $b \leq a + t$, and b is true and a is false.

Proof: By Theorem 4 we can assume that $_*$ coincides with the symmetric difference. This makes the first assertion of the Theorem trivial.

Thus it follows from the Lemma that $b \,_*\, t < a \,_*\, t$ if and only if a and b are different and $a \,.\, t \leq b \leq a + t$. If a is true, this means that b must be true, and $b < a$. If a and b are both false then $b \leq -t$, so that again $b < a$. Otherwise, b is true and a is false and $b \leq a + t$.

COROLLARY. Under the conditions stated, if a is false then $(a + t) \,_*\, t < a \,_*\, t$.

Since our main interest is in being able to compare theories for truthlikeness, Theorem 5 effectively tells us how successful the function $\triangle$ is as a measure of distance from the truth. (It seems that by considering the actual distances $b \,_*\, t$ and $a \,_*\, t$ we can extract marginally more precise results than those we would obtain by asking whether b lies *between* a and the truth; that is, whether $B(a, b, t)$.) There are, it can be seen, three different categories of comparison possible. The first – given by the Corollary – our measure shares even with Popper's original proposal ($\diamondsuit$).

It is that *the truth content of a false theory is closer to the truth than is the false theory itself*. It seems hard to quarrel with this conclusion. The second is also a consequence of (◇): *the stronger of two comparable true theories is the closer to the truth*. This again is unexceptionable. (It is, indeed, a consequence of the condition (5) of upward strict monotony. Nevertheless, measures proposed by Tichý [1976] and Niiniluoto [1976] fail to satisfy this condition, and even violate the much feebler condition of upward weak monotony[1]. See Popper [1976], pp. 148f. and Miller [1976], pp. 365–368). But the third mode of comparison, which instructs us that *the stronger of two comparable false theories is the closer to the truth*, is altogether less amusing. It holds, of course, not just for $\triangle$ but for all normal metrics. The limitation is the more glaring in that amongst false theories *only* those that are logically comparable are comparable with respect to their distance from the truth. Certainly this possibility is not a consequence of Popper's (◇), which permitted no comparisons of false theories at all. I defer to Section 4, however, an assessment of whether this additional capability of our theory is to be welcomed or to be deplored. Before that we must discuss the generalization of these results to Brouwerian algebras, and in particular to the lattice of deductive theories.

3. NORMAL BROUWERIAN GEOMETRY

It is a ready consequence of Gödel's theorem that in most rich languages the set T of all true sentences is not finitely axiomatizable. (There are other arguments that point to the same conclusion.) A general treatment of truthlikeness has therefore to be conducted in the context of Tarski's *calculus of deductive systems*, a deductive system being a set of sentences A which is closed under the operation Cn of logical consequence. Tarski writes $Cn(A) = A$ when A is such a set. Deductive systems are partially ordered, of course, by the relation of set inclusion; and this ordering $\subseteq$ is in fact a lattice ordering, generating the kind of lattice now known as a Heyting algebra. If A and B are systems then their greatest lower bound is just their intersection $A \cap B$ (for this is also a system); their least upper bound is the set $Cn(\{a \,.\, b \mid a \in A \text{ and } b \in B\})$. The greatest system of all, the set S of all sentences, serves as a unit element of the lattice; the poorest, the set L of all tautologies, as the zero. Thus the lattice ordering is the reverse of the usual ordering on the (Boolean) algebra of logic: there the tautology is the unit, the contradiction the zero. To facilitate comparison it is therefore useful to consider not the lattice generated by $\subseteq$, but its dual, the lattice generated by $\supseteq$. Under this ordering (written $\leq$ in this section), which corresponds to the ordering by logical derivability, $A \cap B$ is the least upper bound of A and B; we now write it as $A + B$. Likewise the system $Cn(\{a \,.\, b \,/\, a \in A \text{ and } b \in B\})$ will be written $A \,.\, B$. The lattice zero 0 is what we called S, the unit 1 what we called L. It is possible also to define the complement $-A$ of a system A. (But it must be realized that this complement is not a contradictory; that is, we have $A + -A = 1$, but not always $A \,.\, -A = 0$.) These three operations correspond quite closely to their sentential archetypes. But there is nothing in this lattice to correspond to the conditional $b \to a$ of sentential logic. The conditional is usually defined as the weakest sentence strong enough to satisfy *modus ponens*: $b \to a$ is the weakest c such that $b \,.\, c \leq a$. Yet there exist systems A, B for which no such weakest C can be found[2]. Thus *the generalized lattice of logic* (as I shall call it) does not possess some of the most distinctive features of the (Boolean) algebra of sentences. However, although there is in general no conditional $B \to A$, there is everywhere a *difference* or *remainder* $B - A$, defined as the strongest (that is, narrowest under the ordering $\leq$) system C for which $B \leq A + C$. Accordingly Curry ([1963], p. 162) refers to this sort of lattice as a *subtractive lattice*, though it is more commonly known, following McKinsey and Tarski [1946] p. 124, as a *Brouwerian algebra*. (Popper [1963], p. 321, calls the logical variant the 'dual-intuitionistic calculus'.) It is the purpose of this section to consider autometrizations of Brouwerian algebras, with an eye of course to the generalized lattice of logic $\mathscr{L}$.

In terms of the remainder $B - A$ we can define the *symmetric difference* of two systems A and B:

(6) $$A \mathbin{\triangle} B = (A - B) + (B - A).$$

A function $_*$ from $\mathscr{L} \times \mathscr{L}$ to $\mathscr{L}$ will be called *autometric* if and only if

(7) $A_* B = 0$ if and only if $A = B$,

(8) $A_* B = B_* A$,

(9) $A_* C \leq (A_* B) + (B_* C)$.

We shall show that the symmetric difference $A \mathbin{\triangle} B$ is indeed an autometric operation on the lattice $\mathscr{L}$. The proof is adapted from Nordhaus and Lapidus [1954], p. 218, but it is convenient to start with an easy lemma.

LEMMA. In a Brouwerian algebra $\mathscr{L}$ we have $A + (B - A) = A + B$.

Proof: It will be recalled that $B - A$ is the strongest C such that $B \leq A + C$. Since $B \leq A + B$ it follows that $B - A \leq B$, so that $A + (B - A) \leq A + B$. But $A \leq A + (B - A)$, and by the definition of difference $B \leq A + (B - A)$. Thus $A + B \leq A + (B - A)$, which completes the proof.

THEOREM 6. $A \triangle B$ is an autometric operation on $\mathscr{L}$.

Proof: Axioms (7) and (8) are routine, and we need only concern ourselves with the triangle inequality (9). Since the term $C + ((A \triangle B) + (B \triangle C))$ can be written $C + (B - C) + (A - B) + (C - B) + (B - A)$, because addition associates and commutes, we can use the Lemma to collapse together the three leading summands. They turn out to equal $C + B + A$ (indeed, the whole term is equal to this). Thus $A \leq C + ((A \triangle B) + (B \triangle C))$, and so $A - C \leq (A \triangle B) + (B \triangle C)$. By symmetry the same holds for $C - A$, and by addition the proof is complete.

Again we shall say that an autometric operation $_*$ is *normal* if for every system A we have $A_* 0 = A$. The analogue of Theorem 2 is straightforward.

THEOREM 7. If $_*$ is a normal autometric operation on $\mathscr{L}$ then

$$A \triangle B \leq A_* B \leq A + B.$$

Proof: The triangle inequality applied to 0, A, B, yields at once $A_* B \leq A + B$ as before. Equally similarly we have $A \leq B + (A_* B)$, and so by the definition of difference $A - B \leq A_* B$. Likewise $B - A \leq A_* B$, and consequently $A \triangle B \leq A_* B$.

We can now move on to the special case of distance from the truth. When T is not finitely axiomatizable, the usual case indeed, $-T$ is the unit 1 (that is, L, the set of tautologies; see my [1974a], p. 174). Thus $A \, . -T$ does not differ from A. One could well expect, therefore, in the light of Theorem 3, that $A_* T$ would always and necessarily be A when A is true; and therefore that the distance of A from the truth would always be the same as its truth content $Ct_T(A)$. In fact, this is not so, as Theorem 8 will explain. In view of the general unfamiliarity with Brouwerian algebra, and even with the generalized lattice of logic, we prepare for this theorem with a couple of simple lemmas. In the proofs of these we shall call upon some elementary model-theoretic ideas; they are explained more fully on p. 428 below.

LEMMA. If T is an atom in a Brouwerian algebra $\mathscr{L}$ then if A is false $A \triangle T = A + T$, and if A is true $A \triangle T = A - T$.

Proof: $T - A$ is the strongest B such that $T \leq B + A$. If A is true, then $T \leq A$, so that $T - A = 0$. If A is false, the strongest B required is patently not 0, so that it must be T, since no system lies between T and 0.

$A - T$ is the strongest B for which $A \leq B + T$. Certainly $A \leq A + T$. So suppose that there is a B for which $B < A$ and $A \leq B + T$. If A is false it cannot equal $B + T$, so we have $B < A < B + T$. This last is, however, impossible, for the number of models must *increase* across $<$, yet B differs from $B + T$ in one model only. Thus there is no B stronger than A for which $A \leq B + T$. That is, for *false* A we have $A - T = A$.

Summarizing, we see that for false A the symmetric difference $A \triangle T = A + T$, and for true A it is $A - T$.

LEMMA. $A \triangle T \neq A + T = Ct_T(A)$ if and only if $A = C + T$ for some false theory C. In this case $A \triangle T = C$.

Proof: If $A \triangle T \neq A + T$, then by the previous Lemma A is true and $A - T \neq A + T$. But by the Lemma to Theorem 6 we have $A + T = (A - T) + T$. Thus $A - T$ differs from its own truth content, so must be false. So A is the truth content of the false theory $A - T$.

Conversely, suppose that $A = C + T$ for some false theory C. Then $C < A$, and by the definition of remainder $A - T \leq C$. Since A is true, the previous Lemma establishes that $A \triangle T$ differs from A, which is $A + T$. Moreover, if $A - T \neq C$, then $A - T$ differs from A by more than a single model. So $A \triangle T = C$.

THEOREM 8. If $*$ is a normal autometric operation on $\mathscr{L}$ then $A * T = A \triangle T = A + T$ unless $A = C + T$ for some false theory C. In this case $A * T$ is equal to A or to C.

Proof: If A is not $C + T$ for some false C the result follows at once from the last Lemma and Theorem 7. Otherwise we have $C \leq A * T \leq A$, so that $A * T$ may in general be either C or A.

The Lemma on p. 173 of my [1974a] states that if A is recursively axiomatizable it is not the truth content of any false theory. Thus if A is true and recursively axiomatizable then $A * T = A$.

THEOREM 9. There is no autometric operation on $\mathscr{L}$ that is normal and downward strictly monotone.

Proof: Immediate from the preceding remark.

Thus for axiomatizable A we have generally that $A_* T = A + T$. *The distance from the truth of an axiomatizable theory is equal to its truth content.* But amongst axiomatizable theories truth content increases monotonically with content (see the second corollary to Theorem 5 on p. 174 of my [1974a]).

THEOREM 10. Suppose that $_*$ is a normal autometric operation on $\mathscr{L}$. Then if B and A are axiomatizable,

$$B_* T < A_* T \text{ if and only if } B < A.$$

Proof: Immediate.

Thus if B is closer than A is to the truth T, it is also closer than A is to every unaxiomatizable complete theory X. In other words, *approach to the truth via axiomatizable theories is quite independent of where the truth in truth is.*

This is the objection that originally turned me away from Popper's suggestion to measure the truthlikeness of a theory by its truth content. It is an objection that in the next section we shall have to turn our attention towards.

4. SOME OBJECTIONS ANSWERED

Theorems 5, 9, and 10 encapsulate a number of counterintuitive consequences of the conditions we have laid on autometric functions in the Boolean algebra $\mathscr{B}$ and the Brouwerian algebra $\mathscr{L}$. Of these objections the easiest disarmed is that embodied in Theorem 9, which says that no normal autometric operation on $\mathscr{L}$ is at the same time downward strictly monotone. Although the condition (4) of downward strict monotony was introduced *ad hoc* in Section 2 as a way of disqualifying a clutch of autometric functions that were otherwise admissible, it is undoubtedly a not unattractive condition to have our measures of distance obey. And as it happens we can rather easily arrange that there is a normal metric $_*$ on $\mathscr{L}$ that satisfies (4) provided we relax very modestly the requirement that $_*$ be *auto*metric.

After all, the particular novelty of autometric functions stems very largely from the fact that in each case the *space of values*, or *ground space* (Ellis [1951b]), or *range* of the operation $_*$ is a lattice with least element. That the *space of arguments* (*domain*) be also a Boolean or Brouwerian algebra has been dictated by our problem situation, but otherwise has not played a great role (though the monotony conditions (4), (5) can only be formulated if the domain is at least an ordered set). Indeed, any set whatever may be taken as the domain of a 'Boolean metric space' (Blumenthal [1970], p. 93). Thus there is no real reason why the lattice $\mathscr{L}$ should not be metrized by means of any other lattice[3]; the only reservation one might have is that such a move would insinuate an element of arbitrariness where none was before. The alternative lattice I shall propose, however, is such a natural 'extension' of $\mathscr{L}$ itself that this fear is not likely to survive a single paragraph.

The new lattice of values is going to be a Boolean algebra, in fact the set of all subsets (power set) of the space of all models associated with the language under consideration. More precisely, if we identify any two models in which exactly the same sentences hold (models that, in standard terminology, are *elementarily equivalent*) we obtain a set of the power of the continuum: to each complete theory there corresponds exactly one model, namely the model in which the complete theory holds. By standard methods we can induce a topology on this space, which is called the Stone space of the Lindenbaum/Tarski algebra $\mathscr{B}$ from which it originates. The Stone space is a compact Hausdorff space: closed sets in it can be correlated with deductive systems, singletons with complete systems, and clopen (closed and open) sets with finitely axiomatizable systems (sentences). In every case the system A is correlated with the set $\mathscr{M}(A)$ of models in which it holds. The power set of the Stone space is of course a Boolean algebra, which we shall write $\mathscr{S}$. It seems easiest to continue using $+$, $.$, $-$, $\triangle$, 0, 1, $\leq$, and so on for the Boolean operations and relations in $\mathscr{S}$, in which case it can speedily be confirmed that $\mathscr{M}(A + B) = \mathscr{M}(A) + \mathscr{M}(B)$ and that $\mathscr{M}(A \,.\, B) = \mathscr{M}(A) \,.\, \mathscr{M}(B)$, but that neither $\mathscr{M}(-A) = -\mathscr{M}(A)$ nor $\mathscr{M}(A \triangle B) = \mathscr{M}(A) \triangle \mathscr{M}(B)$. Thus the function $\mathscr{M}$ embeds the lattice $\mathscr{L}$ somewhat heteromorphically in the Boolean algebra $\mathscr{S}$. Now $\mathscr{M}(T)$ is a unit set in $\mathscr{S}$, so is an atom; thus if $_*$ is a normal downward strictly monotone autometric operation on $\mathscr{S}$ then $H {}_* \mathscr{M}(T) = H \triangle \mathscr{M}(T)$, for every H in $\mathscr{S}$, by Theorem 4. In particular, for every system A we have $\mathscr{M}(A) {}_* \mathscr{M}(T) = \mathscr{M}A) \triangle \mathscr{M}(T)$. If we now define an operation $\circ$ on $\mathscr{L}$ by the identity $A \circ B = \mathscr{M}(A) {}_* \mathscr{M}(B)$, then $\circ$ satisfies (1), (2), (3), is 'normal' in the sense that $A \circ 0 = \mathscr{M}(A)$ for all A, and is downward strictly monotone. In this way we are able to impose a Boolean-valued metric on the Brouwerian algebra $\mathscr{L}$.

Indeed, this representation effectively allows us to forget about Brouwerian algebras altogether. For we may simply *identify* A with $\mathscr{M}(A)$, and restrict ourselves to consideration of the autometric $_*$ in the algebra $\mathscr{S}$. Not all elements of $\mathscr{S}$, of course, are theories. But amongst those that are, Theorems 5 and 10 apply with full force: if A is false then $B_* T < A_* T$ if and only if $B < A$; and if A and B are axiomatizable then $B_* T < A_* T$ if and only if $B < A$. As I have indicated above (p.415), it is the latter result that is especially serious. It seems strange, to say the least, that one's approach to the truth should be quite independent of where the truth in fact is located.

Yet suppose that a number of desirable destinations $T, X, Y, Z, \ldots$, are scattered within a short distance of the north pole (0) on the surface of the globe, say within a single degree of latitude of it; and that T is the spot we are particularly anxious to reach. Suppose further that there is some minimum length to each permissible journey, and that this quantum of travel is not too small. Then *from many parts of the globe we are sure to get closer to the terminus T simply by heading north*; or more precisely, if B's latitude is greater than A's (that is, B is closer to 0 than A is) then B will be closer to T than A is. This is not of course true if we are already within striking distance of T, but may well be the case if we are, for example, outside the polar circle. Thus from many parts of the globe *progress towards T is at the same time progress towards the other possible destinations X, Y, Z, ...* Approach to T is quite independent of where T is.

This homely example shows that there can be cases where all possible destinations are, or anyway may be, approached by essentially the same route. The analogy is far from perfect, of course. Representing the elements of a Boolean algebra by the points on a surface of a sphere, even if not all the points are included in the representation, is a fairly crude device. By insisting that there be a minimum length of acceptable journey we go some way towards simulating the discrete character of the ordering $\leq$ in $\mathscr{S}$, to be sure; and by ignoring points A and B too close to T we echo quite faithfully the fact that axiomatizable theories are on the whole considerably weaker than complete ones[4]. But 'B is closer to the north pole than A is' is hardly a satisfactory transcription of '$B \leq A$', since the former is a total pre-ordering, whereas the latter is a partial ordering. In contrast, the relation 'B is *due* north of A' has a geographical meaning (in terms of lines of longitude) that does not reflect anything in the algebra $\mathscr{S}$. It is clear, indeed, that '$B \leq A$' corresponds to some relation less broad than 'B is closer to the north pole than A is' and less narrow than 'B is due north of A'. The model can therefore be refined somewhat by denoting by $-A$ the antipode of A, and for each *destination point* [= complete theory] X (such as T) drawing the great circle that touches X's parallel of latitude at X (and $-X$'s parallel of latitude at $-X$). Then we can say that $X \leq A$ if and only if A lies south of this great circle (see

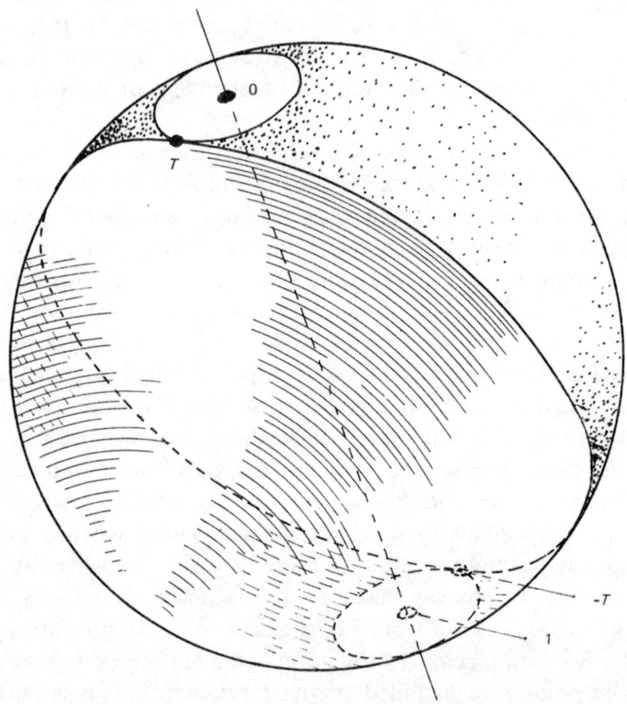

Fig. 1.

Figure 1). More generally we can say that $B \leq A$ if and only if for every destination X, if $X \leq B$ then $X \leq A$. Thus, for example, for half the points A on the globe we shall have $T \leq A$ (these correspond to the truths), and for the other half $T \leq -A$ (these A are false). With this interpretation of $\leq$, however, it is of course no longer the case that B is closer to T than A is *if and only if* $B \leq A$. *If* $B \leq A$, then indeed B is the closer to T (or, anyway, this is no less true now than it was before); but not *only if*. In fact, there will be many cases where B is closer to T than $-B$ is, even though $B \leq -B$ fails for all consistent B. This is not a very astonishing conclusion, given that our measure of distance in this case is a real-valued, rather than a lattice-valued one.

I shall not pursue this Euclidean analogy any further. I do not doubt that it could be significantly refined, were there any point in that. But I suspect that it has already exhausted its usefulness. Elsewhere I shall even offer an alternative Euclidean interpretation of Boolean geometry.

It should perhaps be clear from what has been said that it is the *necessary* conditions contained in Theorems 10 and 5 that make them so subversive. That B can be more truthlike than A *only if* it is logically stronger than A seems to make nonsense of the undisputed fact that most progress in science consists of the replacement of one theory by another with which it is incompatible. The *sufficiency* conditions are much less serious. It may come as something of a shock to discover that whenever B entails A it must also be more truthlike than A; that I would admit. But it does not rule out the possibility that some increases in content are great leaps towards the truth, whilst others are but very small steps. It can be seen therefore that although the autometric $\triangle$ is, on its own, an unsatisfactory measure of distance from the truth, there is nothing to prevent its being incorporated into a richer, and quite satisfactory, theory. Suppose, for example, that we define a real-valued function d on $\mathcal{S} \times \mathcal{S}$ in such a way that $d(A, B) \leq d(C, D)$ whenever $A_* B \leq C_* D$. In that case, as already noted in connection with the Euclidean sphere, $B \leq A$ would be a *sufficient* condition for $B_* T \leq A_* T$, but no longer a *necessary* one. In some such manner, therefore, it might be possible to elude the more pernicious effects of Theorem 10 and Theorem 5.

In other words, to put the matter bluntly, the theory of distance from the truth that we have offered so far may be *true*, but it is just *too weak*, in rather the same way that Popper's condition ($\diamond$) was too weak. It fails to deliver anything like enough comparisons of truthlikeness. Most of the other theories in the literature are, in contrast, demonstrably *too strong*, and therefore *false*: this is the case with Hyland's proposal, and with the variant of it that requires complete independence of the theories' bases (see my [1974b]), with Tichý's theory [1976] (see my [1976], and also Niiniluoto [1977], Sections 4 and 5), and, in my opinion, with all those proposals that hinge on the possibility of all-round improvements in numerical accuracy (see my [1975]; and, for a less sceptical assessment, Popper [1976]). But although a theory that is too strong is undoubtedly not acceptable, it can scarcely be an adequate defence of a weak theory that it *may* be possible to extend it to a satisfactory richer one; one could defend the sublimest trivialities in this way. The very least that we need in addition is some indication of how the richer theory is to be developed[5]. This we have here, since clearly we would wish a real-valued metric d on $\mathscr{S} \times \mathscr{S}$ to satisfy the numerical triangle inequality $d(A, C) \leq d(A, B) + d(B, C)$. The other two metric axioms would of course be consequences of the requirement that all inequalities between $_*$-terms be preserved by d. (If d obeys the requirement, so does d^2; this is why the triangle inequality is not automatically satisfied.) But this indication is not much indication. It cannot be denied that there are few evident signposts to how we should extend the Boolean metric $_*$ to the real-valued metric d – indeed, it was the embarrassing *richesse* of numerical metrics that inveigled us into Boolean geometry in the first place (see p. 416). However, as I shall argue elsewhere, an unexpectedly close relationship exists between *metric operations* and *probabilities*. Indeed, the problem of identifying a numerical metric on the Boolean algebra $\mathscr{S}$ in effect reduces to the problem of identifying a probability measure. I do not wish to pretend that this is an easy task; in fact I do not imagine for a moment that any objective way can be found of selecting such a function uniquely. But this does not mean that we may not be able to narrow a little the field of possibilities in an instructive way[6].

At first glance the theory of distance from the truth that has been promulgated here is, if anything, slightly better placed than is the theory of (logical) probability. The most that the axioms of probability, unaccompanied by any specification of the measure p, can tell us about the relative positions of two elements is that $p(B) \leq p(A)$ if and only if $B \leq A$. If there were no reason for preferring any one probability measure to any other, then logical probability would be no generalization at all of logic; and would therefore be epistemologically quite useless. Theorem 5 assures us that distance from the truth is capable of at least a little more refinement than this. I would argue, therefore, that anyone who holds logical probability to be a useful tool in the analysis of scientific knowledge may well have to accord the same courtesy to distance from the truth. Events might be said, indeed, to have confounded the pessimism of the following remark uttered by Jeffrey in 1973 (Jeffrey [1975], p. 108): 'I see no prospect of a satisfactory explication of *verisimilitude* along the lines Popper has tried so far.' For Jeffrey is a leading investor in the Bayesian trust in probabilities; and if there is no verisimilitude, then there is no probability either[7].

With this I discharge myself from the task of further countering the objections in Theorems 5 and 10. I hope anyway to have shown that these objections, if not innocuous (which they are not), are less disabling than they may at first sight have seemed. It is not yet obviously true that Boolean geometry provides no sort of solution to the problem of verisimilitude.

Department of Philosophy,
University of Warwick

NOTES

[0] The main ideas of Sections 0–3 of this paper were developed in January and February 1975. It was some months, however, before I could convince myself that the theory was anything but a washout; that is, before I could draft Section 4. An early version of the paper was read at Chelsea College in November 1975, and I am grateful to the audience there for the discussion that followed. As always, I am indebted to Sir Karl Popper for his encouragement, and for the interest he has shown in what I had to say. I should also like to thank my father for drawing the figure. Let me further note here that a version of the paper has been published in the *Bulletin of the Section of Logic*, Institute of Philosophy and Sociology, Polish Academy of Sciences, Wrocław. See vol. 6, pp. 15–26.

[1] The conditions of downward and upward weak monotony may be obtained from (4) and (5) by writing $\leqslant$ throughout for $<$. It is of some interest that, although (4) and (5) are not equivalent (as noted in the text), the two conditions of weak monotony are interderivable given the metric axioms (1), (2), (3).

[2] The simplest case is provided by taking for B the system T of all true sentences, and for A any false system. It follows from T's completeness and unaxiomatizability that there is no weakest amongst all false systems. (It would necessarily contain only one false sentence f, which would be a consequence of every false sentence. But then $-f$ would axiomatize T.) Now clearly those systems C for which $T \cdot C \leqslant A$ are just the false systems. Thus there is no weakest such system C, no system to play the part of $T \to A$.

[3] Popper has shown ([1976], p. 152) that if we metrize $\mathscr{L}$ by the (Boolean) algebra of *sets of sentences* then we rather unexpectedly revive condition ($\Diamond$) for increase in truthlikeness. In the algebra in question T, the set of all true sentences, is not an atom; its complement is F, the set of all false sentences. Likewise the complement $-A$ of a theory A is not itself a theory, but the set of sentences that do not follow from A. To say that A is false, therefore, is to say that $T \nleq A$, but not that $T \leq -A$. (Thus Theorem 3 cannot be proved in any useful form.) If we write $A_* T = A \triangle T = A \cdot F + -A \cdot T$, we can see that A's distance from the truth is measured by its *falsity content* ($A \cdot F$),

together with those true sentences that it fails to deliver $(-A \cdot T)$. Consequently $A * T = Ct_F(A) + T - Ct_T(A)$. Since truthlikeness increases as distance from the truth decreases, we have shown that the condition (◇) does indeed follow from this method of defining $*$.

[4] By Gödel's theorem, every axiomatizable extension of Peano's arithmetic P is essentially undecidable. This means that every such extension can itself be extended in continuum many ways. Thus every axiomatizable theory that contains P holds in uncountably many models, and in this clear sense is considerably weaker than any complete theory.

[5] Tichý's and my criticisms of Popper's condition (◇) originally seemed so ruinous because there was no obvious clue to how that theory of truthlikeness might be developed. In consequence, Popper's recent proof (see footnote 3 above) that (◇) characterizes a certain version of Boolean geometry should perhaps help to temper the force of these criticisms. An example of a theory of truthlikeness whose good samaritan is not yet anywhere in sight is the variant of Hyland's proposal that requires maximal independence of the theories' bases (see my [1974b], p. 186). There, at least, the consequence that increase in content is sufficient for increase in truthlikeness is still to be deemed utterly unwelcome.

[6] It is in this light, I suggest, that we should regard the arguments of Popper ([1959], Appendix* vii) and others that purport to show that the probabilities of universal hypotheses must be zero. The uncompromising severity with which Howson [1973] sets about these arguments strikes me as being a little out of place.

[7] It might be argued that the various 'convergence theorems' of probability theory show that most of the applications in this field are largely independent of the initial probability measure: relative probability is what interests the Bayesian faction, not absolute probability. On the other hand, 'relative verisimilitude', though of course definable, is not obviously of any interest. If the said 'convergence theorems' really are theorems, I am prepared to concede a point or two here. But I would note that *some* initial distributions have to be disqualified on other grounds, and these grounds may be sufficient for my purpose.

BIBLIOGRAPHY

Blumenthal, L. M. [1953], *Theory and Applications of Distance Geometry*, Oxford University Press.

Blumenthal, L. M. [1970], 'Lattice Geometries', pp. 1–132 of L. M. Blumenthal and K. Menger, *Studies in Geometry*, W. H. Freeman & Co., San Francisco.

Curry, H. B. [1963], *Foundations of Mathematical Logic*, McGraw Hill, New York.

Elliott, J. G. [1953], 'Autometrization and the Symmetric Difference', *Canadian Journal of Mathematics* **5**, pp. 324–331.

Ellis, D. O. [1951a], 'Autometrized Boolean Algebras I', *Canadian Journal of Mathematics* **3**, pp. 87–93.

Ellis, D. O. [1951b], 'Notes on Abstract Distance Geometry I. The Algebraic Description of Ground Spaces', *The Tôhoku Mathematical Journal*, series 2, **3**, pp. 270–272.

Ellis, D. O. [1951c], 'Notes on Abstract Distance Geometry II. Implications of Basality in Generalized Semimetric Spaces', *Monatshefte für Mathematik* **55**, pp. 185–187.
Howson, C. [1973], 'Must the Logical Probability of Laws be Zero?', *British Journal for the Philosophy of Science* **24**, pp. 153–163.
Jeffrey, R. C. [1975], 'Probability and Falsification: Critique of the Popper Program', *Synthese* **30**, pp. 95–117.
McKinsey, J. C. C. and Tarski, A. [1946], 'On Closed Elements in Closure Algebras', *Annals of Mathematics*, series 2, **47**, pp. 122–162.
Miller, D. W. [1974a], 'Popper's Qualitative Theory of Verisimilitude', *British Journal for the Philosophy of Science* **25**, pp. 166–177.
Miller, D. W. [1974b], 'On the Comparison of False Theories by Their Bases', *British Journal for the Philosophy of Science* **25**, pp. 178–188.
Miller, D. W. [1975], 'The Accuracy of Predictions', *Synthese* **30**, pp. 159–191.
Miller, D. W. [1976], 'Verisimilitude Redeflated', *British Journal for the Philosophy of Science* **27**, pp. 363–381.
Niiniluoto, I. [1977], 'On the Truthlikeness of Generalizations', pp. 121–147 of R. E. Butts and J. Hintikka (eds.), *Basic Problems in Methodology and Linguistics* (Western Ontario Series 11), D. Reidel, Dordrecht.
Niiniluoto, I. [1978], 'Truthlikeness in First-Order Languages', *this volume*, pp. 437–458.
Nordhaus, E. A. and Lapidus, L. [1954], 'Brouwerian Geometry', *Canadian Journal of Mathematics* **6**, pp. 217–229.
Popper, K. R. [1959], *The Logic of Scientific Discovery*, Hutchinson & Co., London.
Popper, K. R. [1963], *Conjectures and Refutations*, Routledge & Kegan Paul, London.
Popper, K. R. [1976], 'A Note on Verisimilitude', *British Journal for the Philosophy of Science* **27**, pp. 147–159.
Tichý, P. [1974], 'On Popper's Definitions of Verisimilitude', *British Journal for the Philosophy of Science* **25**, pp. 155–160.
Tichý, P. [1976], 'Verisimilitude Redefined', *British Journal for the Philosophy of Science* **27**, pp. 25–42.

ILKKA NIINILUOTO

TRUTHLIKENESS IN FIRST-ORDER LANGUAGES

1. INTRODUCTION

Let L be a first-order language, and let $\mathfrak{A}$ be a structure for L. On the basis of the Tarskian definition of truth, the sentences of L are then dichotomized into those which are true in $\mathfrak{A}$ and those which are false in $\mathfrak{A}$. Intuitively, it often seems possible to supplement this two-fold partition of the set S_L of the sentences of L with an additional fine structure. For example, some of the truths in L are 'better' than others at least in the sense that they are more comprehensive, i.e., tell more about the structure $\mathfrak{A}$. Some false sentences of L are more mistaken than others – thus, the statement '3 is the smallest natural number' can be said to be closer to the truth than the statement '5 is the smallest natural number', and the claim that all prime numbers are odd is less erroneous than the claim that all prime numbers are even. Some false sentences of L may even contain (or entail) more true information about the structure than some of the true sentences of L (like tautologies). Similar remarks seem to apply not only to single sentences in L but to theories (i.e., deductively closed subsets of S_L) as well.

But, we may ask, does it really make sense to say that some sentences of L (interpreted in $\mathfrak{A}$) are 'closer to the truth' than some others? In particular, what precisely does it mean to say that one false theory in L is more 'truthlike', or a better 'approximation to the truth', than another theory in L? Can one define a generally applicable measure of 'truthlikeness' with intuitively satisfactory properties, or are there perhaps several viable alternatives? These questions concern what may be called the *logical problem of truthlikeness*. This problem, which is the topic of this paper, should be clearly distinguished from the corresponding *epistemic problem* of truthlikeness which concerns methods of estimating degrees of truthlikeness, or comparing theories for truthlikeness, on the basis of available evidence. Just as truth itself, the notion of truthlikeness is intended to be independent of evidence, that is, independent of our knowledge. Therefore, the systems of so called inductive logic which study the relation between hypotheses and evidence supporting them do not

J. Hintikka, I. Niiniluoto, and E. Saarinen (eds.), Essays on Mathematical and Philosophical Logic, 437–458.

prima facie have anything to do with the logical notion of truthlikeness, even if the inductive probabilities may play an important role within the estimation problem for truthlikeness.[1]

The analysis of truthlikeness does not aim at clarifying the notion of truth *simpliciter*, as this notion is presupposed in our discussion. It does not fall under the scope of many-valued logic, as we operate within the framework of ordinary two-valued logic (see Section 3, however). It is not a branch of the so called fuzzy logic, or of the semantics of vague expressions, as we always assume that the sentences of L have sharp truth conditions within the relevant structures for L.[2]

The methodological import of the notion of truthlikeness is a problem within the philosophy of science rather than within logic, and therefore it will not be discussed in this paper. Here it is sufficient to remark that the main motivation for the study of this notion arises from the semantic analysis of empirical scientific theories: as, in actual fact, many of these theories are false, the idea of scientific progress seems to presuppose the intelligibility of the claim that successive theories in science may come 'closer and closer to the truth'.[3]

The impetus for the present attempts at solving the logical problem of truthlikeness has largely come from the recently noticed failure of the first and so far most influential systematic treatment of this problem. Popper (1962) made the suggestion that two theories T and T' in L may be compared for their truthlikeness (or verisimilitude) as follows: T' is more truthlike than T if and only if

$$T \cap Th(\mathfrak{A}) \subseteq T' \cap Th(\mathfrak{A}),$$
$$T' \cap (S_L - Th(\mathfrak{A})) \subseteq T \cap (S_L - Th(\mathfrak{A})),$$

where at least one of these containments is strict.[4] Miller and Tichý showed, however, that this initially plausible definition of verisimilitude does not work – they proved that whenever T' is more truthlike than T, T' must be true, so that Popper's definition fails to make sense of the comparison of two false theories for their truthlikeness.[5]

2. THE SIMILARITY APPROACH TO TRUTHLIKENESS

Several logicians have suggested that the truth conditions for counterfactual or subjunctive conditionals can be given by using as a primitive notion a relation of *similarity* between possible worlds.[6] Hilpinen observed in 1974, that such similarity relations can be used to define a comparative notion of truthlikeness (Hilpenen, 1976). This idea can be presented in the following way.

Let **M** be the class of all structures for language *L*. For each sentence *h* in S_L there corresponds the class Mod(*h*) $\subseteq$ **M** of those structures in which *h* is true. (The same holds for sets of sentences in S_L.) Let $\mathfrak{A}^*$ be a particular structure in **M** which is taken to represent the actual world. On the basis of a similarity relation (or similarity metric) on **M**, all the structures in **M** are located in a nested system of spheres around $\mathfrak{A}^*$ (see Fig. 1).

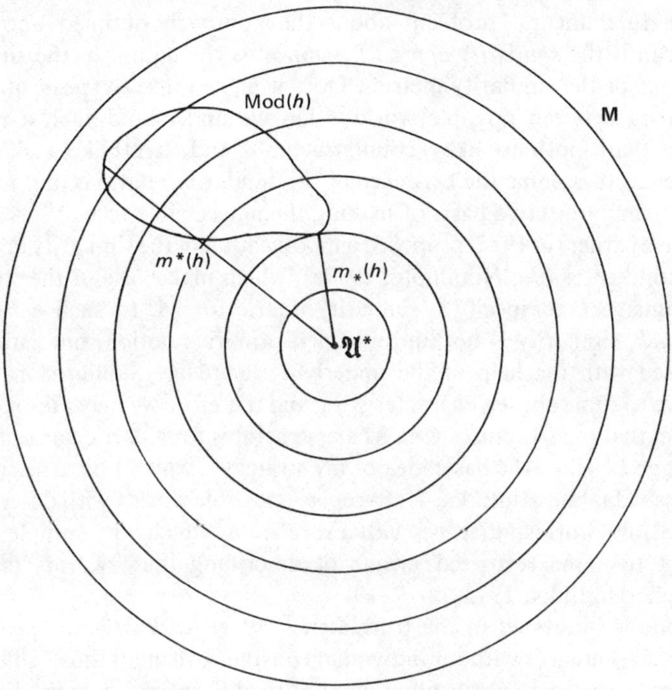

Fig. 1.

If *d* is a similarity metric on **M**, so that $d(\mathfrak{A}, \mathfrak{A}^*)$ gives the distance between structures $\mathfrak{A}$ and $\mathfrak{A}^*$, then we can define for all *h* in S_L

$$(1) \qquad m_*(h) = \inf_{\mathfrak{A} \in \mathrm{Mod}(h)} d(\mathfrak{A}, \mathfrak{A}^*),$$

$$(2) \qquad m^*(h) = \sup_{\mathfrak{A} \subset \mathrm{Mod}(h)} \mathrm{d}(\mathfrak{A}, \mathfrak{A}^*).$$

Hilpinen's definition corresponds now to the following[7]: *h*′ is at least as truthlike than *h* if and only if

(3) $m_*(h') \leqslant m_*(h)$ and $m^*(h') \leqslant m^*(h)$.

This definition has the advantage that it can be immediately generalized to arbitrary theories in L. Moreover, even if many theories will be incomparable according to (3), it is possible that a false theory (that is, false in $\mathfrak{A}^*$) is more truthlike than another false theory. But, on the other hand, definition (3) has the defect that normally no false theory can ever be more truthlike than any true theory.[8]

The fundamental problem about the approach outlined above – we may call it the *similarity approach* – concerns the nature of the similarity relations or the similarity metrics. Does it make sense to speak of overall similarity between possible worlds? Do we understand such a relation better than notions like 'counterfactual' and 'truthlikeness'? What entitles us to assume the existence of the similarity relations and metrics? Is there any objective basis of making them accessible to us?

I have earlier (in 1975) proposed a modification of the similarity approach to truthlikeness (see Niiniluoto, 1977a), which makes use of the structure of language L to specify a similarity metric for **M**. In such a *linguistic approach*, similarity is not an undefined primitive notion, but rather it is analysed with the help of the underlying language. Similarity is always relative to some chosen characteristics, and therefore we may take seriously the fact that the structures $\mathfrak{A}$ in **M** are structures for a particular first-order language L. Thus, the basic idea of my strategy – which I have learnt from Professor Jaakko Hintikka – is to replace possible worlds with descriptions of possible worlds, that is, with expressions which are complete with respect to some restricted means of describing these worlds (see, for example, Hintikka, 1973, pp. 7–10).

If one is interested in the truthlikeness of *generalizations*, i.e., quantificational sentences without individual constants, then all those differences between structures which amount only to differences in their diagrams (that is, in the truth-values of singular sentences) may be considered as irrelevant. In other words, only those dissimilarities between structures which are reflected in the generalizations holding true of them are then considered as significant. (This is not a restrictive assumption in view of the intended applications of the notion of truthlikeness.) Following this suggestion, let us define for structures $\mathfrak{A}$ and $\mathfrak{A}'$ in **M**

(4) $\mathfrak{A} \equiv_g \mathfrak{A}'$ iff for all generalizations h in L:
$$\mathfrak{A} \vDash h \Leftrightarrow \mathfrak{A}' \vDash h.$$

As we are not interested in singular sentences, we may make the simplifying assumption that the non-logical vocabulary V_L of L does not contain individual constants at all. Then the relation $\equiv_g$, defined by (4), is the same as the ordinary relation of *elementary equivalence* with respect to L. Thus, $\equiv_g$ is an equivalence relation in $\mathbf{M}$, and it induces a partition $\mathbf{M}_g$ in $\mathbf{M}$. If V_L consist of a finite number of one-place predicates (i.e., L is a finite monadic language), then this partition $\mathbf{M}_g$ is finite. To obtain a similar partition of $\mathbf{M}$ in the case of finite full first-order languages with n-place predicates ($n > 1$), definition (4) has to be modified as follows:

(5) $\mathfrak{A} \equiv_g^{(d)} \mathfrak{A}'$ iff for all generalizations h of depth $\leqslant d$ in L:
$$\mathfrak{A} \models h \Leftrightarrow \mathfrak{A}' \models h.$$[9]

When the depth d increases, finer and finer finite partitions $\mathbf{M}_g^{(d)}$ of $\mathbf{M}$ are obtained. In the limit, when $d \to \infty$, the partition $\mathbf{M}_g^{(d)}$ approaches the infinite partition $\mathbf{M}_g$ of $\mathbf{M}$ induced by $\equiv_g$.

To each member of $\mathbf{M}_g$ (resp. $\mathbf{M}_g^{(d)}$) there corresponds the strongest generalization in L which is true of each structure in this class. These mutually exclusive generalizations, given in a standard syntactical form, are called the *constituents* (resp. constituents of depth d) of L. Each generalization in L is logically equivalent to a finite disjunction of constituents of L; this disjunction is its *distributive normal form*. In this way, to each consistent generalization in L there corresponds a non-empty subset of the finite set $\mathbf{M}_g$ or $\mathbf{M}_g^{(d)}$.

When this situation is compared with Fig. 1, it is seen that $\mathbf{M}$ has been replaced with a finite set, $\mathbf{M}_g$ or $\mathbf{M}_g^{(d)}$. Moreover, all the relevant differences and similarities between the elements of $\mathbf{M}_g$ (or $\mathbf{M}_g^{(d)}$) are now, as it were, condensed into certain sentences of L.

This achievement suggests the following program for our study.[10] If the truthlikeness of a generalization h in L can be defined as a function of the truthlikeness of the constituents occurring in the normal form of h, then the logical problem of truthlikeness reduces to the problem concerning the distance of the constituents of L from the true constituent. These two problems will be discussed, in the next sections, in the order they have mentioned above.

3. DEGREES OF TRUTHLIKENESS

Let $C_1, \ldots, C_r$ be the constituents of L. (If L is a full first-order language, it is assumed that $C_1, \ldots, C_r$ are the constituents of L at some finite depth d). If h is a consistent generalization in L (of depth $\leqslant d$), then

(6) $\vdash h \equiv \bigvee_{i \in \mathbf{I}(h)} C_i$

for some index set $\mathbf{I}(h) \subseteq \{1, \ldots, r\}$, $\mathbf{I}(h) \neq \emptyset$. Let C_* be the true constituent of L. (C_* is the constituent which corresponds to the equivalence class of the structure $\mathfrak{A}^*$.) Suppose that the distance between constituents is given by measure d.

Following (1) and (2), let us define

(7) $m_*(h) = \min_{i \in \mathbf{I}(h)} d(C_i, C_*)$

and

(8) $m^*(h) = \max_{i \in \mathbf{I}(h)} d(C_i, C_*)$.

The distance of h from the truth (as expressed by C_*) can then be defined as the weighted average of $m_*(h)$ and $m^*(h)$:

(9) $D_1(h) = \gamma m_*(h) + (1 - \gamma) m^*(h)$,

where $0 \leqslant \gamma \leqslant 1$. An important special case of (9) is obtained by choosing $\gamma = 1/2$:

(10) $D_2(h) = \frac{1}{2}(m_*(h) + m^*(h))$.

Tichý (1974) proposed the following measure for the distance of h from the truth:

(11) $D_3(h) = \frac{1}{|\mathbf{I}(h)|} \sum_{i \in \mathbf{I}(h)} d(C_i, C_*)$.

This measure, together with D_1 and D_2, was also considered in Niiniluoto (1977a).

If the maximum of the distances $d(C_i, C_*)$, $i = 1, \ldots, r$, is denoted by K, then three normalized measures for the *degree of truthlikeness* of a generalization h are obtained as follows:

(12) $M_i(h) = 1 - \frac{D_i(h)}{K}$ $(i = 1, 2, 3)$

(see Niiniluoto, 1977a). Here

(13) $$M_3(h) = \frac{1}{|\mathbf{I}(h)|} \sum_{i \in \mathbf{I}(h)} \left(1 - \frac{d(C_i, C_*)}{K}\right).$$

If we define

(14) $$V(h) = 1 - \frac{m_*(h)}{K}$$

and

(15) $$I(h) = 1 - \frac{m^*(h)}{K},$$

then

(16) $$M_1(h) = \gamma V(h) + (1 - \gamma)I(h)$$

and

(17) $$M_2(h) = \frac{1}{2}(V(h) + I(h)).$$

Here $V(h) = 1$ if and only if h is true (in $\mathfrak{A}^*$), and $0 \leqslant V(h) < 1$ for false generalizations h. Function V can thus be said to measure the *degree of truth* in h. On the other hand, $I(h)$ is close to one if and only if the normal form of h does not contain constituents at a great distance from the true one C_*. Function I can thus be said to measure the *information about the truth* in a generalization h. In particular, $V(h) = 1$ but $I(h) = 0$ for a logically true generalization h. Measures M_1 and M_2 thus define truthlikeness as a combination of a 'truth-factor' V and an 'information-factor' I (cf. Hilpinen, 1976).

Measures M_1, M_2, and M_3 satisfy the intuitive desiderata that were mentioned in Section 1 (see Niiniluoto, 1977a). Popper (1976) has pointed out, however, that measure M_3 fails to satisfy the following requirement:

(19) If h and h' are true and $h' \vdash h$, then h does not have a greater degree of truthlikeness than h'.

(19) requires that, among true generalizations, truthlikeness cannot decrease when logical content increases. Fortunately, measures M_1 and M_2 satisfy this reasonable requirement. More generally, if h and h' have the same degree of truth, then $M_2(h) \leq M_2(h')$ if and only if $I(h) \leq I(h')$, where $I(h) \leq I(h')$ is implied by $h' \vdash h$. Measures M_2 and M_3 can therefore be accepted as solutions to the problem of this section.

4. DISTANCE BETWEEN MONADIC CONSTITUENTS

If L is a monadic first-order language (without identity) with the set $V_L = \{O_1, \ldots, O_k\}$ of primitive predicates, then the *constituents* of L are expressions of the following form:

$$(20) \qquad C_i = (\pm)(\exists x)Ct_1(x) \,\&\, \ldots \,\&\, (\pm)(\exists x)Ct_K(x),$$

where $(\pm)$ may be replaced either by the negation sign $\sim$ or by nothing and the *Ct-predicates* Ct_j, $j = 1, \ldots, K$, with $K = 2^k$, are defined by conjunctions of the form

$$(21) \qquad Ct_j(x) = (\pm)O_1(x) \,\&\, \ldots \,\&\, (\pm)O_k(x).$$

If one denotes, for all $i = 1, \ldots, K$,

$$\mathbf{CT}_i = \{j \in \{1, \ldots, K\} | Ct_j \text{ occurs positively in } C_i\},$$

then the constituent C_i in (20) can be expressed as follows:

$$(22) \qquad C_i = \bigwedge_{j \in \mathbf{CT}_i} (\exists x)Ct_j(x) \,\&\, (x)[\bigvee_{j \in \mathbf{CT}_i} Ct_j(x)].$$

A simple way of measuring the distance $d(C_i, C_j)$ between two constituents C_i and C_j $(i, j \leqslant K)$ was proposed in Niiniluoto (1977a). This distance may be defined as the number of Ct-predicates Ct_j, $j = 1, \ldots, K$, which C_i and C_j claim to be different (that is, empty or non-empty). In other words,

$$(23) \qquad d(C_i, C_j) = \mathrm{card}(\mathbf{CT}_i \Delta \mathbf{CT}_j)$$

where Δ is the symbol for symmetric difference. It follows that the maximum distance between two constituents is K, and the number of constituents at the distance of m from a given constituent $(0 \leq m \leq K)$ is $\binom{K}{m}$. The properties of measure (23), combined with measures M_1, M_2, and M_3 of Section 3, were explored in Niiniluoto (1977a).

Definition (23) is illustrated in Fig. 2, where C_j has been chosen as the true constituent C_* (with the corresponding index set $\mathbf{CT}_*$). Relative to the different 'cells' determined by the Ct-predicates, constituent C_i can then make two different kinds of mistakes: it can claim that a non-empty cell is empty (*error of type I*) or that an empty cell is non-empty (*error of type II*). The value of $d(C_i, C_*)$ is, according to (23), simply the total number of errors of type I and II; all the errors of both kinds are thus assumed to have equal weights.

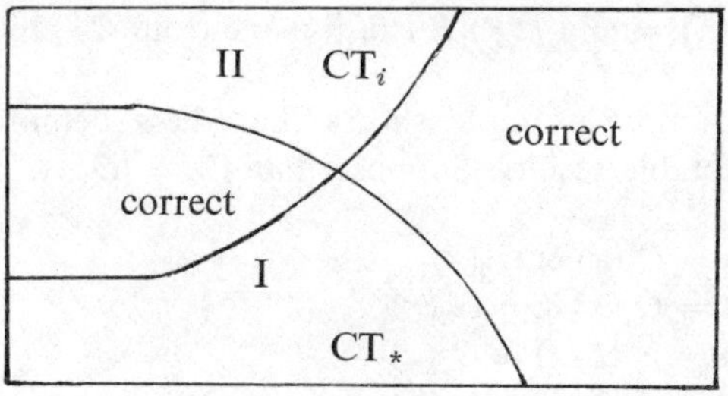

Fig. 2.

Tichý (1976) has proposed a measure of truthlikeness which is generally applicable to the sentences of full first-order languages (see also Section 5 below). He also argues against the acceptability of a putative definition of truthlikeness which he calls 'verisimilitude$_P$' (*ibid.*, p. 39). Measure (23) can be seen to be a special case of 'verisimilitude$_P$'.[11] Tichý's argument amounts to the requirement that, in determining the distance $d(C_i, C_*)$, it is not only the number of errors but also their *seriousness* that counts. The seriousness of these errors depends upon the best way of 'matching' the cells in $\mathbf{CT}_i$ and $\mathbf{CT}_*$ with each other. This, in turn, depends upon the distances between the relevant cells.

Even if Tichý's specific counterexample to 'verisimilitude$_P$', i.e., to measure (23), is not at all convincing (see Example 2 in Section 5 below), his suggestion that the distance between constituents may partly depend upon the pairwise distances between Ct-predicates seems worth exploring. In the special case at hand, Tichý's measure of the distance $\varepsilon(Ct_i, Ct_j)$ between Ct-predicates Ct_i and Ct_j is defined as follows:

(24) $\varepsilon(Ct_i, Ct_j)$ = the number of predicates O_i, $i = 1, \ldots, k$, which have a different sign in Ct_i and Ct_j

(cf. (21)).[12] Tichý says that a relation R is a *linkage* from $\mathbf{CT}_i$ to $\mathbf{CT}_j$ if and only if $\text{dom}(R) = \mathbf{CT}_i$, $\text{rg}(R) = \mathbf{CT}_j$ and $\text{card}(R) \leqslant \text{card}(R')$ for all relations R' with the same domain and the same range. Thus, a linkage R maps the larger of the sets $\mathbf{CT}_i$ and $\mathbf{CT}_j$ onto the smaller one, and $\text{card}(R) = \max\{\text{card}(\mathbf{CT}_i), \text{card}(\mathbf{CT}_j)\}$. The *breath* $\beta(R)$ of a linkage R is defined by

(25) $$\beta(R) = \frac{1}{\text{card}(R)} \sum_{(Ct_u, Ct_v) \in R} \varepsilon(Ct_u, Ct_v),$$

and the *distance* $\delta(C_i, C_j)$ of C_i from C_j by the breath of the narrowest linkage between $\mathbf{CT}_i$ and $\mathbf{CT}_j$:

(26) $\delta(C_i, C_j) = \min\{\beta(R) | R \text{ is a linkage from } \mathbf{CT}_i \text{ to } \mathbf{CT}_j\}$.

The following simple example shows that these definitions lead to intuitively unacceptable results. Suppose that $V_L = \{O_1, O_2\}$, and

$$Ct_1(x) = O_1(x) \,\&\, O_2(x)$$
$$Ct_2(x) = O_1(x) \,\&\, {\sim} O_2(x)$$
$$Ct_3(x) = {\sim} O_1(x) \,\&\, O_2(x)$$
$$Ct_4(x) = {\sim} O_1(x) \,\&\, {\sim} O_2(x).$$

Let the constituents C_1, C_2, and C_* be defined by

$$\mathbf{CT}_1 = \{Ct_1, Ct_3\}$$
$$\mathbf{CT}_2 = \{Ct_1, Ct_2, Ct_3\}$$
$$\mathbf{CT}_* = \{Ct_3, Ct_4\}.$$

(See Fig. 3.) Here C_1 seems to be definitively better than C_2, since C_2 makes all the errors that C_1 makes and some others as well. Yet, $\delta(C_1, C_*)$ is greater than $\delta(C_2, C_*)$, as

$$\delta(C_1, C_*) = \tfrac{1}{2}(1 + 1) = \tfrac{1}{2}(0 + 2) = 1$$

$$\delta(C_2, C_*) = \tfrac{1}{3}(0 + 1 + 1) = \tfrac{2}{3}.$$

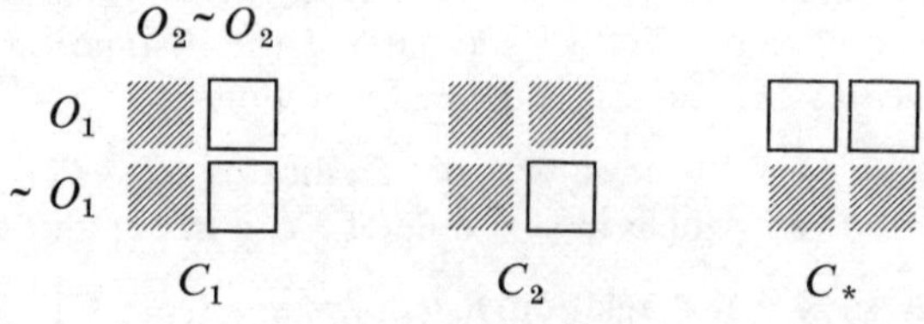

Fig. 3.

This example suggests that Tichý's notion of a linkage does not adequately capture the idea of proximity between constituents. To reconsider the matter, let us first ask the following question: When is an error of type II a serious one? This kind of error about a cell Ct_j is not very serious if there exists a cell in $\mathbf{CT}_*$ which is relatively close to cell Ct_j (cf. Fig. 2). On the other hand, an error of type I about a cell Ct_j is not very serious if there is a cell in $-\mathbf{CT}_*$, the complement of $\mathbf{CT}_*$, which is relatively close to cell Ct_j. To make this idea precise, let us define

(27) $$\varepsilon_{\mathrm{I}}(Ct_j) = \min_{Ct_u \in -\mathbf{CT}_*} \varepsilon(Ct_j, Ct_u), \text{ if } -\mathbf{CT}_* \neq \emptyset$$
$$= k, \text{ otherwise}$$
$$\varepsilon_{\mathrm{II}}(Ct_j) = \min_{Ct_u \in \mathbf{CT}_*} \varepsilon(Ct_j, Ct_u), \text{ if } \mathbf{CT}_* \neq \emptyset$$
$$= k, \text{ otherwise.}$$

ε_{I} and $\varepsilon_{\mathrm{II}}$ thus measure the seriousness of the errors of type I and II, respectively, made concerning a cell Ct_j. They have the value zero if such an error is not made.

Let us correlate with each constituent C_i a binary sequence $x_1x_2 \ldots x_K$, where x_j, $j = 1, \ldots, K$, is the value of the characteristic function of $\mathbf{CT}_i$ at cell Ct_j, i.e.,

$$x_j = 1, \text{ if } Ct_j \text{ occurs positively in } C_i$$
$$= 0, \text{ otherwise}$$

(cf. Niiniluoto, 1977a). Then the distance of the constituent C_i from the true constituent C_* may be defined as follows:

(28) $$d'(C_i, C_*) = \sum_{j=1}^{K} [\varepsilon_{\mathrm{II}}(Ct_j)x_j + \varepsilon_{\mathrm{I}}(Ct_j)(1 - x_j)].$$

Definition (28) is a generalization of the earlier definition (23): measure d of (23) is obtained from (28) by letting ε_{I} and $\varepsilon_{\mathrm{II}}$ have the constant value 1 for all errors.

It was seen above that the measure δ of (25) does not satisfy the following *dominance principle*:

(29) If C_i makes all the errors that C_j makes and some other errors as well, then $d(C_i, C_*) > d(C_j, C_*)$.

In contrast to Tichý's measure δ, the measure d' defined in (28) is immediately seen to satisfy the requirement (29).[13]

An interesting feature of definition (28) is the fact that the function d' is not symmetric. Some comments on this fact are in order here. (I am grateful to Professor Raimo Tuomela for pointing this out to me.) In the first place, David Lewis has indeed argued that degrees of similarity between possible worlds need not be symmetric, since what is important in evaluating the distance of world i from world j may depend upon some facts about world j (Lewis, 1973, p. 51). In our case, the non-symmetry of d' is rooted in the distinction between the two types of errors: if one considers cases (a) and (b), where we evaluate $d'(C_i, C_j)$ and $d'(C_j, C_i)$, respectively, then errors of type I in case (a) are errors of type II in case (b). Secondly, violation of symmetry is by no means necessary for our criticism of Tichý, since there are symmetric measures for the distance between constituents – other than (23) – which satisfy the dominance principle (29). One of them is obtained from (27) and (28) by replacing the set $-\mathbf{CT}_*$ with $\mathbf{CT}_i$ in the definition of ε_{I}. This 'weighted symmetric difference' takes into account the distances between Ct-predicates, which distinguishes it from (23), but in contrast to (27) it treats the two types of errors in a symmetric fashion.

5. MONADIC CONSTITUENTS WITH IDENTITY

If identity $=$ is joined into the monadic language L with the vocabulary $V_L = \{O_1, \ldots, O_k\}$, then the *numerical quantifiers* $\exists^n$ (there exists precisely $n \ldots$) and $\exists^{\geq n}$ (there exists at least $n \ldots$) for all natural numbers n can be expressed in L. In particular, the quantifiers $\exists^n$, for $n = 0, \ldots, d-1$, and $\exists^{\geq d}$ can be expressed by formulas of L which have the quantificational depth d. The constituents of L do not only specify the instantiated and non-instantiated Ct-predicates, but they also indicate (up to their depth) the number of individuals instantiating these Ct-predicates. A constituent of L at the depth d tells, for each Ct-predicate $Ct_1, \ldots, Ct_K$ of L, whether it is instantiated by $0, 1, \ldots$ or $d-1$ individuals or by *at least d* individuals. Thus, to each constituent at depth d there corresponds a unique K-tuple of numbers n_j such that $0 \leq n_j \leq d$ for $j = 1, \ldots, K$.

Let $\langle n_1, \ldots, n_K \rangle$ and $\langle m_1, \ldots, m_K \rangle$ be the K-tuples correlated with two constituents C_i and C_j of L at the same depth d. The simplest way of defining the distance between C_i and C_j is the following:

$$(30) \qquad d(C_i, C_j) = \sum_{t=1}^{K} |n_t - m_t|.$$

Definition (30) is a generalization of formula (23). It can be modified in a similar way as (28) was obtained – e.g., by giving extra weight to mistakes involving errors of type I or type II.

EXAMPLE 1. Let $O_1(x) =$ 'x is a prime number' and $O_2(x) =$ 'x is even'. Let $Ct_1 = O_1 \,\&\, O_2$, $Ct_2 = O_1 \,\&\, {\sim}O_2$, $Ct_3 = {\sim}O_1 \,\&\, O_2$, and $Ct_4 = {\sim}O_1 \,\&\, {\sim}O_2$. In the standard interpretation, the true constituent C_* at depth 3 is then correlated with the sequence $\langle 1, 3, 3, 3 \rangle$. If h is the generalization $(x)(O_1(x) \supset \sim O_2(x))$, i.e., 'all prime numbers are odd', then among the constituents in the normal form of h the closest (according to measure (30)) to C_* is $\langle 0, 3, 3, 3 \rangle$ and the farthest from C_* is $\langle 0, 0, 0, 0 \rangle$. If h' is the generalization $(x)(O_1(x) \supset O_2(x))$, i.e., 'all prime numbers are even', then the corresponding two constituents in the normal form of h' are $\langle 1, 0, 3, 3 \rangle$ and $\langle 3, 0, 0, 0 \rangle$. According to measure M_2, h has a greater degree of truthlikeness than h', since $D_2(h) = \frac{1}{2}(1 + 10) = 5\frac{1}{2}$ and $D_2(h') = \frac{1}{2}(3 + 11) = 7$.

EXAMPLE 2. Tichý (1976) considers the following generalizations:

(B1) $(\exists^1 x)O_1(x) \;\&\; (x) \sim O_2(x)$
(B2) $(\exists^2 x)O_1(x) \;\&\; (x) \sim O_2(x)$
(B3) $(\exists^2 x)O_1(x) \;\&\; (\exists^1 x)O_2(x) \;\&\; (x)(O_1(x) \supset \sim O_2(x))$
(B4) $(\exists^2 x)O_1(x) \;\&\; (\exists^2 x)O_2(x) \;\&\; (x)(O_1(x) \supset \sim O_2(x))$,

where the truth is that the cells Ct_1, Ct_2, Ct_3, and Ct_4 contain 1, 1, 1, and an infinite number ω of individuals, respectively. He argues that, intuitively, "(B1)–(B3) are increasingly better approximations to the truth", while (B3) and (B4) "seem equidistant from the truth" (*ibid.*, p. 26). The same result is given by Tichý's formal definition of truthlikeness (*ibid.*, p. 36). Tichý implicitly assumes that generalizations (B1)–(B4) claim the existence of at least three individuals in the cell Ct_4. With this qualification, (B1)–(B4) are equivalent to the following constituents at depth 3: $\langle 0, 1, 0, 3 \rangle$, $\langle 0, 2, 0, 3 \rangle$, $\langle 0, 2, 1, 3 \rangle$, and $\langle 0, 2, 2, 3 \rangle$. Our measure (30) gives then the result that the distances of (B1)–(B4) from the truth are 2, 3, 2, and 3, respectively, which contradicts Tichý's intuitive and theoretical results.[14] Note that Tichý's counterexample to the notion of 'verisimilitude$_P$' (see Section 4) is based on the claim – false in our view – that (B2) has a greater degree of truthlikeness than (B1).

An interesting feature of measure (30) is the fact that it makes the quantitative notion of truthlikeness relative to the quantificational depth – in the strong sense that h may be more truthlike than h' at the depth d but h' more truthlike than h at some greater depth $d' > d$. However, this may happen only in those cases where the true number of individuals in some cells exceeds the expressive power of our language, that is, is greater than the depth at which we are considering h and h'. For example, if the true number of individuals in some cell is infinite, definition (30) does not allow us to compare a given constituent with the 'whole truth' at one glance, but only with the true constituent at some finite depth. An error made about such a cell then turns out to be greater and greater when deeper and deeper true constituents are considered.[15]

In spite of this relativity to depth, measure (30) can be seen to yield a *comparative* notion of truthlikeness which is insensitive to the increase of depth. Note that each constituent at depth d is equivalent to a finite disjunction of constituents at depth $d+1$. Hence, if h is a generalization with the depth d, then it is equivalent to a finite disjunction $h^{(d')}$ of constituents at depth d', for any $d' > d$. (This disjunction $h^{(d')}$ is called the *expansion* of h to the depth d'.) The following theorem can be easily proved:

(31) If h_1 and h_2 are generalizations of L at depths d_1 and d_2, respectively, then there exists $d_3 \geq \max(d_1, d_2)$ such that, for the expansions $h_1^{(d)}$ and $h_2^{(d)}$ of h_1 and h_2 to depth d, $M_1(h_1^{(d)}) \lesseqqgtr M_1(h_2^{(d)})$ holds for some $d \geq d_3$ if and only if it holds for all $d \geq d_3$.

In particular, Theorem (31) holds also for the measure M_2 of truthlikeness. On the basis of (31), one may define

(32) Generalization h_1 is *more truthlike* than generalization h_2 iff $M_1(h_1^{(d)}) > M_1(h_2^{(d)})$ holds for sufficiently large values of d. Generalizations h_1 and h_2 are *equally truthlike* iff $M_1(h_1^{(d)}) = M_1(h_2^{(d)})$ for sufficiently large values of d.

This comparative notion of truthlikeness is independent of the depth of the generalizations h_1 and h_2. Note that h_1 may be strictly more truthlike than h_2 even if $\lim_{d\to\infty} M_1(h_1^{(d)}) = \lim_{d\to\infty} M_1(h_2^{(d)})$.

6. DISTANCE BETWEEN CONSTITUENTS: GENERAL CASE

Let L be a first-order language (without identity) such that V_L is finite and contains at least one non-monadic predicate. The constituents of L correspond then to finite trees of d-sequences of individuals – they describe, in a systematic way, all the different sequences of d interrelated individuals that can be found in a possible world; in other words, they describe all the different samples of size d which can be drawn (with replacement) from a possible world. More precisely, to each constituent at depth d there corresponds a finite set of finite inverted trees all the branches of which are of length d. First, all the different individuals, as characterized by their properties and their relations to themselves (in the vocabulary V_L), that exist in the world are listed. Secondly, the different kinds of individuals existing in the world are listed relative to each element in the first list, i.e., they are characterized not only by their properties and their relations to themselves but also by their relations to a particular element in the first list. In the third step, the individuals are characterized also by reference to two subsequent elements of the first and the second list. When this operation is repeated for d times, the 'surface structure' corresponding to the constituent is obtained.[16]

For example, a world in which there are two different kinds of people (black and white nodes) and in which marriage between different kinds of people is forbidden may look, at depth 2, as in Fig. 4.

Fig. 4.

A straightforward way of generalizing our treatment of the monadic case (Section 4) is to replace the Ct-predicates (cells) of the monadic language with the different d-sequences of individuals.[16a] Each constituent $C_i^{(d)}$ at depth asserts that certain d-sequences exist; let this set of d-sequences be $\mathbf{D}_i^{(d)}$. The simplest way of measuring the distance between two constituents $C_i^{(d)}$ and $C_j^{(d)}$ is, thus, a generalization of definition (23):

$$(33) \qquad d(C_i^{(d)}, C_j^{(d)}) = \text{card}(\mathbf{D}_i^{(d)} \Delta \mathbf{D}_j^{(d)}).$$

(Cf. Fig. 2 with $\mathbf{CT}_i$ replaced by $\mathbf{D}_i^{(d)}$.) Just like (23), measure (33) does not take into account the distances between different d-sequences. If a measure ε for such a distance is available, definitions (27) and (28) can be immediately generalized to the general situation. Tichý's proposal for such a measure ε is to identify the difference between two d-sequences as the number of atomic sentences of L that are true of one d-sequence and false of the other. This number obviously equals the sum of the distances between those monadic Ct-predicates of L which are satisfied by the corresponding elements of the two d-sequences (see Section 4) plus the number of differences in the relations which hold between the elements of each of the d-sequences.[17]

There is some redundancy in the partial descriptions of possible worlds that constituents give. For example, if marriage between the two kinds of people is allowed, one must add to Fig. 4 not only a sequence of the form

but also the sequence

This means that, in formula (33), some of the mistakes are counted several times. It therefore seems natural to modify (33) by identifying those d-sequences which are the same except the order of elements, and let $\mathbf{D}_i^{(d)}$ in (33) be this *reduced list* of d-sequences. The distance between two elements in reduced lists, i.e., between two equivalence classes of d-sequences, is then defined as the minimum distance between pairs of d-sequences in these classes.

When the distance between two constituents is defined by (33), modified as explained above, the truthlikeness of a constituent $C_i^{(d)}$ at depth d is given by formula

$$(34) \qquad M_1(C_i^{(d)}) = 1 - \frac{d(C_i^{(d)}, C_*^{(d)})}{K(d)},$$

where $C_*^{(d)}$ is the true constituent at depth d and $K(d)$ is the maximum distance between constituents $C_j^{(d)}$ from $C_*^{(d)}$. More generally, formulae (12)–(17) define the degree of truthlikeness for arbitrary generalizations of L at depth d.

For example, the reduced list corresponding to the constituent which allows for all 'mixed' and 'non-mixed' marriages is the following:

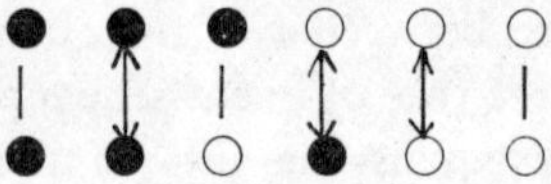

while the reduced list corresponding to the constituent which claims that people represented by white nodes are not allowed to get married at all is the following

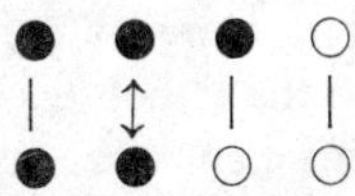

According to our proposal, these two constituents (at depth 2) are equally close to the truth as described in Fig. 4.

Just as in Section 5, each constituent $C_i^{(d)}$ has an expansion to each depth $d + e$, $e = 1, 2, \ldots$ (see Fig. 5).[18] At depth d, the most truthlike generalization is $C_*^{(d)}$, but at depth $d + e$ it (i.e., its expansion to depth $d + e$) is logically weaker and thus less truthlike than $C_*^{(d+e)}$. The relative distances of false generalizations of L from the true constituent may vary when the truth itself becomes 'deeper' – just as in Section 5, our proposal here makes degrees of truthlikeness relative to quantificational depth.

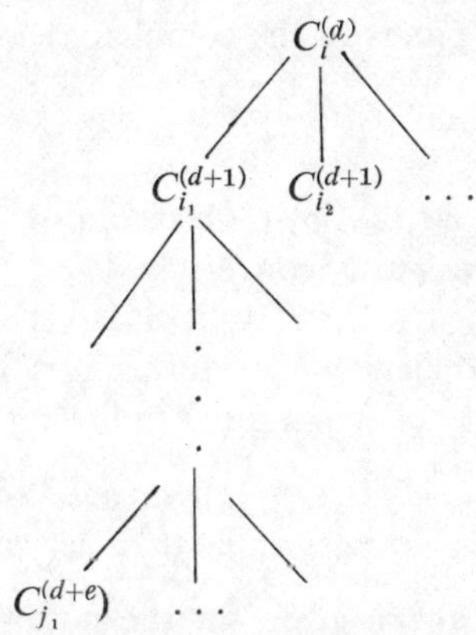

Fig. 5.

7. TRUTHLIKENESS OF THEORIES

We have so far considered the problem of truthlikeness only in the case of first-order sentences. The general account, given in Section 6, is thus, in principle, applicable to any finitely axiomatizable theory which can be formalized in a finite first-order language.

To extend these results, note first that when $d = 0$ and $C_i^{(d)}$ is a propositional tautology, all the infinite branches of Fig. 5 correspond to *complete theories* in L and, conversely, all the complete theories in L correspond to some such branch. If T is a complete theory in L, then $T = \mathrm{Th}(\mathfrak{A})$ for some structure $\mathfrak{A}$ for L. The 'surface structures' described by the constituents $C^{(d+e)}$ in the branch corresponding to T can thus be considered as a sort of finite approximations to the structure $\mathfrak{A}$.

These observations suggest that one may define a comparative notion of truthlikeness for complete theories as follows:

(35) If T_1 and T_2 are complete theories in L and if $\{C_1^{(d)}\}$ and $\{C_2^{(d)}\}$, for $d = 0, 1, \ldots$, are sequences of constituents corresponding to T_1 and T_2, respectively, then T_1 is *more truthlike* than T_2 if, for some d_0,
$$M_1(C_1^{(d)}) > M_1(C_2^{(d)})$$
for all $d \geq d_0$.[19]

A quantitative notion of truthlikeness for complete theories can be defined by the following limit (whenever it exists): If $\{C^{(d)}\}$ is the branch of Fig. 5 which corresponds to the complete theory T, then

(36) $$M_1(T) = \lim_{d \to \infty} M_1(C^{(d)}).$$

Note that if $M_1(T_1) > M_1(T_2)$ by (36), then also T_1 is more truthlike than T_2 by (35), but not necessarily conversely.

As each complete theory T in L is the theory $\mathrm{Th}(\mathfrak{A})$ of some structure $\mathfrak{A}$ (which is unique up to elementary equivalence), definitions (35) and (36) give us the opportunity of making the following definitions:

(37) Structure $\mathfrak{A}$ is *closer to the actual world* $\mathfrak{A}^*$ *than* structure $\mathfrak{A}'$ iff $\mathrm{Th}(\mathfrak{A})$ is more truthlike than $\mathrm{Th}(\mathfrak{A}')$ in the sense of (35).

A similar definition can be given for the notion '$\mathfrak{A}$ is equally close to $\mathfrak{A}^*$ as $\mathfrak{A}'$'.

(38) $$\Delta(\mathfrak{A}, \mathfrak{A}^*) = 1 - M_1(\mathrm{Th}(\mathfrak{A})).$$

The notions defined by (37) and (38) are invariant with respect to elementary equivalence: If $\mathfrak{A}$ is closer to $\mathfrak{A}^*$ than $\mathfrak{A}'$ and if $\mathfrak{B}$ is elementarily equivalent with $\mathfrak{A}$, then $\mathfrak{B}$ is closer to $\mathfrak{A}^*$ than $\mathfrak{A}'$; if $\mathfrak{A}$ is elementarily equivalent to $\mathfrak{B}$, then $\Delta(\mathfrak{A}, \mathfrak{A}^*) = \Delta(\mathfrak{B}, \mathfrak{A}^*)$.

(37) defines a *similarity relation* in the class $\mathbf{M}$ of structures for L, and (38) defines a *similarity metric* Δ in $\mathbf{M}$. We have thus come back where we started in Section 1, but with the achievement that we are now able to *define* those fundamental things – a similarity relation and a similarity metric – which the 'similarity approach' assumes as undefined primitive notions.

It is now possible to extend the definition of truthlikeness to arbitrary first-order theories. By substituting Δ for d in formulae (1) and (2), measures m_* and m^* can be used to define the *degree of truth*, $V(T)$, and the degree of *information about the truth*, $I(T)$, of a theory T in L:

$$(39) \qquad V(T) = 1 - m_*(T)$$
$$(40) \qquad I(T) = 1 - m_*(T).$$

(Note that the maximum value of $\Delta(\mathfrak{A}, \mathfrak{A}^*)$ is 1.) By (38), one obtains

$$(41) \qquad V(T) = 1 - \inf_{\mathfrak{A} \in \mathrm{Mod}(T)} \Delta(\mathfrak{A}, \mathfrak{A}^*) = \sup_{\mathfrak{A} \in \mathrm{Mod}(T)} M_1(\mathrm{Th}(\mathfrak{A}))$$
$$(42) \qquad I(T) = 1 - \sup_{\mathfrak{A} \in \mathrm{Mod}(T)} \Delta(\mathfrak{A}, \mathfrak{A}^*) = \inf_{\mathfrak{A} \in \mathrm{Mod}(T)} M_1(\mathrm{Th}(\mathfrak{A})).$$

The *degree of truthlikeness* of the theory T is then defined by

$$(43) \qquad M_1(T) = \gamma V(T) + (1 - \gamma) I(T)$$
$$= \gamma \sup_{\mathfrak{A} \in \mathrm{Mod}(T)} \Delta(\mathfrak{A}, \mathfrak{A}^*) + (1 - \gamma) \inf_{\mathfrak{A} \in \mathrm{Mod}(T)} \Delta(\mathfrak{A}, \mathfrak{A}^*).$$

An important special case of (43) is again obtained by choosing $\gamma = 1/2$. This quantitative measure of truthlikeness for theories T in L shares the basic properties that M_1 has in a less general situation (see Niiniluoto, 1977a, and Section 3 above).

8. CONCLUSION

This paper shows that the 'linguistic approach' to truthlikeness which was developed for generalizations of a monadic first-order language in Niiniluoto (1977a) can be extended, in a natural way, to arbitrary theories in a finite first-order language. (See note 16a, however.)

By using Hintikka's theory of distributive normal forms as a tool for obtaining finite approximations of ordinary first-order structures, we have been able to define a similarity relation and a similarity metric in the class of structures for a first-order language L. This result may be interesting for those who wish to employ such similarity relations or metrics in dealing with other problems than the logical problem of truthlikeness. The fact that the similarity relation and metric obtained in this paper are invariant with respect to elementary equivalence is a consequence of our decision to deal here only with first-order languages. But, in principle, the same approach can be followed for all extensions of first-order logic where constituents can be defined. In this way, it would be possible to find sharper and sharper similarity metrics up to those which are invariant with respect to isomorphism.[20]

Department of Mathematics and *Department of Philosophy,*
University of Helsinki

NOTES

[1] Inductive probabilities $P(h/e)$ as such do not even serve as reasonable estimates of the degree of truthlikeness of hypothesis h on the basis of evidence e, since h may be highly truthlike even if e gives us strong reasons for believing in the falsity of h. Probabilities $P(h/e)$ can be used in the evaluation of *expected* degrees of truthlikeness, however. (See Niiniluoto, 1977a.)

[2] See, for example, the several papers in April/May issue of *Synthese* **30** (1975).

[3] For the 'realist' theory of scientific progress, see Popper (1963), pp. 114–116, 231–235, and Niiniluoto (1977b).

[4] Th($\mathfrak{A}$), the theory of $\mathfrak{A}$, is defined as the set of all the sentences of L which are true in $\mathfrak{A}$.

[5] See Miller (1974a), Tichý (1974). For generalizations of this result, see Miller (1974b), (1975), and Harris (1974), (1975).

[6] See, for example, Stalnaker (1968) and Lewis (1973).

[7] Note that Hilpinen gives this definition by assuming only a comparative notion of similarity.

[8] The only possible exception to this fact would be the situation where Mod(T) for some false theory T contains structures $\mathfrak{A} \in \mathbf{M}$ which are arbitrarily close to $\mathfrak{A}^*$, so that $m_*(T) = 0$ in spite of the falsity of T.

[9] The (quantificational) depth of a generalization h in L is defined as the maximum number of nested and connected quantifiers in h (see Hintikka, 1973).

[10] This 'program' has also been proposed in Tichý (1974) and (1976). Tichý (1974) develops this suggestion only in the case of propositional logic; his treatment of this case has been effectively criticized by Miller (1975a) and (1975b). See, further, Sections 4–6 below.

[11] In this special case, the language L contains only monadic predicates, and only '1-kinds', in Tichý's sense, are considered in the definition of truthlikeness.

[12] Definition (24) is applicable in the very simple situation where the Ct-predicates are defined by (21). Perhaps a more 'realistic' picture, which is also more applicable, is obtained by using the machinery that Carnap developed for his inductive logic (see Carnap, 1971). Let

$$\mathfrak{F}_1 = \{O_1^1, \ldots, O_{q_1}^1\}$$
$$\ldots$$
$$\mathfrak{F}_r = \{O_1^r, \ldots, O_{q_r}^r\}$$

be *families of predicates* in the sense that, within each family $\mathfrak{F}_i$, the predicates O_j^i, $j = 1, \ldots, q_i$, are jointly exhaustive and mutually exclusive. Examples of such families are the family of colour predicates and all families obtained through finite partitions of the values of some quantity. Let ϵ_i be a *distance* measure on family $\mathfrak{F}_i$, $i = 1, \ldots, r$, normalized so that its maximum value is one. For example, in the family of colour predicates it should be the case that 'green' is closer to 'blue' than 'red'. Instead of (21), the Ct-predicates of the language L with families $\mathfrak{F}_i$, $i = 1, \ldots, r$, are defined by

$$Ct_{i_1 i_2 \ldots i_r} = O^1_{i_1} \;\&\; O^2_{i_2} \;\&\ldots\&\; O^r_{i_r}.$$

Instead of (24), the distance between two Ct-predicates is now defined by

$$\varepsilon(Ct_{i_1 \ldots i_r}, Ct_{j_1 \ldots j_r}) = \frac{1}{r} \sum_{v=1}^{r} \varepsilon_v (O^v_{i_v}, O^v_{j_v}).$$

13 Note that in the example discussed above, we have by (27) and (28)

$$d'(C_1, C_*) = \varepsilon_{\mathrm{II}}(Ct_1) + \varepsilon_{\mathrm{I}}(Ct_4) = 1 + 1 = 2$$
$$d'(C_2, C_*) = \varepsilon_{\mathrm{II}}(Ct_1) + \varepsilon_{\mathrm{II}}(Ct_2) + \varepsilon_{\mathrm{I}}(Ct_4) = 1 + 1 + 1 = 3.$$

14 Cf. Miller (1976), where essentially the same criticism, but in different formulation, against Tichý's example is given.

15 This kind of relativity to quantificational depth is completely analogous to the relativity of truthlikeness to the expressive depth (i.e., the vocabulary V_L) of L which is argued for in Niiniluoto (1977a).

16 For details about constituents, see Hintikka (1971) and (1973).

16a *Added in proof:* A straightforward generalization of the treatment given for the monadic case is obtained by noting that a constituent $C_i^{(d)}$ can be written in the form of (20) where the Ct-predicates Ct_j are replaced by attributive constituents (Ct-predicates) $\mathrm{Ct}_j^{(d-1)}$ of depth $d-1$. Here the attributive constituents $Ct_j^{(d-1)}$ themselves have a tree-like structure. In analogy to formula (23), the distance between two constituents of depth d can then be defined by

$$(*) \qquad d(C_i^{(d)}, C_j^{(d)}) = \mathrm{card}(\mathbf{CT}_i^{(d-1)} \Delta \mathbf{CT}_j^{(d-1)}).$$

where $\mathbf{CT}_i^{(d-1)}$ is the set of attributive constituents $Ct_k^{(d-1)}$ of depth $d-1$ occurring in C_i.

I did not develop this idea in the present paper, since I did not have a solution to the problem of defining the distance between two attributive constituents. Instead, I tried to avoid this problem by considering d-sequences in the place of attributive constituents, since the distance between d-sequences is easy to define. However, I no longer believe that this idea (i.e., formula (33) and its generalizations) yields an entirely adequate general solution, even if it seems to work reasonably in a number of simple situations. (I am grateful to Professor Pavel Tichý's critical remarks.) Measure (33) is in some cases too coarse: it may fail to make a difference between constituents which one would like to keep separate. I shall try to discuss, in another paper, the approach which is based upon (*) and its variants. (See *Synthese* **38** (1978), 281–330.)

17 Cf. note 12.

18 As we are concerned with an objective notion like truthlikeness, it should be assumed that all inconsistent constituents, trivially inconsistent or not, are eliminated from Fig. 5. (Cf. Hintikka, 1973.)

19 I have not been able to decide whether a theorem similar to (31) holds also for this

case. If not, then we may say simply that some complete theories in L are incomparable for truthlikeness.

[20] Cf. the results of Rantala and Karttunen in the present volume about constituents within infinitely deep languages.

BIBLIOGRAPHY

Carnap, R., 'A Basic System of Inductive Logic, Part I', in R. Carnap and R. C. Jeffrey (eds.), *Studies in Inductive Logic and Probability*, vol. I, University of California Press, Berkeley, 1971, pp. 35–165.

Harris, J. H., 'Popper's Definition of "Verisimilitude"', *The British Journal for the Philosophy of Science* **25** (1974), 160–166.

Harris, J. H., 'On Comparing Theories', *Synthese* **32** (1975), 29–76.

Hilpinen, R., 'Approximate Truth and Truthlikeness', in M. Przelecki, K. Szaniawski, and R. Wójcicki (eds.), *Formal Methods in the Methodology of Empiricial Sciences*, D. Reidel and Polish Scientific Publishers, Dordrecht and Warsaw, 1976.

Hintikka, K. J., 'Surface Information and Depth Information', in K. J. Hintikka and P. Suppes (eds.), *Information and Inference*, D. Reidel, Dordrecht, 1970, pp. 263–297.

Hintikka, K. J., *Logic, Language-Games, and Information*, Oxford University Press, Oxford, 1973.

Lewis, D., *Counterfactuals*, Blackwell, Oxford, 1973.

Miller, D., 'Popper's Qualitative Theory of Verisimilitude', *The British Journal for the Philosophy of Science* **25** (1974), 166–177 [1974a].

Miller, D., 'On the Comparison of False Theories by Their Bases', *The British Journal for the Philosophy of Science* **25** (1974), 178–188 [1974b].

Miller, D., 'The Accuracy of Predictions', *Synthese* **30** (1975), 159–191 [1975a].

Miller, D., 'The Accuracy of Predictions: A Reply', *Synthese* **30** (1975), 207–219 [1975b].

Miller, D., 'Verisimilitude Redeflated', *The British Journal for the Philosophy of Science* **27** (1976), 363–381.

Niiniluoto, I., 'On the Truthlikeness of Generalizations', in R. E. Butts and J. Hintikka (eds.), *Basic Problems in Methodology and Linguistics* (Western Ontario Series 11), D. Reidel, Dordrecht, 1977 [1977a].

Niiniluoto, I., 'Notes on Popper as Follower of Whewell and Peirce', *Ajatus* **37** (1977), 272-327 [1977b].

Popper, K. R., 'Some Comments on Truth and the Growth of Knowledge', in E. Nagel, P. Suppes, and A. Tarski (eds.), *Logic, Methodology, and Philosophy of Science*, Stanford University Press, Stanford, 1962, pp. 285–292.

Popper, K. R., *Conjectures and Refutations: The Growth of Scientific Knowledge*, Routledge & Kegan Paul, London, 1963.

Popper, K. R., 'A Note on Verisimilitude', *The British Journal for the Philosophy of Science* **27** (1976), 147–159.

Stalnaker, R. C., 'A Theory of Conditionals', in N. Rescher (ed.), *Studies in Logical Theory* (APQ Monograph No. 2), Blackwell, Oxford, 1968.

Tichý, P., 'On Popper's Definition of Verisimilitude', *The British Journal for the Philosophy of Science* **25** (1974), 155–160.

Tichý, P., 'Verisimilitude Redefined', *The British Journal for the Philosophy of Science* **27** (1976), 25-42.

INDEX OF NAMES

SYNTHESE LIBRARY

Studies in Epistemology, Logic, Methodology,
and Philosophy of Science

1. J. M. Bocheński, *A Precis of Mathematical Logic.* 1959, X + 100 pp.
2. P. L. Guiraud, *Problèmes et méthodes de la statistique linguistique.* 1960, VI + 146 pp.
3. Hans Freudenthal (ed.), *The Concept and the Role of the Model in Mathematics and Natural and Social Sciences. Proceedings of a Colloquium held at Utrecht, The Netherlands, January 1960.* 1961, VI + 194 pp.
4. Evert W. Beth, *Formal Methods. An Introduction to Symbolic Logic and the Study of Effective Operations in Arithmetic and Logic.* 1962, XIV + 170 pp.
5. B. H. Kazemier and D. Vuysje (eds.), *Logic and Language. Studies Dedicated to Professor Rudolf Carnap on the Occasion of His Seventieth Birthday.* 1962, VI + 256 pp.
6. Marx W. Wartofsky (ed.), *Proceedings of the Boston Colloquium for the Philosophy of Science 1961-1962,* Boston Studies in the Philosophy of Science (ed. by Robert S. Cohen and Marx W. Wartofsky), Volume I. 1963, VIII + 212 pp.
7. A. A. Zinov'ev, *Philosophical Problems of Many-Valued Logic.* 1963, XIV + 155 pp.
8. Georges Gurvitch, *The Spectrum of Social Time.* 1964, XXVI + 152 pp.
9. Paul Lorenzen, *Formal Logic.* 1965, VIII + 123 pp.
10. Robert S. Cohen and Marx W. Wartofsky (eds.), *In Honor of Philipp Frank,* Boston Studies in the Philosophy of Science (ed. by Robert S. Cohen and Marx W. Wartofsky), Volume II. 1965, XXXIV + 475 pp.
11. Evert W. Beth, *Mathematical Thought. An Introduction to the Philosophy of Mathematics.* 1965, XII + 208 pp.
12. Evert W. Beth and Jean Piaget, *Mathematical Epistemology and Psychology.* 1966, XII + 326 pp.
13. Guido Küng, *Ontology and the Logistic Analysis of Language. An Enquiry into the Contemporary Views on Universals.* 1967, XI + 210 pp.
14. Robert S. Cohen and Marx W. Wartofsky (eds.), *Proceedings of the Boston Colloquium for the Philosophy of Science 1964-1966, in Memory of Norwood Russell Hanson,* Boston Studies in the Philosophy of Science (ed. by Robert S. Cohen and Marx W. Wartofsky), Volume III. 1967, XLIX + 489 pp.

15. C. D. Broad, *Induction, Probability, and Causation. Selected Papers.* 1968, XI + 296 pp.
16. Günther Patzig, *Aristotle's Theory of the Syllogism. A Logical-Philosophical Study of Book A of the Prior Analytics.* 1968, XVII + 215 pp.
17. Nicholas Rescher, *Topics in Philosophical Logic.* 1968, XIV + 347 pp.
18. Robert S. Cohen and Marx W. Wartofsky (eds.), *Proceedings of the Boston Colloquium for the Philosophy of Science 1966-1968,* Boston Studies in the Philosophy of Science (ed. by Robert S. Cohen and Marx W. Wartofsky), Volume IV. 1969, VIII + 537 pp.
19. Robert S. Cohen and Marx W. Wartofsky (eds.), *Proceedings of the Boston Colloquium for the Philosophy of Science 1966-1968,* Boston Studies in the Philosophy of Science (ed. by Robert S. Cohen and Marx W. Wartofsky), Volume V. 1969, VIII + 482 pp.
20. J.W. Davis, D. J. Hockney, and W. K. Wilson (eds.), *Philosophical Logic.* 1969, VIII + 277 pp.
21. D. Davidson and J. Hintikka (eds.), *Words and Objections: Essays on the Work of W.V. Quine.* 1969, VIII + 366 pp.
22. Patrick Suppes, *Studies in the Methodology and Foundations of Science. Selected Papers from 1911 to 1969.* 1969, XII + 473 pp.
23. Jaakko Hintikka, *Models for Modalities. Selected Essays.* 1969, IX + 220 pp.
24. Nicholas Rescher *et al.* (eds.), *Essays in Honor of Carl G. Hempel. A Tribute on the Occasion of His Sixty-Fifth Birthday.* 1969, VII + 272 pp.
25. P. V. Tavanec (ed.), *Problems of the Logic of Scientific Knowledge.* 1969, XII + 429 pp.
26. Marshall Swain (ed.), *Induction, Acceptance, and Rational Belief.* 1970, VII + 232 pp.
27. Robert S. Cohen and Raymond J. Seeger (eds.), *Ernst Mach: Physicist and Philosopher,* Boston Studies in the Philosophy of Science (ed. by Robert S. Cohen and Marx W. Wartofsky), Volume VI. 1970, VIII + 295 pp.
28. Jaakko Hintikka and Patrick Suppes, *Information and Inference.* 1970, X + 336 pp.
29. Karel Lambert, *Philosophical Problems in Logic. Some Recent Developments.* 1970, VII + 176 pp.
30. Rolf A. Eberle, *Nominalistic Systems.* 1970, IX + 217 pp.
31. Paul Weingartner and Gerhard Zecha (eds.), *Induction, Physics, and Ethics: Proceedings and Discussions of the 1968 Salzburg Colloquium in the Philosophy of Science.* 1970, X + 382 pp.
32. Evert W. Beth, *Aspects of Modern Logic.* 1970, XI + 176 pp.
33. Risto Hilpinen (ed.), *Deontic Logic: Introductory and Systematic Readings.* 1971, VII + 182 pp.
34. Jean-Louis Krivine, *Introduction to Axiomatic Set Theory.* 1971, VII + 98 pp.
35. Joseph D. Sneed, *The Logical Structure of Mathematical Physics.* 1971, XV + 311 pp.
36. Carl R. Kordig, *The Justification of Scientific Change.* 1971, XIV + 119 pp.
37. Milič Čapek, *Bergson and Modern Physics,* Boston Studies in the Philosophy of Science (ed. by Robert S. Cohen and Marx W. Wartofsky), Volume VII. 1971, XV + 414 pp.

38. Norwood Russell Hanson, *What I Do Not Believe, and Other Essays* (ed. by Stephen Toulmin and Harry Woolf), 1971, XII + 390 pp.
39. Roger C. Buck and Robert S. Cohen (eds.), *PSA 1970. In Memory of Rudolf Carnap,* Boston Studies in the Philosophy of Science (ed. by Robert S. Cohen and Marx W. Wartofsky), Volume VIII. 1971, LXVI + 615 pp. Also available as paperback.
40. Donald Davidson and Gilbert Harman (eds.), *Semantics of Natural Language.* 1972, X + 769 pp. Also available as paperback.
41. Yehoshua Bar-Hillel (ed.), *Pragmatics of Natural Languages.* 1971, VII + 231 pp.
42. Sören Stenlund, *Combinators, λ-Terms and Proof Theory.* 1972, 184 pp.
43. Martin Strauss, *Modern Physics and Its Philosophy. Selected Papers in the Logic, History, and Philosophy of Science.* 1972, X + 297 pp.
44. Mario Bunge, *Method, Model and Matter.* 1973, VII + 196 pp.
45. Mario Bunge, *Philosophy of Physics.* 1973, IX + 248 pp.
46. A. A. Zinov'ev, *Foundations of the Logical Theory of Scientific Knowledge (Complex Logic),* Boston Studies in the Philosophy of Science (ed. by Robert S. Cohen and Marx W. Wartofsky), Volume IX. Revised and enlarged English edition with an appendix, by G. A. Smirnov, E. A. Sidorenka, A. M. Fedina, and L. A. Bobrova. 1973, XXII + 301 pp. Also available as paperback.
47. Ladislav Tondl, *Scientific Procedures,* Boston Studies in the Philosophy of Science (ed. by Robert S. Cohen and Marx W. Wartofsky), Volume X. 1973, XII + 268 pp. Also available as paperback.
48. Norwood Russell Hanson, *Constellations and Conjectures* (ed. by Willard C. Humphreys, Jr.). 1973, X + 282 pp.
49. K. J. J. Hintikka, J. M. E. Moravcsik, and P. Suppes (eds.), *Approaches to Natural Language. Proceedings of the 1970 Stanford Workshop on Grammar and Semantics.* 1973, VIII + 526 pp. Also available as paperback.
50. Mario Bunge (ed.), *Exact Philosophy – Problems, Tools, and Goals.* 1973, X + 214 pp.
51. Radu J. Bogdan and Ilkka Niiniluoto (eds.), *Logic, Language, and Probability. A Selection of Papers Contributed to Sections IV, VI, and XI of the Fourth International Congress for Logic, Methodology, and Philosophy of Science, Bucharest, September 1971.* 1973, X + 323 pp.
52. Glenn Pearce and Patrick Maynard (eds.), *Conceptual Change.* 1973, XII + 282 pp.
53. Ilkka Niiniluoto and Raimo Tuomela, *Theoretical Concepts and Hypothetico-Inductive Inference.* 1973, VII + 264 pp.
54. Roland Fraïssé, *Course of Mathematical Logic* – Volume 1: *Relation and Logical Formula.* 1973, XVI + 186 pp. Also available as paperback.
55. Adolf Grünbaum, *Philosophical Problems of Space and Time.* Second, enlarged edition, Boston Studies in the Philosophy of Science (ed. by Robert S. Cohen and Marx W. Wartofsky), Volume XII. 1973, XXIII + 884 pp. Also available as paperback.
56. Patrick Suppes (ed.), *Space, Time, and Geometry.* 1973, XI + 424 pp.
57. Hans Kelsen, *Essays in Legal and Moral Philosophy,* selected and introduced by Ota Weinberger. 1973, XXVIII + 300 pp.
58. R. J. Seeger and Robert S. Cohen (eds.), *Philosophical Foundations of Science. Proceedings of an AAAS Program, 1969,* Boston Studies in the Philosophy of

Science (ed. by Robert S. Cohen and Marx W. Wartofsky), Volume XI. 1974, X + 545 pp. Also available as paperback.

59. Robert S. Cohen and Marx W. Wartofsky (eds.), *Logical and Epistemological Studies in Contemporary Physics*, Boston Studies in the Philosophy of Science (ed. by Robert S. Cohen and Marx W. Wartofsky), Volume XIII. 1973, VIII + 462 pp. Also available as paperback.
60. Robert S. Cohen and Marx W. Wartofsky (eds.), *Methodological and Historical Essays in the Natural and Social Sciences. Proceedings of the Boston Colloquium for the Philosophy of Science 1969-1972*, Boston Studies in the Philosophy of Science (ed. by Robert S. Cohen and Marx W. Wartofsky), Volume XIV. 1974, VIII + 405 pp. Also available as paperback.
61. Robert S. Cohen, J. J. Stachel and Marx W. Wartofsky (eds.), *For Dirk Struik. Scientific, Historical and Political Essays in Honor of Dirk J. Struik*, Boston Studies in the Philosophy of Science (ed. by Robert S. Cohen and Marx W. Wartofsky), Volume XV. 1974, XXVII + 652 pp. Also available as paperback.
62. Kazimierz Ajdukiewicz, *Pragmatic Logic*, transl. from the Polish by Olgierd Wojtasiewicz. 1974, XV + 460 pp.
63. Sören Stenlund (ed.), *Logical Theory and Semantic Analysis. Essays Dedicated to Stig Kanger on His Fiftieth Birthday*. 1974, V + 217 pp.
64. Kenneth F. Schaffner and Robert S. Cohen (eds.), *Proceedings of the 1972 Biennial Meeting, Philosophy of Science Association*, Boston Studies in the Philosophy of Science (ed. by Robert S. Cohen and Marx W. Wartofsky), Volume XX. 1974, IX + 444 pp. Also available as paperback.
65. Henry E. Kyburg, Jr., *The Logical Foundations of Statistical Inference*. 1974, IX + 421 pp.
66. Marjorie Grene, *The Understanding of Nature: Essays in the Philosophy of Biology*, Boston Studies in the Philosophy of Science (ed. by Robert S. Cohen and Marx W. Wartofsky), Volume XXIII. 1974, XII + 360 pp. Also available as paperback.
67. Jan M. Broekman, *Structuralism: Moscow, Prague, Paris*. 1974, IX + 117 pp.
68. Norman Geschwind, *Selected Papers on Language and the Brain*, Boston Studies in the Philosophy of Science (ed. by Robert S. Cohen and Marx W. Wartofsky), Volume XVI. 1974, XII + 549 pp. Also available as paperback.
69. Roland Fraïssé, *Course of Mathematical Logic* – Volume 2: *Model Theory*. 1974, XIX + 192 pp.
70. Andrzej Grzegorczyk, *An Outline of Mathematical Logic. Fundamental Results and Notions Explained with All Details*. 1974, X + 596 pp.
71. Franz von Kutschera, *Philosophy of Language*. 1975, VII + 305 pp.
72. Juha Manninen and Raimo Tuomela (eds.), *Essays on Explanation and Understanding. Studies in the Foundations of Humanities and Social Sciences*. 1976, VII + 440 pp.
73. Jaakko Hintikka (ed.), *Rudolf Carnap, Logical Empiricist. Materials and Perspectives*. 1975, LXVIII + 400 pp.
74. Milič Čapek (ed.), *The Concepts of Space and Time. Their Structure and Their Development*, Boston Studies in the Philosophy of Science (ed. by Robert S. Cohen and Marx W. Wartofsky), Volume XXII. 1976, LVI + 570 pp. Also available as paperback.

75. Jaakko Hintikka and Unto Remes, *The Method of Analysis. Its Geometrical Origin and Its General Significance,* Boston Studies in the Philosophy of Science (ed. by Robert S. Cohen and Marx W. Wartofsky), Volume XXV. 1974, XVIII + 144 pp. Also available as paperback.
76. John Emery Murdoch and Edith Dudley Sylla, *The Cultural Context of Medieval Learning. Proceedings of the First International Colloquium on Philosophy, Science, and Theology in the Middle Ages – September 1973,* Boston Studies in the Philosophy of Science (ed. by Robert S. Cohen and Marx W. Wartofsky), Volume XXVI. 1975, X + 566 pp. Also available as paperback.
77. Stefan Amsterdamski, *Between Experience and Metaphysics. Philosophical Problems of the Evolution of Science,* Boston Studies in the Philosophy of Science (ed. by Robert S. Cohen and Marx W. Wartofsky), Volume XXXV. 1975, XVIII + 193 pp. Also available as paperback.
78. Patrick Suppes (ed.), *Logic and Probability in Quantum Mechanics.* 1976, XV + 541 pp.
79. Hermann von Helmholtz: *Epistemological Writings. The Paul Hertz/Moritz Schlick Centenary Edition of 1921 with Notes and Commentary by the Editors.* (Newly translated by Malcolm F. Lowe. Edited with an Introduction and Bibliography, by Robert S. Cohen and Yehuda Elkana), Boston Studies in the Philosophy of Science (ed. by Robert S. Cohen and Marx W. Wartofsky), Volume XXXVII. 1977, XXXVIII+204 pp. Also available as paperback.
80. Joseph Agassi, *Science in Flux,* Boston Studies in the Philosophy of Science (ed. by Robert S. Cohen and Marx W. Wartofsky), Volume XXVIII. 1975, XXVI + 553 pp. Also available as paperback.
81. Sandra G. Harding (ed.), *Can Theories Be Refuted? Essays on the Duhem-Quine Thesis.* 1976, XXI + 318 pp. Also available as paperback.
82. Stefan Nowak, *Methodology of Sociological Research: General Problems.* 1977, XVIII + 504 pp.
83. Jean Piaget, Jean-Blaise Grize, Alina Szeminska, and Vinh Bang, *Epistemology and Psychology of Functions,* Studies in Genetic Epistemology, Volume XXIII. 1977, XIV+205 pp.
84. Marjorie Grene and Everett Mendelsohn (eds.), *Topics in the Philosophy of Biology,* Boston Studies in the Philosophy of Science (ed. by Robert S. Cohen and Marx W. Wartofsky), Volume XXVII. 1976; XIII + 454 pp. Also available as paperback.
85. E. Fischbein, *The Intuitive Sources of Probabilistic Thinking in Children.* 1975, XIII + 204 pp.
86. Ernest W. Adams, *The Logic of Conditionals. An Application of Probability to Deductive Logic.* 1975, XIII + 156 pp.
87. Marian Przełęcki and Ryszard Wójcicki (eds.), *Twenty-Five Years of Logical Methodology in Poland.* 1977, VIII + 803 pp.
88. J. Topolski, *The Methodology of History.* 1976, X + 673 pp.
89. A. Kasher (ed.), *Language in Focus: Foundations, Methods and Systems. Essays Dedicated to Yehoshua Bar-Hillel,* Boston Studies in the Philosophy of Science (ed. by Robert S. Cohen and Marx W. Wartofsky), Volume XLIII. 1976, XXVIII + 679 pp. Also available as paperback.
90. Jaakko Hintikka, *The Intentions of Intentionality and Other New Models for Modalities.* 1975, XVIII + 262 pp. Also available as paperback.

91. Wolfgang Stegmüller, *Collected Papers on Epistemology, Philosophy of Science and History of Philosophy*, 2 Volumes, 1977, XXVII + 525 pp.
92. Dov M. Gabbay, *Investigations in Modal and Tense Logics with Applications to Problems in Philosophy and Linguistics.* 1976, XI + 306 pp.
93. Radu J. Bogdan, *Local Induction.* 1976, XIV + 340 pp.
94. Stefan Nowak, *Understanding and Prediction: Essays in the Methodology of Social and Behavioral Theories.* 1976, XIX + 482 pp.
95. Peter Mittelstaedt, *Philosophical Problems of Modern Physics,* Boston Studies in the Philosophy of Science (ed. by Robert S. Cohen and Marx W. Wartofsky), Volume XVIII. 1976, X + 211 pp. Also available as paperback.
96. Gerald Holton and William Blanpied (eds.), *Science and Its Public: The Changing Relationship,* Boston Studies in the Philosophy of Science (ed. by Robert S. Cohen and Marx W. Wartofsky), Volume XXXIII. 1976, XXV + 289 pp. Also available as paperback.
97. Myles Brand and Douglas Walton (eds.), *Action Theory. Proceedings of the Winnipeg Conference on Human Action, Held at Winnipeg, Manitoba, Canada, 9-11 May 1975.* 1976, VI + 345 pp.
98. Risto Hilpinen, *Knowledge and Rational Belief.* 1979 (forthcoming).
99. R. S. Cohen, P. K. Feyerabend, and M. W. Wartofsky (eds.), *Essays in Memory of Imre Lakatos,* Boston Studies in the Philosophy of Science (ed. by Robert S. Cohen and Marx W. Wartofsky), Volume XXXIX. 1976, XI + 762 pp. Also available as paperback.
100. R. S. Cohen and J. J. Stachel (eds.), *Selected Papers of Léon Rosenfeld,* Boston Studies in the Philosophy of Science (ed. by Robert S. Cohen and Marx W. Wartofsky), Volume XXI. 1978, XXX + 927 pp.
101. R. S. Cohen, C. A. Hooker, A. C. Michalos, and J. W. van Evra (eds.), *PSA 1974: Proceedings of the 1974 Biennial Meeting of the Philosophy of Science Association,* Boston Studies in the Philosophy of Science (ed. by Robert S. Cohen and Marx W. Wartofsky), Volume XXXII. 1976, XIII + 734 pp. Also available as paperback.
102. Yehuda Fried and Joseph Agassi, *Paranoia: A Study in Diagnosis,* Boston Studies in the Philosophy of Science (ed. by Robert S. Cohen and Marx W. Wartofsky), Volume L. 1976, XV + 212 pp. Also available as paperback.
103. Marian Przełęcki, Klemens Szaniawski, and Ryszard Wójcicki (eds.), *Formal Methods in the Methodology of Empirical Sciences.* 1976, 455 pp.
104. John M. Vickers, *Belief and Probability.* 1976, VIII + 202 pp.
105. Kurt H. Wolff, *Surrender and Catch: Experience and Inquiry Today,* Boston Studies in the Philosophy of Science (ed. by Robert S. Cohen and Marx W. Wartofsky), Volume LI. 1976, XII + 410 pp. Also available as paperback.
106. Karel Kosík, *Dialectics of the Concrete,* Boston Studies in the Philosophy of Science (ed. by Robert S. Cohen and Marx W. Wartofsky), Volume LII. 1976, VIII + 158 pp. Also available as paperback.
107. Nelson Goodman, *The Structure of Appearance,* Boston Studies in the Philosophy of Science (ed. by Robert S. Cohen and Marx W. Wartofsky), Volume LIII. 1977, L + 285 pp.
108. Jerzy Giedymin (ed.), *Kazimierz Ajdukiewicz: The Scientific World-Perspective and Other Essays, 1931 - 1963.* 1978, LIII + 378 pp.

109. Robert L. Causey, *Unity of Science*. 1977, VIII+185 pp.
110. Richard E. Grandy, *Advanced Logic for Applications*. 1977, XIV + 168 pp.
111. Robert P. McArthur, *Tense Logic*. 1976, VII + 84 pp.
112. Lars Lindahl, *Position and Change: A Study in Law and Logic*. 1977, IX + 299 pp.
113. Raimo Tuomela, *Dispositions*. 1978, X + 450 pp.
114. Herbert A. Simon, *Models of Discovery and Other Topics in the Methods of Science*, Boston Studies in the Philosophy of Science (ed. by Robert S. Cohen and Marx W. Wartofsky), Volume LIV. 1977, XX + 456 pp. Also available as paperback.
115. Roger D. Rosenkrantz, *Inference, Method and Decision*. 1977, XVI + 262 pp. Also available as paperback.
116. Raimo Tuomela, *Human Action and Its Explanation. A Study on the Philosophical Foundations of Psychology*. 1977, XII + 426 pp.
117. Morris Lazerowitz, *The Language of Philosophy. Freud and Wittgenstein*, Boston Studies in the Philosophy of Science (ed. by Robert S. Cohen and Marx W. Wartofsky), Volume LV. 1977, XVI + 209 pp.
118. Tran Duc Thao, *Origins of Language and Consciousness*, Boston Studies in the Philosophy of Science (ed. by Robert S. Cohen and Marx. W. Wartofsky), Volume LVI. 1979 (forthcoming).
119. Jerzy Pelč, *Semiotics in Poland, 1894 - 1969*. 1977, XXVI + 504 pp.
120. Ingmar Pörn, *Action Theory and Social Science. Some Formal Models*. 1977, X + 129 pp.
121. Joseph Margolis, *Persons and Minds, The Prospects of Nonreductive Materialism*, Boston Studies in the Philosophy of Science (ed. by Robert S. Cohen and Marx W. Wartofsky), Volume LVII. 1977, XIV + 282 pp. Also available as paperback.
122. Jaakko Hintikka, Ilkka Niiniluoto, and Esa Saarinen (eds.), *Essays on Mathematical and Philosophical Logic. Proceedings of the Fourth Scandinavian Logic Symposium and of the First Soviet-Finnish Logic Conference, Jyväskylä, Finland, 1976*. 1978, VIII + 458 pp. + index.
123. Theo A. F. Kuipers, *Studies in Inductive Probability and Rational Expectation*. 1978, XII + 145 pp.
124. Esa Saarinen, Risto Hilpinen, Ilkka Niiniluoto, and Merrill Provence Hintikka (eds.), *Essays in Honour of Jaakko Hintikka on the Occasion of His Fiftieth Birthday*. 1978, IX + 378 pp. + index.
125. Gerard Radnitzky and Gunnar Andersson (eds.), *Progress and Rationality in Science*, Boston Studies in the Philosophy of Science (ed. by Robert S. Cohen and Marx W. Wartofsky), Volume LVIII. 1978, X + 400 pp. + index. Also available as paperback.
126. Peter Mittelstaedt, *Quantum Logic*. 1978, IX + 149 pp.
127. Kenneth A. Bowen, *Model Theory for Modal Logic. Kripke Models for Modal Predicate Calculi*. 1978, X + 128 pp.
128. Howard Alexander Bursen, *Dismantling the Memory Machine. A Philosophical Investigation of Machine Theories of Memory*. 1978, XIII + 157 pp.
129. Marx W. Wartofsky, *Models: Representation and Scientific Understanding*, Boston Studies in the Philosophy of Science (ed. by Robert S. Cohen and Marx W. Wartofsky), Volume XLVIII. 1979 (forthcoming). Also available as a paperback.
130. Don Ihde, *Technics and Praxis. A Philosophy of Technology*, Boston Studies in

the Philosophy of Science (ed. by Robert S. Cohen and Marx W. Wartofsky), Volume XXIV. 1979 (forthcoming). Also available as a paperback.

131. Jerzy J. Wiatr (ed.), *Polish Essays in the Methodology of the Social Sciences,* Boston Studies in the Philosophy of Science (ed. by Robert S. Cohen and Marx W. Wartofsky), Volume XXIX. 1979 (forthcoming). Also available as a paperback.
132. Wesley C. Salmon (ed.), *Hans Reichenbach: Logical Empiricist.* 1979 (forthcoming).
133. R.-P. Horstmann and L. Krüger (eds.), *Transcendental Arguments and Science.* 1979 (forthcoming). Also available as a paperback.

SYNTHESE HISTORICAL LIBRARY

Texts and Studies
in the History of Logic and Philosophy

Editors:

N. KRETZMANN (Cornell University)
G. NUCHELMANS (University of Leyden)
L. M. DE RIJK (University of Leyden)

1. M. T. Beonio-Brocchieri Fumagalli, *The Logic of Abelard.* Translated from the Italian. 1969, IX + 101 pp.
2. Gottfried Wilhelm Leibniz, *Philosophical Papers and Letters.* A selection translated and edited, with an introduction, by Leroy E. Loemker. 1969, XII + 736 pp.
3. Ernst Mally, *Logische Schriften,* ed. by Karl Wolf and Paul Weingartner. 1971, X + 340 pp.
4. Lewis White Beck (ed.), *Proceedings of the Third International Kant Congress.* 1972, XI + 718 pp.
5. Bernard Bolzano, *Theory of Science,* ed. by Jan Berg. 1973, XV + 398 pp.
6. J. M. E. Moravcsik (ed.), *Patterns in Plato's Thought. Papers Arising Out of the 1971 West Coast Greek Philosophy Conference.* 1973, VIII + 212 pp.
7. Nabil Shehaby, *The Propositional Logic of Avicenna: A Translation from al-Shifā: al-Qiyās,* with Introduction, Commentary and Glossary. 1973, XIII + 296 pp.
8. Desmond Paul Henry, *Commentary on De Grammatico: The Historical-Logical Dimensions of a Dialogue of St. Anselm's.* 1974, IX + 345 pp.
9. John Corcoran, *Ancient Logic and Its Modern Interpretations.* 1974, X + 208 pp.
10. E. M. Barth, *The Logic of the Articles in Traditional Philosophy.* 1974, XXVII + 533 pp.
11. Jaakko Hintikka, *Knowledge and the Known. Historical Perspectives in Epistemology.* 1974, XII + 243 pp.
12. E. J. Ashworth, *Language and Logic in the Post-Medieval Period.* 1974, XIII + 304 pp.
13. Aristotle, *The Nicomachean Ethics.* Translated with Commentaries and Glossary by Hypocrates G. Apostle. 1975, XXI + 372 pp.
14. R. M. Dancy, *Sense and Contradiction: A Study in Aristotle.* 1975, XII + 184 pp.
15. Wilbur Richard Knorr, *The Evolution of the Euclidean Elements. A Study of the Theory of Incommensurable Magnitudes and Its Significance for Early Greek Geometry.* 1975, IX + 374 pp.
16. Augustine, *De Dialectica.* Translated with Introduction and Notes by B. Darrell Jackson. 1975, XI + 151 pp.

17. Arpád Szabó, *The Beginnings of Greek Mathematics.* 1979 (forthcoming).

18. Rita Guerlac, *Juan Luis Vives Against the Pseudodialecticians. A Humanist Attack on Medieval Logic.* Texts, with translation, introduction and notes. 1978, xiv + 227 pp. + index.

SYNTHESE LANGUAGE LIBRARY

Texts and Studies
in Linguistics and Philosophy

Managing Editors:

1. Henry Hiż (ed.), *Questions.* 1977, xvii + 366 pp.
2. William S. Cooper, *Foundations of Logico-Linguistics. A Unified Theory of Information, Language, and Logic.* 1978, xvi + 249 pp.
3. Avishai Margalit (ed.), *Meaning and Use. Papers Presented at the Second Jerusalem Philosophical Encounter, April 1976.* 1978 (forthcoming).
4. F. Guenthner and S. J. Schmidt (eds.), *Formal Semantics and Pragmatics for Natural Languages.* 1978, viii + 374 pp. + index.
5. Esa Saarinen (ed.), *Game-Theoretical Semantics.* 1978, xiv + 379 pp. + index.
6. F. J. Pelletier (ed.), *Mass Terms: Some Philosophical Problems.* 1978, xiv + 300 pp. + index.